STRUCTURES AND NORMS IN SCIENCE

SYNTHESE LIBRARY

STUDIES IN EPISTEMOLOGY,

LOGIC, METHODOLOGY, AND PHILOSOPHY OF SCIENCE

VOLUME 260

STRUCTURES AND NORMS IN SCIENCE

Volume Two of the Tenth International Congress of Logic, Methodology and Philosophy of Science, Florence, August 1995

Edited by

MARIA LUISA DALLA CHIARA
University of Florence

KEES DOETS
University of Amsterdam

DANIELE MUNDICI
University of Milan

and

JOHAN VAN BENTHEM
University of Amsterdam
and *Stanford University*

KLUWER ACADEMIC PUBLISHERS
DORDRECHT / BOSTON / LONDON

A C.I.P. Catalogue record for this book is available from the Library of Congress

ISBN 0-7923-4384-0 (Vol. 2)
ISBN 0-7923-4383-2 (Vol. 1)
ISBN 0-7923-4385-9 (Set)

Published by Kluwer Academic Publishers,
P.O. Box 17, 3300 AA Dordrecht, The Netherlands.

Kluwer Academic Publishers incorporates
the publishing programmes of
D. Reidel, Martinus Nijhoff, Dr W. Junk and MTP Press.

Sold and distributed in the U.S.A. and Canada
by Kluwer Academic Publishers,
101 Philip Drive, Norwell, MA 02061, U.S.A.

In all other countries, sold and distributed
by Kluwer Academic Publishers Group,
P.O. Box 322, 3300 AH Dordrecht, The Netherlands.

Printed on acid-free paper

Printed in the Netherlands

TABLE OF CONTENTS

EDITORIAL

These volumes continue a long tradition. They document part of the scientific program of the 10th International Congress of Logic, Methodology and Philosophy of Science, held at Florence, August 1995. The founding fathers of the Congress, including Alfred Tarski, Evert Beth, and Abraham Robinson, left us a legacy of cooperation between logicians and philosophers of science, on an international scale which is still unique. At the time, this alliance seemed natural. Philosophers and mathematicians in the fifties still remembered the heighdays of logical positivism, and their interests were broad. Not accidentally, Tarski's best-known textbook is a joint introduction to logic and methodology of the exact sciences. In the meantime, many things have changed. Logicians have developed new contacts with computer science and linguistics, philosophers of science also look toward history and sociology. Nevertheless, our Congress is still there, like a Grand Hotel, with perhaps a somewhat faded splendour, but with a distinctive style, elegance and comfort. And it serves an important purpose.

The invited lectures published here demonstrate much of what goes on in the fields of the Congress. Through an extensive selection process, General and Sectional Program Committees have chosen speakers whose excellence in research deserves recognition. How much this honour is appreciated is shown by a remarkably high acceptance rate. The sectional organisation covers the traditional subdisciplines of mathematical logic, philosophical logic, as well as their interfaces with computer science, linguistics, philosophy. Philosophy of science is broadly represented, too, including both general issues of methodology and ethics, and specific foundational issues of natural sciences, social sciences and humanities. In addition to this planned part, the Congress also serves as the meeting place for an international community from all over the world. There have been a host of interesting contributed talks, often leading to osmosis through the sectional barriers of the program. Selections of these papers will appear in a number of satellite volumes, edited by the Florence local organizers.

The program of the Congress has further special features. Traditionally, the opening and closing lectures are given by eminent scholars and scientists, who give us a broad perspective on the intellectual environment that we are working in. We are pleased to include the lectures by professors Beltrametti and van Fraassen, who did just that. Also, the general Program Committee has organized a number of special Symposia, to high-light some topics of current interest, and to enhance the connectivity of our field. In Florence, these were “Semantics and Semiotics”, “Logic and Philosophy of Science”, “Logic in Eastern Europe” and “Logic in the Far East”. These symposia are represented here, too. Not included are the so-called Affiliated Meetings on “Model Theory” (organized jointly by the Kurt Gödel Society and the Italian Association of Logic and its Applications, AILA) and “Proof Theory for Natural Language” (European Association for Logic, Language and Information), whose materials will be published elsewhere by the responsible organizers.

There remains the pleasant task of expressing thanks where these are due. We have been assisted very generously by Mrs. Annie Kuipers (Kluwer Academic Publishers). Most of the editorial logistics have been taken care of by Kees Doets at the University of Amsterdam, assisted by Jeroen Haak and Joeri van Ruth. The four editors themselves see this Volume as the visible finished product of several years of effective and pleasant collaboration in a joint Dutch-Italian team. Readers with even the slightest knowledge of contemporary Europe will know that this is indeed a winning combination ...

Maria Luisa Dalla Chiara (Firenze)
Kees Doets (Amsterdam)
Daniele Mundici (Milano)
Johan van Benthem (Amsterdam and Stanford)

SECTIONS, PLENARY LECTURES AND SPECIAL SYMPOSIA

LOGIC

Section 1. Proof Theory and Categorical Logic

Section Program Committee:
G.E. MINTS, Stanford (chair)
V. OREVKOV, St. Petersburg
J. LAMBEK, Montréal

Invited speakers:
A. CANTINI, Proof-theoretic aspects of self-referential truth
Università degli Studi di Firenze
A. JOYAL, Free lattices, communication and money games
Université du Quebec à Montréal
J. KRAJÍČEK, On methods for proving lower bounds in propositional logic
Mathematical Institute of Science, Prague
V.YU. SAZONOV, On bounded set theory
Russian Academy of Sciences, Pereslavl-Zalessky

Section 2. Model Theory, Set Theory and Formal Systems

Section Program Committee:
W.A. HODGES, London (chair)
X. CAICEDO, Bogota
C.A. DI PRISCO, Caracas

Invited speakers:
PH.G. KOLAITIS, Infinitary logic in finite model theory
University of California, Santa Cruz
A. SCEDROV, Decision problems for second-order linear logic
(with P.D. LINCOLN and N. SHANKAR)
University of Pennsylvania, Philadelphia
and
SRI International

S. Todorčević,
Comparing the continuum with the first two uncountable cardinals
University of Toronto
and
Matematički Institut, Belgrade

Section 3. Recursion Theory and Constructivism

Section Program Committee:
W. Sieg, Pittsburgh (chair)
P. Odifreddi, Torino
K. Amboss-Spies, Heidelberg

Invited speakers:
K. Kelly, Church's thesis and Hume's problem
(with O. Schulte)
Carnegie Mellon University, Pittsburgh

Y.N. Moschovakis, The logic of functional recursion
Department of Mathematics, University of Los Angeles

H. Schwichtenberg, Computability for partial continuous functionals
(with K. Stroetman)
Universität München

R. Soare, Computability, enumerability, macros, and meaning
University of Chicago

G. Tamburrini,
Mechanistic theories in cognitive science: the import of Turing's thesis
Instituto di Cibernetica del C.N.R.

Section 4. Logic and Computer Science

Section Program Committee:
C. Böhm, Roma (chair)
J.A. Bergstra, Amsterdam and Utrecht
P.G. Clote, Boston

Invited speakers:
P. Martin-Löf, The semantic significance of computation
Academy of Sciences, Stockholm, Sweden

G.D. Plotkin, Type theory and recursion
University of Edinburgh

D.S. SCOTT, 25 years of domain theory

Carnegie Mellon University, Pittsburgh

Section 5. Philosophical Logic

Section Program Committee:

P. GÄRDENFORS, Lund (chair)
G. SAMBIN, Padova
M. DETLEFSEN, Notre Dame

Invited speakers:

G. BOOLOS (†), Frege's theorem and the Peano postulates

MIT, Cambridge MA

E. CASARI, Conjoining and disjoining on different levels

Centro Fiorentino di Storia e Filosofia della Scienza

K. DOŠEN, Logical consequence, a turn in style

Matematički Institut, Belgrade
and
IRIT, University of Toulouse III

D. MAKINSON, Applying normative rules with restraint (with S.O. HANSON)

Ville d'Avray, France
and
Uppsala University

GENERAL PHILOSOPHY OF SCIENCE

Section 6. Methodology

Section Program Committee:

A.E. MUSGRAVE, Otago (chair)
C.A. HOOKER, Newcastle
N.C.A. DA COSTA, Sao Paulo

Invited speakers:

P. ACHINSTEIN, The empirical character of evidence

Johns Hopkins University, Baltimore

S. FRENCH, Partiality, pursuit and practice

University of Leeds

Section 7. Probability, Induction and Decision Theory

Section Program Committee:

R.C. Jeffrey, Princeton (chair)
T. Seidenfeld, Pittsburgh
H. Gaifman, New York

Invited speakers:

M. Goldstein, Prior inferences for posterior judgements
Science Labs, Durham

B. Skyrms, Game theory, rationality and evolution
University of California, Irvine

Section 8. History of Logic, Methodology and Philosophy of Science

Section Program Committee:

N. Cartwright, London (chair)
P. Maddy, Irvine
L. Laudan, Hawaii

Invited speakers:

U. Gähde, Anomalies and the revision of theory-elements: notes on the advance of Mercury's perihelion
Universität Bayreuth

G. Moore, The prehistory of infinitary logic: 1885–1955
McMaster University, Hamilton, Ontario

Section 9. Ethics of Science and Technology

Section Program Committee:

K. Shrader-Frechette, Tampa (chair)
B. Hansson, Lund
R. Qiu, Beijing
W. Shea, Montréal

Invited speakers:

S. Beder, Engineers, ethics and sustainable development
University of Wollongong

L. Jonathan Cohen, Are there ethical reasons for being, or not being, a scientific realist?
Oxford

D. Fan, Rights and autonomy of scientists: cases in P.R. China
Beijing

I. Hronszky, On ethics and changing expertise in technology assesment
University of Budapest

M. Kaiser,
Unpopular and precautionary science: towards a new ethics of science
The Research Council of Norway, Oslo

N. Oreskes, Testing models of natural systems: can it be done?
Darthmouth College, Hanover

PHILOSOPHICAL AND FOUNDATIONAL PROBLEMS ABOUT THE SCIENCES

Section 10. Philosophy of Logic, Mathematics and Computer Science

Section Program Committee:
H. Field, New York (chair)
G. Longo, Paris
J. Etchemendy, Stanford

Invited speakers:
Ch.D. Parsons, What can we do "in principle"?
Harvard University, Cambridge MA

J. Pearl, Causation, action, and counterfactuals
University of California, Los Angeles

Section 11. Philosophy of the Physical Sciences

Section Program Committee:
J.S. Earman,
and
K.L. Manders, Pittsburgh (chairs)
G. Toraldo di Francia, Firenze

Invited speakers:
D. Albert, On the character of statistical-mechanical probabilities
Columbia University, New York City

P. Galison, unable to attend
Harvard University, Cambridge MA

G. Ghirardi, Macroscopic reality and the dynamical reduction program
Universita degli Studi di Trieste

Section 12. Philosophy of Biological Sciences

Section Program Committee:

M. RUSE, Guelph and Montpellier (chair)
P. KITCHER, Santa Dominga
W.J. VAN DER STEEN, Amsterdam

Invited speakers:

R. BURIAN, On conflicts between genetic and developmental viewpoints and their attempted resolution in molecular biology
Virginia Polytechnic Institute and State University, Blacksburg VA

J. GAYON, The "paramount power of selection": from Darwin to Kauffman
Université de Bourgogne, Dijon
and
Institut d'Histoire et de Philosophie des Sciences et des Techniques, Paris

C.N. VAN DER WEELE,
Environmental causes in ontogeny and an ethics of attention
University of Utrecht

Section 13. Philosophy of Cognitive Science and AI

Section Program Committee:

N. BLOCK, Cambridge MA (chair)
M. DAVIES, Oxford
A. CLARK, St. Louis

Invited speakers:

D. CHALMERS, The extended mind
University of California, Santa Cruz

F. JACKSON, Mental causation without the language of "thought"
The Australian National University

J. KIM, What is the problem of mental causation?
Brown University, Providence RI

B.P. MCLAUGHLIN, Classical constituents in Smolensky's ICS architecture
Rutgers University, New Brunswick NJ

Section 14. Philosophy of Linguistics

Section Program Committee:

S. SOAMES, Princeton (chair)
H. SCHNELLE, Bochum
G. CHIERCHIA, Milano

Invited speakers:

D.R. DOWTY, Non-constituent coordination, wrapping, and multimodal categorical grammars

Ohio State University, Columbus OH

H. KAMP,

Cognitive structure, shared cognitive structure and lexical semantics

Universität Stuttgart

T.D. PARSONS, Meaning sensitivity and grammatical structure

University of California at Irvine

Section 15. Social Sciences

Section Program Committee:

PH. PETTIT, Canberra (chair)
J. ELSTER, Oslo and Chicago
A. ROSENBERG, Riverside

Invited speakers:

R.M. DAWES, Qualitative consistency masquerading as quantitative fit

Carnegie Mellon University, Pittsburgh

D.M. HAUSMAN, Why does evidence matter so little to economic theory?

University of Wisconsin

D. SPERBER, Selection and attraction in cultural evolution

CNRS and École Polytechnique, Paris

PLENARY SPEAKERS

Opening

E.G. BELTRAMETTI, Trends in the foundations of physics

Department of Physics
University of Genoa

Closure

B.C. VAN FRAASSEN,

Structure and perspective: philosophical perplexity and paradox

Princeton University

SPECIAL SYMPOSIA

Symposium 1. Logic and Philosophy of Science: current interfaces

Section Program Committee:
T.A.F. KUIPERS, Groningen (chair)
J. BUTTERFIELD, Cambridge UK
G. HELLMAN, Minneapolis

Invited speakers:
K. KELLY, Reliable belief revision
(with O. SCHULTE and V. HENDRICKS)
Carnegie Mellon University, Pittsburgh
and
University of Copenhagen

V. RANTALA,
Explanatory translation: beyond functionalism and reductionism
Tampere University

P. THAGARD, Abductive reasoning: logic, visual thinking, and coherence
(with C. SHELLEY)
University of Waterloo

P. WEINGARTNER, Can the laws of nature (physics) be complete?
Salzburg University

Symposium 2. Semantics and Semiotics

Section Program Committee:
J. PELC, Warsaw (chair)
G. CHIERCHIA, Milano
P. LEONARDI, Venezia

Invited speakers:
U. ECO, On meaning, logic and verbal language
Milano

D. FØLLESDAL, Semantics and semiotics
Stanford University
and
University of Oslo

Symposium 3. Logic and Philosophy of Science in the Far East

Section Program Committee:
R. QIU, Beijing (chair)

T. Fujimoto, Tokyo
H. Hsiu Hwang, Hong Kong
S. Xiaoli, Beijing

Invited speakers:

W. Lu, Logic in China

Beijing

H. Sakamoto, Recent develepments of logic, computer science and philosophy of science in Japan

Nihon University, Tokyo

S-Y. Song, Philosophy of science in Korea, 1950–1990

Hallym University, South Korea

C-H. Lin, Philosophy of science in Taiwan
(with D. Fu)

National Taiwan University, Taipei
and
National Tsing-Hua University, Hsinchu

Symposium 4. Logic in Central and Eastern Europe

Section Program Committee:

P. Hájek, Prague,
and
S. Artemov, Moscow (chairs)
L. Tondl, Prague

Invited speakers:

L. Bukovský, Logic in Czechoslovakia and Hungary

Katedra matematickej informatiky, Kosice

V.A. Uspensky, Mathematical logic in the former Soviet Union: brief history and current trends

Moscow State University

D. Vakarelov, Logic in Central and Eastern Europe: Balkan region

Sofia University

R. Wójcicki, The postwar panorama of logic in Poland

PAN, Warszawa

Executive committee of the Division of Logic, Methodology and Philosophy of Science, International Union of History and Philosophy of Science, 1991–1995:

President J.E. FENSTAD (Norway), 1st Vice-President D. PRAWITZ (Sweden), 2nd Vice-President R. WÓJCICKI (Poland), Secretary General E. SOBER (USA), Treasurer A. PRELLER (France), Past President L.J. COHEN (UK), Assessors M. BOFFA (Belgium), R. CHUAQUI (Chile, †), R.P.H. FORREST (Australia), M. HALLETT (Canada), H. ISHIGURO (Japan), V. LEKTORSKY (Russia), G.E. MINTS (USA), G. SAMBIN (Italy).

General Program Committee:

J. VAN BENTHEM (Chair, the Netherlands and USA), R. CHUAQUI, (Chile, †), P. HÁJEK (Czech Rep.), D. MUNDICI (Italy), R. QIU (China), E. SOBER (USA).

Local Organizing committee:

M.L. DALLA CHIARA (chair), L. AMADUCCI, T. ARECCHI, A. BENEDETTI, S. BERNINI, M. BRESCIANI, S. CALIFANO, A. CANTINI, E. CASARI, E. CASTELLANI, G. CORSI, P. CORSI, F. DESSI-FULGHERI, A. FONNESU, P. GALLUZZI, R. GIUNTINI, E. GIUSTI, F. LAUDISA, P. MANGANI, A. MARCJA, P. MINARI, M. MUGNAI, A. PAGNINI, E. PANCONESI, P. PARRINI, A. PERUZZI, G. PERUZZI, S. POGGI, M. POLSINELLI, P. ROSSI, D. SEARS, G. TORALDO DI FRANCIA.

Financial sponsors of the congress:

The 10th International Congress of Logic, Methodology and Philosophy of Science was arranged by the International Union of History and Philosophy of Science (IUHPS), Division of Logic, Methodology and Philosophy of Science (DLMPS), in collaboration with the University of Florence and the Region of Tuscany. Other supporting institutions are: the European Union (European Commission, Directorate General XII for Science, Research and Development), the National Research Council (the Presidence and the Committees for Philosophy and Mathematics), the University of Siena, the Italian Association for Logic and its Applications (AILA), the Italian Society for Logic and Philosophy of Sciences (SILFS), the Kurt Gödel Society (KGS), the Carlo Marchi Foundation (Florence), the Florence Center for the History and Philosophy of Science, the Italian Institute for Philosophical Studies (Naples), the Center for Epistemology and History of Science F. Enriques (Bologna, Cesena), the Norwegian Research Council, the City of Florence and the Province of Florence. The Congress was held under the auspices of the President of the Italian Republic.

Inaugural Address

E.G. BELTRAMETTI

TRENDS IN THE FOUNDATIONS OF PHYSICS

ABSTRACT.

We review some among the many trends in the foundations of physics. Reductionism is a traditional one: new milestones have been achieved in the recent past but puzzling aspects are accompanying the confinement properties of quarks. New achievements and persisting open problems characterize unified theories of the known basic interactions. An impressive accumulation of experimental results on the fundamental aspects of quantum behaviour came in the last few years; the variety of domains where quantum phenomena are currently investigated is extending beyond expectations. Remarkable developments in the mathematical foundations of quantum mechanics have accompanied the last decades and still characterize many branches of intense research work.

1. INTRODUCTION

After the two great scientific revolutions of our century, carried by relativity and quantum theory, there has been an impressive accumulation of new knowledge in physics. This growth of knowledge called into play many and deep foundational issues, though not needing new conceptual revolutions. It would be presumptuous to try an account of all the relevant contemporary trends in foundations of physics: we shall only deal with a few of them.

The great development of subnuclear physics is one of the contexts where foundational problems, and even puzzling ones, are encountered: they refer for instance to the outcomes of the reductionism approach, a matter that we shall consider in section 2.

In the last decades there has been an impressive accumulation of new quantum phenomena; they encompass new effects in condensed matter physics, spectacular individual-particle experiments, tests on typical quantum correlations (in particular the so-called Einstein-Podolsky-Rosen correlation), all involving fundamental aspects of the quantum behaviour. They will be shortly reviewed in section 3.

Section 4 is devoted to quote a number of problems and ideas that emerge from the research trends in a more theoretical and mathematical context.

2. REDUCTIONISM AND UNIFICATION

Reductionism is an old trend, a guideline along the historical path of physics (as well as of many other sciences). Loosely speaking, it embodies the aspiration to reduce the number of independent concepts necessary to formulate the fundamental laws; it supports questions like "what is a physical system made of?" It urges to explain and interpret a class of phenomena in terms of categories and objects pertaining to an underlying level.

M.L. Dalla Chiara et al. (eds.), Structures and Norms in Science, 3–19.

But the reductionist procedure has recoursive aspects. Take for instance the descriptive level of standard thermodynamics. It admits an underlying descriptive level consisting of the so-called kinetic theory that adopts as building blocks the atoms, the about ninety elements of the Mendeleev table. But if we call into play the internal degrees of freedom of atoms a next level opens, the atomic physics level, which uses electrons and atomic nuclei as new building blocks. One can go down further and take into account the internal structure of nuclei thus coming the a new level characterized by a dramatic reduction of the building blocks: just electrons, protons, and neutrons. And again one can realize that protons and neutrons exhibit an internal structure, made of "quarks", thus opening a new level where the building blocks become electrons and a couple of quarks.

Thus we see that a stratification of different layers emerges, and it is important to notice that these layers are to a large extent independent, decoupled. Why this decoupling? One reason is that each level has its own energy threshold and when the energy is below that threshold the level is "frozen". So, e.g., the atomic physics level has an energy threshold of about 1 ev, so that an atom undergoing a dynamical process (think of the collision with another atom) at lower energies behaves as an uncuttable object, its internal degrees of freedom are "frozen". Similarly, the nuclear physics level has an energy threshold of about 1 Mev (and the subnuclear level an even higher one): at lower energies a nucleus behaves as an unbreakable object and its internal structure does not came into play. Since the existence of these energy thresholds is a typical quantum phenomenon we see that the above stratification of decoupled layers has a quantum root.

Reductionism works, is readable, in physics just thanks to the mentioned stratification: in absence of a decoupling among the different layers a reductionist approach would remain hidden into too big a complexity. The traditional reductionist program of reconstructing the universe from simple elementary laws sounds somewhat unrealistic; the above remarks might justify P. Anderson's more skeptical view [1]: *The more the elementary-particle physicists tell us about the nature of the fundamental laws, the less relevance they seem to have to the very real problems of the rest of science.*

A major aspect of reductionism is the search of "elementary building blocks". When, in 1932, the discovery of the neutron offered the way out to understand the internal structure of atomic nuclei, only three building blocks - namely electrons, protons and neutrons- appeared necessary to form all matter around us. But almost contextually the discovery of new "elementary" particles started up, and the epocal revolution carried out, around the fifties, by the development of accelerator physics (from the age of cosmic-ray hunting to the age of farming new particles in laboratories) strongly enhanced what became the elementary particle physics.

In the sixties the zoo of "elementary" particles appeared embarassingly crowded: less and less likely to look at them as the ultimate building blocks.

But a new descriptive level grew up, conjecturing new elementary constituents (for an account of this matter we refer to [2]). Besides the four "leptons" known at that time -that is the electron (e) with its neutrino (ν_e), and the muon (μ) with its neutrino (ν_μ) (and their antiparticles)- it turned out that all the "hadrons" known since then could be modeled as made of three "quarks", named *up* (u), *down* (d) and *strange* (s). It is worth recalling that these new building blocks were assumed to have fractional electric charge (in the usual unit of the electron charge): the u quark with charge $+\frac{2}{3}$, the d and s quarks with charge $-\frac{1}{3}$. As an example, the proton is thought of as made of two u's and one d, the neutron as made of two d's and one u, the positive pion as made of one u and one anti-d.

But in the seventies a new particle (the ψ or J particle) was discovered not admitting an interpretation in terms of the above quarks: it required a fourth quark named *charm* (c), and experimental evidence for it was found. Somewhat later another new meson (the Y meson) was discovered which needed the existence of a fifth quark whose simbol b stands for *beauty* or for the less poetic *bottom*: again, experimental evidence for this new quark was found. In the meantime also a new heavy lepton, denoted τ, was discovered, and its neutrino ν_τ conjectured.

The whole pattern found a very satisfactory fitting inside the so-called Standard Model developed since 1967 [2]. This model, which has acquired great credit thanks to a number of successful predictions -among others the discovery in 1973 of the "neutral currents" and then the discovery in the early eighties of the Z_0 and $W^\pm$ particles- needed one more quark, the sixth one, named *top* (t), whose experimental evidence finally came quite recently. According to this model the building blocks of everything are the six quarks u, d, s, c, b, t, and the six leptons e, ν_e, μ, ν_μ, τ, ν_τ, which are viewed as elementary, as point particles. Are we at the end of the reductionism route? The question "what is made of" no longer applies to the quarks and leptons? And no more quarks and leptons?

No matter what the answer to these question will be, a puzzling novelty emerges: quarks seem unable to have an autonomous existence, unable to live as isolated particles. Since the sixties an intense experimental research started up aimed at the observation of individual quarks [2]. The experiments were focused on the fractional electric charge of quarks. In ionization experiments the signature of an isolated quark would be a ionization power $\frac{1}{9}$ or $\frac{4}{9}$, relative to a "normal" particle of unit charge (indeed the ionization power is a quadratic function of the charge). Also Millikan-like experiments were performed, on the ground of the fact that, due to the fractional charge, the decay of a quark would include another quark among the decay products (otherways no charge conservation would be possible), so that the lightest quarks must be stable and it would be possible to find sometimes them mixed up with ordinary matter, thus giving some chance to the detection of samples of matter having fractional charge. But the result of these experiments was negative, despite some "announcements" on the contrary: no isolated quark has ever been reliably observed.

Why such an unexpected result? One might guess that we have not yet reached sufficiently high energies to "extract" an isolated quark from a nucleon, thus postponing the appearence of isolated quarks to the availability of higher energies (some troubles would however come from what is believed about the cosmological evolution). But today the favored attitude among physicists is more radical [2]: it is generally accepted that it is, and it will be, impossible to extract a quark, that quarks cannot exist isolated, that they can exist only confined with other quarks. One could support such a confinement destiny of quarks by picturing some unusual force acting among them, a force which is weak at low distances (and indeed quarks appear almost free inside a nucleon) but becomes increasingly strong when the distance increases. A lot has been studied and written about this, but it seems that behind all that there is something we are not able to prove.

In 1976, at a meeting of the German Physical Society, Heisenberg expressed the opinion that until isolated quarks are not observed the question of composition of "elementary" particles is just philosophical, devoid of physical meaning: better to give up. But the explanatory and predictive power of the quark model is well established. In high energy scattering experiments the existence of scattering centers inside nucleons (much smaller than the nucleon size) comes out much in the same way as in the historical Rutherford experiments on the atomic structure. Summing up, we are presently faced with the fact that quarks require a generalized notion of observability, and that the success of a model based on certain building blocks does not entail an autonomous existence of these building blocks.

Connected with reductionism is another traditional trend of physics: the effort of unifying within a single theory different phenomenologies, different interactions. There are many examples of unification in the history of physics: think for instance of the 17-th century unification of the forces causing the falling down of bodies on the ground with the forces governing the motion of planets around the Sun; or think of the 19-th century unification of magnetic with electric phenomena, of the unification of optics with electromagnetism, of thermodynamics with mechanics. More recently other impressive unifications occurred, like the understanding of the dynamics of stars in terms of nuclear physics phenomena, or like the intimate connection of cosmology with elementary particle physics.

A main unification issue refers to the four basic interactions that are operating in our phenomenological world: the gravitational interaction, the weak interaction (responsible, for instance, of the β-decay of nuclei), the electromagnetic interaction, and the strong one (responsible, for instance, of the binding of protons and neutrons inside a nucleus). The above mentioned Standard Model provided a new milestone in unification: weak and electromagnetic interactions appear unified inside that model [2]. Actually, the Standard Model embodies also the ingredients to describe the strong interactions (the so-called Quantum Cromodynamics is inside the model) but the purpose of achieving a wider unification including also the strong interaction goes beyond its purpose. This is the goal of the so-called Great Unification Theories which are beeing the object of intense research activity, and represent by themselves an important trend of contemporary physics. The simplest models of Great Unification Theories predicted a mean life of the proton of about 10^{30} years, contrary to recent experimental evidence that points at larger values. The models presently under consideration predict the unification of the electroweak and the strong interaction at so high an energy (perhaps attainable by the future Large Hadron Collider project at CERN) that also the gravitational interaction might be called into play. This makes the problem very complicated since the way to quantize the gravitational field has still mysterious aspects.

The main credit related to the recent achievements along the stream of reductionism and unification goes to the elementary-particle physicists. But a sense of crisis, of disease, is now readable inside their community. Besides some slowing down in the rhythm of major cognitive novelties during the last decade, the crisis manifests itself at the organizational and social level. Nuclear physics emerged from the second World War with national states openly recognizing its value and power; subnuclear physics then inherited much of this credit. Now it is difficult to convince people that particle physics promises practical outcomes, while present day societies seem more and more shifted toward instrumentalism: knowledge, curiosity is valued less than usefulness. On the other hand, the costs of high-energy accelerator physics have reached levels that the governments do not consider negligible on their budget scale (though accelerator physics is often organized within internationl cooperations). Thus the scientific projects are not handled solely inside the scientific community: they enter the political sphere. Paradigmatic of this situation is the cancellation of the Superconducting Super Collider Project by the USA Congress in 1993, when the construction (in Texas) was already in progress.

A folkloristic example of anomalous organizational trends in particle physics can be found by just looking at the papers published by the teams working at the big accelerator laboratories. For instance, papers by the teams working at LEP (CERN, Geneva) have commonly some 500 Authors, and it is expected that when LEP will be transformed into a proton-antiproton collider (the LHC project) the above figure will be about four times larger. It is clear that the parameters enabling someone to become a (co)author are not precisely the ones traditionally adopted by the scientific community. Parallel haphazard growths obviously accompany the costs of an experiment, the complexity of its experimental apparata, and the needed technical support. Under these conditions the number and variety of performed experiments tends to decrease, while their factual repeatability becomes more and more improbable.

3. ADVANCES IN QUANTUM PHENOMENA

While particle physics is knowing some jurassic problems, a fresh "low scale" physics has grown, producing an impressive accumulation of new experimental results on fundamental aspects of physics [3]. Experiments once deemed as *gedankenexperiments* now become routinely feasible; far ranging progresses in the practical mastering of quantum phenomena have been accomplished in the last years. The variety of domains where quantum phenomena are currently investigated is extending beyond expectations, while their essential unity is ever more clear.

Many different chapters might be singled out such as: (i) new achievements in interferometry, (ii) experiments with single electrons, atoms, molecules, or photons, (iii) quantum correlation experiments, (iv) high temperature superconductivity, (v) quantum Hall effect, (vi) quantum optics, (vii) quantum criptography, and so on. In the sequel of this section we shall limit ourselves to some features of the first three ones.

(i) Interferometry phenomena emerged historically as wave phenomena. Imagine (see Fig. 1) a light beam impinging on a beam splitter BS_1 where it is splitted into a transmitted and a reflected wave that, after reflection on the mirrors M_1 and M_2, converge on a second beam splitter BS_2 behind which two detectors C, C' receive the light. The interferometry phenomenon consists in the fact that the light intensity measured by C (or by C') is in general not the sum of the intensities that correspond to leaving open only path I (hence stopping path II) and then leaving open only path II (hence stopping path I).

When we come to the quantum description, we better speak of a collection of photons impinging on the beam splitter BS_1. Any *which way* observation (e.g. by putting detectors on each path) would reveal that each photon is detected either on path I or on path II: so we are led to think that upon impinging on BS_1 each photon has some chance of being transmitted and some chance of being reflected. According to quantum mechanics even a maximal information about the state of the incoming photon is not sufficient to determine whether it will be transmitted or reflected, what outlines the typical indeterministic nature of quantum theory. Interferometry is still there, as a paradigmatic quantum phenomenon. It is indeed an empirical fact that the probability that a photon will be counted by C is not the sum of the probabilities of reaching C via path I plus the probability of reaching C via path II (similarly for C'). The two paths "interfere".

Interferometry phenomena have been observed long ago not only for photons but also for electrons (one can still imagine an arrangement as in Fig. 1 with proper realisations of the beam splitters and "mirrors"); since a few decades also for neutrons [4], thanks to efficient interferometers made of perfect silicon crystals; more recently also for atoms [5], i.e. for more and more massive and complex objects.

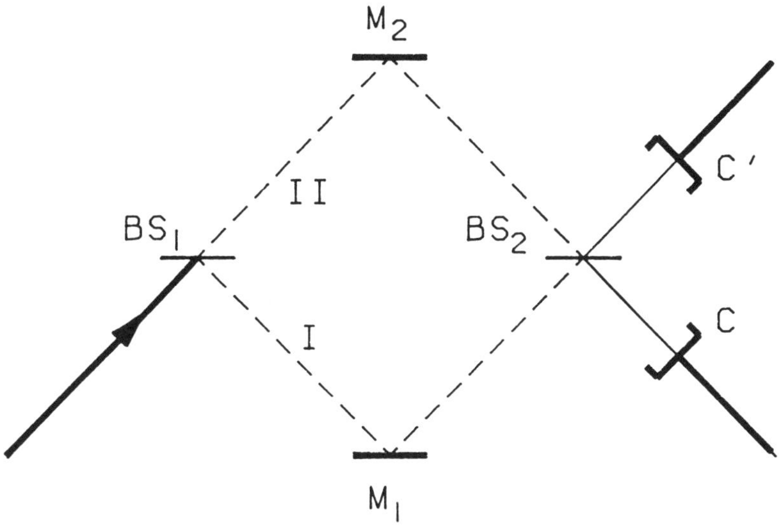

Fig. 1.

The advances in interferometry have allowed checks of many sophisticated quantum effects. Let us mention just a few.

1) Referring to the interferometry with massive particles, suppose that a constant force field acts on the particle. Then a change in the flux of this force through the interferometer loop (think of the loop formed by paths I and II of Fig. 1, and imagine that the loop plane is tilted with respect to the acting force) causes a phase shift between the two paths, hence a change in the interferometry outcome. This effect has been observed for the static gravitational field acting on neutrons [6].

2) Quantum mechanics predicts that, when we deal with charged particles, the insertion of a magnetic flux through the interferometer loop causes a phase shift between the two paths - hence an observable change in the interferometry outcome- even when the magnetic field is confined in a region well inside the interferometer loop so that there is no magnetic field along the two paths and the particle does not experience any magnetic field. This effect, predicted in the late fifties by Aharonov and Bohm has been unambiguously tested in the eighties [7], after a series of previous experimental attempts that were subject to questionable aspects. It appears to be the prototype of a full family of new quantum phenomena, called "topological effects" [8].

3) According to our common intuition the state of a physical system is left unchanged by a full-turn rotation of the system itself: a rotation of 2π is the same as no rotation. But according to quantum mechanics it is not precisely so when we deal with spin-$\frac{1}{2}$ particles, like electrons, protons, neutrons: for them a rotation of 2π changes the sign of the wave function while only a rotation of 4π is the same as no rotation. This effect has been experimentally observed with neutron interferometry, by causing rotations of the particle spin on one path of the interferometer [8].

(ii) Loosely speaking, the recent experiments with single electrons, atoms, molecules, or photons can be splitted into two main families: a) Condensed matter experiments, in which the observed individual particle is surrounded by, and interacts with, the neighbouring atoms, and b) Atomic-physics type experiments, in which the observed individual particle is trapped into a small, though macroscopic, volume and is usually far from any other perturbing matter.

A major example of family a) is related to novel microscopy techniques that allow to observe and manipulate matter with atomic resolution. We refer in particular to the Scanning Tunnelling Microscope, whose discovery motivated the 1986 Nobel prize awarded to Binnig and Roher. A very thin metallic tip scans above a conducting surface, and records the tunnelling electron current (a typical quantum effect) which flows across the gap when a voltage is applied between the tip and the metallic surface. This current is extremely sensitive to the distance between the tip and the surface, thus allowing a high resolution surface topograph [9]. This technique allows also to move and bring along on the metallic surface individual impurity atoms adsobed on the surface. In May 1993, Crommie, Lutz and Eigler [10], of the IBM Almaden Center, succeeded in constructing a corral consisting of 48 iron atoms arranged on a circle with diameter of 14.3 nanometers on a copper surface. Then one electron was added into the enclosure and the charge density measured. The result [11] is a picture of the squared modulus of the wave function of this additional individual electron: it appeared on the cover of the November 1993 issue of *Physics Today.*

Among other spectacular advances within family a) one might quote what might be called the single electron electronics [12]. It became possible to count electrons one at a time and control their flow in low temperature turnstile microscopic gates. Structures of this sort, that have typical sizes of about 10^{-6} cm, are also referred to as quantum dots, or zero dimensional electron gases, or Coulomb islands. These small electronic systems have a quantum behaviour, e.g., they have discrete energy levels: they behave as artificial atoms [13]. Several such artificial atoms can be coupled together to form artificial molecules or artificial solids. The age of artificial atoms has only just begun and it might open the way to fascinating applications.

Come now to some examples of experiments that can be classified within the family b) said above. In 1984-85 H. Dehmelt (he was awarded the Nobel prize in 1989) and his group stored one and the same electron in a trap [14] during 10 months! The trap was a Penning trap: a device in which a proper combination of an electrostatic and a magnetostatic field can keep a single electron bounded inside a ultrahigh-vacuum volume having linear size of a few millimeters. This electron constitutes a pseudo-atom (it was called "geonium"), endowed with a typical structure of energy levels depending on the shape and the strength of the electric and magnetic fields. The trapped electron was monitored by exciting jumps from different energy levels, the involved electromagnetic frequencies being radio frequencies.

One of the main purposes in building up the "geonium" was a very precise measurement of the intrinsic magnetic moment μ_e of the electron. The impressive resolution allowed the extremely sensitive result

$$\mu_e = 1.001159652188$$

in the usual Bohr magneton unit. If one views the electron as a point particle then heroic quantum electrodynamical calculations would give $\mu_e = 1.001159652133$, hence a deviation from the experimentally observed value of one part over 10^{11}. Extrapolation techniques [14] can then be used to evaluate to what extent an electron is not a point particle: the estimated electron radius turns out to be about 10^{-20} cm., the smallest geometrical size presently occurring in physics.

Trapping techniques are efficient also for more complex objects, for instance for a single ion [15]. In this case the trapped particle has its own internal degrees of freedom that can be excited and observed. Think, e.g., of an ion having a short living excited state: then a fluorescence detection can be conceived. A laser beam resonant with the transition between the ground state and this excited state impinges on the ion causing transitions to the excited state which then decays again to the ground state, thus re-emitting a photon of the same wavelength in some other direction. In these conditions the ion acts as a fluorescence scatterer. Typically, $10^7 - 10^8$ photons can be scattered per second; even with low photon detection efficiency, $10^3 - 10^4$ photons per second can be observed into a small solid angle, making a single ion visible by the naked eye. Even more subtle phenomena can be observed: suppose for instance that the ion has a metastable (i.e., long living) excited state at an excitation energy lower than the energy of the short living state considered before. Then sometimes a jump from the higher excited state to the metastable state can occur. These quantum jumps, that occur at random times, can be directly observed because the transition to the metastable state is associated with sudden interruption of the fluorescence signal. Indeed, when the ion is in the metastable state it is no longer able to absorb and scatter photons of the laser beam. When the ion returns to the ground state then the fluorescence goes back to the initial value. One can thus follow directly these jumps on an individual ion [15].

It is also possible to trap together more than one ion and detect them by analogous fluorescence techniques. When several ions are trapped together, the combined effect of Coulomb repulsion and trapping forces can give rise to molecule-like or solid-like structures in which, however, the inter-ion distances are quite larger than in usual molecules or solids. It is possible to observe closely the progressive construction of these new physical structures and detect phenomena concerning them, like phase transitions to random motion when the temperature overcomes some critical value [15].

(iii) Quantum mechanics is intrinsically indeterministic: even a maximal information about the initial conditions allows only probabilistic predictions. It is not so in classical mechanics where every physical quantity has a definite value when the state of the physical system carries a maximal information, that is when the state is "pure". Thus in the quantum description the probability enters at a basic level [16]; it is not just a daughter of ignorance, as it is the case in the classical description. A simple example of indeterministic behaviour can be picked up from what said in (i) above, where beam splitters were considered as elements of interferometry devices. Indeed, even a maximal information about the state of the impinging particle (think of a photon whose direction, energy and polarization are fully specified) is not enough to predict whether the particle will be transmitted or reflected by the beam splitter.

Indeterminism has been often considered as an unpleasant feature of quantum mechanics, it has been an aching pain in the heart of many people, even of some founding fathers of quantum theory. Think of the celebrated Einstein's sentence *God does not play dice.* So, since the beginning of quantum theory the question of its completeness was raised. Could we imagine that the quantum description of a physical system is not complete, admitting a deeper level of description where determinism is restored? Does quantum mechanics appear indeterministic simply due to present unability to "seeing" some additional variables that would make complete the description of the physical system? Coming again to the example of photons impinging on a beam splitter, could we imagine that each photon carries a hidden variable, a label that we are presently unable to read, which deterministically specifies whether it will be reflected or transmitted by the beam splitter? If so, the indeterministic behaviour of photons would evaporate, since we would just be faced with some level of ignorance about the initial state of the impinging photons. Theories embodying such an hypothetical incompleteness of quantum mechanics are usually called hidden-variable (HV) theories.

But no empirical inadequacy of quantum mechanical predictions has ever been observed: thus HV theories are committed to be noncontradictory with the quantum mechanical level of description. In other words, when averag-

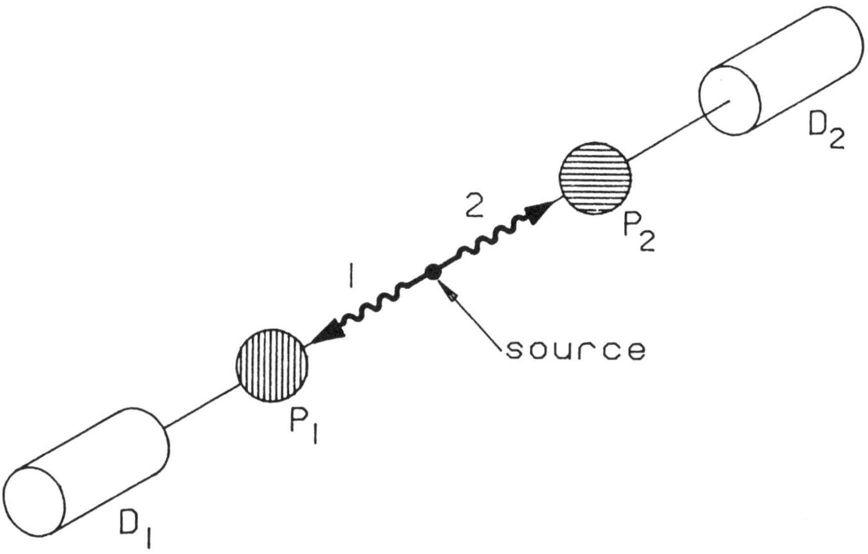

Fig. 2.

ing over the unobserved hidden variables one should recover the quantum predictions. Can this compatibility commitment be fulfilled?

Before 1964 this question was tackled mainly from the mathematical side: no-go theorems (among them the fundamental ones of J. von Neumann and of A.M. Gleason) were proved which restricted the admissible forms of HV theories [16]. In 1964 J. Bell added a milestone by proving that for a wide class of HV theories (the so called local ones) it is impossible to reproduce all the predictions of quantum mechanics [17]. In other words, Bell proved that there exist experiments, not yet performed at that time, which could test HV theories against quantum mechanics.

The difficulty of the proposed experiments forbidded unambiguous tests before the eighties, but in 1981 the famous optical experiment of Aspect, Grangier, and Roger [18,19] arrived. In short, the experiment makes use of a source emitting pairs of photons and the correlation between the polarizations of the two photons is what is measured. The source consisted of a beam of calcium atoms crossed by laser light: the atoms become excited and jump back to the ground state by emission of two photons in cascade. Before reaching their detectors the two photons have to pass analizers of linear polarization, as schematically shown in Fig. 2. According to Bell's discovery the rate of coincidences between the counts of the two detectors, as a function of the relative orientation of the polarization analysers P_1, P_2 cannot be the same for quantum mechanics and for local HV theories. Here "local" means essentially that changing the setting of P_2 has no effect on the probability with which photon 1 passes P_1. The outcome of the experiment agrees with quantum mechanics ruling out local HV theories (despite some persisting discussion on the interpretation of the experiment).

The novelty carried about by Aspect's experiment can also be expressed by saying that the observed correlation between the counts of the two detectors does not admit an interpretation in terms of a "common cause": we are faced with a correlation that does not possess a "causal" explanation" [20]. Or, from another point of view, we can say that the empirical probabilities emerging from the mentioned experiment do not find a classical Kolmogorovian representation within a single probability space [21,22].

It is remarkable the fact that already in 1935, thirthy years before Bell and almost half a century before Aspect's experiment, Einstein, Podolsky and Rosen [23] pointed out at correlations like the one above as the paradigm of the conflict between quantum mechanics and classically minded, deterministic, realistic views: what is somewhat emphatically called the EPR paradox.

4. ON THE RELATIONSHIP BETWEEN CLASSICAL AND QUANTUM BEHAVIOUR

The subject matter that might be covered by the heading of this section includes many trends of contemporary research. We shall just mention a few.

(i) A first question can be related to the situation left by Aspect's experiment as summarized in the last part of the previous section: is there some residual place where the advocates of HV theories can hold the flag of determinism? A corner is left: the one of contextual, nonlocal HV theories. Bohmian mechanics is the paradigmatic example [24]. It is an elaboration of De Broglie's old idea of pilot waves: it is based on the conjecture that a complete description of a physical system needs both the quantum wave function ψ and a hidden variable which is identified with the position coordinate. Take, e.g., the simplest case of one particle of mass m moving in one dimension: then Bohmian mechanics says that a complete description is provided by the pair $(x, \psi(x))$ formed by the coordinate x and by the wave function $\psi(x)$, obeying the time evolution equations [25]

$$i\hbar\frac{\partial}{\partial t}\psi(x) = H\psi(x) \quad , \quad \frac{dx}{dt} = \frac{\hbar}{m}\Im(\frac{1}{\psi}\frac{\partial\psi}{\partial x})$$

(where H is the Hamiltonian operator and $\Im$ stands for "the imaginary part of"). The first of these equations is recognized as the usual Schroedinger equation, while the second one says how the position is driven by ψ.

The position has a privileged role in Bohmian mechanics, what suggests the question of how one can handle those physical situations in which the position is hardly seen to be a natural candidate for a hidden variable (think, e.g., of the intrinsic spin of a particle). But the headquarters of Bohmian mechanics are able to state that also these situations can be dealt with. It appears that Bohmian mechanics is in full agreement with quantum mechanics: no departure between them is envisaged. This might raise the question whether Bohmian mechanics carries mainly an epistemological interest, without aiming at a challenge to quantum mechanics.

(ii) A traditional way of featuring the branching point between classical and quantum mechanics is the study of the algebraic structure formed by the two-valued observables (also called yes-no experiments, or events, or properties, or propositions) [16]. In the classical case they are represented by the subsets of the phase space associated to the physical system; in the quantum case they are represented by the closed subspaces of the (separable, complex) Hilbert space associated with the system. In both cases the collection of two-valued observables is naturally endowed with an order relation (set theoretical containment). But if one looks deeper into these partially ordered structures the branching point comes in view: in the classical case one has to do with Boolean algebras while in the quantum case weaker structures emerge, namely orthomodular lattices, whose study goes back to the seminal 1936 paper of G. Birkhoff and J. von Neumann [26]. The names of many other distinguished mathematicians mark the itinerary of the study of orthomodular structures. Orthomodularity became a guideline of intense research work not only in mathematics but also in logics and in foundations of quantum mechanics. One might speak of the saga, or epopee of orthomodularity.

One particular aspect of the orthomodularity issue has been the "coordinatization problem": characterize lattice theoretically those orthomodular lattices that are the lattices of the closed subspaces of a separable, complex (or real, or quaternionic) Hilbert space. In other words, the solution of the coordinatization problem would provide a reconstruction of the Hilbert-space formulation of quantum theory starting from mathematically simple and physically plausible assumptions.

In the sixties and seventies the coordinatization program made a good part of its path [16,27]: conditions were found that make an orthomodular lattice representable by the closed subspaces of an Hermitean space $\{V, K, \langle\cdot,\cdot\rangle\}$, where V is a vector space, K a field with an involution (a *-field) and $\langle\cdot,\cdot\rangle$ an Hermitean form from $V \times V$ into K. But the question remained open of how to force this Hermitean space into a Hilbert space over $\mathbf{R}$ (the reals), $\mathbf{C}$ (the complexes), or $\mathbf{Q}$ (the quaternions). For a while it was conjectured that the orthomodularity condition might be enough; in this context orthomodularity reads as the condition that the sum of any closed subspace M of V and its orthogonal $M^\perp$ equals V. But in 1980 H. Keller [28], a mathematician of the Zurich school, put a decisive end to this speculation: he constructed a "nonclassical" Hilbert space, that is an infinite-dimensional orthomodular Hermitean space over a *-field different from **R**, **C**, **Q**. Quite recently, M.P. Soler [29], a young researcher of the same Zurich group, found the answer to the above question by the following sharp theorem: If the orthomodular Hermitean space has an infinite orthonormal sequence then K=**R**, **C**, or **Q** and $\{V, K, \langle\cdot,\cdot\rangle\}$ is the corresponding Hilbert space.

This result puts a fundamental stone in the program of recostructing the Hilbert space formulation of quantum theory. It outlines that the saga of orthomodularity, though half a century old, is still producing fresh results and fresh scholars.

(iii) The relationship between quantum and classical behaviour still involves open problems. Atoms exhibit the quantum behaviour but when a large number of them are brought together to form a compound system then the quantum behaviour progressively disappears; when the macroscopic scale is attained the physical system as a whole (e.g., the center-of-mass motion) forgets about quantum mechanics. Why the composition of quantum systems progressively creates classical objects? Why we never experience a macroscopic object in a nonlocalised state (think of the state propagating along the two arms of an interferometer) while its atoms, if isolated, can live in such a state? Why quantum mechanics allows many more states than we seem to encounter for macroscopic objects? Too crude an attitude toward these questions can lead to paradoxes (Schrödinger's cat is a popular example); the difficulty to tackle these questions supports the puzzling aspects of the quantum measurement theory, based, as it is, on the attempt to use a quantum-like description for both the measured system and the (macroscopic) measuring instrument.

Mesoscopic physics, that is the study of physical systems that are intermediate between the atomic and the macroscopic scale, appears to be the natural framework to explore the transition between quantum and classical behaviour. It is indeed a very promising and fast developing research field [30].

From the theoretical side, some ideas have called much attention. Ghirardi, Rimini and Weber (GRW) conjectured modifications of quantum dynamics, assuming that at random times any microscopic system is affected by a process of spontaneous localisation in position [31]. This localisation mechanism would be hardly observable on a microscopic scale, since it is expected to occur with an average period comparable with the age of the universe, but it is cumulative for a compound system thus producing the desired classical behaviour of macro-objects. The GRW model is not merely interpretational: it carries in principle testable consequences, though the feasibility of these tests is still somewhat out of reach. The future will perhaps say whether Nature has chosen the GRW mechanism.

Another idea to explain why macroscopic objects always appear to us in a small classical subset of a much larger quantum menu of states goes back to the differences in predictions of quantum theory for systems which are closed (i.e., isolated) or open (i.e., interacting with the environment). The quantum theory of open system was developed some decades ago [32] but the fact that *openess* can be responsible for transition from quantum to classical was fully recognized more recently. A "decoherence" effect [33] emerges which results in a selection process that dynamically eliminates nonclassical states: a generic initial state from the rich quantum menu will evolve so that only certain classical states will be left on the scene.

(iv) The observables of standard classical mechanics are real functions on the phase space of the physical system; the observables of standard quantum mechanics are self-adjoint operators on the Hilbert space associated with the physical system. Since some years the usefulness of some generalisation of the notion of observable became apparent. Actually, both definitions can be seen as particular cases of the following more general one: An observable is an affine mapping from the convex set of states into the convex set of probability measures on some measurable space. This space is understood as the one where the observable takes values, typically the real line $\mathbf{R}$. The above notion of observable is physically natural: it translates the fact that a state of the physical system determines the probability of each possible outcome of an observable.

The relevant novelty carried by this notion of observable lies in the fact that also unsharp, or fuzzy, observables are encompassed. To visualize this fact, consider an observable A, think of a measuring instrument for A and focus attention on some subset X of the reading scale of the instrument. Then consider the property (A, X) expressed by the statement "the value of A lies in X", or in other words "the pointer of the instrument appears in X". When one deals with sharp observables, as the ones of standard classical and quantum mechanics, this property admits only two outcomes, say "yes" and "no", or 0 and 1 (notice that we are not claiming that one of these outcomes will occur with certainty: an indeterministic context is admitted). But it is conceivable that the property (A, X) might be not two-valued, having a more complex array of outcomes. This possibility captures an idea of unsharpness or fuzzyness. As an example we might imagine that a noise makes the pointer of the measuring instrument not stable, so that one is not faced with either seeing the pointer stopping inside or outside X.

The usefulness of going beyond sharpness is a known fact in set theory: the construction of a fuzzy-set theory goes back to the sixties [34]. Similarly, the usefulness of going beyond sharpness can manifest itself in logic where the study of fuzzy logics represents a fruitful and promising research tool [35]. The study of the mathematical structures that can host the notion of fuzzy properties represents a new chapter of mathematics: having in mind that orthomodular partially ordered sets are the typical carriers of sharp properties, the problem arises of how these structures have to be relaxed to accomodate fuzzy properties [36,37].

When the notion of observable formulated above is applied to the set of states of a quantum system (the set of all density operators in a Hilbert space) it produces, besides the standard sharp observables represented by the self-adjoint operators, also a new class of observables, the POV (Positive Operator Valued) measures. The Hilbert space edifice of quantum mechanics implemented by the additional "floor" of POV measures is a consistent context hosting also fuzzy observables and fuzzy properties [36]. Its study, and the analysis of the associated experimental situations, received much attention in the last years [38].

When the mentioned notion of observable is applied to a classical-like set of states, that is to a simplex, something surprising is waiting us: the fuzzy observables can exhibit typical quantum features [39]. For instance, they can exhibit uncertainty relations though being comeasurable (i.e., admitting a joint observable). The old tenet according to which Heisenberg's uncertainty relations can only refer to non-comeasurable observables has to be revisited. As another example, fuzzy observables in a classical context can give rise to probabilities not admitting a classical Kolmogorovian representation, hence violating Bell-type inequalities [40]. Again, the traditional tenet according to which Bell-type inequalities cannot be violated in classical contexts has to be revisited. These facts fit with the possibility recently proved [39] of constructing a classical extension of quantum mechanics: we refer to a framework in which a classical-like set of states is determined that can be surjectively mapped into the usual quantum set of states so that every quantum observable finds a classical representative preserving all statistical features.

5. SUMMING UP

We have mentioned some among the many trends in foundation of physics. We hope they are enough to show the vitality of the research work in this field as well as the fact that many open problems are still there. The quantum behaviour, about 70 years after having been discovered, is still dominating the scene. The reliability of quantum theory is all the more surprising when one considers the extension of its range of validity: from the atomic scale down to the subnuclear one, and up to the astronomical scale (white dwarfs, neutron stars, ...). No experimental fact has emerged up to now casting doubts about the main concepts of quantum mechanics. As nobody can seriously believe that "quantum" is the last word of Nature, future research work on the limits of validity of quantum theory represents an intriguing challenge.

Department of Physics
University of Genoa

REFERENCES

1. P.W. Anderson, *Science* **177**, 393 (1972)
2. Y. Ne'eman and Y. Kirsh, *The Particle Hunters*, Cambridge University Press, (1986)
3. E.G. Beltrametti and J.M. Levy-Leblond (eds.), *Advances in Quantum Mechanics*, Plenum (NATO ASI Series B:Physics, vol. 347) (1995)
4. H. Rauch, *Achievements in neutron interferometry*, in [3]
5. J. Mlynek, V. Balykin and P. Meystre (eds.), *Optics and Interferometry with Atoms*, May 1992 issue of *Applied Physics* **B**
6. J.L. Staudenmann, S.A. Werner, R. Colella and A.W. Overhauser, *Phys. Rev.* **A21**, 1419 (1980)
7. A. Tonomura, *Electron interferometry and holography*, in [3]
8. G.I. Opat, *Interferometry with particles of non-zero mass: topological experiments*, in [3]
9. J. Tersoff and D.R. Haniann, *Phys. Rev.* **B31**, 805 (1985)
10. M.F. Crommie, C.P. Lutz and D.M. Eigler, *Science* **262**, 218 (1993)
11. G.P. Collins, *Physcs Today* **46**, n.11, p.17 (1993)
12. M. Devoret, *Single-electron effects in nanostructures*, in [3]
13. M.A. Kastner, *Physics Today* **46**, n.1, p.24 (1993)
14. H. Dehmelt, *Rev. Mod. Phys.* **62**, 525 (1990)
15. S. Haroche, *Experiments with single atoms, molecules or photons*, in [3]
16. E.G. Beltrametti and G. Cassinelli, *The Logic of Quantum Mechanics*, Addison-Wesley, Reading (1981)
17. J. Bell, *Physics* **1**, 195 (1964)
18. A. Aspect, P. Grangier and G. Roger, *Phys. Rev. Lett.* **47**, 460 (1981)
19. A. Aspect and P. Grangier, *Experimental tests of Bell's inequalities*, in [3]
20. B.C. van Fraassen, *Synthese* **52**, 25 (1982)
21. I. Pitowsky, *Quantum Probability-Quantum Logic*, Lecture Notes in Physics **321**, Springer-Verlag (1989)
22. E.G. Beltrametti and M. Maczynski, *J. Math. Phys.* **32**, 1280 (1991); **34**, 4919 (1993)
23. E. Einstein, B. Podolsky and N. Rosen, *Phys. Rev.* **47**, 777 (1935)
24. D. Bohm, *Phys. Rev.* **85**, 166 (1952); **85**, 180 (1952); **89**, 458 (1953)
25. D. Durr, S. Goldstein and N. Zanghi, *J. Stat. Phys.* **67**, 843 (1992)
26. G. Birkhoff and J. von Neumann, *Ann. Math.* **37**, 823 (1936)
27. S.S. Holland, *Bull. Amer. Math. Soc.* **32**, 205 (1995)
28. H. Keller, *Math. Z.* **172**, 41 (1980)
29. M.P. Soler, *Comm. Algebra* **23**, 219 (1995)
30. C. Di Castro, F. Guerra and G. Jona-Lasinio (eds.), *Mesoscopic Physics and Fundamental Problems in Quantum Mechanics*, May-June 1995 issue of *Il Nuovo Cimento* **110**
31. G.C. Ghirardi, A. Rimini and T. Weber, *Phys. Rev.* **D34**, 470 (1986)
32. E.B. Davies, *Quantum Theory of Open Systems*, Academic Press (1976)
33. W.H. Zurek, *Physics Today* **44**, n.10, p.36 (1991)
34. L.A. Zadeh, *Information and Control* **8**, 338 (1965)
35. M.L. Dalla Chiara, *Int. J. Theor. Phys.* **34**, 1331 (1995)
36. G.P. Cattaneo, *Int. J; Theor. Phys.* **32**, 1709 (1993)
37. R.J. Greechie and D.J. Foulis, *Int. J. Theor. Phys.* **34**, 1369 (1995)
38. P. Busch, M. Grabowski and P. Lahti, *Operational Quantum Physics*, Springer-Verlag (1995)
39. E.G. Beltrametti and S. Bugajski, *J. Phys. A: Math. Gen.* **28**, 3329 (1995)
40. E.G. Beltrametti and S. Bugajski, *J. Phys. A: Math. Gen.* **29**, 247 (1996)

6
Methodology

PETER ACHINSTEIN

THE EMPIRICAL CHARACTER OF EVIDENCE[1]

We philosophers of science have long debated the issue of what counts as evidence for a scientific theory or hypothesis. Today there is as much disagreement as ever. We argue about whether evidence or confirmation should be understood by reference to probability, whether it is quantitative, whether it is inductive, hypothetico-deductive, or neither, whether predictions of new phenomena provide stronger evidence than explanations of old one, whether there are universal standards of evidence or whether such standards vary from one person or community to the next, and even over whether scientists believe hypotheses on the basis of evidence or solely because of interactions with peers.

No doubt these debates keep our juices flowing. But this state of affairs is discouraging, not only for its seeming lack of progress and agreement, but because of something else as well. You would think that scientists, who, after all, are very much concerned with what counts as evidence for what, would take an interest in, and benefit from, philosophical accounts of evidence. Yet most scientists pay little heed to philosophical theories of evidence. Partly this is a result of specialization; it is hard enough to keep up in your own field. Partly this may be the result of the lack of general agreement in philosophy about this or any other issue. But I want to suggest also that scientists believe that they, not philosophers, are in the best position to judge whether, and to what extent, some body of information confirms some hypothesis. They may view this as something other than a philosophical question. They may view it as an empirical, not an a priori, issue. In what follows I want to ask whether such an empirical view of confirmation can be defended.

Many, though by no means all, philosophical accounts of evidence subscribe to the following two theses:

> *Thesis 1*: **Objectivism.** Whether a body of information is evidence for some hypothesis, and how strong that evidence is, is an objective fact. It does not depend on, or vary with, what one person or group knows or believes.

Suppose I have just swallowed a quart of lead paint. We might say that this fact is evidence — quite strong evidence — that I will become ill. If it is, the objectivist will say, it is evidence for this even if no one, including

[1] A more detailed version of the argument in the present paper can be found in my "Are Empirical Evidence Claims A Priori?" *British Journal for the Philosophy of Science*, 46 (1995), 447–473.

M.L. Dalla Chiara et al. (eds.), Structures and Norms in Science, 23–34.

myself, realizes that I drank a quart of paint or that doing so usually has disasterous consequences. In this respect the concept of evidence functions like the concepts "sign" and "symptom." Something can be a sign or symptom of something else without anyone realizing this, indeed, without anyone knowing or believing that the sign or symptom is present.

> *Thesis 2*: **A Priorism.** Whether, and the extent to which, some body of information, if true, confirms an hypothesis is an a priori, not an empirical, question. It can be settled completely by logical or mathematical computation.

To be sure, whether the proposition reporting the evidence is true, and whether the hypothesis is true, are both empirical issues. But whether, and to what extent, the evidence report, if true, supports or confirms the hypothesis is a matter to be determined completely a priori.

Both the objective and the a priori theses are held by a range of philosophers with different views about evidence. Let me mention just these: Carnap (who holds a view defining evidence as logical probability), Hempel (whose satisfaction theory of confirmation in the 1940's offers one solution to his famous paradox of the ravens), Glymour (with the well-known bootstrap theory of evidence), hypothetico-deductivists (who define evidence in terms of logical entailment), and at least some inductivists who claim that e confirms h if and only if h is an inductive generalization of e or if h is entailed by an inductive generalization of e. In all of these cases whether, and to what extent, e if true supports h is an objective fact that has nothing to do with what any person or group knows or believes. Moreover, it is an a priori fact whose truth is to be determined completely by calculation, not by observation or experiment. Carnap draws an analogy between inductive and deductive logic. Whether e deductively implies h is an objective fact that is settleable a priori. An analogous claim is made for inductive logic, where the question is whether and to what extent e supports h.

Now, in order to generate the problem I want to deal with I will introduce a third thesis that seems very obvious. Let me first state it in a quite general way.

> *Thesis 3*: **Flawed Evidence.** If the procedures for gathering evidence for an hypothesis are flawed, then this affects whether and how much the evidence gathered supports the hypothesis.

Consider a very simple example. Suppose a drug company is testing the hypothesis that a new drug D for relieving ordinary headaches is effective in approximately 95% of the cases, i.e.,

h: Approximately 95% of those who take drug D to relieve headaches obtain relief from using D.

Furthermore, suppose a study is made to test this hypothesis. The results are reported as follows:

e_1: 1000 patients with headaches were given drug D; 950 of these had relief, while 50 did not.

You might be tempted to say that e_1 is pretty good evidence for h. But wait! Suppose there is another study of 1000 headache patients who were similar to those in the first study, with one significant exception. Those in the second study were not given drug D or anything else. Despite this 990 headache sufferers (40 more than in the first study) experienced relief of their symptoms and did so more quickly than those in the first study. We ought to conclude that the first study was flawed with respect to the drug's causal efficacy. No control group was reported in the first study. And when, in effect, the second study functions as a control group there is the strong suggestion that the drug has no causal efficacy. We ought to conclude that this information about the lack of a control group — about the absence of any determination of how headache sufferers fare without taking drug D — should affect whether and to what extent the results of the first study (i.e., e_1) support the hypothesis h about the effectiveness of the drug.

Consider some other possible evidential flaws. Suppose all of the headache sufferers in the first study turned out to be three year old girls with very mild symptoms. Since the hypothesis h does not confine itself to that select group of headache sufferers, this revelation concerning the lack of variety in the evidence e_1 will affect the extent to which e_1 supports h. Suppose, finally, it turns out that all of those in the study were also taking some other drug D' in addition to D, where D' relieves headaches in 95% of the cases and prevents D from doing so (if it can). Such a fact, like the previous one, will affect whether, or at least the extent to which, evidence report e_1 supports hypothesis h.

How does, or should, this Flawed Evidence thesis affect the other two theses about evidence — the Objectivity thesis and the A Priori thesis? It appears that the three are inconsistent. Consider again our drug hypothesis h (that the drug is approximately 95% effective), and the reported results e_1 (that of the 1000 headache sufferers who took drug D, 950 got relief). According to the Objectivity thesis, we are to suppose that whether, and to what extent, result e_1 supports hypothesis h is an objective fact. According to the A Priori thesis, we are to suppose that whether, and to what extent, e_1 if true supports h is to be determined completely a priori — by some type of arm-chair calculation. But how can this be so given the Flawed Evidence thesis? If the procedure used to gather evidence e_1 was flawed — if, for example, no control group was tested, if subjects were unvaried in age, sex, and severity of symptoms, if the subjects were all taking another drug that relieves headaches — this will affect whether, or the extent to which, result e_1 supports h. But whether such evidential flaws exist is an empirical not an a priori issue. Given just the test results reported in e_1 we cannot know a priori, but only empirically, whether a control group was also tested, whether those tested in e were varied in age, sex, and severity of symptoms, and whether they were taking drugs in addition to D.

Hold on, you may say, trying to be as sympathetic as possible to the a priorist. Just look at the evidential report e_1. It says nothing at all about whether or not a control group was tested, whether the subjects were sufficiently varied, or whether they were taking other drugs. So, a priori, you would not accord h a very high degree of confirmation based just on the evidential report e_1. And this is something you can say a priori. To quantify matters, suppose that on the basis of the report e_1 alone — absent information about control groups, variation, and other drugs — you calculate a priori that the degree of support e_1 confers upon h is significant, say, greater than 1/2, but not extremely high. Now suppose that new information e_2 is discovered which indicates the aforementioned evidential flaws — no control group, no variation, and the fact that other drugs were taken. You now determine that the degree of support given to h by e_1 together with this new information e_2 is considerably less than 1/2. And again you do so by a priori calculation. What we are forgeting, the a priorist will insist, is to apply what Carnap calls the "Requirement of Total Evidence," which he states as follows:

> **Requirement of total evidence**: In the application of inductive logic to a given knowledge situation, the total evidence available must be taken as a basis for determining the degree of confirmation.[2]

So if, to begin with, e_1 by itself is our total (relevant) evidence available, and we calculate a priori that the degree of confirmation of h on e_1 is, say, .55, then that is the appropriate number to use in determining what to believe. If later we acquire new information e_2, and we calculate a priori that the degree of confirmation of h on $e_1 \& e_2$ is, say, .35, then to be sure there is a change in the confirmation number we are to use. But this is all to be decided a priori. The question of whether we know e_1 together with e_2, or just e_1 alone, to be true is an empirical issue. But whether and to what extent e_1 by itself, if true, confirms h is a priori and is not affected by whether or not e_2 is true. Similarly, whether and to what extent e_1 together with e_2, if true, confirms h is a priori, and is not affected by any other empirical facts.

Carnap's requirement of total evidence seems to save the day for those espousing objectivity, a priorism, and the thesis about evidential flaws. Any flaws once discovered and incorporated into the evidence statement will affect degree of confirmation. But is that enough? Is the thesis about evidential flaws fully satisfied, and if not should it be? This needs more discussion, in view of a crucial fact to which I now turn, viz. the fact that evidential flaws may be entirely unexpected.

Let us consider a new and more sophisticated study for drug D. In this study a large sample of patients with headache symptoms is tested by being

[2] Rudolf Carnap, *Logical Foundations of Probability* (University of Chicago Press, 1962), p. 211.

given drug D. There is an equally large control group of headache sufferers who are not given D but a placebo. Both groups are varied with respect to properties that we have some reason to believe might be causally relevant with respect to relief of headaches, such as age, sex, dosage, and severity of symptoms. Neither group is taking other drugs. Given the total information available to us, there is no reason at all to believe the study is flawed. Let b (for background information) contain the information I have just cited concerning how the study was conducted, and let e_3, the results of the study, be the following:

e_3: 95% of those in the study taking drug D got relief of their headaches, while none in the control group did.

Suppose, however, that despite our best efforts there is a flaw in the study, although we do not know this, nor is it reasonable for us, given what we do know, to expect such a flaw. What might such a flaw be? Well, suppose that unbeknownst to us, all of those in the study who took drug D also happen to eat some exotic fruit after the onset of a headache, while no one in the control group does. And it is this exotic fruit, not drug D, that is relieving the headaches. But, given what we know, we have no reason to believe that one group eats this fruit while the other does not; nor do we have any reason to believe that this fruit can relieve headaches.

Here is the question I want to raise. Should we say that, given our background information concerning how the test was conducted, the fact e_3 that 95% of those taking drug D got relief while none in the control group did constitutes (strong) evidence for the hypothesis h that approximately 95% of those who take drug D to relieve headaches obtain relief from the drug? Or should we say that although, in the light of our background information, e_3 seems to be (strong) evidence for h, nevertheless it is not because the study is flawed?

The Flawed Evidence thesis, as formulated earlier, requires that we draw the latter conclusion. It requires that if there is a flaw in gathering evidence for an hypothesis then this affects whether, and how much, the evidence supports the hypothesis. And it does so whether or not such a flaw is known to us or could reasonably have been expected. The Flawed Evidence thesis, as formulated, just speaks about flaws. It does not restrict such flaws only to ones known or reasonably expected. And surely a flaw of the sort I have described above is a genuine flaw. If it had been known or expected that those in the study taking drug D also eat the exotic fruit which relieves headaches, while none in the control group eat the fruit, then results e_3 would not be considered particularly strong evidence for the efficacy of drug D.

Suppose, then, we retain the Flawed Evidence condition, as earlier stated. What implications will this have? If we also retain the Objectivity thesis, then we will have to abandon the A Priori thesis. This is because whether, and to what extent, in the light of background information b, test results e support an hypothesis h depends not just on information reported in e and b. It depends in addition on whether there are any flaws in the evidence gathering that are unexpected given e and b. But whether there are such flaws is an empirical fact, not an a priori one. It is not a fact that can be determined a priori given information in our possession. If such an empirical fact can affect the truth of claims of the form "e is evidence for h, given b" or "given b, e confirms h to degree r," then, contrary to the a priori thesis, the latter evidence claims are empirical, not a priori.

What is a poor a priorist to do, assuming that he wants to retain objectivity? Obviously abandon the Flawed Evidence thesis, or modify it appropriately. Here is how he can modify it. Say that only known or expected flaws in evidence gathering procedures can affect the strength of the evidence. Suppose there is an evidential flaw in the way our results e were obtained to test hypothesis h. But suppose that, given our background information b, together with e, there was no reason to expect such a flaw to exist. Then the fact that it does exist has no effect on whether, or the extent to which, test results e support hypothesis h. So in my previous example, since the flaw involving the exotic fruit is completely unexpected, given what we know, its existence has no effect on whether or the extent to which the test results involving drug D confirm the hypothesis about the drug's efficacy. If we determine, a priori of course, that our test results e provide strong evidence for hypothesis h given our background information, then this is true whether or not the unexpected flaw exists. If we discover the flaw, and add this to our background information b, then, in the light of this modified b, the test results e will no longer provide strong evidence for h. But now we have changed the background information. And when we do so we can determine a priori whether this changes the degree of confirmation.

Accordingly, to retain the A Priori thesis, we can modify the Flawed Evidence thesis so that only those flaws that are known or expected, given our total information, can affect whether and to what extent that information supports an hypothesis. Obviously there are other strategies for avoiding the inconsistency besetting the three theses about evidence. Here I want to consider one alternative, which may seem radical to some but to me seems reasonable. It is to retain the original Flawed Evidence thesis, retain the Objectivity thesis, and reject the A Priori thesis. What concept of evidence will result if we do these things?

First, the resulting concept will be objective in this sense: Whether e if true confirms h, and how much it does so, is an objective fact that is not affected at all by who believes or knows what.

Second, the present concept is an empirical one in this sense. A claim that information e, if true, is evidence for an hypothesis h, or that it supports h to such and such an extent, is an empirical claim, not an a priori one. It can be defeated by an empirical discovery concerning how the information was gathered. So in our drug example, the claim that the results of the test constitute strong evidence that the drug is 95% effective is defeated by the empirical discovery about the exotic fruit, which showed the test to be flawed.

Third, despite what might otherwise be thought, there is a relationship between the present empirical concept of evidence and that of belief, though this relationship is different from the one exemplified by the a priori concept. The difference is between (i) there being a reason to believe something and (ii) someone's being justified in believing something. Let us begin with the a priori concept. For Carnap and other a priorists there is an important relationship between evidence and what one is justified in believing. The relationship, which is implied by Carnap's formulation of the "requirement of total evidence," is this:

> If e is evidence for h (or if e confirms h to degree r), and if e represents one's total (relevant) empirical knowledge, then one is (to some extent) justified in believing h on the basis of e (or one is justified in believing h to the degree r on the basis of e).

By contrast, in the case of the empirical concept of evidence, the relationship between evidence and belief is this:

> If e is evidence for h (or if e confirms h to degree r), then e is at least some good reason for believing h (to the degree r).

This concept of a "good reason to believe" is not the same as the a priorist's idea of what one is justified in believing. The latter depends on what else is believed by the person or group. The former does not. The fact that I have just swallowed a quart of paint is a very good reason to believe I will become ill. It is a good reason to believe this even if no one, including me, happens to be justified in believing this — because, say, no one knows it is paint I have swallowed, or that paint, or that much of it, is poisonous.

Suppose, then, we agree that it is at least logically possible to reject the A Priori thesis while retaining the Objectivity and Flawed Evidence theses. What emerges is an objective, empirical concept of evidence that provides an objective reason for belief that is not tied to what anyone in particular knows or believes. It is one thing to urge that such a concept is possible. Is it ever actually employed? And even if it is, should it be?

Let me urge — perhaps with reckless abandon — that the answer to both questions is Yes. As for actual employment, my argument is this. Numerous evidential statements of the form "e is evidence for h" are in fact defended or criticized by appeal to empirical claims. Instead of manufacturing fictitious examples, let me give a real one. In 1897, J.J. Thomson discovered the electron.[3] He did so on the basis of a series of experiments with cathode rays. In the first of these experiments he connected a cathode tube to a bulb containing a magnet capable of deflecting the rays into cylinders containing an electrometer. He obtained the following experimental result:

[3]J. J. Thomson, "Cathode Rays," *Philosophical Magazine* 44 (1897), 293–316.

Thomson's experimental result e_4: Whenever the cathode rays were deflected by a magnetic field, an electrometer in their path showed a large negative charge of electricity; when the rays were not deflected into the electrometer, the latter showed no such large charge.

Thomson took this experimental result e_4 to be evidence for the following hypothesis:

Thomson's hypothesis: Cathode rays carry with them a negative charge.

However, immediately following his claim that e_4 is evidence for his hypothesis Thomson noted another experimental fact:

e_5: As the cathode rays impinge on the electrometer the deflection of the electrometer increases up to a certain value and then remains the same, even though the cathode rays continue to impinge on it.

If, as Thomson claims, the deflection of the electrometer is evidence that cathode rays carry a negative charge, then, it might be objected, the longer the cathode rays strike the electrometer the greater the charge should be — which is contrary to what is observed. In other words, the experimental result e_5 might be thought to cast doubt on the statement that Thomson's initial experimental result e_4 is evidence for the hypothesis that cathode rays carry a negative charge. In response to such an objection Thomson makes the following empirical claim:

C: When the cathode rays pass through the cylinders containing the electrometer, the air between the inner and outer cylinders becomes an electrical conductor, and the charge escapes from the inner cylinder to the earth; so the charge does not continuously increase.

In short, Thomson is defending his original evidential statement that his initial experimental results reported in e_4 are evidence for his hypothesis. He is doing so by saying that e_4 is evidence for this hypothesis despite the new experimental result e_5. He is defending this by appeal to the empirical claim C regarding the conductivity of the air betweeen the two cylinders. His defense of the statement that e_4 is evidence for his hypothesis is clearly not a priori. This seems to fit much better with an empirical concept of evidence rather than an a priori one. (How an a priorist might reply I will take up in a moment.)

Not only does Thomson offer an empirical defense of a claim that his experimental results confirm his hypothesis, he also rejects another evidential claim on empirical grounds — a claim that was made 14 years earlier by Heinrich Hertz.[4] In 1883 Hertz conducted experiments on cathode rays in which he tried to deflect them by means of an electric field. Hertz took his experiments, which showed no electrical deflection, to provide strong evidence that

[4]Heinrich Hertz, *Miscellaneous Papers* (London, 1896), p. 254.

cathode rays are not charged particles, but some type of ether wave. In 1897 Thomson challenged Hertz's evidential claim, not by producing some a priori calculation, but by making an empirical claim, which he later verified experimentally. His empirical claim was that when cathode rays pass though the gas present in a cathode tube they make the gas an electrical conductor, which then screens off the electric force from the charged particles. This screening off effect should be reduced, Thomson predicted, if the gas in the tube is more completely evacuated. Thomson then conducted new experiments with more fully evacuated tubes in which an electric field deflected the cathode rays. In effect Thomson was claiming that there was an evidential flaw in Hertz's earlier experiments — viz. the lack of sufficient evacuation of gas in Hertz's cathode tubes. Moreover, he seems to be saying, given Hertz's knowledge and techniques at the time, this was a flaw that Hertz might not reasonably have expected. Before his own experiments with more evacuated tubes Thomson considered his empirical claim that Hertz's cathode tubes were not sufficiently evacuated at least to cast some doubt on the statement that Hertz's experimental results constitute strong evidence that cathode rays are not charged particles. After his own experiments with more evacuated tubes Thomson took the results of these experiements to refute Hertz's evidential claim.

In short, Thomson defended one statement of the form "e is evidence for h" empirically, and he attacked another statement of this form empirically. This seems to fit much better with an empirical rather than an a priori concept of evidence. Or does it? An impatient and determined a priorist will surely have a reply. Here it is.

The evidence statements that Thomson appears to be defending or attacking on empirical grounds are incompletely formulated, since they fail to bring in background assumptions being made. If such background assumptions are explicitly introduced into the evidence statements themselves, then such statements will readily be seen to be a priori — a priori true or false, as the case may be. Recall Thomson's initial evidence statement:

> T: The fact that the electrometer showed a large negative charge when cathode rays were deflected into it but not otherwise is strong evidence that cathode rays carry a negative charge.

According to the a priorist, evidential claim T is incomplete. Part of what is missing in that statement is a background assumption that the only known cause of the separation of leaves in an electroscope is the presence of an electric charge; another missing element is a statement concerning the limited deflection of the leaves and Thomson's claim C regarding a cause of that limitation. Put all of this information into the evidence statement, the a priorist will insist, and you make it complete and a priori.

A supporter of the empirical concept of evidence — whom I shall hereafter call an empiricist — need not accept such ideas. First, why should one say that Thomson's initial evidential statement T is incomplete as stated? On what grounds does the a priorist claim it is incomplete? Is it just that T is not true a priori? That begs the question. Is is that T does not contain all the background information that could be cited in its defense? But then all empirical statements would be incomplete.

Second, an empiricist will want to distinguish the evidential statement a speaker is making from any empirical defense of such a statement that the speaker or others might offer. This, the empiricist will urge, reflects better than the a priorist's total evidence condition what scientists and others actually do when they make evidential claims. In practice they defend them empirically. This practice is better reflected with an empirical concept of evidence than with an a priori one.

Third, the empiricist can argue, what type of defense of an evidential claim is given, or called for, depends on what has been or might be questioned in the particular context of inquiry. Not all appropriate defenses of an evidential claim will, if added to that claim, transform it into an a priori statement. For example, if we add Thomson's defense of his evidential claim T to T itself an a priori statement does not emerge.

Finally, and most importantly, an empiricist will reject the a priorist's claim that by incorporating all the known background information into the evidential statement we can make the latter a priori. Recall the study involving the headache drug D that, given our information, looked perfectly designed to us. It employed a control group; it exhibited variety; none of those tested was taking other drugs known to relieve headaches; and so forth. The a priorist now claims that if such background information about how the test was conducted were incorporated into the evidence statement then that statement would be a priori. An empiricist concerning evidence, however, will remain unconvinced. The empiricist will say that even when such background information is included the resulting evidential statement is not a priori. It can be shown to be false by empirical means, viz. by discovering an unexpected flaw in the evidence-gathering procedure. This is what happened in the exotic fruit example. And the empiricist will say, it is also what happened with Hertz's evidential claim about cathode rays, when Thomson produced his later experiments.

So, I believe, what an a priorist needs to argue is this. Even if an empirical concept of evidence is sometimes, or perhaps usually, employed in the sciences, an a priori concept is superior. Why is it superior? Let me suggest two reasons an a priorist might offer; and let me reject them one by one.

First, an a priorist such as Carnap will say that when scientists and others gather evidence to test an hypothesis what they seek, what they care about, is that they be justified in believing (or disbelieving) the hypothesis on the basis of that evidence. And they will note that an a priori concept of evidence, but not an empirical one, ties evidence to what we are justified in believing.

An empiricist can make two responses. First, an a priori concept of evidence such as the ones proposed by Carnap, Glymour, and hypothetico- deductivists do not in fact tie evidence to what one is justified in believing. Suppose we take Carnap's approach, and suppose that in accordance with his preferred confirmation function $c^{\star}$, the degree of confirmation of hypothesis h on evidence e equals .95. Suppose further that e represents my total (relevant) empirical knowledge. Am I therefore justified in believing h on e to the degree .95? Not on your life. For suppose that neither I nor anyone I know is able to perform the logical, mathematical, and linguistic computatations in Carnap's system required to determine that the degree of confirmation of h on e is .95; and suppose that neither I nor anyone I know has any idea whether there is a calculation that will enable even Carnap himself to determine this. Then clearly I am not justified in believing h on e to the degree .95. This is a problem facing all of the other a priori systems mentioned.

The second response of an empiricist cuts philosophically deeper. Recall that the a priorist claims that when scientists and others gather evidence to test an hypothesis what they seek, what they care about, is being justified in believing (or disbelieving) the hypothesis on the basis of the evidence. An empiricist may agree that this is one of the things they may want or care about, but there is something deeper. What evidence-seekers want most fundamentally is to discover a good reason to believe (or disbelieve) the hypothesis. This is an objective notion of good reason that does not depend on what particular individuals or groups believe or know.[5] When Thomson sought evidence for the hypothesis that cathode rays are negatively charged particles he was not simply trying to show that he (but not say Hertz) was justified in believing this. He was trying to present a good reason (for anyone) to believe this, even if others were not yet in a position to believe it. This objective, person-independent concept of good reason for belief is, according to the empiricist, what we most fundamentally want when we seek evidence. And it is something provided by the empirical concept of evidence, but not the a priori one.

An empiricist, then, need not be at all moved by the a priorist's first reason for thinking an a priori concept of evidence is superior to an empirical one. The empiricist will simply deny that the most important reason for seeking evidence is to obtain person-relative justification. Rather it is to obtain person-independent reasons for believing. This, an empiricist can claim, is more in keeping with the objective thesis about evidence which both the a priorist and the empiricist espouse.

I turn briefly to a second reason the a priorist may offer for the claim that an a priori concept of evidence is superior to an empirical one. With an a priori concept the issue of whether, and to what extent, e if true confirms h can be settled with the highest form of certainty, a priori certainty. With an empirical concept of evidence this is impossible. To old-fashioned rationalists

[5]See Laura J. Snyder, "Is Evidence Historical?" in Peter Achinstein and Laura J. Snyder (eds.) *Scientific Methods: Conceptual and Historical Problems* (Krieger, 1994), 95–117.

about scientific statements generally this is no doubt an advantage. To empiricists about scientific statements generally this should not be an advantage. Carnap, Hempel, Glymour, and hypothetico-deductivists who are empiricists in other respects do not regard propositions in physics as inferior to those in mathematics on the ground that statements in physics being empirical lack the a priori certainty of statements in mathematics. Why should they be allowed to make the analogous claim in the case of evidence? I don't think they should be allowed to do so. Whether and to what extent Hertz's experiments supported his hypothesis that cathode rays are not charged particles should be open to as much empirical doubt as the hypothesis itself. The problem that J.J. Thomson found with Hertz's evidential claim was not that it lacked a priori certainty. The problem was that it was criticizable on empirical grounds.

A final very brief comment. If I am right, then if you want to determine whether or how much an evidential report confirms some scientific hypothesis, ask the scientist, not the philosopher.[6]

Department of Philosophy
Johns Hopkins University
Baltimore

[6]Special thanks are due to Laura J. Snyder for helping me see the light about evidence. My good colleague Robert Rynasiewicz put up his usual formidable obstacles to my vision.

STEVEN FRENCH

PARTIALITY, PURSUIT AND PRACTICE*

INTRODUCTION

My intention in this paper is to draw connections between three areas of recent interest in the philosophy of science, broadly characterised: the introduction of partial structures into the model-theoretic or 'semantic' approach, the underlying structure of pursuit and heuristics and the recent emphasis on scientific 'practice'. The connections are, perhaps, obvious, but worth highlighting nonetheless and they have not to my knowledge yet been fully explored.

I shall begin by considering the third vertex of my conceptual triangle, practice, and work backwards though my title.

PRACTICE MAKES SCIENCE

During the 1950s and 1960s there emerged, in the philosophy of science, an emphasis on the nature and role of models in science. This emphasis took two forms: on the one hand there was the development of what is now generally known as the 'semantic' or 'model-theoretic' approach, lead by Suppes and Beth and, later, Suppe and van Fraassen, according to which theories are regarded as families of mathematical models; on the other, there was the work of Achinstein, Hesse and Kuhn, which focused on models of physical systems and the role of analogy. Although proponents of the semantic view of theories have often drawn on the wide ranging use of models of the latter kind in support of their approach, the relationship between the two senses of 'model' was and still is, unclear.

Kuhn, in particular, placed models at the heart of scientific practice, where a model was to be understood in his sense of an 'exemplar' which effectively structures solutions to the problems of 'normal' science. More than the implicit relativism, more than the eschewing of any talk of truth, more even than the rather confused but according to the later Kuhn, central idea that occupants of different paradigms live in 'different worlds', it is this notion of

*I would like to thank James Ladyman for numerous discussions of these topics and for prompting many of the ideas presented here. Whatever is right or interesting about them is largely down to him. I would also like to thank Otávio Bueno, whose email conversations helped me get clear on a number of points. Finally, a version of this paper was presented at a meeting of the British Society for the Philosophy of Science and I would like to thank Jim Cushing, Colin Howson, David Papineau, Stathis Psillos, Mauricio Suárez and John Worrall for their interesting and useful comments. Of course, the responsibility for all errors and infelicities, philosophical or otherwise, is entirely mine.

M.L. Dalla Chiara et al. (eds.), Structures and Norms in Science, 35–52.

exemplars which was taken as fundamental by the advocates of the sociology of scientific knowledge, perhaps because it indicated a way in which science could be taken as interestingly rule-free. It is the exemplars which drive the conceptual extensions of the scientific 'net' through the modeling process: 'the production of new scientific knowledge entails seeing new situations as being relevantly like old ones.' (Pickering 1992 p. 4). This process in turn is regarded as radically open-ended — in the sense that '*nothing* within the net fixes its future development' (ibid.; my emphasis) — with 'closure' achieved via the mediation of socio-political interests. It is these two aspects — of rule-free extensions via modeling and analogy and a sense of openness — that I am primarily interested in here.

More recently, an alternative view has begun to be articulated which differs from the above in two crucial respects: i) a shift away from science-as-knowledge to science-as-practice, with an attendant elaboration of concepts which, it is claimed, speak directly to the latter; and ii) an insistence on the multiplicity of elements which can account for closure, linked to a view of the 'net' as 'patchy' and heterogeneous[1] (the beginnings of this swing away can be seen in Pickering 1990 and it is fully evident in his 1992 and Pickering and Stephanides (1992); with regard to ii) Hacking's influence cannot be understated and is made explicit in Pickering 1992[2]). Nevertheless, and despite these differences, this new approach retains the dual emphasis on the role of models in driving practice and the open-ended nature of this process.

Indeed, the importance of models is elevated even further in this account, since, with interests shoved aside, closure must be achieved in some other way, namely by the bringing together of various disparate 'extensions', or models — theoretical, phenomenological, experimental, or whatever — into 'association'. This is, of course, a non-trivial achievement and the difficulties involved fall under the heading of 'resistances'. These arise when models are brought into interaction in some way, when aspects of one model — certain fundamental relationships, say — are imported into another, for example. Resistance, then, although this is typically not made at all explicit, is a manifestation of structural discord.

These elements of the sociology of scientific practice, and, in particular, the emphasis on models and their extensions, are clearly laid out in Pickering's co-authored analysis of Hamilton's development of quaternions:

'A given model can be extended in an indefinite number of ways, nothing within the model itself foreshadows which should be chosen. Thus part of the problem of getting to grips with practice is that of understanding closure, of understanding why some individual or group extends particular models in particular ways. [This is the fundamental problem of

---

[1] Gremmen refers to this as the 'bucket' view of practice, in the sense that everything is thrown into the bucket! (Gremmen 1995).

[2] And interestingly enough, given the model-theoretic context of my paper, this particular influence with regard to the 'disunity of science' can in turn be traced back to Suppes and his 1974 set of lectures which resulted in the book *Probabilistic Metaphysics* (Suppes 1984).

the present paper on which Pickering and I differ so completely.] The solution to this problem appears to lie in the observation that models are not extended in isolation. Modeling typically aims at producing *associations* in which a plurality of projected elements hang together in some way. And the important point here is that the achievement of such associations is not guaranteed in advance - particular modeling sequences readily lead to mismatches in which intended associations are not achieved. *Resistances*, that is, arise in practice to the achievement of goals. Encounters with resistance set in train a process of *accommodation*, in which the openness of modeling is further exploited in trial-and-error revisions and substitutions of models, modeling sequences, and so on, aimed at proceeding further toward the intended association.' (Pickering and Stephanides, op. cit., pp.140–141)

There are claims here that are strikingly obvious, and there are claims here that are strikingly problematic, or at least so they seem to me[3]. That models are not extended in isolation will hardly come as a shock to the system of any but the most naive student of the field. Of course we may disagree as to the nature and composition of the embedding context but, at least in this example — and I take it to be representative of this view of scientific practice — the primary concern is with the contextual role of other models.

Likewise, there is little that is controversial in the assertion, as it stands, that 'resistances' arise in practice and are then dealt with by a process of accommodation. One obvious, and rather stubborn, piece of resistance would be the material world, a.k.a. reality, although Pickering himself could hardly be expected to accept the latter statement without some qualification (in his more recent work he has advocated a kind of 'pragmatic realism' which 'recognizes that the production and transformation of scientific knowledge in accommodation to resistance is inseparable from a larger process of the production and transformation of complex and heterogeneous forms of life.' (1990, p. 688)).

However, as Pickering notes, the more interesting case is that of conceptual practice, as in the mathematical example he explores with Stephanides, where the question as to what might count as 'resistance' is considerable less clear. This particular example is used to exemplify the decomposition of the modeling process into the triad 'bridging', 'transcription' and 'filling'. Thus, Hamilton's move from the geometrical representation of complex algebra in terms of points in a plane, to a three-place system '...established a bridgehead to a possible three-dimensional extension of complex algebra.' (Pickering and Stephanides, op. cit., p.148) As Pickering and Stephanides note, the latter is not produced out of nothing but can be regarded as a creative extension of the two-place system already known. In constructing this extension, Hamilton attempted to preserve the association between complex algebra and its geometrical representation in terms of a relation of one-to-one correspondence '...between the algebraic notation $x + iy + jz$ and suitably defined three-dimensional line segments' (ibid p.149). Relatedly, he also tried to preserve the rules of algebraic multiplication and here we see the operation of 'transcription' in action. As a consequence, terms appeared in the

[3]The term 'closure' is also to be found in Kuhn (1970).

relevant equations that were undefined, and the assignment of values to such terms represents what Pickering and Stephanides call 'filling' (the emergence of such 'gaps' being another general feature of the modeling process in their view).

Resistance arose in this process as the association between algebraic and geometrical representations of a one-to-one correspondence between elements and operations broke down and accommodation was sought by abandoning commutativity ($ij = -ji$). Further resistances arose as the two representations led to different results, and a further accommodation was made by introducing a third imaginary quantity (the product of i and j). This in turn resulted in a further bridge from three-place to four-place systems, those of quaternions.

As it stands, (and I agree that the above is a rather superficial gloss, although I believe my point can be carried forward) there seems little here that would be unacceptable to even the most traditional of philosophers of science. The resistance, in particular, comes not from any kind of socio-political reaction but from the inherent nature of the mathematical systems themselves; that is, from the mathematical *structures*. What the bridging process does, of course, is to import the basic structural elements successfully applied in one domain into another. Note that what we are talking about here is *importation* rather than creation; the 'target' domain must be structured to at least some minimal extent before bridging is attempted, otherwise resistance would simply never arise. In the kind of conceptual practice discussed above, the source of this minimal structure will be principally mathematical in nature, whereas in scientific practice I would claim that it is ultimately empirical, deriving perhaps from low lying phenomenological models. Given, then, that these domains are at least minimally structured in this way, it is not that surprising that further elaboration of the imported structure will encounter 'resistance' due to structural incompatibilities. Of course, if these incompatibilities are sufficiently extensive, they will prevent the given importation in the first place[4].

Seeking 'some accommodation to resistance' then involves 'tinkering' with various extensions that are available in the given conceptual space in a series of 'free' and 'forced' moves. Thus transcription can be regarded as a sequence of forced moves, insofar as it follows 'from what is already established concerning the base model' (Pickering and Stephanides, ibid p.150). Bridging and filling in, on the other hand, are 'free' moves in that they '...mark tentative choices within the *indefinitely open* space of cultural extension.' (ibid p.142; my emphasis)

'The outcome of particular modeling sequences is thus at once structured by choice but not determined by it; it is something genuinely to be found in practice.' (ibid)

[4]For both the platonist and empiricist philosophers of mathematics, these 'resistances' might be traced back to external 'reality', understood in some respective form, while for the formalist it would ultimately have to do with the rules for manipulation of symbols.

For Pickering the existence of 'free' moves is crucial to his conclusion that the products of 'practice' — quaternions, in Hamilton's case — are to be viewed as the contingently successful end points of the dialectic of resistance and accommodation. It is this aspect of contingency that is central to this kind of approach[5].

The earlier and much discussed model of 'opportunities-in-context' is now seen as covering the process of goal-formation in science and this is taken to be inherently contingent:

> 'I believe that one can see an underlying general structure to goal-formation in science, and that modeling and coherence are important concepts in unraveling that structure. But I do not think that one can offer causal explanations of goal-formation on this basis. To the contrary, there seems to be an ineradicable element of chance here. There is an explanatory gap that I cannot see how to bridge between possessing a given range of resources and assembling them into a coherent goal.' (Pickering 1990, p. 684).

As an example he gives Zweig's interpretation of the fundamental representation of the SU(3) group as referring to quarks, rather than purely mathematical entities, a move which Pickering claims cannot be understood or justified at all in terms of the context and resources available at the time. All that we can say by way of explanation is that 'it just happened'. (ibid., p. 721, fn 16).

This is heavily redolent of the 'flash of genius' view[6] and it is precisely on this point that Pickering's characterisation of the modeling process above is so problematic, in particular his claims that '[a] given model can be extended in an indefinite number of ways ... ' and '... nothing within the model itself foreshadows which should be chosen.' It is on the latter claim, in particular, that I wish to focus attention here.

THE PRACTICE OF PURSUIT

As discussed in the context of scientific practice, pursuit has been described as the 'nether region' between discovery and justification (Laudan 1980, p. 174). I should perhaps emphasise right away that I am not convinced that the boundaries between these domains can be sharply drawn, not on the grounds that sociological considerations emerge on both sides of the line, but rather that the appropriate epistemic attitude is the same across them. I shall return to this point below. With the positivist claim, famously repeated by Popper, that all that is epistemically interesting can be found in the domain of justification only, the domain of 'discovery', taken to cover both heuristics per se and pursuit, was left in a methodological vacuum. Obviously one way to fill this vacuum is with assorted interests and opportunities-in-context and if the

[5]It also provides one of the more fundamental connections between Pickering's earlier and later work.

[6]Again, we find this view expressed in Kuhn, op. cit. p. 119, as noted by Nersessian in a paper which addresses many of the issues touched on here from the perspective of a 'cognitive' approach (1993, p.11).

wall between domains is toppled also, these pour over into justification as well (see Koertge 1993). But this is to disregard the claim that there is actually no vacuum to be filled, that the domain of discovery has a rich and interesting and primarily *epistemic* structure of its own. This sort of claim has, of course, been made and elaborated by a number of philosophers of science. In the present discussion I shall draw in particular on the work of Post, who emphasises that the structure of the domain of heuristics derives from that of the 'heuristic objects' — the models themselves (Post 1971, p. 217).

Take the example that Pickering himself has referred to — Zweig's interpretation of the representation of SU(3). An alternative understanding of this move can be achieved via Redhead's account of supplying a physical interpretation for surplus mathematical structure within a theory (Redhead 1975). This is related to Post's heuristic guideline of 'Adding to the Interpretation' (closely connected with 'Taking Models Seriously'): 'We take a hitherto incompletely interpreted part of the abstract formalism of the theory, and give it a tentative interpretation of our own, at some level.' (Post, op. cit., p. 241) This interpretation, tentative as it is, is still some way from being put forward in the state of unfettered freedom that Pickering suggests — Zweig himself was not operating in a vacuum. Indeed, Pickering's own earlier work spells out just how detailed the context actually was:

> The idea of reducing the number of elementary particles by explaining some of them as composite was by no means new. As early as 1949, C.N. Yang and Enrico Fermi had written a paper entitled 'Are Mesons Elementary Particles?', in which they speculated that the pion was a nucleon-antinucleon composite. When the strange particles began to be discovered, the Fermi-Yang approach was notably extended by the Japanese physicist S. Sakata in 1956. (1984, p. 86)

Suitably formulated in group-theoretic terms, the Sakata model became the main rival to Gell-Mann's Eightfold Way, until the 1964 discovery of the Ω-particle resolved the competition in the latter's favour. As Pickering notes, representing the hadronic symmetries in group-theoretic terms 'positively invited' the view that hadrons were composite and the most well known and well developed position along these lines was the SU(3) tradition, whose fundamental representation — from which all others can be mathematically derived — contained three members[7]. 'From there it was but a small step to the identification of the fundamental triplet representation of SU(3) with a triplet of fundamental entities: the quarks.' (ibid., p. 87)

Zweig studied both the Sakata model and the reviews of SU(3) and, independently of Gell-Mann, came to the view that the observed multiplet hadronic structure could be accounted for in terms of two or three hadronic constituents. The next step was not only 'simple and direct' but obvious: 'As one would expect for a neophyte in theoretical physics, he treated quarks as

[7]The importance of the group-theoretic representation of such symmetries will be returned to below.

physical constituents of hadrons and thus ... derived all of the predictions of SU(3) plus more besides.' (ibid., p. 89)

Coming back to more general considerations regarding the structure of the domain of heuristics, the kinds of criteria that have been suggested can be taken over into theory pursuit; indeed, I think that an obvious case can be made for claiming that the theoretic guidelines of theory construction and development also function as reasons for theory choice in non-epistemic situations (or at least situations that are not explicitly epistemic — certain of these reasons may involve reference to other theories, suitably supported), that is, conferring appropriate warrant in theory pursuit. Such criteria include the much maligned General Correspondence Principle, simplicity (problematic, granted) and the retention of well confirmed universal invariance principles (Post op. cit.).

A nice example of pursuit in a situation where explicitly empirical factors had little or no role to play is that of the competition between the paraparticle and colour models of quarks in the mid-1960's. This episode is one of the more important in the history of modern physics as it resulted in the birth of quantum chromodynamics (QCD). It began with a statistics problem: there was good reason to believe that quarks should be assigned spin 1/2, and, therefore, according to the spin-statistics theorem, should be regarded as fermions. On the other hand, there were cases in which the most successful model was one in which the wave-function of the three quark collective was symmetric under interchange of any two quarks. The problem, then, was to reconcile the apparently fermionic nature of quarks with symmetric wave-functions. Greenberg proposed a particularly elegant solution to this problem which was derived from his long standing investigation of the possibility of particles with mixed-symmetry wave-functions: he suggested that quarks just were examples of these theoretical possibilities, or paraparticles (another example of according the formalism ontological significance). Shortly afterwards, Han and Nambu put forward an alternative solution to the problem, in which quarks were given an extra degree of freedom, subsequently denoted by 'colour'. The rest, as they say, is history.

Pickering has considered this episode from the perspective of opportunities-in-context and concluded that the colour model won the day because paraparticle theory was simply too obscure, too complex and too 'out of the ordinary' for most physicists (1984, pp. 94–95; private correspondence). Elsewhere I have argued (French 1995) that this kind of analysis simply cannot bear the weight of historical circumstance. Rather than turning to external factors to do with opportunities-in-context and the comparative unfamiliarity of the paraparticle model (a specific claim which I would dispute) in order to explain the scientific community's preference, we should focus on the objective characteristics of the models themselves[8]. In fact, the decisive factor, I claim, is that the colour model was able to be gauged, whereas the paraparticle one was not. Gauge invariance, of course, was incorporated within quantum electrodynamics (QED), and 'inherited', as it were, the immense empirical success of the latter. There was thus good reason to pursue a model which could be shown to similarly incorporate this invariance principle, over one which could not (pursuit being thus differential)[9]. The precise nature of this similarity between QED and QCD and, relatedly, that of the 'objective characteristics' of the models involved, will be touched on below[10].

However, my intention is not to bury the 'science-as-historically-contingent' approach under the weight of historical counter-example — the present work has a different axe to grind. The point I want to emphasise is two-fold: i) that the space of possibilities might be drastically more limited than proponents of this approach seem to appreciate, once we accept that even apparently 'free' moves in pursuit are structured by objective considerations; ii) that, relatedly, the structure of the model itself plays a crucial role in 'foreshadowing' which is chosen.

This is not to say that such foreshadowing is completely determined by the structural characteristics of the models concerned. As Post remarks, 'Following Wigner, we may look forward to the day when a new theory is uniquely given by a conjunction of invariance requirements. In most cases at present we are faced with a looser situation.' (op. cit., p. 227) How to characterize this 'looseness' or 'openness to further development' in a more formal manner is a fundamental problem in the philosophy of science. One possible way of doing this has been suggested in the context of that other current of thought regarding models, the semantic or model-theoretic approach. This also offers a suitable context for considering the structural characteristics of models and thus for accommodating the points noted above. Thus this paper can also be viewed as an attempt to extend this approach to considerations of theory pursuit.

PARTIALITY MAKES PRACTICE

As is by now well known, according to the model-theoretic approach theories are represented as families of models, rather than sets of axioms and theorems

[8]In private correspondence Pickering has acknowledged that, on this point, he relied upon Nambu who described Greenberg's work as 'very formal'. Nevertheless, he maintains that the choice to pursue colour was made in purely contingent terms regarding the routine techniques which could be applied.

[9]'Gauge invariance ... is used as a guide in constructing new field theories.' (Post op. cit., p. 227)

[10]As is very well known, the gauge invariance of quantum electrodynamics is inherited, in turn, from classical electromagnetism, although it acquires rather more ontological significance in the former than the latter as the existence of photons 'drops out' of such a gauge invariant theory.

given a partial interpretation via appropriately formulated correspondence rules. Typically the advantages of the former are touted as compared to the latter, although there is no reason to be dogmatic here. Typically, again, the approach is presented against the foil of the syntactic view, with the deficiencies in the latter laid bare, particularly the role of the correspondence rules and their status as internal components of the theory. Alternatively, and some might say, more appropriately, it is presented in terms of capturing the various uses of and relationships between, models in scientific practice. It is this way of going about things that I will take here, although the two are obviously not unrelated: showing how the semantic view can account for the employment of conjunctions of theories in biological explanations, for example, is undertaken in the context of a demonstration of the failure of the syntactic view in this regard, principally because of the nature of the correspondence rules (Lloyd 1988, Thompson 1989).

The rubric 'theories are to be represented in terms of their (mathematical) models' leaves open the nature of this representation: for van Fraassen, following Beth, it is expressed in terms of state spaces, for Suppe the chosen mode is that of relational systems, while Suppes (and Sneed) adopt a more general line, where by defining an appropriate set-theoretical predicate, one can specify either a state space, a relational system, or some other mathematical structure (Suppe 1989, p. 4). The precise details are not so important here, although in the work to be described below, we have chosen to follow Suppes.

What *is* important is that a theory is to be represented in terms of some extra-linguistic structure, such as a model. It is here in the (meta-level) specification of the structure that we can appropriately modify the semantic approach so as to accommodate the looseness or openness referred to above.

Informally (mathematically speaking) we can represent a model in terms of a first-order structure of the following form:

$$\mathcal{M} = \langle A, R_i \rangle_{i \in I}$$

where A is the set of appropriately designated individuals of the domain under consideration and R_i, $i \in I$, is a family of partial relations defined on A (Mikenberg, da Costa and Chuaqui 1986; da Costa and French 1990). Given our lack of complete knowledge of the domain concerned, it seems appropriate that $\mathcal{M}$ be a partial structure in the sense that the relevant properties and relations among the members of A are captured by a family of partial relations R_i, $i \in I$, which are 'partial' in the sense that any relation R_i, $i \in I$, of arity n_i is not necessarily defined for all n_i-tuples of elements of A[11].

[11]More rigorously, a partial relation defined on a set A, for example, a binary relation R, can be introduced as follows: R is an ordered triple $\langle R_1, R_2, R_3 \rangle$, where R_1, R_2 and R_3 are mutually disjoint sets such that $R_1 \cup R_2 \cup R_3 = A^2$; R_1 is the set of ordered pairs which satisfy R, R_2 is the set of ordered pairs which do not satisfy R, and R_3 is the set of ordered pairs for which it is left open whether they satisfy R or not. When R_3 is empty, R constitutes a normal binary relation and can be identified with R_1.

There are two fundamental points to note immediately. The first is simply that introducing partial structures into the model-theoretic approach allows one to accommodate the sense of openness that I have highlighted as crucial to theory pursuit. The process of theory development can be seen as one of defining and re-defining the relationships between the appropriate set of elements, be they genes, elementary particles or whatever. In the initial stages, some sub-set of the R_i will be laid down according to the sorts of heuristic criteria suggested by Post, for example. Perhaps the most important concerns the Generalised Correspondence Principle, which can be taken as expressing the idea that this sub-set, or portions of it, will be bequeathed by the theory's predecessor(s). Adherence to well-confirmed symmetry and invariance principles will then constrain this set of relations in further ways. In all cases, it is the objective qualities of the models that are important, as expressed in terms of their relational structure.

Thus, whereas logically the space of opportunity may be infinitely wide, in practice it is severely delimited in terms of heuristic structure[12]. There is, of course, still an element of freedom in the choice of which criterion is taken to dominate. Perhaps here, at this more elevated level, it might be argued that non-epistemic considerations come back into play, directing the choice of methodological element rather than the theoretical. This opens up that whole can of worms concerning the justification of scientific methodology and here, at least, I think I want to slam the lid back down before things get too messy. Suffice to say that, certainly with regard to heuristics and pursuit, a case can be made for the iterative view that, in the early stages of the development of a science, rather crude models are tentatively put forward, are seen to be successful and criteria for future developments thereby extracted. Of course, this hinges on a theory independent notion of 'success' and here the differences with the sociological approach yawn wide again. Ultimately, I believe, it is empirical considerations that have the final say (at least when it comes to science, rather than mathematics), producing the ultimate 'forced' move through an act of final 'resistance'.

The second point to note is that in the practice of theory development these criteria may be applied sequentially in different order on different occasions, with empirical considerations also added to the mixture. It is here that we see the distinction between the kind of reasons adduced under justification and those put forward as heuristic criteria begin to blur and crumble[13]. (Of course, the sociologists of science also typically reject the distinction, albeit for very different reasons.)

[12]This of course accounts for the observation that underdetermination, while a conceptual possibility, so rarely occurs in practice where, as scientists themselves are quick to point out, it is hard enough to obtain even one model.

[13]Nersessian makes a similar point and replaces 'justification' and 'discovery' with the unitary context of 'development' (Nersessian 1993, p. 6)

The introduction of partial structures thus allows a treatment of evolving theories within the semantic approach, as Suppe notes (op. cit., p 427). There is a problem of theory individuation here, or more specifically, whether a change in a theory is regarded as a further development of that theory, or as leading to a different, although related, one. Characterising theories as partial structures allows for changes in the family of relations R_i without thereby necessitating that we regard the structure as referring to a different theory entirely.

It is because of this open nature of scientific theories that the appropriate doxastic attitude should be characterised as weaker than that of belief-as-true (in the correspondence sense). Modifying the Tarskian account of truth by introducing partial structures then permits the elaboration of a structural account of partial or pragmatic truth suitable for capturing such an attitude (Mikenberg, da Costa and Chuaqui op. cit. and da Costa and French 1993). That this account is also appropriate for capturing the requisite attitude during pursuit should be clear. Here talk of acceptance and belief in 'truth' is explicitly eschewed in the literature, to be replaced, typically, by the notion of merely 'entertaining' a theory. But 'entertainment' implies far too frivolous an activity, as if there were no commitment whatsoever to the theory being pursued, and suggests just the sort of view which I am trying to oppose here. There are, I would claim, good reasons for rejecting the construction of separate doxastic frameworks for pursuit and justification — a suitable notion of acceptance tied to that of 'partial' truth can do it all[14] (da Costa and French op. cit.). Again the barrier collapses. Of course, it is this sense of openness which, by featuring in both 'domains', erodes the foundations[15].

Pulling back from the view that theories are to be regarded as true in the correspondence sense opens up a new perspective on the relationship between theories, regarded in terms of mathematical models by the semantic approach, and scientific models of phenomena or physical systems. As is becoming in-

[14]Even, or perhaps particularly, if the theory is inconsistent. It is in such cases where talk of 'truth' seems so obviously inappropriate and where the notion of entertainment is entertained. But shifting to partial truth and representational belief allows one to retain an appropriate sense of doxastic commitment, whilst acknowledging that, heuristically, the inconsistency can be regarded as a 'signpost' pointing the way to a consistent successor (da Costa and French 1993).

[15]One can also accommodate this openness within other forms of the semantic approach. Thus Suppe notes that '... theories undergoing active development frequently are not put forward as literally or even counterfactually true, but rather as "promising", "worth pursuing", "approximately correct", "being on to something", "being in a certain respect importantly true, but in other respects profoundly wrong as well", and so on. And in their developmental states, such theories typically are importantly incomplete in their accounts of phenomena.' (Suppe 1989, p. 347). This is taken to support Suppe's 'quasirealism', articulated in the context of his 'relational systems' version of the semantic approach, according to which theories purport to describe how the world would behave if it were "nice and clean".

creasingly clear, this relationship is problematic[16]. Typically the two have been viewed as distinct, although, as I have mentioned above, the ubiquitous nature of scientific models is often cited as motivating the semantic approach. This distinction is often taken to be underpinned by the realist sentiment that theories are supposed to provide true descriptions of the phenomena, whereas models most definitely are not. From the perspective of partial truth, however, the distinction becomes blurred[17]. By their very nature, scientific models — regarded perhaps as the units of pursuit — are considered to be only partially true, but then, I would argue, so are theories. But, of course, there are models and models! What about iconic models, impoverished models and idealisations in general? With regard to the latter, as Suppe notes (p. 426), the use of theories as idealisations or approximations causes the distinction to crumble further. In the 1990 paper with da Costa we claimed, with the impetuousness, if not of youth, then at least of someone writing their first paper on a topic, that all models — idealisations, iconic, impoverished — could be equally treated from the partial structures perspective. Consider the infamous billiard ball model, for example. This, it is usually asserted, is strictly false of the gas atoms whose behaviour it is supposed to model. Yet for the model to have any pragmatic value, to be useful, it must be true in some minimal, partial sense. But in what sense, precisely? Clearly the elements of the model — the billiard balls — do not correspond to the gas atoms. But it is not the elements of the model per se that we are interested in because they do not actually do any of the analogical work — rather it is the properties and relationships that are asserted to hold between the balls that are important. Certain of these relationships — certain of the R_i — are taken to correspond to certain of the relationships holding between the atoms, as represented by the kinetic theory of gases. One way of understanding this relationship between relationships (!) is in terms of an isomorphism holding between some members of the R_i of one model and some members of the R_i of the other[18]. Thus we might say that there is a partial isomorphism between the sets of relations concerned (French and Ladyman, forthcoming), and this is another way of explicating what Hesse meant by a 'positive' analogy[19]. The 'negative'

[16]Expressions of disquiet concerning this relationship can be found in Olivé (1985) and Pérez Ransanz (1985), responding to an account of theory identification by van Fraassen (1985) and, more recently, in Collier (1992), Downes (1992) and Suárez (forthcominga).

[17]This mirrors the blurring of the distinction in actual scientific practice.

[18]As Collier notes, the problem is not that of formally establishing isomorphisms between models but rather that of ruling out those which are uninteresting (op. cit., pp. 294–295). What I am about here, of course and at least in part, is delineating the sorts of considerations that are typically drawn upon to rule out the uninteresting and unworkable isomorphisms in pursuit.

[19]cf Nersessian (op. cit., p. 13) who credits Sellars with the insight that '... analogical reasoning creates a bridge from existing to new conceptual frameworks through the mapping of relational structures from the old to the new.' She goes on to claim that '... most cognitive theories of analogy agree that the creative heart of analogical reasoning is a modeling process in which relational structures from existing modes of representation and problem solutions

analogy of course refers to the dissimilarities between the R_i in the two cases and the 'neutral' analogy takes in those relations which are not yet known to either hold or definitely not hold in both cases. Again, the partial structures approach would seem to offer a natural way of accommodating these distinctions. This, I think, gives a way of treating, in fairly general terms, theories, models[20] and the diverse relationships between them[21], embracing idealisations, approximations, iconic models[22] and, of course, the Kuhnian exemplars with which I began this discussion[23].

Our concern, then, is with the explicit representation of the structural relationships between the elements of the domain under consideration. The various relationships that hold between models (and, more generally, theories)

---

are abstracted from a source domain and are fitted to the constraints of the new problem domain.' (p. 20) Pertinent features of this process as revealed by psychological studies include the preservation of relational systems and the isomorphic mapping of relationships (ibid.).

[20]If we focus on structure, there would seem to be no fundamental difference between theories and models as such (see Suppe op. cit., p. 425). This approach would therefore mesh nicely with those views which emphasise the heuristic use of models and suggest that theories might be regarded as clusters of (scientific) models, for example (ibid., p. 426).

[21]Of the recent contributions to the study of these relationships, some of the more interesting include those of Suárez (forthcominga and b) and Huggett (1995), which offer alternative perspectives to that presented here.

[22]It has recently (very recently!) been argued that the most natural way to understand idealization and approximation within the semantic approach is by means of the process of 'deidealisation', which takes one from the theory to the model of the phenomena (Suarez forthcominga). It is then claimed that at least one example from scientific practice — the construction of the first model of superconductivity — runs counter to this, with negative consequences for the semantic approach as a whole. However, this argument fails to engage the sort of approach being presented here. Setting aside the question whether the semantic approach is indeed committed to this notion of 'deidealisation' or not, the above account of the relationships between models is an atemporal one, in which we are concerned with the logical connections, rather than temporal ones concerning the provenance and construction of the models. But the supposed counter-example is concerned precisely with the latter; indeed it is a criticism of the 'theory driven' view of model construction in general. With regard to the construction of models, I am quite prepared to accept that this may proceed in a variety of directions, from theories to phenomenological models or vice versa, and, of course, I would emphasise the role of heuristic guidelines in this construction. Whatever the view that is adopted on the latter, it is quite independent of the nature of the analysis of the inter-relationships between the models once they have been constructed.

[23]Sticking my neck out even further, I believe that all examples of scientific 'representation', broadly construed, might be treated in this way. Consider for example, diagrams of experimental apparatus: at a recent Leeds seminar on the difficulties in reproducing Joule's famous paddle-wheel experiment it was pointed out that drawings culled from Joule's papers could not be relied upon for the purposes of acccurate reconstruction. However, this is to presume that accurate reconstruction was the ultimate purpose of such diagrams. Rather, I would hazard, their function was to indicate the most pertinent relationships between the various components of the apparatus, these relationships revealing — at least in this case — the lines of force involved. Nersessian, again, has drawn attention to the role of images and 'imagistic reasoning' in theory development and claims that, in the quantitative phase of development, the image '...serves primarily to make certain structural relationships visualizable ...' (op. cit., p. 24)

represented in this way can be analysed in terms of the relationships that hold between such families of relations. And, of course, it is precisely the nature and characteristics of these relationships that we are interested in from the point of view of theory pursuit.

Thus, to return to our earlier example, the analogy between QED and QCD which both encouraged and permitted the expression of the latter in gauge invariant form can be expressed in terms of an appropriate relationship of similarity holding between the respective R_i in the two cases. Leaving it at that, of course, adds little to the discussion, either historically or philosophically.

Taking the former perspective first, particles which interact via the strong nuclear force can be grouped into multiplets of 'isospin', a spin-like vector property. The isospin symmetry of the strong nuclear force can then be represented by the SU(2) group. Regarding the gauge invariance of QED from a group theoretical perspective[24], Yang attempted to construct a similar theory for the strong interactions which would be invariant under local gauge transformations of SU(2). This attempt led to the famous Yang-Mills gauge theory of the strong interactions which, crucially, differed from its electrodynamic analog in containing non-linear terms referring to the self-interaction of particles 'carrying' the strong force. Formally speaking, this difference derives from the difference in group structure involved, which, in turn, can be traced to the difference in the physical properties represented.

Yang-Mills gauge theory came to be regarded as the exemplar for gauge theories in general and was extended to apply to the electroweak interaction[25]. However, the negative analogy with QED turned out to be fundamental as Yang-Mills theory could not be shown to be renormalisable, a feature which allows calculations to be made to arbitrarily high orders of approximation. As is well known, the crucial breakthrough came with the work of 't Hooft[26] and his invocation of spontaneous symmetry breaking, and the similarity with QED was restored. With the availability of renormalisable gauge theories of the strong interactions, the choice facing physicists was now that of selecting the appropriate gauge group. In the search for a 'realistic'[27] theory of the strong interactions, it was the SU(3) group of quark colours mentioned previously which was chosen and QCD was the result. What we have here is a complex web of similarities and dissimilarities existing between models which, indeed, were constructed on the basis of the former, the ultimate ground for this incorporation of similarity and also for the dissimilarities that arise, being empirical.

[24]The relevant group here is U(1) and, corresponding to the symmetry represented by this group, the property that is conserved is, of course, charge.

[25]The history of these developments is given with great detail and clarity, as ever, in Pickering (1984).

[26]Pickering ibid., pp. 173–180.

[27]The term is Pickering's; ibid., pp. 207 and 223.

Turning to the philosophical perspective, 'similarity', as it stands, is a rather vague and imprecise notion[28]. An obvious way of getting a tighter grip on it is to decompose the similarity relation in turn into that of partial isomorphism holding between certain members of the families of relations concerned (again see French and Ladyman op. cit.)[29]. The 'degrees and respects' characteristic of the former notion can then be captured in terms of the number and kinds of relations included in the relevant sub-sets. In the case of QED and QCD, it is in terms of the correspondence between certain members of the R_i that refer to the relationships between the particles expressed by the relevant Lagrangians, that we can say that the theories are similar with respect to gauge invariance[30]. The R_i may refer not only to 'physical' relationships in this manner but also to mathematical ones[31]. Thus the fundamental role of group theory, both heuristically, as I have indicated, and formally, can be accommodated within this approach: the group-theoretic similarities between U(1) and SU(2), say, establish the bridge between domains, heuristically, and the partial isomorphism between models, formally[32].

Likewise the more particular notion of 'association' can now be understood in terms of such structural analogies, holding between the algebraic and geometric models of complex numbers in the case of quaternions or between QED and Yang-Mills gauge theories in the history outlined above. Notice that in the latter case, as an example of scientific rather than conceptual practice per se, the ultimate source of the analogy, partial as it is, lies in the properties of the particles described by the theories, these properties, such as charge and isospin, being introduced in order to accommodate *experimental* results. Where the analogy breaks down, 'accommodation' might be sought for in terms of changes in the relevant family R_i. Thus, in Hamilton's case, the fundamental relationship of commutativity, previously taken to hold between the elements concerned, was abandoned and a new relation effectively defined and introduced. The move from a positive to a negative analogy in the face of resistance gets expressed in a shift of the partial isomorphism across the R_i.

[28] See Giere (1988), for example.

[29] All analogies must surely ride piggy back on such similarities, for how else are they to be motivated? Likewise, the notion of 'family resemblance', famously appealed to by Kuhn in his account of the way exemplars drive practice, can be considered in terms of a 'partial isomorphism' holding between members of the R_i of the respective parts of the 'scientific corpus'.

[30] For an analysis of the supposed metaphysical nature of gauge invariance, see Liu (1995)

[31] '... there is no theoretical way of drawing a sharp distinction between a piece of pure mathematics and a piece of theoretical science. The set-theoretical definitions of the theory of mechanics, the theory of thermodynamics, the theory of learning, to give three rather disparate examples, are on all fours with the definitions of the purely mathematical theories of groups, rings, fields, etc. From the philosophical standpoint there is no sharp distinction between pure and applied mathematics, in spite of much talk to the contrary.' (Suppes 1967, pp. 29–30)

[32] From the perspective of the role of mathematics in science, there is of course much more that could be said here.

The 'resistance', then, lies in the structures or, even more explicitly, in the characteristics of the R_i[33]. In the case of conceptual practice, this structural resistance is regarded as reducible only under some further philosophical interpretation of that practice, such as an empiricist approach to the philosophy of mathematics, whereas in that of scientific practice, the source of the resistance is again, ultimately, empirical[34]. Thus, the resistance represented by the occurrence of non-linear terms in the field equations of Yang-Mills theory can also, of course, be expressed structurally and, to repeat, derives ultimately from the properties of the interaction particles which, in turn, were introduced to accommodate the well supported phenomenological model of the strong force. The renormalisability problem was then overcome through the introduction of spontaneous symmetry breaking; that is, through an accommodation expressed in terms of changes in the R_i. New members of this family were introduced in order to overcome the problem, thereby restoring the partial isomorphism between the relational predicates in terms of which the renormalisability of QED and QCD can be expressed[35].

What we have then, is the following sort of process: on the basis of some perception of structural correspondence between elements of different domains, broadly and perhaps crudely understood, (such as that holding between the properties of sets of elementary particles, as in our example above) a family of (partial) relations, found to hold good in one domain, typically well understood, is projected into another, about which knowledge is even less complete or, rather, even more partial. Certain of these relations may fail to hold in the new domain, as resistances arise. Others may come to be established as accommodations are made. The correspondence, then, is with only a sub-set of the original set but nevertheless a (positive) analogy is established. The initial bridge may then serve as the basis for further exploration, as further relations are brought into the correspondence. The bridge itself is an 'open' structure, projecting into a domain about which knowledge is only partial[36]. What we have when we compare different models as to their similarity is a time-slice of a heuristic process: the family of relations projected into the new domain forms the basis of the partial isomorphism formally holding between

[33]Thus 'resistance' can be equated with the absence of the appropriate isomorphism holding.

[34]Although, as I have said, the resistance is typically mediated by a network of relationships between models.

[35]Again the problem of theory individuality arises: in what sense is gauge theory with spontaneous symmetry breaking the 'same' theory as Yang-Mills gauge theory? Given that the latter was, if not constantly, then certainly frequently, evolving, it seems appropriate to give an initial identification of the theory in terms of some fundamental sub-set of the R_i. The possible connection with Lakatosian notions of a 'hard core' is obvious.

[36]Nersessian, again: 'According to Maxwell, a physical analogy provides both a set of mathematical relationships and an imagistic representation of the structure of those relationships drawn from a "source" domain to be applied in analyzing a "target" domain about which there is only partial knowledge.' (op. cit., p. 17)

the relevant models. And the nature of this set — *which* relations to project — is delineated by the sorts of heuristic criteria outlined by Post.

One of these criteria that I have already mentioned is that of 'Taking models seriously'; to which I would add, yes, of course, but not just in terms of the elements A. The set of relations R_i should also be 'taken seriously' and the structural characteristics expressed by these relations may have enormous heuristic significance. Indeed, it might perhaps be said that, underlying all of the above considerations is the most fundamental heuristic guideline, which is also the final 'message' of this paper — take the structure seriously.

Department of Philosophy
University of Leeds

REFERENCES

Collier, J.D. (1992), 'Critical Notice: Paul Thompson, *The Structure of Biological Theories*', *Canadian Journal of Philosophy* **22**, pp. 287–298

da Costa, N.C.A. and French, S. (1990) 'The Model-Theoretic Approach in the Philosophy of Science', *Philosophy of Science* **57** pp. 248–265

da Costa, N.C.A.and French, S. (1993), 'Towards an Acceptable Theory of Acceptance', in S. French and H. Kamminga, *Correspondence, Invariance and Heuristics*; D. Reidel, pp. 137–158.

Downes, S. (1992) 'The Importance of Models in Theorizing: A Deflationary Semantic View', *PSA 1992* Vol. 1, pp. 142–153

French, S. (1995), 'The Esperable Uberty of Quantum Chromodynamics', *Studies in History and Philosophy of Modern Physics* 26 pp.87–105.

French, S. and Ladyman, J. (forthcoming), 'Reinflating the Semantic Approach'; submitted.

Giere, R.N. (1988), *Explaining Science*; University of Chicago Press.

Gremmen, B. (1995), 'The Structure of Scientific Practice', paper read at the *10th International Congress of Logic, Methodology and Philosophy of Science*, August 19–25 1995, Florence, Italy.

Huggett, N. (1995), 'A New Criterion for Theory Equivalence', paper read at the *10th International Congress of Logic, Methodology and Philosophy of Science*, August 19–25 1995, Florence, Italy.

Koertge, N. (1993), 'Ideology, Heuristics and Rationality in the Context of Discovery', in S. French and H. Kamminga, *Correspondence, Invariance and Heuristics*; D. Reidel, pp. 125–136.

Kuhn, T.S. (1970), *The Structure of Scientific Revolutions*; University of Chicago Press, 2nd ed.

Laudan, L. (1980), 'Why was the Logic of Discovery Abandoned?', in T. Nickels (ed.) *Scientific Discovery, Logic and Rationality*; D. Reidel, pp. 173–183.

Liu, C. (1995), 'Gauge Invariance and its Metaphysical Nature', paper read at the *10th International Congress of Logic, Methodology and Philosophy of Science*, August 19–25 1995, Florence, Italy.

Lloyd, E. (1988), *The Structure and Confirmation of Evolutionary Theory*, Greenwood Press

Mikenberg, I., da Costa, N.C.A. and Chuaqui, R. (1986), 'Pragmatic Truth and Approximation to Truth', *Journal of Symbolic Logic* **51** pp. 201–221.

Nersessian, N. (1993), 'How Do Scientists Think? Capturing the Dynamics of Conceptual Change in Science' in R.N. Giere (ed.), *Cognitive Models in Science*, Minnesota Studies in the Philosophy of Science, Vol. XV, University of Minnesota Press, pp. 3–44

Olivé, L. (1985), 'Realismo e antirrealismo en la concepción semántica de las teorías', *Crítica* **17** pp. 31–38; English summary pp. 39–40

Pérez Ransanz, A.R. (1985), 'El concepto de teoría empírica según van Fraassen', *Crítica* **17** pp. 3–12; English summary pp. 13–19

Pickering, A. (1984), *Constructing Quarks*; Edinburgh University Press.

Pickering, A. (1990), 'Knowledge, Practice and Mere Construction'; *Social Studies of Science* **20** pp. 682–729.

Pickering, A. (1992), 'From Science as Knowledge to Science as Practice', in A. Pickering (ed.), *Science as Practice and Culture*; University of Chicago Press, 1992, pp.1–26.

Pickering, A. and Stephanides, A. (1992), 'Constructing Quaternions: On the Analysis of Conceptual Practice', in Pickering (ed.), op. cit., pp.139–167.

Post, H.R. (1971), 'Correspondence, Invariance and Heuristics'; *Studies in History and Philosophy of Science* **2** pp. 213–255.

Redhead, M.L.G. (1975), 'Symmetry in Intertheory Relations', *Synthese* **32** pp. 77–112.

Suárez, M. (forthcominga), 'Idealization and the Semantic View of Scientific Theories', preprint.

Suárez, M. (forthcomingb), 'How Theories save Phenomena: A Case Against Embedding', preprint.

Suppe, F. (1989), *The Semantic Conception of Theories and Scientific Realism*; Univ. of Illinois Press.

Suppes, P. (1967), *Set-Theoretical Structures in Science; mimeograph*, Stanford University.

Suppes, P. (1984), *Probabilistic Metaphysics*, Blackwell.

Thompson, P.(1989), *The Structure of Biological Theories*, State University of New York Press.

van Fraassen, B. (1985), *The Scientific Image*, Oxford University Press.

7
Probability, Induction and Decision Theory

MICHAEL GOLDSTEIN

PRIOR INFERENCES FOR POSTERIOR JUDGEMENTS

ABSTRACT.

We consider the minimal assumptions of temporal consistency that will allow us to make meaningful prior statements about our posterior judgements. These consistency conditions suggest a natural geometric interpretation of 'prior inference' which we develop and interpret. We discuss various applications of the geometric construction.

1. INTRODUCTION

In the subjectivist description of inference, judgements made at certain time points are modified by acquisition of relevant information to yield revised judgements. The most familiar form of such revision is conditioning, using Bayes theorem, and indeed, in some accounts, conditional and posterior judgements are used interchangeably.

However, it is difficult, in general, to identify those circumstances under which formal belief revision, of any kind, may be identified with actual posterior judgements. Informally, if we write down now explicit rules under which future beliefs are to be determined, and these rules are not consistent with certain standard transformations, such as those based on probabilistic conditioning or probability kinematics, then, subject to certain technical qualifications, we are susceptible to being made into sure losers, provided bets may be accumulated over time. While such results may place constraints upon our current efforts to predetermine our future beliefs, they can say nothing as to whether it is possible for us to predetermine rules for our future beliefs at all.

More strongly, given our current uncertainties about the nature of our future observations, the difficulties in describing exhaustive partitions over all such possible observations and of specifying, a priori our conditional beliefs over such partitions, and our uncertainties as to whether, by the time that further information has been received, we might purely by introspection have changed our prior description of the problem, then it is clear that, in general, there can be no such logical identification between any formal prior evaluation and the corresponding posterior belief. Instead, as we shall argue, the relationship between the formal evaluation and the actual posterior judgement is that the former, under certain conditions, is a 'prior inference' for the latter, in a way which we shall make precise below. This relationship is a consequence of the temporal sure preference principle, which is a minimal coherence requirement relating beliefs at different time points. This minimal requirement is, in particular, appropriate when we are dealing with large and complex problems, which are beyond our ability or our interest to fully specify our beliefs

M.L. Dalla Chiara et al. (eds.), Structures and Norms in Science, 55–71.

over. Indeed, we shall argue that there is nothing special or privileged about a full probabilistic prior specification, but that this is simply a special case of a hierarchy of levels of detail at which beliefs may be considered and revised.

In this paper, we describe the sure preference principle, deduce the property of prior inference, and develop a natural geometric formulation within which the prior inference may be developed in full generality, giving various applications of this formulation.

2. TEMPORAL SURE PREFERENCE

What do our current beliefs imply about our future beliefs? As an extreme example, consider Doctor Jekyll, who knows that he will turn into Mister Hyde. Today, Doctor Jekyll makes certain collections of probabilistic judgements. Tomorrow, as Mister Hyde, he will again make some such collection of judgements. However, while these preferences may be rational at each individual time point, there need be no linkage whatsoever between the two collections of judgements. Such an example suggests that, in order to establish links between our judgements at different time points, we need ways of describing 'temporal rationality' which go beyond being internally rational at each time point.

Our description will concern preferences between certain rewards, as assessed at different time points. Such preferences involve both probability and utility judgements, while our interest here is only with changes in probabilistic judgements. Therefore, all preferences shall concern small penalties that are in some scale whose relative values are unchanging over time. The simplest such scale for our purpose is to suppose that each penalty is in units of probability currency, i.e. tickets in a lottery with a single prize, or a continuous analogue, for example when the penalties are fractions of the circumference of a probability wheel, where the prize is awarded to the contestant who has the ticket for the region in which the pointer lands (see, for example, the discussion in [11]). With this penalty scale, our preferences obey the **expectation preference property**, namely that preferring penalty A to penalty B is equivalent to assigning $\mathrm{E}(A) < \mathrm{E}(B)$, as the expectation for the penalty equals the probability of the reward. Therefore, changes in preferences between A and B over time correspond to changes in probability, rather than utility (to the extent that these can be separated).

Current preferences for penalty A over penalty B, even when tightly constrained by conditional statements about preferences given possible future evidential outcomes, cannot require you to hold certain future preferences, if for no other reason than that you may have further, hitherto unsuspected, information or insights into the problem before you come to make your future judgements. However, while current preferences cannot determine future preferences, it is more compelling to suggest that future preferences may determine prior preferences. Suppose that you know that tomorrow you will prefer penalty A to B. Should you prefer A to B now? The reasons that we have suggested as to why current preferences cannot determine future preferences, based on unpredictable outcomes and unanticipated probabilistic insights, do not apply when the actual future preference is known. What is left are disagreements of a fundamentally different nature, for example, that Doctor Jekyll may consider that his judgements when he turns into Mister Hyde will be intrinsically inferior. Such views are not inherently contradictory, which is why we need operational temporal criteria to determine whether and to what extent your prior analysis may be of value in determining your future judgements.

For our argument to be generally applicable, any principles of temporal consistency must be weak enough to be compelling in many situations. We shall suggest a particular temporal consistency principle, which is both sufficient to construct our interpretation of the inferential process and also weak enough to be widely applicable.

In order for your future preferences to influence your current preferences, you must firstly know what your future preference will be. For definiteness, suppose that you must choose between two (probability currency) random penalties, A and B. Suppose that at some future time the values of A and B will be revealed, and you will pay the penalty that you have chosen. We say that you have a **sure preference** for A over B at (future) time t, if you know now, as a matter of logic, that at time t you will not express a strict preference for penalty B over penalty A.

The temporal consistency principle that we impose is that future sure preferences are respected by preferences today. We call this the **temporal sure preference principle**, as follows.

The temporal sure preference principle *Suppose that you have a sure preference for A over B at (future) time t. Then you should not have a strict preference for B over A now.*

The temporal sure preference principle is similar in spirit to the reflection principle as given in [13], which provides constraints on current probability judgements over events conditional on future probability judgements about these events. Much of the argument for and against the reflection principle is also relevant to temporal sure preference, and discussion of various possible criticisms of a generalised version of reflection is given in [14].

The reasons why our development is based upon the temporal sure preference principle are as follows.

(1) The principle applies in a great many problems concerning assessments of uncertainty. Indeed, I have yet to be involved in an inferential problem in the particular field which is my primary concern, namely statistical inference, where I would be unhappy to follow temporal sure preference. The main reason that you might have for not following the principle is that you might not now trust your future ability to make sensible judgements. This might be the case, for example, if, like Jekyll and Hyde, you knew that at this future time you would be under the influence of some mind-altering substance. We envisage rather more prosaic applications, for which you currently believe that further thought and observation would place you in a stronger rather than a weaker position to make good judgements.

(2) This is a very weak requirement on our current preferences, as sure preference is a very strong requirement for our future preferences. Without some temporal principle, we know of no reason why beliefs specified at different time points should bear any relation, except accidentally, and we know of no weaker temporal coherence principle which leads to a subjectivist account of the inferential process. The principle is implied by, but substantially weaker than, the conditions imposed in most other accounts of temporal coherence which lead to applicable methodology. For example, the usual, albeit implicit, Bayesian temporal assumption is firstly that you can now anticipate and specify numerical beliefs for all possible combinations of future outcomes, and secondly that such conditional evaluations will, with no further consideration on your part, become your actual posterior evaluations given the evidence.

(3) We do not claim that temporal sure preference, or any other temporal principle, should be viewed as a basic rationality criterion. Instead, we view temporal sure preference as an operationally testable property of your current beliefs about your future beliefs, which, we suggest, you will usually find it reasonable to accept. However, on those occasions where you judge, for whatever reason, that your future judgements do not currently satisfy temporal sure preference, then you will need to consider afresh whether there are any links between your current and future beliefs, as the relationships that we shall develop below need no longer apply.

(4) Coherence is not a property which automatically applies to collections of beliefs. Any large collection of judgements assessed at a particular time may contain many inconsistencies, and the role of coherence is to suggest criteria which we may apply with the intention of identifying and resolving such inconsistencies. In the same way, temporal sure preference is not an automatic feature of the development of beliefs over time, but rather it is a condition which you may voluntarily impose upon aspects of your beliefs in the sense that when you see that a particular collection of beliefs will violate this condition, then you may take steps to avoid this conflict. As a simple example, there are many statements which you now know to be true, i.e. give probability 1 (for example, concerning what you had for breakfast this morning). You also know that, within a year, you will doubtless forget the truth of various of these statements and thus your future probability assignments will be less than 1, and you now consider these future judgements less reliable than your current judgements. Thus, you now see that without further action on your part temporal sure preference is violated. If it is now important to you to make informative judgements about these quantities in the future, then you may recover temporal sure preference, for example by recording sufficient information concerning your current beliefs and the reasons why you hold these beliefs to ensure that you may have confidence in your future judgements again. Just as you cannot hope to make all of your beliefs at a particular time coherent, but instead apply coherence to particular subcollections of beliefs, you cannot make all of your beliefs cohere over time, but reserve the temporal sure preference principle, and the resulting inferential structures, for those particular sub-collections of beliefs for which you want to develop a satisfactory inferential account.

(5) The principle avoids certain types of temporal sure loss, and may be viewed as a natural temporal extension to the sure loser argument based on accumulating bets over time. You are a 'temporal sure loser' if you would now pay c to switch penalty A for penalty B, given that you will, at time t, accept $c/2$ to switch B for A, in the sense that you could program an automaton to place bets, using the same knowledge that you have, in such a way that the automaton would be bound to win against you over time. The force of temporal sure loss arguments is a matter of some controversy; for discussion of criticisms of temporal sure loss arguments, see [1], [10]. Our view is that, as the various penalty comparisons that we will consider are derived directly from the natural elicitation method for our probabilistic judgements, we should usually be concerned that honest elicitations of belief at two time points will lead us into sure loss. We therefore view the temporal sure preference principle as a minimal, and often very plausible, extension of the sure-loser argument, which if, unlike Jekyll and Hyde, you are trying to perform well in your overall collection of judgements, you will often find it reasonable to obey.

(6) The temporal sure preference principle is sufficient to act as the basis of our theory in that we may derive from it an operational account of the inferential content of the subjectivist theory, for individuals with limited abilities to enumerate future possibilities and to specify beliefs over such possibilities. This account unifies our inferences over a wide variety of types of object, by revealing the underlying geometric form for all such inferences.

3. EXPECTATION AND TEMPORAL PREFERENCE

We are concerned with the relationships between our current and future beliefs. In most circumstances, it will be very difficult to describe fully such beliefs. Therefore, we choose a formulation within which we may treat as primitive and specify directly those limited aspects of our beliefs which are needed for our argument. Thus, we treat expectation as the primitive quantification for our approach. We follow the development of de Finetti in [2], and define the expectation of a random quantity, X, as the value $\overline{x}$ that you would choose for x, if faced with the penalty $L = k(X - x)^2$, where k is a constant defining the units of loss. The conditional expectation of X given event H is the corresponding quantity when the penalty L is replaced by $L_H = kH(X - x)^2$, where H is here used as the indicator function for the corresponding event. Again, we suppose that the penalty is paid in units of some scale, such as probability currency, which obeys the expectation preference property.

Now suppose that, for a particular random quantity X, you specify a current expectation $\mathrm{E}(X)$ and you intend to express a revised expectation $\mathrm{E}_t(X)$ at time t. As $\mathrm{E}_t(X)$ is currently unknown to you, you may express beliefs about this quantity, and in particular make a current specification for the quantity $\mathrm{E}((X - \mathrm{E}_t(X))^2)$. A consequence of the temporal sure preference principle is the following.

Theorem 1 *Suppose that F is a random quantity whose value you will surely know by time t. If you assess a current expectation for $(X - F)^2$, then to satisfy the temporal sure preference principle you must now assign*

$$\mathrm{E}((X - \mathrm{E}_t(X))^2) \leq \mathrm{E}((X - F)^2). \tag{1.1}$$

Proof At time t, $\mathrm{E}_t(X)$ is your preferred choice x for penalty $k(X - x)^2$. Therefore, at time t, as F will be a known constant, you will not prefer penalty $L_t^* = k(X - F)^2$, to the penalty $L_t = k(X - \mathrm{E}_t(X))^2$. Given temporal sure preference, you therefore do not prefer penalty L_t^* to L_t now. Therefore, (1.1) follows by the expectation preference property. □

An immediate corollary follows by setting $F = \mathrm{E}_t(X) + b$ for a general constant b. From theorem 1, we have

$$\min_b \mathrm{E}((X - \mathrm{E}_t(X) - b)^2) = \mathrm{E}((X - \mathrm{E}_t(X))^2),$$

so that $(X - \mathrm{E}_t(X))$ has zero expectation, giving the following corollary.

Corollary 1 *For any random quantity, X, and any future time t, given temporal sure preference, we must have that*

$$\mathrm{E}(\mathrm{E}_t(X)) = \mathrm{E}(X). \tag{1.2}$$

This result was argued for directly in [4], but is better understood as a consequence of the temporal sure preference principle. Without temporal sure preference, there is no particular reason why this type of conglomerability should hold; for example Doctor Jekyll has no reason to be conglomerable over the beliefs of Mister Hyde.

Note that theorem 1 concerns relationships between collections of expectations which you have already specified. It would be wrong to say, by extension, that your current value for $\mathrm{E}((X - \mathrm{E}_t(X))^2)$ establishes a strict lower bound for the expectation that you could give for any quantity $(X - F)^2$, for which F will be known by t, as your considerations, when specifying $\mathrm{E}((X - F)^2)$ might yield further insights which could lead you to change your value for $\mathrm{E}((X - \mathrm{E}_t(X))^2)$. Indeed, such insights might be precisely the reason that you carried out your evaluation of $\mathrm{E}((X - F)^2)$. In many problems, there will be a very large collection of quantities F which, if you so decide, you may know the value of by t, and theorem 1 applies to that subcollection of quantities over which you have currently specified expectations.

In particular, we will apply the theorem to various combinations of the elements of a general finite data vector $D = (D_1, D_2, \ldots, D_k)^T$, where each element of D will be known by t. The particular combination of most interest in our account is the **adjusted expectation**, $\mathrm{E}_D(X)$, which is defined as

$$\mathrm{E}_D(X) = \mathrm{E}(X) + \mathrm{Cov}(X, D)(\mathrm{Var}(D))^{\dagger}(D - \mathrm{E}(D)). \tag{1.3}$$

where $(\mathrm{Var}(D))^{\dagger}$ is the Moore-Penrose generalised inverse of $\mathrm{Var}(D)$. $\mathrm{E}_D(X)$ is the linear combination of the elements of D which minimises, over all choices of coefficients α, the quantity $\mathrm{E}((X - \alpha_0 - \alpha_1 D_1 - \ldots - \alpha_k D_k)^2)$, or equivalently, it is the choice of coefficient vector α under penalty $L(\alpha) = (X - \alpha_0 - \alpha_1 D_1 - \ldots - \alpha_k D_k)^2$, thus extending our definition of expectation to cover linear combinations of random choices. Further

$$\mathrm{E}((X - \mathrm{E}_D(X))^2) = \\ \mathrm{Var}_D(X) = \mathrm{Var}(X) - \mathrm{Cov}(X, D)(\mathrm{Var}(D))^{\dagger}\mathrm{Cov}(D, X). \tag{1.4}$$

$\mathrm{Var}_D(X)$ is termed the **adjusted variance** of X given D. Observe that adjusted expectations and variances depend only on the prior second order structure, namely on specification of the mean vector, variance matrix and covariance vector for X, D. Thus, we may use these quantities to simplify the task of prior specification and belief updating in the usual Bayes analysis; for an overview of the resulting methodology, see [3]. The foundational interpretation of these quantities will be developed below.

In the particular case where the elements of D form an event partition, i.e. each D_i equals zero or one and $\sum_i D_i = 1$, then adjusted expectation is equivalent to conditional probability, as, if A is the indicator function for an event,

$$\mathrm{E}_D(A) = P(A|D) = \sum_i P(A|D_i)D_i \tag{1.5}$$

namely the random quantity which takes value $P(A|D_i)$ if event D_i occurs. Therefore, our general analysis of adjusted expectation will also offer an interpretation of the foundational role played by probabilistic conditioning.

Note from theorem 1 the following corollary.

Corollary 2 *If vector D will be known by time t, and you have made the appropriate second order prior specification to determine $\mathrm{E}_D(X)$, then, under the conditions of theorem 1,*

$$\mathrm{E}((X - \mathrm{E}_t(X))^2) \leq \mathrm{E}((X - \mathrm{E}_D(X))^2) = \mathrm{Var}_D(X).$$

Therefore, joint second order prior specification over X and D places upper bounds on the variance of $(X - \mathrm{E}_t(X))$. We shall make these relationships more precise in the following sections.

4. BAYES LINEAR SUFFICIENCY

The link between temporal preference and posterior judgements comes through the general notion of Bayes linear sufficiency. We now summarise the relevant results on linear sufficiency, from [9]. Given a finite collection of random quantities $X = (X_1, X_2, \ldots, X_r)$, each with finite variance, we denote by $\langle X \rangle$ the set of linear combinations, $Y = \sum_i \alpha_i X_i$, of the elements of X, where the α_is are any real constants. A random linear functional $A(.)$ on $\langle X \rangle$ is one for which, for each element $Y \in \langle X \rangle$, $A(Y)$ is a random quantity and $\forall Y, Z \in \langle X \rangle$, $A(\alpha Y + \beta Z) = \alpha A(Y) + \beta A(Z)$. We denote by $\mathcal{V}(X)$ the collection of random linear functionals G on $\langle X \rangle$ for which $\mathrm{Var}(G(X_i))$ is finite for each i. Given a finite collection $H = (H_1, H_2, \ldots, H_s)$ of elements of $\mathcal{V}(X)$, and a further element $G \in \mathcal{V}(X)$, we say that G is **Bayes linear sufficient** for H, written $G \succeq^* H$, if, for each $Y \in \langle X \rangle$,

$$\mathrm{E}((\mathrm{Y} - \mathrm{G}(\mathrm{Y}))^2) = \inf_{Z, Z_1, Z_2, \ldots Z_s \in \langle X \rangle} \mathrm{E}((\mathrm{Y} - \mathrm{G}(\mathrm{Z}) - \textstyle\sum_{\mathrm{i}=1}^{\mathrm{s}} \mathrm{H_i}(\mathrm{Z_i}))^2). \tag{1.6}$$

We may establish sufficiency by assessing certain orthogonality relations. We say that random linear functionals G, K on $\langle X \rangle$ are **orthogonal**, written $G \perp K$ if $\mathrm{E}(\mathrm{G}(\mathrm{W})\mathrm{K}(\mathrm{Y})) = 0$, $\forall W, Y \in \langle X \rangle$. A necessary and sufficient condition for sufficiency is that $G \succeq^* H$, if and only if both $(I - G) \perp G$ and $(I - G) \perp H_i$, $i = 1, 2 \ldots$, where I is the identity functional, $I(Z) = Z$. We say that G is in **standard form** if $G \succeq^* G$. We denote the collection of standard elements of $\mathcal{V}(X)$ by $\mathcal{S}(X)$.

We say that $Y_1, Y_2, \ldots, Y_m$ **separate** Z, written $Z = Y_1 \oplus Y_2 \oplus \ldots \oplus Y_m$, if $Z = Y_1 + Y_2 + \ldots + Y_m$, where $Y_i \in \mathcal{S}(X)$, $Y_i \perp Y_j, \forall i, j$. We have the following decomposition. Suppose $Z, U_1, \ldots, U_m \in \mathcal{S}(X)$. Then

$$\begin{aligned} Z &\succeq^* U_m \succeq^* U_{m-1} \succeq^* \ldots \succeq^* U_2 \succeq^* U_1 \Leftrightarrow \\ Z &= (Z - U_m) \oplus (U_m - U_{m-1}) \oplus \ldots \oplus (U_2 - U_1) \oplus U_1 \end{aligned} \tag{1.7}$$

In particular, for any finite data vector $D = (D_1, D_2, \ldots, D_k)^T$, let $\Delta(X)$ denote the collection of linear functionals on $\langle X \rangle$ of form $A(Y) = \mathrm{E}(Y) + f(Y)^T D$ where $f(Y)$ is a linear transform from $\langle X \rangle$ to $\Re^k$. It is straightforward to demonstrate that both $\mathrm{E}_D(.) \in \mathcal{S}(X)$, $\mathrm{E}_t(.) \in \mathcal{S}(X)$, and also that

$$\mathrm{E}_D(.) \succeq^* \Delta(X), \tag{1.8}$$

while, from theorem 1, we have, for each $Y, W, Z \in \langle X \rangle, V \in \Delta(X)$ and constant c,

$$\mathrm{E}((Y - \mathrm{E}_t(Y))^2) \leq \mathrm{E}((Y - c\mathrm{E}_t(W) - V(Z))^2), \tag{1.9}$$

so that

$$\mathrm{E}_t(.) \succeq^* \mathrm{E}_D(.). \tag{1.10}$$

We may therefore apply (1.7) to adjusted and posterior expectations to give the property of prior inference which we now describe.

5. PRIOR INFERENCES

Suppose that you specify a current mean vector and variance matrix for a random vector $X = (X_1, ..., X_r)$. You also specify your current expected value for the variance matrix that you will specify for X at time t. From (1.2), this is equivalent to specifying your current value for each $\mathrm{E}((X_i - \mathrm{E}_t(X_i))(X_j - \mathrm{E}_t(X_j)))$.

Suppose that by time t you shall observe the values of $D = \{D_1, D_2, ..., D_k\}$, and you have already specified the mean vector and variance matrix for D and the covariance vector between D and X. From (1.7), (1.8), (1.10), we have the following theorem.

Theorem 2 *Given temporal sure preference, we have, for each* $Y \in \langle X \rangle$,

$$Y = \mathrm{E}_t(Y) + R_t(Y), \;\; \mathrm{E}_t(Y) = \mathrm{E}_D(Y) + S_t(Y), \tag{1.11}$$

where $R_t(Y), S_t(Y)$ *have prior expectation zero, and*

$$R_t(.) \perp \{\mathrm{E}_t(.), \mathrm{E}_D(.), S_t(.), \Delta(X)\},$$
$$S_t(.) \perp \{\mathrm{E}_D(.), \Delta(X)\}, \;\; \mathrm{E}_D(.) \perp \Delta(X). \tag{1.12}$$

We say that a collection of adjusted expectations $\mathrm{E}_D(.)$, and posterior expectations $\mathrm{E}_t(.)$ which satisfy (1.11) forms a **prior inference** for the collection $\langle X \rangle$. These relations were derived informally for an individual quantity and discussed in [5] and [6]. Applications to the subjectivist account of the inductive argument were given in [8]. We have now shown that the property of prior inference follows from a single temporal principle within a framework which will extend naturally to infinite collections of posterior judgements about general collections of quantities.

Theorem 2 summarises the notion of prior inference for posterior judgements. When you evaluate $\mathrm{E}_D(Y)$, this furnishes certain probabilistic information, which is revealing about your posterior judgements $\mathrm{E}_t(Y)$ in precisely the same way that $\mathrm{E}_t(Y)$ is informative for Y, as follows. Just as you replace the relation $Y = \mathrm{E}(Y) + R(Y)$, with $Y = \mathrm{E}_t(Y) + R_t(Y)$, where $\mathrm{Var}(R_t(Y)) \leq \mathrm{Var}(R(Y))$, quantifying your current view that $\mathrm{E}_t(Y)$ is informative for Y, so you replace the relation $\mathrm{E}_t(Y) = \mathrm{E}(Y) + S(Y)$, with $\mathrm{E}_t(Y) = \mathrm{E}_D(Y) + S_t(Y)$, where $\mathrm{Var}(S_t(Y)) \leq \mathrm{Var}(S(Y))$, quantifying your current view that $\mathrm{E}_D(Y)$ is informative for $\mathrm{E}_t(Y)$. Thus, we interpret and exploit beliefs relating adjusted expectations to posterior expectations in precisely the same manner as for beliefs relating posterior expectations for quantities to the values of the quantities themselves. Informally, we have decomposed Y as the sum of three uncorrelated quantities, namely $Y = \mathrm{E}_D(Y) + S_t(Y) + R_t(Y)$, so that $\mathrm{Var}(Y) = \mathrm{Var}(\mathrm{E}_D(Y)) + \mathrm{Var}(S_t(Y)) + \mathrm{Var}(R_t(Y))$. The first term, $\mathrm{Var}(\mathrm{E}_D(Y))$, is resolved automatically, by second order belief specifications over Y and D, and observation of D. The second term, $\mathrm{Var}(S_t(Y))$, is automatically resolved at time t by your specification of $\mathrm{E}_t(X)$, and is in no way linearly predictable from D, while the final term $\mathrm{Var}(R_t(Y))$, is the current expected remaining uncertainty at t and is not linearly predictable from any judgements or observations made by t.

Note that theorem 2 operates under the same conditions as theorem 1, namely that it imposes consistency conditions between judgements that you have currently made. As prior inferences utilise those aspects of your beliefs over future observations which you have currently quantified, and each such prior quantification may be difficult and time consuming, it is essential that you may limit the number of prior specifications that the theory requires to a manageable level. We use adjusted expectations precisely because these allow you to exploit whatever collection of second order belief statements you are currently prepared to make rather than requiring you to make some, possibly extremely large, probabilistic prior specification over some potentially enormous sample space of possible outcomes before you make your inference.

Probabilistic conditioning is the special case of adjusted expectation, where the data vector consists of indicator functions for a finite partition D, so that one and only one of the quantities D_i will be one, and the rest zero. Therefore, conditioning is a prior inference as follows. We have, from theorem 2, that the posterior probability that you assess at time t for a general event A satisfies

$$\mathrm{P}_t(A) = P(A|D) + S_t(A), \tag{1.13}$$

where, if $D_i = 1$ then $P(A|D) = P(A|D_i)$, and $\mathrm{E}(R_t(A)|D_j) = \mathrm{E}(S_t(A)|D_j) = 0, \forall j$.

Thus, conditioning on a finite partition is a valid prior inference, which is why it is not usually a priori incoherent to pretend that you intend to use Bayes conditioning to update your beliefs, though there is no reason, in general, to equate a collection of zero conditional expectations for S_t with a prior belief that S_t will be zero.

Interpreting conditioning as a prior inference resolves many of the troubling aspects of the usual Bayes formalism where conditional and posterior probabilities are considered, by default, to be the same. We may have doubts about the meaning of a conditioning statement: for example, what does it mean to say that you are conditioning on an event and nothing else, when the fact that you have observed nothing else is not considered to be part of the conditioning event? We have confusions in application: for example, how can you uniquely identify your observations with a conditioning event when there are many alternative descriptions of what you have seen with different probabilistic quantifications? We have problems of practicality, namely how can you hope to carry out the Bayes program for realistically complex problems, where you cannot list all of the possible outcomes, let alone quantify all your beliefs over these outcomes, particularly as almost all of this effort will be rendered irrelevant by the particular data outcomes. There are problems of circularity, in that the strongest possible statement that we could logically make in trying to identify conditional with posterior probabilities is that you might judge the two types of probability to be equal unless you change your mind, which you are free to do at any moment simply by claiming fresh insights into the problem, so that there can be no operational force to such an identification.

All of these obscurities disappear when we view conditioning as a form of prior inference, based on whatever aspects of future observations you now feel able and willing to quantify uncertainties for. Such quantifications reduce your current assessments of future uncertainties in just the same way as for any other adjusted expectations. What we let go of, in this interpretation, is the view that you can or must, by exhaustive thought now, logically determine your future judgements precisely. In a particular application, you may choose to set the prior variances for the residual terms S_t in (1.13) equal to zero, if this is appropriate for the case at hand, but this is no more special than any other assessment for these variances, and the evaluation of these quantities, in any case, may only be made at the future time t.

6. APPLICATION: ANALYSING EXPERT JUDGEMENTS

An example of the way in which prior inferences may be used to structure a complex analysis is described in [9] as follows. A water company wants to estimate the cost for renovation of various assets that are in poor condition in each of n zones. (There are many such assets and many zones.) In practice, the most accurate way to estimate renovation costs for a particular asset in a given zone is based on a careful study of the values of various key variables within that zone. For example, an assessment of the approximate cost for each length of sewer needing renovation in a zone might use studies of key variables such as diameter, material, depth and location within that zone. Less accurate estimates can be obtained more cheaply, using typical values for the key variables of all sewers in the zone. An even simpler alternative is to classify the zones, for example into urban and rural, giving the same estimate of for renovation costs for each zone with a given classification.

Typically, a company will have detailed estimates in a few zones, estimates based on a typical sewer in other zones, plus rough estimates under some simple classification for every zone. The data are in the form of expert posterior expectation assessments, based on a variety of detailed evidence, for which it would be effectively impossible to list a priori all of the possible evidential outcomes. Suppose that there are r levels of accuracy to which an assessment may be made. We denote the values of the asset in each of the n zones as $X = (X_1, ..., X_n)$. The expert assessment at accuracy level j for X_i, the value of the asset in zone i, is elicited as a posterior expectation $\mathrm{E}_j(X_i)$. For notational convenience, we add a further level of accuracy, namely observing the actual value of X_i, and denote $\mathrm{E}_{r+1}(X_i) = X_i$.

Therefore, the data consists of various individual assessments, made at different levels of accuracy, for different zones. Each assessment gives information both about the corresponding asset value, and about the values of other assessments that may be carried out in the same and in different zones. The data arrives sequentially, and must be processed in a coherent manner. We now describe how temporal sure preference was used to impose a natural and tractable statistical description upon the problem.

As an assessment of the value of an asset in a particular zone at a particular level of accuracy makes use of all of the information which is available for assessment of that value at lower levels of accuracy, we first suppose that full assessments at each level of accuracy are made in sequence, so that we may consider the vectors $\mathrm{E}_j(X)$ as expectation vectors made at a series of time points t_j over the whole collection X. Under temporal sure preference, we view each assessment $\mathrm{E}_j(X)$ as Bayes linear sufficient for all assessments at level j or lower, so that each $\mathrm{E}_j(X)$ forms a prior inference for each vector $\mathrm{E}_k(X)$, $r+1 \geq k > j$.

We introduce the increments $F_j = \mathrm{E}_j(X) - \mathrm{E}_{j-1}(X)$, $j = 1, ..., r+1$. By application of theorem 2, we have the following stochastic description of the development of our beliefs, namely that

$$X = E(X) + \sum_{k=1}^{r+1} F_k, \tag{1.14}$$

and for each level j

$$\mathrm{E}_j(X) = \mathrm{E}(X) + \sum_{k=1}^{j} F_k. \tag{1.15}$$

Each F_j has zero expectation. The elements of these vectors are uncorrelated across levels of assessment j but not across zones i. Therefore, the vectors F_j are mutually uncorrelated, and all of the second order structure in the problem is described by specifying the prior expectation vector for X and, by specifying the variance matrices $V_j = \mathrm{Var}(F_j)$, as each $\mathrm{Cov}(F_j, F_k) = 0, j \neq k$.

Equations (1.14), (1.15) define all of the relationships between the quantities of interest X and the data which are any subcollection of expert judgements $\mathrm{E}_j(X_i)$. The model is fully specified for a second order analysis, in which each expert judgement $\mathrm{E}_j(X_i)$ is a prior inference both for the corresponding quantity X_i, and also for the further expert judgements $\mathrm{E}_k(X_i)$, $k > j$. Thus, the description is sufficient for a Bayes linear analysis of the problem, details of which are given in [9]. There is an element of subjective judgement as to whether we should modify the above argument to take into account that the various expectations are specified by a collection of individuals, and are only partially observed, but this was not felt to substantially change the modelling or the prior specifications.

7. GEOMETRIC FORMULATION FOR PRIOR INFERENCES

We now describe the geometric argument underlying the notion of prior inference. This will allow a natural extension of prior inferences to general classes of random objects. We begin by expressing geometrically the definition of expectation through quadratic scoring rules, as follows. For any finite or infinite collection X of random quantities, we construct the linear space, $\langle X \rangle$, of finite linear combinations of the elements of X. We view $\langle X \rangle$ as a vector space, where each random quantity is a vector and linear combinations of vectors are the corresponding linear combinations of random quantities. We construct an inner product space $[X]$ over the closure of the equivalence classes of $\langle X \rangle$, with respect to the inner product $(Y, Z) = \mathrm{E}(YZ)$, with corresponding norm $\|Y\|^2 = \mathrm{E}(Y^2)$. This structure is motivated and discussed in [7]. We denote the orthogonal projection of $Y \in [X]$ into the subspace $[D]$ as $\mathrm{P}_D(Y)$.

$\mathrm{E}(Y)$ is the value y chosen with respect to penalty $(Y - y)^2$, in probability currency, so that y is the value minimising $\mathrm{E}((Y - y)^2)$, i.e. minimising $\|Y - yC\|^2$, where C denotes the unit constant, i.e. $C = 1$. Therefore, we have $\mathrm{P}_C(Y) = \mathrm{E}(Y)C$. Similarly, if D is a collection of elements of X, including the element C, then, for any $Y \in [X]$, it follows from the definition of adjusted expectation that

$$\mathrm{E}_D(Y) = \mathrm{P}_D(Y). \tag{1.16}$$

Note in particular that $\mathrm{E}_C(Y)$, where C is the space spanned by all constant quantities, which in this case is simply a single unit constant, is numerically equal to $\mathrm{E}(Y)$.

We denote by $\mathrm{E}_t(X)$ the collection of posterior expectations $\mathrm{E}_t(Y), Y \in X$. We now construct the enlarged collection $X(t) = X \cup \mathrm{E}_t(X)$, and the corresponding inner product space $[X(t)]$. Within this space, from theorem 1, in order to satisfy the temporal sure preference principle you must specify

$$\|Y - \mathrm{E}_t(Y)\|^2 \leq \|Y - F\|^2, \tag{1.17}$$

for any quantity F which will definitely be known by time t. Therefore, creating the collection $D(t) = \mathrm{E}_t(X) \cup D$, where D is any collection of quantities, including the unit constant, which will definitely be known by t, we have, for each $Y \in [X]$, that

$$\mathrm{E}_t(Y) = \mathrm{P}_{D(t)}(Y). \tag{1.18}$$

We may therefore derive theorem 2 directly from the properties of the corresponding projections, which will allow us to immediately generalise the property of prior inference to any inner product for which the various expectations may be similarly represented as projections. The argument is as follows.

A linear operator A is positive, written $A \geq 0$, if A is self adjoint with $(AY, Y) \geq 0, \forall Y \in [X]$. We write $A \geq B$ if $A - B \geq 0$. If P_1, P_2 are orthogonal projections with ranges M_1, M_2 respectively, then $P_1 \geq P_2$ if and only if $M_2 \subseteq$

M_1 or equivalently if and only if $\|P_1(Y)\| \geq \|P_2(Y)\| \, \forall Y \in [X]$. Further $P^- = P_1 - P_2$ is an orthogonal projection if and only if $P_1 \geq P_2$ and $P^+ = P_1 + P_2$ is an orthogonal projection if and only if $P_1 \perp P_2$. All of the properties of Bayes linear sufficiency described in section (4) follow from the above properties of the corresponding projections.

From (1.16), (1.18), we therefore have

$$I \geq \mathrm{E}_t \geq \mathrm{E}_D \geq \mathrm{E}_C, \tag{1.19}$$

where I is the identity operator, so that $I - \mathrm{E}_t, \mathrm{E}_t - \mathrm{E}_D, \mathrm{E}_D - \mathrm{E}_C$ are each orthogonal projections with mutually orthogonal ranges. We may therefore write $I - \mathrm{E}_C = (I - \mathrm{E}_t) + (\mathrm{E}_t - \mathrm{E}_d) + (\mathrm{E}_d - \mathrm{E}_C)$ and all of the orthogonalities in relations (1.11), (1.12) follow. We have proved the following geometric version of theorem 2.

Theorem 3 *For any inner product satisfying (1.16), (1.18), we have the relations (1.11), (1.12), where the orthogonalities in (1.12) are with respect to the inner product.*

For which inner products will relations (1.16), (1.18) be satisfied? Our development depends on the property that the squared norm $\|Y - yC\|^2 = \mathrm{E}(L(Y,y))$ is the expectation of quadratic loss, $L(Y,y) = (Y-y)^2$, which is a proper scoring rule for $\mathrm{E}(Y)$. Therefore, we may equate expectation with the projection into $[C]$, as the expectation of the quadratic penalty is minimised by $\mathrm{E}(Y)$, so that $\mathrm{E}(Y)\mathrm{C}$ is the projection of Y onto the unit constant. Further, we may equate posterior expectation with projection into $[D(t)]$, as, at t, we will prefer penalty $L(Y, \mathrm{E}_t(Y))$ to $L(Y,F)$ for any F known by t, so, by temporal sure preference, relation (1.18) follows, from which we may deduce (1.19) so that theorem 3 follows.

We may similarly apply the notion of prior inferences to general random objects for which we have a notion of expectation, whenever we can construct an inner product from a norm that is derived from the expectation of a proper scoring rule. The general representation is as follows. We have a finite or infinite collection X of random objects of a particular type. We construct the linear space $\langle X \rangle$, as before. We say that a norm on $\langle X \rangle$ is proper if, for any constant y and random Y, $\|Y - y\|^2 = \mathrm{E}(L(Y,y))$, where $L(Y,y)$ is a proper scoring rule for $\mathrm{E}(Y)$, namely the preferred choice of constant y for penalty $L(Y,y)$ is $\mathrm{E}(Y)$, or equivalently $\mathrm{E}(L(Y,y))$ is minimised at $\mathrm{E}(Y)$. We construct the inner product space $[X]$ as before, but with respect to the inner product corresponding to this norm. Adjusted expectation, for such a norm, may be defined as a projection through relation (1.16). We may construct the geometric interpretation given by Theorem 3 from the temporal sure preference principle, as the projection of an element of $\langle X \rangle$ into the constant subspace corresponds to the expectation of the random object, and relation (1.18) follows from (1.17), which follows for any squared norm which is the expectation of a proper scoring rule. We therefore have the following corollary.

Corollary 3 *Whenever the inner product is based on a proper norm, then given temporal sure preference, the prior inferences of theorem (2) hold.*

So far, we have illustrated our development for random scalar quantities under quadratic loss. As a second example, suppose that X is a collection of random $m \times m$ matrices. If $A = (A_{ij})$ is a random matrix and you must choose matrix $a = (a_{ij})$ with loss function $L(A, a) = \text{trace}((A-a)(A-a)^T) = \sum_{i,j}(A_{ij} - a_{ij})^2$, then the choice is $a = \text{E}(A)$, so that $L(A, a)$ is a proper scoring rule for $\text{E}(A)$.

The corresponding norm for any random matrix A is given as $\|A\|^2 = \text{E}(\text{trace}(AA^T))$, with inner product $(A, B) = \text{E}(\text{trace}(AB^T))$. $\langle X \rangle$ is the collection of linear combinations of the random matrices in X and the space of constant objects, C^*, is of dimension m^2, and is spanned by the orthogonal collection of matrices C_{ij} where C_{ij} has a 1 in the $(i, j)^{th}$ position and zero elsewhere. This construction is derived, motivated and illustrated in [12], as a natural setting for learning about variance matrices, allowing us to make inferences over possibly very large random matrices based on a comparatively small number of prior quantifications of belief, for example allowing us to make simple inferences about population variance matrices based on sample variance matrices, exploiting natural exchangeability judgements.

Projection of the random matrix $A = (A_{ij})$ into C^* is equivalent to choosing the matrix $a = (a_{ij})$ to minimise $\|A - \sum_{i,j} a_{ij} C_{ij}\|^2 = \sum_{i,j} \text{E}((A_{ij} - a_{ij})^2)$ which minimum is attained at $a = \text{E}(A)$. Temporal sure preference similarly requires $\text{E}(\text{trace}((A - \text{E}_t(A))(A - \text{E}_t(A))^T)) \leq \text{E}(\text{trace}((A - F)(A - F)^T))$ for any matrix F which will be known by t. From corollary 3, we therefore have the prior inference for random matrices, namely that, for any matrix A, and collection of matrices D, including the constant collection C which will be known by t, we have

$$Y = \text{E}_t(Y) + R_t(Y), \;\; \text{E}_t(Y) = \text{E}_D(Y) + S_t(Y),$$

where $R_t(Y), S_t(Y)$ each have prior expectation zero, and we have $R_t(.) \perp \{\text{E}_t(.), \text{E}_D(.), S_t(.), \Delta(X)\}$, and $S_t(.) \perp \{\text{E}_D(.), \Delta(X),\}$ where in this case $A \perp B$ is given by $\text{E}(\text{trace}(AB^T)) = 0$.

We therefore have a similar interpretation of the relationship between linear fitting and posterior expectations for matrices to that which we developed for expectations of scalar quantities, which itself is a generalisation of our interpretation of conditioning. Each generalisation offers the opportunity to further restrict the number of belief specifications upon which we base our prior inference. Formally, probabilistic specification over some set Ω corresponds to specifying the Hilbert space $\text{L}^2(\Omega, P)$ of square integrable functions over Ω under some prior measure P, while expectation specification corresponds to specifying whichever subspace of $\text{L}^2(\Omega, P)$ appears to be appropriate to the task. If Ω is very large, then typically we will specify and analyse small subspaces of the whole space, but this is simply a subjective choice in each problem. In each such case the interpretation of conditioning or of adjusted expectation will be the same, but the larger the subspace of $L^2(\Omega, P)$ which is specified, then the more informative will be the projections on which the prior inference is based and the more stringent will be the the implications of the various orthogonality conditions imposed in theorem 3. However, this is simply a difference in degree, reflecting the differing amounts of prior information and data which have been brought to bear.

The generalisation to random matrices allows further restriction of the prior specification from belief statements about individual random quantities to belief statements about collections of random quantities. Again, this may reduce, and so simplify, the detail required of the prior specification, and correspondingly it may be that the resulting projections will be less informative and the residual orthogonalities may impose weaker constraints. Again, these are simply differences of degree, and, in large, complex problems, where full belief specification would be extremely difficult and time consuming, we must, in each problem, decide which level of analysis is within our abilities and is adequate to the problem.

In each case, whether probabilistic conditioning for events, adjusted expectations for random quantities, or more general projections, for example for random matrices, the inferential argument is the same. Temporal sure preference establishes certain stochastic relations which you now assert between the various formal rules that you may now determine and the actual posterior evaluations that you will make. You may use these relations now to guide your prior reasoning and choice of observations, and you now expect that these considerations will be useful to you at the future time in resolving part of the uncertainty about your future judgements.

Department of Mathematical Sciences
University of Durham

REFERENCES

1. Armendt, B., 1992, Dutch strategies for diachronic rules: when believers see the sure loss coming, PSA 1992, vol 1, ed. D. Hull, M.Forbes, K.Okruhlik, East Lansing, 217–229, Philosophy of Science Association.
2. de Finetti, B., 1974, Theory of probability, vol.1 , London, Wiley
3. Farrow, M. and Goldstein, M., 1993, Bayes linear methods for grouped multivariate repeated measurement studies with application to crossover trials, Biometrika, 80, 39–59
4. Goldstein, M., 1983, The Prevision of a Prevision, J. Amer. Statist. Ass., 78, 817–819
5. Goldstein, M., 1985, Temporal Coherence, in Bayesian Statistics 2, J.-M. Bernardo et al. eds, 231–248, North Holland
6. Goldstein, M. 1986, Separating beliefs, in Bayesian Inference and Decision Techniques, P.K. Goel and A. Zellner, eds, 197–215, North Holland
7. Goldstein, M., 1988, Adjusting belief structures, J. R. Statist. Soc., 50, 133–154.
8. Goldstein, M., 1994, Revising exchangeable beliefs: subjectivist foundations for the inductive argument, in Aspects of uncertainty. A tribute to D. V. Lindley, Freeman, P. and Smith, A.F.M., eds., 201–222, Wiley

9. Goldstein, M and O'Hagan, A, 1995, Bayes linear sufficiency and systems of expert posterior assessments, J. Roy. Statist. Soc. B, to appear.
10. Skyrms, B., 1993, Discussion: a mistake in dynamic coherence arguments?, Philosophy of Science, 60, 320–328.
11. Walley, P., 1991, Statistical inference with imprecise probabilities, Chapman and Hall.
12. Wilkinson, D. and Goldstein, M., 1995, Bayes linear adjustment for variance matrices, in Bayesian Statistics 5, J. M. Bernardo et al eds, to appear.
13. vanFrassen, B., 1984, Belief and the will, Journal of Philosophy, 81, 235–256.
14. vanFrassen, B., 1995, Belief and the problem of Ulysses and the sirens, Philosophical Studies, 77, 7–37.

BRIAN SKYRMS

GAME THEORY, RATIONALITY AND EVOLUTION*

1. INTRODUCTION

The Theory of Games was conceived as a theory of interactive decisions for rational agents. The basic solution concept — that of Nash equilibrium — is built on rationality assumptions whose strength has only recently been made explicit. Refinements of Nash equilibrium, such as subgame perfection, require even stronger rationality assumptions. Here, it seems, is the pure domain of rational choice theory, where rational decision makers reflect on each others rational decisions, and rationality assumptions are multiplied by strategic interaction.

In Evolutionary theory there is also strategic interaction in the form of frequency dependent selection, but there is no presumption of rationality — let alone common knowledge of rationality — operative. Indeed the organisms that are evolving may not even be making decisions at all. Nevertheless, game theoretic ideas have been fruitfully applied to strategic interaction in evolution. And the key equilibrium concept is almost the same.

The most striking fact about the relationship between evolutionary game theory and economic game theory is that at the most basic level, a theory built of hyperrational actors and a theory built of possibly non-rational actors are in fundamental agreement. This fact has been widely noticed, and its importance can hardly be overestimated. Facile criticisms of game theory along the lines "people just aren't rational" must be reconsidered. There are many roads to the Nash equilibrium concept, only one of which is based on highly idealized rationality assumptions.

However, as we look more closely at the theory in the evolutionary and rational choice settings, differences begin to emerge. At the onset, a single population evolutionary setting imposes a symmetry requirement which selects Nash equilibria which might appear implausible in other settings. Furthermore, refinements of the Nash equilibrium are handled differently. Standard evolutionary dynamics, the replicator dynamics, does not guarantee elimination of weakly dominated strategies. In a closely related phenomenon, when we consider extensive form games, evolutionary dynamics need not eliminate strategies which fail the test of sequential rationality. Going further, if we generalize the theories to allow for correlation, we find that the two theories

*A fuller discussion of issues raised in this paper together with more complete references to the literature may be found in my book *Evolution of the Social Contract*, which is forthcoming with Cambridge University Press.

M.L. Dalla Chiara et al. (eds.), Structures and Norms in Science, 73–86.

can diverge dramatically. Correlated evolutionary game theory can even allow for the fixation of strongly dominated strategies.These are strategies which fail under even the weakest theory of rational choice.

The situation is therefore more complicated than it might at first appear. There are aspects of accord between evolutionary game theory and rational game theory as well as areas of conflict. This is as true for cultural evolution as for biological evolution. The phenomena in question thus have considerable interest for social and political philosophy, and they can be displayed in strategic situations which illustrate some of the central issues in those areas.

2. RATIONAL EQUILIBRIUM

Suppose a number of rational decision makers must each independently make decisions which jointly determine the payoff to each player. A combination, or profile, of players acts is a Nash equilibrium just in case if each knows the others' acts, he cannot increase his payoff by unilateral deviation. If randomized acts (probability distributions over pure acts) count as possible acts then in a finite game (a finite number of players and a finite number of pure acts) there is always a Nash equilibrium.

Why should we expect rational players to play a Nash equilibrium? Well, if they were not at a Nash equilibrium and each knew what the others were doing, then by the definition of Nash equilibrium, some player would not be maximizing her expected payoff thus she would not be rational. But if players are deciding independently, why should we assume that each knows what the others will do?

One sort of answer is that the theory of rationality will specify uniquely what players will do, and each player — not only being rational but also knowing that the other players are also rational — will be able to deduce from the theory what the other players will do. This sort of defense of the Nash equilibrium concept in the context of zero sum games is, in fact, given by von Neumann and Morgenstern (1947 p. 148). If, however, "rationality" is taken to mean Bayesian Rationality — that is, maximizing expected utility — then it will not do the job. Typically, maximizing expected utility does not, by itself, determine a player's strategy uniquely. The act that she takes to maximize expected utility will depend on her beliefs. All that we can really conclude in ignorance of her beliefs is that it will not be a strongly dominated act — one such that there is a specific alternative that gives a higher payoff no matter what the other players do.

Suppose we assume not only that the players are Bayesian rational, but also that this is common knowledge — all know that all know that ... that all players are rational. Then what we get is the "rationalizable strategic behavior" of Pearce (1984) and Bernheim (1984). Nash equilibria are rationalizable profiles, but there are also rationalizable profiles that are not Nash equilibria. In the case of two-person games, rationalizable strategies are what remain after iterated elimination of strongly dominated strategies. Any pair of such rationalizable strategies is a rationalizable profile. In the well-known game of Rock, Scissors, Paper no strategy is strongly dominated and all strategies are rationalizable. In the game of Prisoner's Dilemma, the strategy Defect strongly dominates the strategy Cooperate, so the only rationalizable profile is that in which both players defect.

On the other hand, one might attempt a more stringent theory of rationality, which does in fact specify a unique choice of action for every player in every game. This is the heroic project of Harsanyi and Selten (1988). In between Rationalizability and A General Theory of Equilibrium Selection in Games, we have various refinements. One of the most straightforward is that players should never use weakly dominated strategies, ones such that some alternative can do better but can't do worse. These more modest attempts at narrowing down the range of possible options have their own interest, but unlike the heroic approach, they cannot keep the fundamental rationale for the Nash equilibrium itself from unraveling. For if rational players cannot deduce how others will act, there is no general reason to expect them to play a Nash equilibrium.

3. EVOLUTIONARY GAME THEORY

Let us consider the case of individuals who are paired at random from a large population to engage in a strategic interaction. Reproduction is asexual and individuals breed true. We measure the payoffs in terms of evolutionary fitness — expected number of offspring. Payoff to a strategy depends on what strategy it is paired against, so we have frequency dependent selection in the population. We write the payoff of strategy A when played against strategy B as $U(A|B)$. Under these assumptions, the expected Fitness for a strategy is an average of its payoffs against alternative strategies weighted by the population proportions of the other strategies:

$$U(A) = SUM_i U(A|B_i)P(B_i)$$

The expected fitness of the population is the average of the fitnesses of the strategies with the weights of the average being the population proportions:

$$UBAR = SUM_j U(A_j)P(A_j)$$

We postulate that the population is large enough that we may safely assume that you get what you expect.

The population dynamics is then deterministic. Assuming discrete generations with one contest per individual per generation we get:

$$\text{DARWIN MAP:} \qquad P'(A) = P(A)U(A)/UBAR$$

where $P(A)$ is the proportion of the population today using strategy A and P' is the proportion tomorrow. Letting the time between generations become small, we find the corresponding continuous dynamics:

$$\text{DARWIN FLOW:} \quad dP(A)/dt = P(A)[U(A) - UBAR]/UBAR$$

The Darwin flow has the same orbits as the simpler dynamics gotten by discarding the denominator, although the velocity along the orbits may differ:

$$\text{REPLICATOR FLOW:} \quad dP(A)/dt = P(A)[U(A) - UBAR]$$

In the following discussion we will concentrate on this replicator dynamics.

The replicator flow was introduced by Taylor and Jonker (1978) to build a foundation for the concept of evolutionarily stable strategy introduced by Maynard-Smith and Price (1973). The leading idea of Maynard Smith and Price was that of a strategy such that if the whole population uses that strategy, it cannot be invaded. The definition offered is that A is evolutionarily stable just in case both:

(i) $U(A|A) \geq U(B|A)$ (for all B different that A) and
(ii) If $U(A|A) = U(B|A)$ then $U(A|B) > U(B|B)$.

The leading idea only makes sense for mixed strategies if individuals play randomized strategies. But the replicator dynamics is appropriate for a model where individuals play pure strategies, and the counterparts of the mixed strategies of game theory are polymorphic states of the population. We retain this model, but introduce the notion of an evolutionarily stable state of the population. It is a polymorphic state of the population, which would satisfy the foregoing inequalities if the corresponding randomized strategies were used. We use ESS to mean evolutionarily stable state in this sense, an evolutionarily stable strategy corresponding to the special case of an evolutionarily stable monomorphic state.

4. NASH FROM NATURE

Condition (i) in the definition of ESS looks a lot like the definition of Nash equilibrium. It is, in fact, the condition that $< A, A >$ is a Nash equilibrium of the associated two person game in which both players have the payoffs specified in by the fitness matrix, $U(-|-)$. The second condition adds a kind of stability requirement. The requirement is sufficient to guarantee strong dynamical stability in the replicator dynamics:

EVERY ESS is a STRONGLY DYNAMICALLY STABLE (or ATTRACTING) EQUILIBRIUM in the REPLICATOR DYNAMICS

This is more that sufficient to guarantee Nash equilibrium in the corresponding game:

IF A is a DYNAMICALLY STABLE EQUILIBRIUM in the REPLICATOR DYNAMICS then $< A, A >$ is a NASH EQUILIBRIUM of the CORRESPONDING TWO-PERSON GAME.

(The converse of each of the foregoing propositions fails.)

Evidently, Nash equilibrium has an important role to play here, in the absence of common knowledge of rationality or even rationality itself. The reason is quite clear, but nevertheless deserves to be emphasized. The underlying dynamics is adaptive: it has a tendency towards maximal fitness. Evidently many alternative dynamics — of learning as well as of evolution — share this property. There is a moral here for philosophers and political theorists who have attacked the theory of games on the basis of its rationality assumptions. Game theory has a far broader domain of application than that suggested by its classical foundations.

5. SYMMETRY

For every evolutionary game there is a corresponding symmetric two person game, and for every ESS in the Evolutionary game there is a corresponding symmetric Nash equilibrium in the two person game. Symmetry is imposed on the Nash equilibrium of the two person game because the players have no identity in the evolutionary game. Different individuals play the game. The things which have enduring identity are the strategies. Evolutionary games are played under what I call The Darwinian Veil of Ignorance. For example, consider the game of Chicken. There are two strategies: Swerve; Don't. The fitnesses are:

$U(S|S) = 20$
$U(S|D) = 15$
$U(D|S) = 25$
$U(D|D) = 10$

In the two person game there are two Nash equilibria in pure strategies: player 1 swerves and player 2 doesn't, player 2 swerves and player 1 doesn't. There is also a mixed Nash equilibrium with each player having equal chances of swerving. In the evolutionary setting there are just swervers and non-swervers. The only equilibrium of the two person game that corresponds to an ESS of the evolutionary game is the mixed strategy. It corresponds to an evolutionary stable polymorphic state where the population is equally split between swervers and non-swervers. Any departure from the state is rectified by the replicator dynamics, for it is better to swerve when the majority don't and better to not swerve when the majority do.

The evolutionary setting has radically changed the dynamical picture. If we were considering learning dynamics for two fixed individuals, the relevant state space would be the unit square with the x-axis representing the probability that player one would swerve and the y-axis representing the probabilities that player 2 would swerve. With any reasonable learning dynamics, the mixed equilibrium of the two person game would be highly unstable and the two pure equilibria would be strongly stable. The move to the evolutionary setting, in effect restricts the dynamics to the diagonal of the unit square. The probability that player one will encounter a given strategy must be the same as the probability that player two will. It is just the proportion of the population using that strategy. On the diagonal, the mixed equilibrium is now strongly stable.

For another example which is of considerable importance for social philosophy, consider the simplest version of the Nash bargaining game. Two individuals have a resource to divide. We assume that payoff just equals amount of the resource. They each must independently state the minimal fraction of the resource that they will accept. If these amounts add up to more than the total resource, there is no bargain struck and each players gets nothing. Otherwise, each gets what she asks for.

This two person game has an infinite number of Nash equilibria of the form: Player 1 demands x of the resource and player 2 demands $(1 - x)$ of the resource $[0 < x < 1]$. Each of these Nash equilibria is strict — which is to say that a unilateral deviation from the equilibrium not only produces no gain; it produces a positive loss. Here we have the problem of multiple Nash equilibria in especially difficult form. There are an infinite number of equilibria and, being strict, they satisfy all sorts of refinements of the Nash equilibrium concept. [See van Damme (1987)].

Suppose we now put this game in the evolution context that we have developed. What pure strategies are evolutionarily stable? There is exactly one: Demand half! First, it is evolutionarily stable. In a population in which all demand half, all get half. A mutant who demanded more of the natives would get nothing; a mutant who demanded less would get less. Next, no other pure strategy is evolutionarily stable. Assume a population composed of players who demand x, where $x < 1/2$. Mutants who demand $1/2$ of the natives will get $1/2$ and can invade. Next consider a population of players who demand $x > 1/2 > 1$. They get nothing. Mutants who demand $0 < y < (1 - x)$ of the natives will get y and can invade. So can mutants who demand $1/2$, for although they get nothing in encounters with natives, they get $1/2$ in encounters with each other. Likewise, they can invade a population of natives who all demand 1. Here the symmetry requirement imposed by the evolutionary setting by itself selects a unique equilibrium from the infinite number of strict Nash equilibria of the two person game. The "Darwinian Veil of Ignorance" gives an egalitarian solution.

This is only the beginning of the story of the evolutionary dynamics of bargaining games. Even in the game I described, there are evolutionarily stable polymorphic states of the population which are of considerable interest. And in more complicated evolutionary games we can consider individuals who can occupy different roles, with the payoff function of the resource being role dependent. However, at the most basic level, we have an powerful illustration of my point. The evolutionary setting for game theory here makes a dramatic difference in equilibrium selection, even while it supports a selection of a Nash equilibrium.

6. WEAKLY DOMINATED STRATEGIES

The strategic situation can be radically changed in bargaining situations by introducing sequential structure. Consider the Ultimatum game of Güth, Schmittberger and Schwartze (1982). One player —the Proposer — demands some fraction of the resource. The second player — the Responder is informed of the Proposer's proposal and either takes it or leaves it. If he takes it the first player gets what she demanded and he gets the rest. If he leaves it, neither player gets anything.

The are again an infinite number of Nash equilibria in this game, but from the point of view of rational decision theory they are definitely not created equal. For example, there is a Nash equilibrium where the Proposer has the strategy "Demand half" and the Responder has the strategy "Accept half or more but reject less." Given each players strategy, the other could not do better by altering her strategy. But there is nevertheless something odd about the Responder's strategy. This can be brought out in various ways. In the first place, the Responder's strategy is weakly dominated. That is to say that there are alternative strategies which do as well against all possible Proposer's strategies, and better against some. For example, it is weakly dominated by the strategy "Accept all offers". A closely related point is that the equilibrium at issue is not subgame perfect in the sense of Selten (1965). If the Proposer were to demand 60% this would put the responder into a subgame, which in this case would be a simple decision problem: 40% or nothing. The Nash equilibrium of the subgame is the optimal act: Accept 40%. So the conjectured equilibrium induces play on the subgame which is not an equilibrium of the subgame.

For simplicity, let's modify the game. There are 10 lumps of resource to divide; lumps are equally valuable and can't be split; the proposer can't demand all 10 lumps. Now there is only one subgame perfect equilibrium — the Proposer demands 9 lumps and the Responder has the strategy of accepting all offers. If there is to be a Bayesian rational response to any possible offer, the Responder must have the strategy of accepting all offers. And if the Proposer knows that the Responder will respond by optimizing no matter what she does, she will demand 9 lumps.

It is worth noting that the subgame perfect equilibrium predicted by the foregoing rationality assumptions does not seem to be what is found in experimental trials of the ultimatum game — although the interpretation of the experimental evidence is a matter of considerable controversy in the literature. Proposers usually demand less than 9 and often demand 5. Responders often reject low offers, in effect choosing zero over 1 or 2. I do not want to discuss the interpretation of these results here. I only note their existence. What happens when we transpose the ultimatum game into an evolutionary setting?

Here the game itself is not symmetric — the proposer and responder have different strategy sets. There are two ways to fit this asymmetry into an evolutionary framework. One is model the game as an interaction between two disjoint populations. A proposer is drawn at random form the proposer population and a responder is drawn at random from the responder population and they play the game with the payoff being expected number of offspring. The alternative is a single population model with roles. Individuals from a single population sometimes are in the role of Proposer and sometimes in the role of Responder. In this model individuals are required to have the more complex strategies appropriate to the symmetrized game, for example:

If in the role of Proposer demand x;
If in the role of Responder and proposer demands z, take it;
Else if in the role of Responder and proposer demands z', take it;
Else if in the role of responder and the proposer demands z'',leave it;
ETC.

The evolutionary dynamics of the two population model was investigated by Binmore, Gale and Samuelson (1995) and that of the one population symmetrized model by myself for the replicator flow and by Harms (1994) for the Darwin map.

Despite some differences in modeling, all these studies confirm one central fact. Evolutionary dynamics need not eliminate weakly dominated strategies; evolutionary dynamics need not lead to subgame perfect equilibrium. Let me describe my results for a small game of Divide 10, where proposers are restricted to two possible strategies: Demand 9; Demand 5. Responders now have four possible strategies depending on how they respond to a demand of 9 and how they respond to a demand of 5. The typical result of evolution starting from a population in which all strategies are represented, is a polymorphic population which includes weakly dominated strategies.

One sort of polymorphism includes FAIRMEN types who demand 5 and accept 5 but reject greedy proposals; together with EASY RIDERS who demand 5 and accept all. The FAIRMAN strategy is weakly dominated by the EASY RIDER strategy, but nevertheless some proportion of FAIRMEN can persist in the final polymorphism. Another sort of polymorphism consists of GAMESMEN who demand 9 and accept all, together with MAD DOGS who accede to a demand of 9 but reject a FAIRMAN's demand of 5. MAD DOG is weakly dominated by GAMESMAN but nevertheless some proportion of MAD DOGS can persist in the final polymorphism. Which polymorphism one ends up with depends on what population proportions one starts with. However, in either case one ends up with populations which include weakly dominated strategies.

How is this possible? If we start with a completely mixed population — in which all strategies are represented — the weakly dominated strategies must have a smaller average fitness than the strategies which weakly dominate them. The weakly dominating strategies by definition do at least as well against all opponents and better against some. Call the latter the Discriminating Opponents. As long as the discriminating opponents are present in the population the weakly dominating do better than the weakly dominated, but the Discriminating Opponents may go extinct more rapidly than the weakly dominated ones. This leaves a polymorphism of types which do equally well in the absence of Discriminating Opponents. This theoretical possibility is, in fact, the typical case in the ultimatum game. This conclusion is not changed, but rather reinforced, if we enrich our model by permitting the Proposer to demand 1,2,3,4,5,6,7,8,9. Then we get more complicated polymorphisms which typically include a number of weakly dominated strategies.

It might be natural to expect that adding a little mutation to the model would get rid of the weakly dominated strategies. The surprising fact is that such is not the case. The persistence of weakly dominated strategies here is quite robust to mutation. It is true that adding a little mutation may keep the population completely mixed, so that weakly dominated strategies get a strictly smaller average fitness than those strategies which weakly dominate them, although the differential may be very small. But mutation also has a dynamical effect. Other strategies mutate into the weakly dominated ones. This effect is also very small. In polymorphic equilibria under mutation these two very small effects come into balance.

That is not to say that the effects of mutation are negligible. These effects depend on the mutation matrix, Mij, of probabilities that one given type will mutate into another given type. We model the combined effects of differential reproduction and mutation by a modification of the Darwin Flow:

$$dP(A_i)/dt$$
$$= (1-e)[P(A_i)(U(A_i) - UBAR)/UBAR] + e[SUM_j P(A_j)M_{ij} - P(A_i)]$$

[See Hofbauer and Sigmund (1988) p. 252]

There is no principled reason why these probabilities should all be equal. But let us take the case of a uniform mutation matrix as an example. Then, in the ultimatum game, introduction of mutation can undermine a polymorphism of FAIRMEN and EASY RIDERS and lead to a polymorphism dominated by GAMESMEN and MAD DOGS. With a uniform mutation matrix and a mutation rate of .001, there is a polymorphism of about 80% gamesmen and 20% MAD DOGS — with other types maintained at very small levels by mutation. Notice that the weakly dominated strategy MAD DOG persists here at quite substantial levels. Other mutation matrices can support polymorphisms consisting mainly of FAIRMEN and FREE RIDERS. In either case, we have equilibria which involve weakly dominated strategies.

A crack has appeared between game theory based on rationality and game theory based on evolutionary dynamics. There are rationality based arguments against weakly dominated strategies and for subgame perfect equilibrium. We have seen that evolutionary dynamics does not respect these arguments. This is true for both one population and two population models. It is true for both continuous and discrete version of the dynamics. It remains true if mutation is added to the dynamics. I will not go into the matter here, but it also remains true if variation due to recombination — as in the genetic algorithm — is added to the dynamics. We shall see in the next section that the crack widens into a gulf if we relax the assumption of random pairing from the population.

7. STRONGLY DOMINATED STRATEGIES

Consider a two person game in which we assume that the players are Bayesian rational and know the structure of the game, but nothing more. We do not assume that the players know each others strategies. We do not assume that Bayesian rationality is Common Knowledge. In general, this assumption will not suffice for Nash equilibrium or even Rationalizability. But one thing that it does is guarantee that players will not play strongly dominated strategies.

In certain special games, this is enough to single out a unique Nash equilibrium. The most well-known game of this class is the Prisoner's Dilemma. In this game both players must choose between cooperation $[C]$ and defection $[D]$. For each player:

$U(C|C) = 10$
$U(D|C) = 15$
$U(C|D) = 0$
$U(D|D) = 5$

Defection strongly dominates Cooperation and if players optimize they both defect.

If we simply transpose our game to an evolutionary setting, nothing is changed. The game is symmetric; players have the same strategy sets and the payoff for one strategy against another does not depend on which player plays which. Evolutionary dynamics drives the cooperators to extinction. Defection is the unique evolutionarily stable strategy and a population composed exclusively of defectors is the unique evolutionarily stable state. So far evolution and rational decision theory agree.

Let us recall, however, that our evolutionary model was based on a number of simplifying assumptions. Among those was the assumption that individuals are paired at random from the population to play the game. This may or may not be plausible. There is a rich biological literature discussing situations where it is not plausible for biological evolution. With regard to cultural evolution I believe that many social institutions exist in order to facilitate non-random pairing. A more general evolutionary game theory will allow for correlated pairing.

Here the fundamental objects are conditional pairing probabilities, $P(A|B)$, specifying the probability that someone meets an A-player given that she is a B-player. These conditional pairing proportions may depend on various factors, depending on the particular biological or social context being modeled.The expected fitness of a strategy is now calculated using these conditional probabilities:

$$U(A) = SUM_i U(A|B_i)P(B_i|A).$$

Now suppose that nature has somehow — I don't care how — arranged high correlation between like strategies among individuals playing the Prisoner's Dilemma. For instance, suppose $P(C|C) = P(D|D) = .9$ and $P(C|D) = P(D|C) = .1$. Then the fitness of cooperation exceeds that of defection and the strategy of cooperation takes over the population. Correlated evolutionary game theory permits models in which a strongly dominated strategy is selected.

We can, of course, consider correlation in two person games between rational decision makers. This was done by Aumann (1974) (1987) in his seminal work on correlated equilibrium. Aumann takes the natural step of letting the "coin flips" of players playing mixed strategies be correlated. The resulting profile is a joint correlated strategy. Players know the joint probability distribution and find out the results of their own "coin flips". If, whatever the results of those flips, they do not gain in expected utility by unilateral deviation, they are at a correlated equilibrium. However, this natural generalization is quite different than the generalization of mixed strategies as polymorphic populations that I have sketched. In particular, there is only one Aumann correlated equilibrium in Prisoner's Dilemma. It has each player defecting. More generally, Aumann correlated equilibria do not use strongly dominated strategies.

If fact, evolutionary game theory can deal with two kinds of mixed strategy. The first kind arises when individuals themselves use randomized strategies. The second kind interprets a population polymorphism as a mixed strategy. I have focused on the second kind in this paper. We have assumed that individuals play pure strategies in order to preserve the rationale for the replicator dynamics. If one drops the independence assumption from the first kind of mixed strategy one gets Aumann correlated equilibrium. If one drops the independence assumption from the second kind, one gets the kind of correlated evolutionary game theory I am discussing here. In this setting new phenomena are possible — the most dramatic of which include the fixation of strongly dominated strategies.

Prisoner's Dilemma is so widely discussed because it is a simple paradigm of the possibility of conflict between efficiency and strict dominance. Everyone would be better off if everyone cooperated. Cooperation is the efficient strategy. Whatever the other player does, you are better off defecting. Defection is the dominant strategy. In the game theory of rational choice dominance wins. But in correlated evolutionary game theory, under favorable conditions of correlation, efficiency can win.

The point is general. If there is a strategy such that it is best if everyone uses it, then sufficiently high autocorrelation will favor that strategy. Perfect correlation imposes a kind of Darwinian categorical imperative for under these conditions, Others do unto you as you do unto them. Then a strategy, A, such that $U(A|A) > U(B|B)$ for all B different from A, will be carried to fixation by the replicator dynamics from any initial state in which it is represented in the population.

On the other hand, in some strategic situations anticorrelation may promote social welfare. Suppose that there are two types of a species that do not do well when interacting with themselves, but each of which gains a large payoff when interacting with the other type. Then the payoff to each type and the average payoff to the species might be maximized if the types could somehow anticorrelate. The situation hypothesized is one where mechanisms for detection might well evolve to support pairing with the other type. If we consider species such as our own which reproduce sexually, the idea should not seem too farfetched.

The introduction of correlation in the manner indicated, takes evolutionary game theory into largely uncharted waters — unexplored by the traditional game theory based on rational choice. The basic structure of the theory is relaxed, opening up novel possibilities for the achievement of efficiency. One of the simplest and most dramatic examples is the possibility of the evolution of the strongly dominated strategy of cooperation in the one-shot Prisoner's Dilemma.

8. CONCLUSION

From the fundamental insight that in the random pairing model Darwin supports Nash, we have moved to an appreciation of ways in which game theory based on rational choice and game theory based on evolution may disagree. In the Ultimatum game evolution does not respect weak dominance or subgame perfection. In the Prisoner's Dilemma with correlation evolution may not respect strong dominance.

Is this an argument for the evolution of irrationality? It would be a mistake to leap to this conclusion. Irrationality in Bayesian terms does not consist in failing to maximize Darwinian fitness, but rather in failing to maximize expected utility, where subjective utility and probability are determined by one's preferences. One can conjecture utility functions which save Bayesian rationality of prima facie irrational behavior in putative Ultimatum or Prisoner's Dilemma games. Whether this is the best way to explain observed behavior remains to be seen. But if it were, we could talk about the evolution of utility functions which disagree with Darwinian fitness rather than the evolution of irrationality.

However that may be, when we look at the structure of game theory based of rational choice and evolutionary game theory, we find that beyond the areas of agreement there are also areas of radical difference.

Department of Philosophy
University of California at Irvine

REFERENCES

Aumann, R.J. (1974) "Subjectivity and Correlation in Randomized Strategies" *Journal of Mathematical Economics* 1:67–96.

Aumann, R.J. (1987) "Correlated Equilibrium as an Expression of Bayesian Rationality" *Econometrica* 55:1–18.

Binmore, K., Gale, J. and Samuelson, L. (1995) "Learning to be Imperfect: The Ultimatum Game" *Games and Economic Behavior* 8: 56–90.

Bernheim, B.D. (1984) "Rationalizable Strategic Behavior" *Econometrica* 52:1007–1028.

Bomze, I. (1986) "Non-cooperative Two-person Games in Biology: A Classification" *International Journal of Game Theory* 15: 31–57.

Dawes, R. and Thaler, R. (1988) "Anomalies: Cooperation" *Journal of Economic Perspectives* 2: 187–197.

Güth, W., Schmittberger, R. and Schwarze, B. (1982) "An Experimental Analysis of Ultimatum Bargaining" *Journal of Economic Behavior and Organization* 3: 367–388.

Güth, W., and Tietz, R. (1990) "Ultimatum Bargaining Behavior: A Survey and Comparison of Experimental Results" *Journal of Economic Psychology* 11, 417–449.

Hamilton, W. D. (1964) "The genetical evolution of social behavior" *Journal of Theoretical Biology* 7: 1–52.

Harms, W. (1994) "Discrete Replicator Dynamics for the Ultimatum Game with Mutation and Recombination" Technical Report: University of California, Irvine.

Hirshliefer, J. (1987) "On the Emotions as Guarantors of Threats and Promises" in *The Latest on the Best: Essays on Evolution and Optimality* ed. J. Dupré. Cambridge Mass.: M.I.T. Press.

Hofbauer, J. and Sigmund, K. (1988) *The Theory of Evolution and Dynamical Systems* Cambridge: Cambridge University Press.

Maynard Smith, J. and Price, G.R. (1973) "The Logic of Animal Conflict" *Nature* 146:15–18.

Maynard Smith, J. and Parker, G.R. (1976) "The Logic of Asymmetric Contests" *Animal Behavior* 24, 159–175.

Kitcher, P. (1993) "The Evolution of Human Altruism" *The Journal of Philosophy* 10, 497–516.

Milgrom, P., North, D. and Weingast, B. (1990) "The Role of Institutions in the Revival of Trade: The Law Merchant, Private Judges, and the Champagne Fairs", *Economics and Politics* 2:1–23.

Pearce, D.G. (1984) "Rationalizable Strategic Behavior and the Problem of Perfection" *Econometrica* 52:1029–1050.

Samuelson, L. (1988) "Evolutionary Foundations of Solution Concepts for finite two-player normal form games" in *Proceedings of the Second Conference on Theoretical Aspects of Reasoning About Knowledge* ed. M. Vardi. Los Altos, California: Morgan Kaufmann, 211–226.

Selten, R. (1975) "Reexamination of the Perfectness Concept of Equilibrium in Extensive Games" *International Journal of Game Theory* 4: 25–55.

Selten, R. (1965) "Spieltheoretische Behandlung eines Oligopolmodells mit Nachfragetragheit" *Zeitschrift für die gesamte Staatswissenschaft* 121: 301–324, 667–689.

Skyrms, B. (forthcoming) *Evolution of the Social Contract Cambridge*: Cambridge University Press.
Skyrms, B. (forthcoming) "Mutual Aid" in *Modeling Rationality, Morality and Evolution* ed. Peter Danielson Oxford: Oxford University Press.
Skyrms, B. (forthcoming) "Evolution of an Anomaly" in *Protosoziologie* vol. 8 "Rationalität II"
Skyrms, B. (1995) "Introduction to the Nobel Symposium on Game Theory" *Games and Economic Behavior* 8 3–5.
Skyrms, B. (1994) "Darwin Meets 'The Logic of Decision': Correlation in Evolutionary Game Theory" *Philosophy of Science* 61, 503–528.
Skyrms, B. (1994) "Sex and Justice" *The Journal of Philosophy* 91, 305–320. Also to be reprinted in *The Philosopher's Annual* XVII Atascadero, Ca.: Ridgeview.
Skyrms, B. (1990) *The Dynamics of Rational Deliberation* Cambridge, Mass.: Harvard University Press.
Sober, E. (1992) "The Evolution of Altruism: Correlation, Cost and Benefit" *Biology and Philosophy* 7: 177–187.
Thaler, R. (1988) "Anomalies: The Ultimatum Game" *Journal of Economic Perspectives* 2, 195–206.
van Damme, E. (1987) *Stability and Perfection of Nash Equilibria* Berlin: Springer.

8
History of Logic, Methodology and Philosophy of Science

ULRICH GÄHDE

ANOMALIES AND THE REVISION OF THEORY-ELEMENTS: NOTES ON THE ADVANCE OF MERCURY'S PERIHELION

The history of anomalies shows that empirical theories are by no means defenseless against conflicts with observational data. By contrast, they have refined immunization mechanisms at their disposal, which can be utilized to protect their fundamental principles. This lecture is devoted to some considerations concerning these immunization mechanisms. I will argue that a special variant of the semantic view of empirical theories, namely the structuralist approach, is especially well suited for the treatment of conflicts of this type. As an illustrative example, I will use one of the most famous anomalies discovered in the nineteenth century: the anomaly of Mercury's perihelion. First, let me present a brief sketch of Mercury's anomaly.

1. THE DISCOVERY OF MERCURY'S ANOMALY

In the middle of the nineteenth century, celestial mechanics was a highly developed field of science. In particular, the orbits of most planets were known with considerable accuracy. It was common knowledge that Kepler's laws hold only approximately: the planets would move in elliptical orbits only, if there was no mutual interaction between them. In that case, the perihelion — the point on the orbit that is nearest to the sun — would not move (Fig. 1).

If, however, the mutual gravitational interaction between the planets is taken into account, this elliptical orbit turns into a rosette (Fig. 2). In this case, the perihelion is not stationary, but moves in a circular orbit itself. The movement of the perihelion is slow compared to the planet's motion along the elliptical orbit: in the case of Mercury, the planet takes about three months to complete its elliptical orbit once. By contrast, it takes about 260 000 years for the perihelion to complete one circular motion.

These data concerning Mercury's orbit were well known already in the middle of the nineteenth century. Nevertheless, Mercury has always been a problematic case for astronomers: It is both the smallest planet and the planet nearest to the sun, which has made it difficult to observe. In 1849 Le Verrier, a young French astronomer, published a fairly sophisticated treatment of Mercury's motion.[1] The values predicted by his theory, however, were not in agreement with the observational data: Mercury's perihelion was moving

[1] Cf. Le Verrier (1859).

M.L. Dalla Chiara et al. (eds.), Structures and Norms in Science, 89–104.

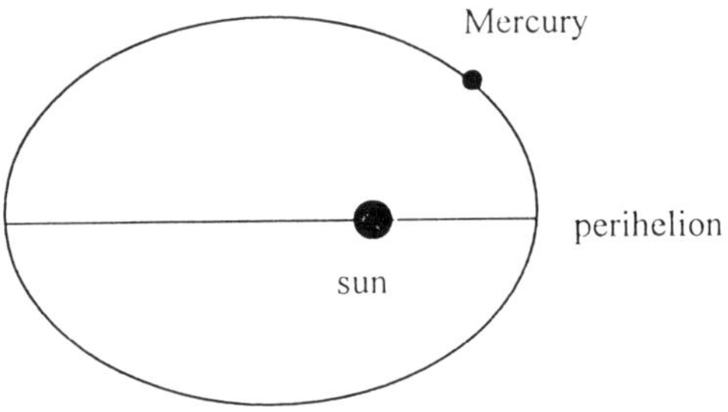

Fig. 1.

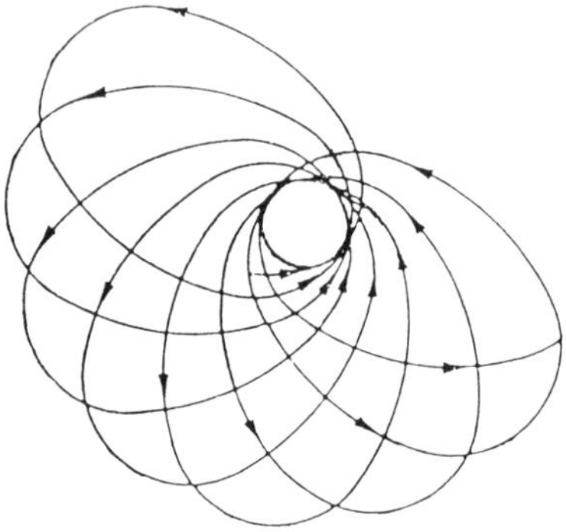

Fig. 2.

faster than predicted. It was a minor difference — about 39 seconds of arc per century according to Le Verrier. Refined measurements by Newcomb (1882) showed that the anomaly amounted to about 43 seconds of arc. However, contemporary astronomers were able to determine the planets' orbits with an observational error of only about 1 second of arc, which was more than necessary to test Le Verrier's predictions. Thus, there was a small, but nevertheless noticeable conflict between *theory and experience* which immediately prompted attempts at correction.

These attempts have drawn considerable attention from within philosophy of science.[2] The philosophical contributions have, however, concentrated mainly on Einstein's revolutionary solution of the anomaly. In this lecture, though, I will focus exclusively on the earlier phase of attempts to remove the anomaly — on revisional attempts that were undertaken within the framework of classical physics. I will try to show that a study of these correctional attempts provides numerous insights into the struggle of an old, well-established theory — namely Newtonian mechanics — against a persistent conflict with observational data. What seemed to be a minor conflict in the beginning turned out to be full-fledged threat to the theory's most fundamental principles in the end. This phase has been the object of intense studies by several historians of science. In particular, I would like to mention the monograph on Mercury's perihelion published by Roseveare (1983). I will refer to his results throughout this lecture.

[2]E.g. Glymour ([2]1981), pp. 277–290.

2. ANOMALIES AND CRITERIA FOR THE CHOICE OF AN APPROPRIATE METATHEORETICAL APPROACH

Mercury's anomaly confronted astronomers with a problem that showed two typically holistic features. First, the theoretical description of Mercury's orbit involved a vast number of hypotheses. These hypotheses included not only Newton's axioms and his law of universal gravitation, but a variety of further assumptions. Among other things, they concerned the masses of the sun and of Venus, which was known to have a major impact on Mercury's motion. This comprehensive complex of hypotheses was confronted with the observational data as a whole. When it failed, it was by no means clear exactly where revisions had to start.

The second holistic aspect consisted in the fact that the treatment of Mercury's motion was not carried out in isolation, but was correlated in numerous ways with the treatment of other systems. Thus, for example, data were used which originated from the analysis of lunar motion or of Venus. In the description of these systems, however, numerous assumptions were once again involved, and here faults might have slipped in just the same. This second holistic feature thus enforced the first: the number of options where a possible revision might start increased exponentially.

According to Quine's well-known position, numerous options may be available of how to eliminate conflicts between *theory and experience*. This does not mean, however, that the choice between different alternatives for correction is an arbitrary one. By contrast, Quine claims that this choice is guided by two main pragmatic rules: On the one hand, one will try to disturb the whole system of knowledge as little as possible. On the other hand, the aim is to obtain a system of knowledge which is as simple as possible. In particular, one will attempt to exclude two types of statements from possible revision: statements close to what Quine calls the "sensory periphery", and statements expressing the most general and fundamental principles of our knowledge.[3]

Although Quine's remarks undoubtely point in the right direction, they nevertheless remain basically vague. The primary aim of my lecture is to show how this deficit of Quine's position can be overcome within a special variant of the semantic view of empirical theories. For that purpose, however, this metatheoretical approach must show certain essential features. Let me list three of them:

1. If one wants to make sense of Quine's notion that some alternatives for revision cut deeper into our system of knowledge than others, then empirical theories have to be interpreted as structured entities: Fundamental parts have to be distinguished from less central, peripheral parts. As is well known, Imre Lakatos has proposed a fairly simple model of this type. Instead of considering a single empirical theory in isolation, he analyzes series of the-

[3] Cf. Quine (1961), pp. 20–46, and Quine ([3]1974), pp. 2–4.

ories, so-called *research programmes.*[4] The theories which belong to one and the same research programme agree with respect to their *hard core*, which contains the theories' fundamental principles. By contrast, they differ with respect to the protective belt surrounding it, which consists of special laws, auxiliary hypotheses, boundary conditions etc. By modifications of this protective belt, the theories' basic principles are immunized to a certain degree against conflict with observational data. One can easily see, however, that Lakatos' model provides insufficient information for an adequate analysis of Mercury's anomaly. It fails to serve this purpose for the following reason: In our treatment of the anomaly, we will focus exactly on those revisions that were carried out within the framework of classical physics. These revisions tried to leave the theory's basic laws — and thus its *hard core* — untouched. By contrast, they were carried out by continuous modifications of the theory's protective belt. However, this protective belt remains without any structure. In order to analyze these processes in detail, we need a more sophisticated metatheoretical concept — one that provides detailed information about the structure also of the specialized parts of an empirical theory. In other words: we need some kind of hierarchical ordering for the theory's laws.

2. In order to remove Mercury's anomaly, the physicists of the nineteenth century developed numerous special variants of Newton's theory. Normally, these specializations were generated by replacing Newton's law of universal gravitation with some other force law. One and the same system — namely the system Mercury/sun — was linked to these different specializations in succession. In order to mirror these processes, we need both an explicit notion of what is meant by the term "application of an empirical theory" , and formal tools to mirror the assignment of different applications to different specializations of Newton's theory.

3. Applications of empirical theories typically relate to *limited* segments of reality. Sometimes, they are limited in a spatiotemporal way — like in the case of the Mercury/sun system. Sometimes they are limited in the sense that only a selected set of features of the system in question is considered. In most cases, they will be limited in both senses. A major part of the research is devoted to the task of fitting these theoretical descriptions of limited segments of reality into one coherent, comprehensive picture. As we will see later on, in the case of Mercury this turned out to be an exceptionally difficult task. A metatheoretical approach, which — implicitly or explicitly — uses the concept of *one global* application from the start cannot mirror these processes adequately. By contrast, we will argue for the use of *local* applications which refer to the limited segments of reality mentioned before. With this proposal, however, formal tools are required to mirror the correlation between the theoretical descriptions of local applications and the data transfer between them.

[4]Cf. Lakatos & Musgrave (1970).

3. LOCALIZING AN ANOMALY: TREELIKE THEORY-NETS

In order to meet these three requirements, we will use the structuralist notion of a treelike theory-net. A theory-net consists of one *basic element* and a considerable number of *specialized theory-elements*. In a simplified version, each theory-element consists of a set of models M_k and a corresponding set of intended applications I_k. Let us turn to the basic element T_0 first. It corresponds to the *hard core* of a research programme. Its set of models M_0 is singled out by a set theoretic predicate which contains the fundamental laws of the theory in question. In the case of classical electrodynamics, for example, the essential constituents of this predicate are Maxwell's equations. In the case of Newtonian mechanics, they consist of Newton's three axioms. The corresponding set I_0 of intended applications contains those systems to which these fundamental laws are to be applied.

From this basic element more specialized theory-elements can be obtained. Let us illustrate this procedure by specifying one branch of the net of Newtonian mechanics that will turn out to be particularly important for an understanding of Mercury's anomaly (cf. the middle branch of Fig. 3). The basic idea behind the construction of this branch is the following: The force function is underdetermined by Newton's axioms. This provides the necessary amount of scope for the formulation of special laws that hold in addition to the theory's basic principles. Thus we may require that all forces operating between Mercury and sun be *conservative* forces. These forces are characterized by the fact that for them the work done around any closed path is zero. Examples of conservative forces are Coulomb's electrostatic forces. Examples of nonconservative forces are velocity-dependent frictional forces. By adding this requirement, the set theoretic predicate that singles out the models of the net's basic element is reinforced. Thus, the corresponding set of models will be a subset of M_0. Trivially, the more specialized theory-element will apply only to a more restricted set of intended applications. Consequently, the corresponding set of intended applications will be a subset of I_0 as well.

This process of specialization may be iterated. A special case of conservative forces are the so-called *central* forces. The magnitude of these forces depends solely on the separation between the interacting particles and is directed along the line that joins them. Among these central forces one may further differentiate with respect to the dependence of the force function on the distance between the interacting particles. Here, the *inverse square* laws played a crucial role in the discussion about Mercury's anomaly. With one more step of specialization, we finally arrive at Newton's law of universal gravitation. This law was used in the first attempts to give a theoretical description of Mercury's orbit which failed, however: Mercury's calculated orbit did not match the observational data.

I will now proceed by arguing that the specialized theory-elements worked as an analogon to the protective belt: For several decades they managed to

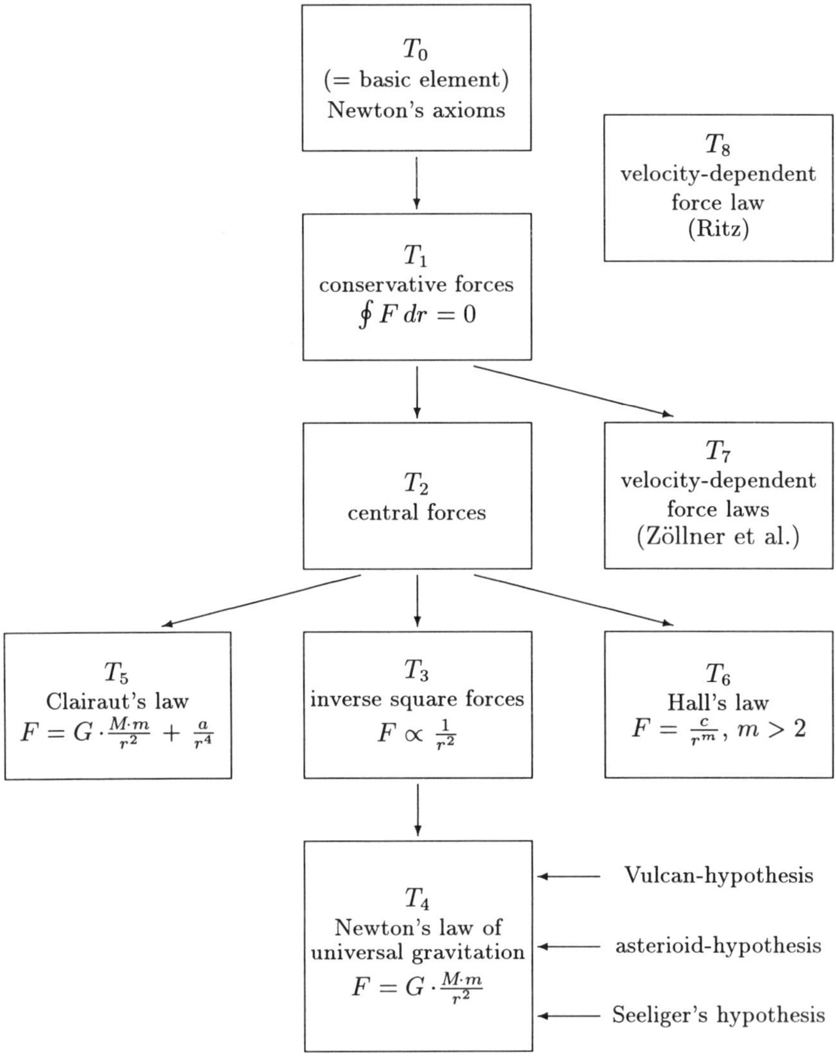

Fig. 3

immunize the basic principles of Newton's theory against possible falsification. This can be shown by following the different attempts to remove the anomaly.

4. "NONTHEORETICAL" REVISION ATTEMPTS: IMMUNIZING THE LAWS OF CLASSICAL PHYSICS

The first proposals were published as early as 1859, the very same year in which the anomaly was discovered. These proposals may rightfully be called *nontheoretical* revision attempts for the following reason: They tried to get rid of the anomaly without touching upon any of the well-established laws of classical physics. In particular, they tried to solve the conflict without questioning Newton's law of universal gravitation. One historical reason for this consisted in the fact that ten years before the discovery of Mercury's anomaly, Le Verrier had used the very same law to predict the existence and orbit of a new planet. When, in 1846, Neptune was discovered, this was seen as a striking confirmation of Newton's law. Consequently, in 1859 the validity of this law was beyond question.[5]

Instead of changing this or any other law of classical physics, the astronomers followed a different strategy: They claimed that the description of the physical system to which Newton's gravitational theory was to be applied was incomplete. The most famous of these proposals was the so-called *Vulcan-hyothesis*. More than a decade before the discovery of Mercury's anomaly, the French astronomer Babinet had claimed the existence of planetary matter between Mercury and the sun, which he baptized Vulcan.[6] In 1859, Le Verrier took up the idea and showed that either a single — so far undiscovered — planet or several asteroids might well cause the perturbing effects on Mercury. Le Verrier at first favored the second hypothesis. He argued as follows: A planet will be more difficult to observe the nearer it comes to the sun. If there was one single object, and if this object had an orbit near the sun, it would be rather far away from Mercury. In that case, it would have to be of considerable size in order to generate the actually observed perturbations. An object of that size, so Le Verrier argued, could hardly escape observation. If, by contrast, Vulcan had an orbit near Mercury, a smaller mass would do. In that case, however, the observational conditions were much better. Thus, in either case it seemed unlikely that the planet had been overlooked by professional astronomers. For that reason, Le Verrier at first favored the hypothesis that several smaller asteroids caused the anomaly. When, however, in the same year an amateur astronomer claimed that he had seen the transit of an unknown planet before the solar disk, Le Verrier himself checked the observational conditions and was convinced that Vulcan had been discovered.[7]

The attempts to confirm the existence of Vulcan lasted for well over two decades. Several amateurs claimed they had seen the object, while many professional astronomers refuted this claim. By 1880, the precision of astronomical measuring apparatus had improved to such an extent that it could be excluded

[5] Cf. Roseveare (1982), p. 19.
[6] Babinet (1846).
[7] Roseveare (1982), p. 25.

that the planet had been overlooked, and the Vulcan hypothesis had to be dropped. It was replaced with Le Verrier's original hypothesis that, instead of one single planet, several asteroids caused the effect. As these asteroids could be considerably smaller than the single planet Vulcan, they were more difficult to observe. By the middle of the 1880s, however, the asteroid hypothesis had to be dropped as well due to lack of observational evidence. The nontheoretical attempts to remove the anomaly continued, however. The most refined and influential of these was the proposal put forward by the Munich astronomer Hugo von Seeliger. Seeliger (1906) tried to show that the anomalous advance of Mercury was caused by the same clouds of fine-grained particles that are responsible for the zodiacal light. It took more than forty years of painstaking work to show that the brightness distribution of the zodiacal light did not fit in with any distribution of matter that could cause Mercury's anomaly.

All *nontheoretical* attempts to remove the anomaly have one thing in common: They use some kind of a matter-hypothesis to protect the well-established laws of classical physics against possible revision. The more fine-grained that putative matter distribution was, the more difficult it became to show that it either did not exist at all or could not be responsible for the anomaly. The matter-hypothesis thus proved to be a powerful tool for the immunization of the fundamental laws of classical physics. In the rather crude form of the Vulcan-hypothesis, it was used to protect Newton's gravitational law against rivals. In the version of Seeliger's hypothesis, it was used until about 1920 to protect Newton's axioms themselves against general relativity.

Most metatheoretical approaches will have some difficulty in providing an adequate account of these nontheoretical revision attempts. The reason lies in the fact that they devote little attention to the question of the entities to which empirical theories are to be applied. If they say anything about that issue at all, they normally identify intended applications with certain *kinds* of systems, like *the harmonic oscillator* or *the free fall.* This may be a useful strategy for some purposes. It is definitely not useful, however, if one wants to analyze the correction attempts described earlier. For this purpose, we need a concept of *intended applications* that refers to physical systems which are characterized by a concrete data base, and are localized in space and time. Structuralism is one of the few metatheoretical approaches which provides a concept of *limited, local intended applications.* Let me first sketch this concept and then point out some of its deficiencies.

According to this concept, intendend applications cannot be interpreted as "segments of reality" , which are given independently of any theorizing. In most cases, the description of intended applications will be theory-laden itself: It will include theoretical assumptions that are taken from empirical theories other than the one to be applied. The example of classical electrodynamics illustrates this point: Here the mechanics of the continua serves as a pre-theory for the description of those systems to which Maxwell's equations are then to be applied. The theory-ladenness is even more obvious in case of theories that belong to the field of microphysics. Any adequate account of intended applications has to mirror this theory-ladenness.

The structuralist concept tries to meet this demand by interpreting intended applications as *partial models*. It starts from the assumption that one can meaningfully distinguish between theoretical and nontheoretical terms with respect to some empirical theory, and then requires that intended applications be described exclusively by using nontheoretical terms. In the case of the Mercury/sun system the corresponding partial model would have to contain information concerning the objects involved — which is a crucial question, as we have seen before —, about the position functions of these objects, and the interval of observation.

This concept of intended applications has some rather obvious deficits, however. First of all, the distinction between theoretical and nontheoretical terms is notoriously unclear. But even if we leave aside this principal problem for a moment, the structuralist concept has two further deficits that make it inadequate for the Mercury case.

First: The concept of partial models presupposes that all values of the nontheoretical functions are known. This assumption is fictitious, however: Although the nontheoretical functions will in general be defined for an infinite number of arguments, the concrete data base is always finite. The available data set may be further restricted for practical reasons. In the case of Mercury, for example, we can observe only part of the orbit: We cannot observe the planet when it is either behind or very near the sun. In view of this fact, we have to generalize the concept of partial models in such a way that it allows for a finite data base.

Second: The structuralist concept of partial models is based on the assumption that for the description of intended applications only those functions may be used that are nontheoretical with respect to the theory in question. This is trivially true at the beginning of a theory's development: When a new theory is applied for the first time, its intended applications have to be described either in terms of everyday language or by means of some pre-theory. Let us call these intended applications *primary applications*. We may face a different situation, though, when a theory is applied to a system, aspects of which have already been described before with the help of the very same theory. Let me illustrate this point by referring to the calculation of Mercury's orbit: When Newton's theory of gravitation was applied to Mercury, values for the masses of both the sun and Venus were used that had been determined in previous applications of that theory. Within the structuralist framework, this fact can be accounted for by liberalizing the concept of partial models a second time: One may distinguish between *primary* and *secondary* applications and allow for the use of values of theoretical functions in describing the latter.[8]

[8]Cf. Gähde (1995) and the literature cited therein.

Let us refer to partial models generalized in these two respects by calling them *data structures*. In structuralist terms the task of providing a theoretical description for some intended application with the help of a certain variant of an empirical theory can then be reformulated as follows: The corresponding data structure has to be extended to a model of the corresponding theory-element. In 1859, the description of Mercury's orbit failed because the data structure then available could not be extended successfully to a model of the theory-element that contains Newton's gravitational law in addition to the axioms of his theory. All nontheoretical attempts of revision rather than trying to solve this problem by shifting the intended application to a different theory-element, searched for a more appropriate and complete data structure for which this task could be fulfilled.

5. THEORETICAL REVISION ATTEMPTS

By 1885, however, it had become obvious that both the Vulcan hypothesis and the asteroid hypothesis had finally failed. At that time, the first phase of theoretical attempts to remove the anomaly began. These attempts started with slight modifications to Newton's gravitational law. Within the framework of theory-nets, they may be described as follows: The intended application corresponding to Mercury's motion is moved to other theory-elements. Some of them were already available within the net, while others were newly created for that purpose (cf. Fig. 3).

An alternative to Newton's gravitational law had already been proposed by Clairaut (1745). He added a small term of the inverse fourth power to Newton's law in order to cope with a putative anomaly in connection with the lunar motion:

$$F = G \cdot \frac{m_1 \cdot m_2}{r^2} + \frac{a}{r^4} \quad .$$

It can be easily shown that the inclusion of an additional term of the inverse third or higher power in the gravitational law leads to an advance of Mercury's perihelion. Furthermore, this advance could be made to fit the observational data then available, if the constant a was chosen properly. There is, however, a straightforward argument against a gravitational law of this type: For small distances of the interacting bodies, the second term would lead to a significant increase in the force acting between them. This effect should be visible in experiments like the one proposed by Cavendish, in which two metal spheres were positioned near each other and the force between them was measured. No such effect was observed, however. For this reason, the astronomers of the late nineteenth century did not consider laws of the Clairaut type as suitable candidates for the substitution of Newton's law in the description of Mercury's motion.

A more serious candidate was proposed by Hall (1894), taking the following form:

$$F = \frac{c}{r^m} \quad .$$

Hall was able to demonstrate that, choosing m slightly greater than 2, this law leads to an advance of Mercury's perihelion that was in accordance with the observational data as well. Furthermore, in 1895, Simon Newcomb showed that Hall's hypothesis was in accordance with the data for the other inner planets within the error of measurement.[9]

Obviously, Hall's law was not exposed to the objection raised against laws of the Clairaut type: It did not lead to significant changes with respect to objects fairly near to each other. Hall's hypothesis, however, had to face another problem: If it was to hold for all planets and satelites in the solar system, it had to apply to the lunar motion in particular. The motion of the moon, however, was the object of intense studies at that time. Brown (1903) published a new theory of the lunar motion which was in conflict with Hall's law: It showed that there was no consistent treatment of the moon's motion in accordance with both this law and the observational data within the error of measurment. Thus, Hall's hypothesis had failed.

Both Clairaut's and Hall's law did not fail, because they could not provide an adequate description of the system sun/Mercury when regarded in isolation: the available data base could be described by both laws law within the error of measurement, if the parameters were properly adjusted. By contrast, they did fail because this description could not be made to fit in with a more comprehensive picture of physical reality which included other physical systems as well. In the case of Hall's law, for example, the parameter m was first successfully adapted to fit the observational data in the case of the system Mercury/sun (Fig. 4). This value for m was then transferred to the theoretical description of the lunar motion. Here, however, it failed to provide a description which was in accordance with the observational data. Similarly, the parameter a in Hall's law was transferred to the Cavendish experiment, where it failed to provide a correct description of the forces acting between the two metal spheres.

Note, that whenever a new value for the advance of Mercury's perihelion was measured, the values of m and a had to be newly adapted; thus they were undergoing continuous change. What remained unchanged, however, was the basic requirement that in each intended application the same value for m or a, respectively, should be used — whatever that value turned out to be. Obviously, this does not constitute a requirement that is imposed on each model in isolation, but on multielemental sets of models. Within the structuralist framework, requirements of this kind are called *constraints*. They describe the correlations that exist between different models of one and the same theory-element. The elements of constraints are those sets of models between which

[9]For details, see Newcomb (1895) and Roseveare (1982), pp. 51–52

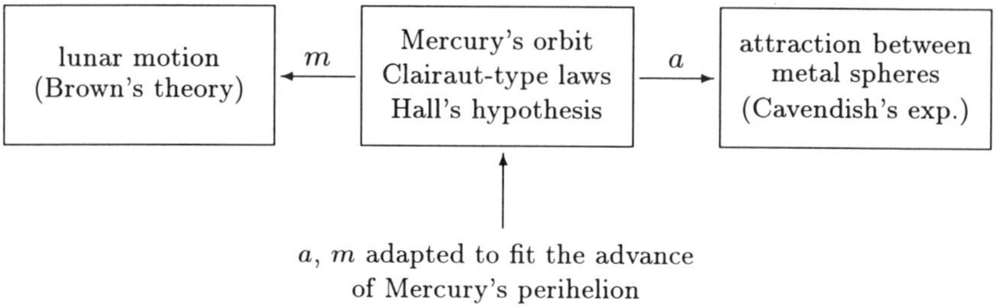

Fig. 4

these correlations exist; thus, they are a subset of the power set of M:

$$(1) \qquad C = \mathcal{P}(M_k) \quad .$$

The fact that constraints impose no restrictions whatsoever on models when regarded in isolation is expressed by the following requirement:

$$(2) \qquad \forall x \, [x \in M_k \rightarrow \{x\} \in C] \quad .$$

The necessity to introduce constraints is an immediate consequence of the use of *local* intended applications that are extended to local models. Each of these local models describes a limited segment of reality. Constraints mirror the correlations and the data transfer that exist between these partial descriptions, and guarantee that they can be fit into a consistent and comprehensive model of reality.

Let us return to a different line of attempts to remove Mercury's anomaly which has to be seen against the background of electrodynamic theory. Due to the success of that theory, physicists tried to extend it in such a way that it would cover not only electromagnetic but gravitational phenomena as well. For that purpose, for each of the electrodynamic force laws proposed by Weber, Gauss, Lorentz and others analogous laws were formulated for the treatment of gravitational phenomena. According to these laws, the gravitational force resembles the electrodynamic force in being velocity-dependent (cf. Fig. 3). This brought in an additional term with respect to Newton's gravitational law, and it was shown that would lead to an advance in the planets' perihelia. Most of the attempts to utilize these force laws for the explanation of Mercury's orbit were carried out just before the turn of the century, although a predecessor of these studies was published as early as 1872.[10] These

[10]Cf. Zöllner (1872), p. 334.

velocity-dependent force laws abandoned the idea that gravitational forces were central forces. Thus, their application to the Mercury motion cut deeper into the net of Newton's theory than the proposals of Clairaut's and Hall's law discussed earlier. They failed, however, because they led to an advance of the perihelion which was far too small to account for the observational data. The first velocity-dependent force law to succeed in this task was proposed by Ritz (1908). Ritz' law was embedded in ballistic theory, in which fictitious particles were responsible for the gravitational force. When adapted to fit the data of Mercury's orbit, it provided values for the motion of the other planets as well as for the lunar motion that largely accorded with the observational data. Ritz' theory was refuted, however, for another reason: It rested on the fallacious assumption that the velocity of light depends on the velocity of the emitting source. Within this theory, the principle according to which actio equals reactio does not hold. If this principle is seen as a fundamental law of Newton's theory, than Ritz' proposal constitutes a challenge to the net's basic element itself. In this sense, it is a forerunner of Einstein's revolutionary solution to the Mercury problem.[11] As is well known, the discovery of general relativity was by no means triggered by the Mercury anomaly. By contrast: the very fact that it had not been designed for that purpose but still succeeded in cutting this Gordian knot, was one of the reasons why Einstein's solution was viewed as a sweeping success of his theory. As I said in the beginning, though, I will confine myself here to those attempts to remove the anomaly that were carried out within the framework of classical physics and not go into any detail concerning Einstein's solution.

6. SUMMARY

Let me instead summarize some advantages of the concept of *theory-nets* for the analysis of anomalies:

1. Unlike Lakatos' unstructured *protective belt*, this concept provides a close-up of the hierarchical system of special laws and auxiliary hypotheses that are used to immunize the theory's basic principles. Theory-nets differ from other forms of axiomatizations on account of the fact that they do not only contain an axiomatic reformulation of the theory's fundamental principles, but of specialized variants of that theory as well.

2. Consequently, this concept allows us to localize exactly where a conflict between *theory and experience* first occurs. This is important, because conflicts with observational data claimed to challenge an empirical theory as a whole often turn out to be directed only against special variants of that theory.

3. The concept of theory-nets allows us to analyze the different attempts that are made in order to eliminate an anomaly in great detail. One can closely follow the path along which the problematic intended application is moved

[11] Cf. Einstein (1915).

through the net, which modifications are carried out during that process in which order, and how deeply these modifications cut into the net.

4. This allows us to state hypotheses about the dynamics of theories more precisely, and to test these reformulated hypotheses by using historical data. Thus, Quine's thesis that we will try to carry out those revisions first that disturb our system of knowledge as little as possible may be translated into the language of theory-nets as follows: In case of conflict *between theory and experience*, revision will start at the lower parts of the net containing highly specialized variants of the theory. Only after a failure of these attempts, corrections will move upward, thereby afflicting successively more fundamental laws. However, our analysis of the Mercury case shows that Quine's statement at best describes the general trend of the different revision attempts. This trend led from the nontheoretical revision attempts, carried out immediately after the discovery of the anomaly, via more or less far-reaching modifications of special laws of classical physics to Einstein's revolutionary solution to the anomaly, which finally challenged the net's basic element itself.

5. The more differentiated the analysis becomes, though, the more obvious the deviations from Quine's thesis. Thus, for example, several decades before the most refined version of the nontheoretical revision attempts - Seeliger's matter-hypothesis — had finally failed, modifications of well-established laws of classical physics were put forward on a trial basis (example: Clairaut-type laws or Hall's hypotheses as substitutes for Newton's law of universal gravitation). This fact contradicts Quine's thesis that comparatively far-reaching corrections of our system of knowledge will not be considered *before* less drastic attempts have definitely failed.[12]

6. The early predecessors of the velocity-dependent force laws, in particular, seem to contradict Quine's position. With respect to these force laws, however, it should be noted that Quine's position comprises two aspects: Not only do we search for those revisions that disturb the total system of our knowledge as little as possible, but also for a system of knowledge that is as simple as possible. Our account of the reasons that motivated the formulation of velocity-dependent force laws shows that they were introduced in order to meet the second demand: With their help, the physicist of the nineteenth century tried to develop a comprehensive theory that covered both electrodynamic and gravitational phenomena. If it had been successful, this theory would have led to a considerable gain in simplicity — at the cost of more radical intrusions into the system of physical knowledge. Thus, an adequate

[12]This fact indicates that Quine's all too vague and informal claims have to be substituted with more subtle and detailed hypotheses concerning the dynamics of empirical theories in case of conflict between *theory and experience* which can be confronted with the empirical (historical) data base. These refined hypotheses, too, might be explicable within the structuralist framework. Thus, it constitutes an interesting question for future research, if the so-called *bootstrap-method* developed in Glymour (1981) and subsequent publications can be translated and, perhaps, be stated more precisely in the framework of theory-nets.

account of these revision attempts would require a precise reformulation of both aspects of Quine's position. In the preceding sections, I have tried to show how a precise interpretation can be given of Quine's first thesis within the framework of theory-nets. The task of providing a formal explication of his second thesis (which concerns the struggle for simplicity) is much more ambitious even if analysis is restricted to the special case of Mercury's anomaly, and is beyond the scope of this lecture.[13]

7. Our analysis of the Mercury case demonstrates that the underdetermination of some functions by the theory's fundamental laws is crucial for the shaping of its protective belt. In the case of Newton's theory, for example, the force function is underdetermined within the net's basic element. This provides the scope for the formulation of a series of different special laws all of which can be consistently combined with Newton's axioms. Each combination is represented by a specialized theory-element within the net. If the extension of some intended application to a model of a certain specialized theory-element fails, other specializations may be considered instead: the problematic application is moved to another theory-element.

8. The theoretical description of Mercury's orbit turns out to be correlated with the treatment of numerous other intended applications in a complex way. In the treatment of these other systems, faults might also have slipped in. The correlation between different intended applications thus leads to a significant increase in the number of options of where possible revisions might start. In view of this fact, the efficiency of the protective belt is considerably reinforced.

The structuralist approach allows us to describe and analyze these correlations in great detail. This is due to the fact that it starts from a concept of *local* applications and *local* models and provides technical tools to describe the data transfer that exists between them. (In this lecture, I have only mentioned constraints that correlate models of one and the same theory-element. More important, however, are so called *links* that are used to describe the data transfer between models of different theory-elements.) Furthermore, the structuralist concept allows us to describe how these correlations change over time as the theory develops. It can thus provide some first insights into the dynamics of holistic phenomena.[14]

Kulturwissenschaftliche Fakultät, Philosophie
Universität Bayreuth

[13]In the Mercury case, this task would presuppose an analysis of the intertheoretical relations that exist between the net of Newtonian mechanics and the net of classical electrodynamics. The structuralist notion of *intertheoretical links* can be used to describe in detail the correlations that exist between different theory-nets. Thus, the structuralist framework might turn out to be a powerful tool for the formulation and analysis of the second aspect of Quine's position as well.

[14]Cf. Gähde (1989).

REFERENCES

Babinet, M. (1846), "Mémoire sur les nuages ignés du soleil considérés comme des masses planétaires", *Acad. Scie. Paris* **22**, 281–186.

Brown, E.W. (1903), "On the Verification of the Newtonian Law", *Mon. Not. r. astr. Soc.* **64**, 530.

Clairaut, A.C. (1745), "Du système du monde dans les principes de la gravitation universelle", *Hist. Mem. Acad. r. Sci.* **32**, 329–64

Einstein, A. (1915), "Erklärung der Perihelbewegung des Merkur aus der allgemeinen Relativitätstheorie", *Sitzungsber. k. preuss. Akad. Wiss.* Pt. 2, 831–9

Gähde, U. (1989), "Bridge Structures and the Borderline Between the Internal and External History of Science", in: K. Gavroglu et. al. (eds.), *Imre Lakatos and Theories of Scientific Change*, Boston Studies in the Philosophy of Science, vol. 111, Dordrecht — Boston — London: Kluwer.

Gähde, U. (1996), "Holism and the Empirical Claim of Theory Nets", in: W. Balzer, C.U. Moulines (eds.), *Structuralist Theory of Science: Focal Issues, New Results*, Berlin — New York: de Gruyter.

Glymour, C. (1981), *Theory and Evidence*, second printing, with corrections, Princeton, N.J.: U.P.

Hall, A. (1894), "A Suggestion in the Theory of Mercury", *Astr.* J. **14**, 49–51

Lakatos, I. & Musgrave, A. (eds.) (1970), *Criticism and the Growth of Knowledge*, London: Cambridge University Press.

Le Verrier, U.J.J. (1859), "Theorie du mouvement de Mercure", *Ann Observ. imp. Paris (Mém.)* **5**, 1–196.

Newcomb, S. (1882), "Discussion and Results of Observations on Transits of Mercury From 1677 to 1881", *Astr. Pap. am. Ephem. naut. Alm.* **1**, 367–487, U.S. Govt. Printing Office, Washington, D.C.

Newcomb, S. (1895), "The Elements of the Four Inner Planets and the Fundamental Constants of Astronomy", *Astr. Pap. am. Ephem. naut. Alm.* **6**, U.S. Govt. Printing Office, Washington, D.C.

Quine, W.V.O. (1961), *From a Logical Point of View*, 2nd. edition, New York: Harper and Row.

Quine, W.V.O. (1974), *Methods of Logic*, 3rd. edition, London.

Ritz, W. (1908), "Recherches critiques sur l'éléctrodynamique générale", *Ann. chim. Phy. Ser.* 8 **13**, 145–275. Reprinted in: *Gesammelte Werke*, pp. 317–422, Paris: Gauthier-Villars (1911).

Roseveare, N.T. (1982), *Mercury's Perihelion. From Le Verrier to Einstein*, Oxford: Clarendon Press.

Seeliger, H. von (1906), "Das Zodiakallicht und die empirischen Glieder in der Bewegung der inneren Planeten", *Sitzungsber. Math. Naturwiss. Kl. Bayer. Akad. Wiss. München* **36**, 595–622.

Zöllner, F. (1872), "Ueber die Natur der Cometen", Leipzig: Engelmann.

Zöllner, F. (1882), "Erklärung der universellen Gravitation aus den statistischen Wirkungen der Elektricität und die allgemeine Bedeutung des Weber'schen Gesetzes", Leipzig: Fock.

GREGORY H. MOORE

THE PREHISTORY OF INFINITARY LOGIC: 1885–1955

> History is difficult ... because the connections that matter are usually numerous, often hidden, and then subsequently neglected.
>
> S. Mac Lane [33, 441]

1. INTRODUCTION

Traditionally, logic was restricted to proofs having a finite number of steps and to expressions of finite length. Around 1954–56, infinitely long formulas entered the mainstream of mathematical logic through the work of Henkin, Karp, Scott, and Tarski. Soon Hanf and Tarski used such logics to settle negatively the 30-year-old problem of whether the first strongly inaccessible cardinal is measurable, a result Tarski communicated to the first LMPS congress in 1960. Infinitary logic continues to be fertile in unexpected ways, as shown by Kolaitis at the present congress.

Yet infinitary logic has a long prehistory. Infinitely long formulas were introduced by C.S. Peirce in the 1880s, used by Schröder in the 1890s, developed further by Löwenheim and Lewis in the 1910s, explored by Ramsey and Skolem in the 1920s, extended by Zermelo and Helmer in the 1930s, studied by Carnap, Novikov, and Bochvar (and explicitly rejected by Gödel) in the 1940s, and exploited by A. Robinson (1951). But there was no ongoing tradition of infinitary logic before the mid-1950s, and most of the researchers mentioned were unaware of their predecessors' work on infinitary logic.

What makes a logic "infinitary"? Before 1954, infinitary logic took two main forms: (1) infinitely long formulas (see §4 below), and (2) rules of inference with infinitely many premises, which began with Hilbert, Tarski, and Zermelo circa 1930. (For lack of space, we leave such infinitary rules of inference to another article.)

The continuous development of infinitary logic that began in 1954 was part of a broader effort during the mid-1950s to extend first-order logic to logics that hopefully retained valuable features of first-order logic but had greater expressive power (e.g. weak second-order logic, generalized quantifiers). Foremost among those features was completeness (every valid sentence is provable) and strong completeness (every consistent set of sentences has a model).

Yet the work of Henkin and Tarski (and their students Karp and Scott) in the mid-1950s was motivated at first by algebraic concerns and especially by Boolean algebras, sometimes with additional operators (e.g. cylindric algebras). The crucial connection—between Boolean algebras and logics—is in-

M.L. Dalla Chiara et al. (eds.), Structures and Norms in Science, 105–123.

triguing because it was within the tradition begun by Boole and continued by Peirce, Schröder, Löwenheim, and the early Skolem that infinitely long formulas were first used. In 1961 Henkin gave one perspective on the earlier work [19, 167]:

> The notions of a *calculus of formulas* and of a *formal deductive system* originated in the efforts of mathematicians to attack the ancient logical problem of explaining what constitutes a valid deduction .... But ... there subsequently appeared new problems, and new points of view, ... and these have led to a progressive incursion of infinitistic methods in the study of the foundations of mathematics .... This evolution has been marked first by the consideration of infinitistic rules of inference[1], then by the introduction of systems with nondenumerably many symbols[2], and finally by allowing formulas of infinite length[3].
>
> [1] See Carnap 1935 [7], Rosser 1937 [42], for example.
>
> [2] See Maltsev 1936 [34], Henkin 1949 [17], Robinson 1951 [40], for example.
>
> [3] See Henkin 1955 [18], Tarski 1958 [55], for example.

Uncountable languages and infinitary rules of inference arose more or less as Henkin stated. Nevertheless, as we shall see below, the use of formulas of infinite length did not come as the final stage, but as the very first.

Before turning to the prehistory of infinitary logic, let us consider the work by Henkin and Tarski that led to the continuous development of infinitary logic.

2. HENKIN AND INFINITELY LONG RELATION SYMBOLS: 1954–55

During his first year at Berkeley (1953–54), Henkin was introduced to Tarski's theory of cylindric algebras (CAs), whose the principal result was Tarski's representation theorem for locally finite, infinite-dimensional CAs. Tarski invited Henkin to extend this theorem to other CAs.

The theory of locally finite, infinite-dimensional CAs was an algebraic analogue of classical first-order logic. Thus Henkin found it natural to look for an extended first-order logic analogous to the ω-dimensional CAs. In the spring of 1954 he found such a logic with infinitary relation symbols, and in September he presented his results to a conference in Amsterdam [20].

Henkin's paper gave a representation theorem for cylindric algebras, and his proof was metamathematical, owing something to his well known proof for the completeness of first-order logic. In the 1954 proof, he wrote that he would

> extend the ordinary notion of a formula and speak of formulas of infinite length. More specifically we shall consider formalisms in which

> the primitive symbols are the same as in the usual first order formalisms, but in which the rules of formula formation provide that a basic formula may be constructed by placing a sequence of individual symbols of type η after a predicate symbol—where η may be either a finite or transfinite ordinal number. [18, 93]

Formulas were constructed from basic formulas as in first-order logic, and so only finitely many connectives and quantifiers appeared in a formula, which "has the form of a well-ordered sequence of symbols".

Henkin then discussed a possible objection to infinitely long formulas: formulas are physical objects and so infinitely long formulas do not really exist. He replied that, mathematically, it does not matter what a symbol really is, but only that the rules are stated clearly. With a sufficiently rich set theory as a metalanguage,

> we can develop a theory of symbols in a precise way so as to admit infinitely long formulas. One way to do this is simply to identify symbols with certain sets—say the ordinal numbers—so that the whole theory of expressions simply becomes a part of the theory of transfinite sequences of ordinal numbers. Alternatively we could axiomatize our theory, taking symbols and expressions as undefined objects and providing suitable axioms .... In this case we would need ... a new primitive operation, infinitary in character, which would permit us to construct infinitely long formulas. [18, 93–94]

He added that his new infinitary logic was very much like first-order logic, and the new metatheory much like that for a first-order calculus. One difference was that in the new logic it was not always possible to construct the closure of a formula. Then he gave an example showing that the Compactness Theorem failed for his new infinitary logic [18, 95].

Henkin's 1955 article marks the end of the prehistory of infinitary logic. In February 1956 he encouraged his graduate student Karp to use infinitely long formulas, and at Berkeley in the fall of 1956 he conducted a seminar with Tarski on such formulas. At that seminar Karp presented her results. The way was then open for the continuous development of infinitary logic.

3. SCOTT, TARSKI, AND INFINITELY LONG CONJUNCTIONS: 1956–58

Henkin's first paper using infinitary logic [18] was influenced by Stone's 1936 representation theorem for Boolean algebras [50]. Likewise, Stone's theorem was fundamental to the work of Scott and Tarski that led to their joint paper [46] on infinitary propositional logic. A weakened form of Stone's theorem stated that every Boolean algebra is a homomorphic image of a field of sets, and in 1947 Loomis strengthened this by showing that every σ-complete Boolean algebra is σ-representable, i.e. is a σ-homomorphic image of a σ-field of sets [30]. In 1955 Tarski's student Chang [9] had found a necessary and sufficient condition for an α-complete Boolean algebra to be α-representable, where $\alpha \geq \omega$. Then Scott [45] and Tarski [53] each wrote an abstract on conditions for such representability. These abstracts formed the starting point for their joint paper on infinitary logic [46], which they presented at the Summer Institute of Symbolic Logic held in 1957 at Cornell University.

Scott and Tarski pointed out that their results essentially concerned "identities in Boolean algebras with infinitary operations .... These remarks, however, are restated here in terms of the syntax of the sentential calculus with infinitely long formulas" [46, 83]. They added that the syntax of first-order logic as extended by infinitely long expressions was being studied in detail by Karp.

What was Tarski's earlier attitude to infinitely long formulas? In his 1933 paper on truth, he wrote: "Whenever a language contains infinitely many sentences, the definition [of truth] constructed automatically according to the above scheme would ... consist of infinitely many words, and such sentences cannot be formulated either in the metalanguage or in any other language" [51, 188–189]. This was an unambiguous denial of the possibility of languages having infinitely long sentences.

Nevertheless, in 1937 Tarski used a device, equivalent to such an infinitely long sentence, in a paper by Kuratowski on definable order-types. In 1936 Tarski had shown that well-ordering is not a first-order property, i.e. there is no set M of first-order sentences such that the models of M are precisely the well-orderings [52, 382]. By using descriptive set theory, Kuratowski gave a new proof of this result and extended it to first-order logic supplemented by a new kind of quantifier: "there exists a positive integer n such that $\varphi_n(x)$". He also treated this quantifier as a countably infinite disjunction: $\varphi_1(x) + \varphi_2(x) + \ldots$ where $+$ represented "or" [28, 100]. Kuratowski remarked that Tarski had found a way to introduce this new quantifier; namely, postulate that to each infinite sequence $\varphi_1(x), \varphi_2(x), \ldots$ of propositional functions there exists a propositional function $P(x, n)$ of two variables, which, for each positive integer n, is equivalent to $\varphi_n(x)$, and define the new quantifier as "there exists an n such that $P(x, n)$". Thus, countably infinite conjunctions and disjunctions were definable in a finitary language having a quantifier varying over the positive integers. Kuratowski was willing to accept a language with such infinitely long formulas, whereas Tarski dispensed with them by an ostensibly finitary device.

Two decades later, Tarski apparently had no qualms about using infinitely long formulas. Scott and Tarski gave an explicit definition for the formulas and theorems of a propositional logic having conjunctions and disjunctions of length less than any infinite regular cardinal α [46, 84]. Their first result was a completeness theorem for this logic when $\alpha = \omega_1$. In a revised version of the paper [47, 168], they remarked that this metatheorem, if expressed in terms of Boolean algebras, showed that the free Boolean σ-algebra is isomorphic to a σ-field of sets, a result equivalent to Loomis's theorem. They then gave a completeness theorem for infinitary propositional logics with $\alpha > \omega_1$, and this they regarded as new [47, 169].

At Cornell, Tarski also presented a paper on infinitary logics that extend the usual first-order logic P_0. He mainly treated the logic, now called $L_{\omega_1\omega_1}$, that he named P_1. Its disjunctions and conjunctions were of countable length, as were its strings of quantifiers, but its relation symbols were of finite length. Allowing relational systems to have denumerably many relations, he extended various notions from first-order model theory to the model theory of P_1. His main result gave necessary and sufficient conditions for a class of relational systems to be axiomatized by universal sentences, and extended his recent first-order theorem on this matter. In particular, he remarked, the class of all well-ordered relations is the class of all models of a set of P_1-sentences, but not of any set of first-order sentences [54, 161–163].

Tarski added that, in order to study algebras with operations having denumerably many arguments, one needed a logical system with denumerably long atomic formulas and with conjunctions and disjunctions whose length was the power of the continuum [55, 176].

Given the shift in Tarski's attitude toward infinitely long formulas from the 1930s to the mid-1950s, it is of interest to see whom he cites as having used them. In 1957 he cited only Jordan [25] as his precursor and regarded Jordan (1949) as using, in effect, the logic P_∞ with conjunctions and strings of quantifiers having a length of arbitrarily large cardinality [1957, 163]. In 1958, Tarski also mentioned Krasner (1938) as his precursor in infinitary logic.

To what extent did Jordan and Krasner actually use infinitely long formulas? Jordan introduced infinite conjunctions and disjunctions quite explicitly [25, 57], but his concerns were mainly with axioms leading to distributive lattices, and logical matters were all but absent. Likewise Krasner, except for referring briefly to "logical functions" [27, 377], is essentially an exercise in generalizing Galois theory. The logical content is minimal. Nevertheless there were many explicit uses—in logic—of infinitely long formulas prior to the mid-1950s, and it is time to explore them.

4. BEGINNINGS OF A TRADITION: PEIRCE AND SCHRÖDER

The earliest hint of an infinitely long formula in logic occurs in Boole's *Mathematical Analysis of Logic* (1847). There he expanded an arbitrary Boolean function into an infinite series, treating this Maclaurin series in a purely formal way and not asking whether the derivatives appearing in it were well defined [4, 60]. In his *Laws of Thought* (1854), he again used such an infinite logical series [5, 72]. But on both occasions the aim was merely to show that an arbitrary Boolean function $f(x)$ was equal to $f(1)x + f(0)(1 - x)$, from which the infinitary formula had disappeared.

The Peirce–Schröder tradition, growing out of the work of Boole, began to use infinitely long formulas almost half a century before there emerged rules of inference with infinitely many premises. In 1885 Peirce stated infinitely long formulas quite explicitly when introducing his symbols for quantifiers:

> Here ... we may use $\sum$ for *some*, suggesting a sum, and $\prod$ for *all*, suggesting a product. Thus $\sum_i x_i$ means that x is true of some one of the individuals denoted by i or $\sum_i x_i = x_i + x_j + x_k +$ etc. In the same way, $\prod_i x_i$ means that x is true of all these individuals, or $\prod_i x_i = x_i x_j x_k$ etc. ... $\sum_i x_i$ and $\prod_i x_i$ are only *similar* to a sum and a product; they are not strictly of that nature, because the individuals of the universe may be innumerable. [38, 194–195]

Thus Peirce introduced the quantifier "there exists an x, $P(x)$" in the form "$P(i)$ or $P(j)$ or $P(k)$ or ..." where i, j, k, etc. were names for all individuals in the universe of discourse under discussion. This universe could be infinite, as he explicitly allowed by the word "innumerable", and in that case the formula was infinitely long.

No doubt a role was played in Peirce's thinking by the use of infinite series in mathematical analysis, and perhaps he recalled Boole's use of such series in logic. Peirce's understanding of quantifiers differs that of Frege (1879) and Peano (1889), both of whom treated quantifiers in the way, now standard, that involves no infinitely long expressions.

Schröder's 1895 book, which extended Peirce's work on relations, contains some very sophisticated uses of infinitely long formulas. Thus, in what he described as "a procedure that possesses a certain boldness", he considered an infinitary proposition that had a universal quantifier for uncountably many (in fact, continuum many) variables [43, 512]. Yet often there was an ambiguity, due to his notation, that made it difficult to determine when he was using an infinitary proposition and when it was merely a union or intersection.

5. END OF A TRADITION: LÖWENHEIM AND SKOLEM

Despite this ambiguity, Schröder's successors understood that at times he expanded quantifiers into infinite conjunctions or disjunctions. After Schröder died in 1902, E. Müller reconstructed a book on logic, *Abriss der Algebra der Logik*, from Schröder's unpublished writings. Its second volume (1910) contained a brief treatment of such infinite conjunctions and disjunctions. If $\alpha(x)$ was a propositional function considered in a countably infinite domain of individuals, Müller treated the existential proposition $\sum_x \alpha(x)$ as equivalent to the infinite disjunction $\alpha(x_1) + \alpha(x_2) + \alpha(x_3) + \cdots$, where x_1, x_2, x_3,... were the names of all the individuals of the domain. He expanded universal and existential quantifiers in this way even when they were nested inside each other, as in his example $\prod_x \sum_y \alpha(x, y) = \sum_y \alpha(x_1, y) \cdot \sum_y \alpha(x_2, y) \cdot \ldots$ When discussing this example, Müller was more explicit about infinitary propositions than Schröder had tended to be: "If infinitely many domains belong to the system of domains, then we are dealing here with products and sums of infinitely many propositions" [44, §101]. Müller stressed the importance of determining the conditions under which it was legitimate to use such infinite sums and products of propositions.

While reviewing Müller's edition of Schröder's *Abriss*, Löwenheim focused on the role of infinite sums (disjunctions) in Schröderian logic. He criticized Müller for not explicitly defining such infinite sums and products [31, 72]:

> Admittedly the reader can easily extend himself the definition in §91 [for finite disjunctions]. However, the rules for calculating infinite sums of propositions remain unproved. In each case the extension of the axioms and proofs to infinite products and sums [of propositions] is unconditionally required for the general calculus of domains, since otherwise the most important applications become impossible ....

Until 1915, logicians working in the Peirce–Schröder tradition had used infinitely long propositions that contained infinitely many conjunctions or disjunctions. But in that year Löwenheim extended Schröder's and Müller's work by considering infinitely long prefixes of quantifiers as well.

Löwenheim made several distinctions that Schröder lacked. The most important was between a *Relativausdruck* (relational expression) and a *Zählausdruck* (individual expression). Moreover, Löwenheim adopted the convention that relational expressions can contain only finitely many quantifiers, conjunctions, and disjunctions. But he temporarily dropped this convention when he wished to transform a proposition in which a universal quantifier preceded an existential quantifier into an equivalent proposition in which the quantifiers were in reverse order. As an example, he discussed the case when the domain under consideration was countably infinite. He expanded the formula $\prod_i \sum_k A_{ik}$ to the equivalent infinitary formula $\prod_i (A_{i1} + A_{i2} + A_{i3} + \cdots)$ and finally to an infinite sequence of existential quantifiers followed by a quantifier-free infinitary formula

$$\sum_{k_1=1,2,3,\ldots} \sum_{k_2=1,2,3,\ldots} \sum_{k_3=1,2,3,\ldots} \cdots (A_{1k_1} \cdot A_{2k_2} \cdot A_{3k_3} \cdot \ldots).$$

Löwenheim gave the above example while proving Löwenheim's Theorem that every satisfiable first-order formula has a countable model [32, §2]. His proof also involved what van Heijenoort [57, 231] later called the law of infinite conjunction: If the propositions Q_i are true for all positive integers i, then the infinitary conjunction $Q_1 Q_2 Q_3 \ldots$ is also true.

The original proof of Löwenheim's Theorem used an infinitary second-order logic. When in 1920 Skolem, who had previously adopted Schröder's logic, first discussed Löwenheim's Theorem, he found it odd that Löwenheim had proved this theorem about first-order expressions by using second-order notions. It was a desire for purity of proof methods, as well as for greater simplicity, that first led Skolem to bypass those second-order notions. Skolem quantified only over individuals and not over relations.

Yet in 1920 Skolem felt no need to drop the infinitary expansions of finitely long formulas that were so much a part of Löwenheim's work. On the contrary, Skolem made infinitary logic even more explicit than Löwenheim had. This occurred when Skolem gave Löwenheim's Theorem a new proof, which depended strongly on converting a first-order proposition into an equivalent proposition in Skolem normal form.

Skolem immediately generalized Löwenheim's Theorem, thereby obtaining the Löwenheim–Skolem Theorem: If a countable sequence of first-order propositions is satisfiable, it is satisfiable in a countable domain [48, Theorem 2]. His next generalization was to take such a countable sequence and to consider its conjunction: If a countable conjunction of first-order propositions is satisfiable, this conjunction is satisfiable in a countable domain [48, Theorem 3]. But he went further and proved the infinitary theorem: If a proposition P is constructed from first-order propositions by means of countable conjunction and countable disjunction applied any countable number of times, then P is satisfiable in a countable domain. Here Skolem was no longer in first-order logic but in $L_{\omega_1\omega}$, as we now call it.[1] But Skolem's final generalizations were in an even stronger logic, namely $L_{\omega_1\omega_1}$. That is, Skolem now allowed his propositions P, in Skolem normal form, to begin with an infinitely long string of quantifiers. He did not allow an arbitrary infinite string in P, but only a countably infinite string that began with finitely many universal quantifiers followed by infinitely many existential quantifiers; the quantifier-free matrix was allowed to have just the variables found in the quantifiers. Finally, he gave the theorem: If a countable conjunction of propositions P of that form is satisfiable, then that conjunction is satisfiable in a countable domain [48, Theorem 8].

Skolem's 1920 theorems were the most sophisticated results in infinitary logic until the 1940s. But no one chose to take up his various infinitary logics. Moreover, Skolem himself abandoned infinitary logic (1923) in an article criticizing Zermelo's axiomatic set theory. This was the first time set theory was formulated in first-order logic, giving the Skolem paradox: Set theory has a countable model although set theory proves that there are uncountable sets [49]. For the next four decades, Skolem advocated restricting logic to first-order logic (stripped of any infinitary propositions), and advocated as well the set-theoretic relativism expressed in the Skolem paradox.

6. ECHOES OF A TRADITION: HILBERT AND LEWIS

Hilbert was very eclectic in his work on logic and well aware of his predecessors. In August 1904, at Heidelberg, Hilbert delivered his first paper devoted

[1]The recognition that in 1920 Skolem worked in $L_{\omega_1\omega}$ is due to Vaught [58, 166].

entirely to mathematical logic and his first paper containing logical symbols. One apparently Schröderian influence was Hilbert's introduction of quantifiers as formulas that can be infinitely long. For the fixed domain of the positive integers, Hilbert defined "for some x, $A(x)$" to be an abbreviation for the infinitary formula $A(1)$ *o.* $A(2)$ *o.* $A(3)$ *o.*..., where *o.* stood for "oder" (or), and analogously for the universal quantifier with respect to "und" (and) [22, 178]. He used $x^{(o)}$ and $x^{(u)}$ to symbolize "for some x" and "for every x", respectively.

During 1905 Hilbert gave a lecture course, *Logical Principles of Mathematical Thinking* [23], which developed his Heidelberg paper. Even more than at Heidelberg, Hilbert now saw salvation from the paradoxes in what he called "die Axiomenlehre" [23, 250], a formalized version of the axiomatic method. Yet he now relied on a more restricted underlying logic than before: propositional logic [23, 228].

Hilbert omitted quantifiers from his logic in that course because he had become suspicious of the universal quantifier. In his formalized arithmetic, he used only quantifier-free formulas. He acknowledged the need to include laws for "all" but admitted that at present he could not formulate them rigorously [23, 254]. Thus the infinitary quantifiers, so striking in his Heidelberg paper, had disappeared.

In 1923 Hilbert still regarded quantifiers as involving the infinite in an essential way, while he wanted his proof theory to be finitary: "Where does there appear for the first time something going beyond the concretely intuitive and the finitary? Obviously already in the use of the concepts '*all*' and '*there exists*'" [24, 154]. For finite collections, he noted, the universal and existential quantifiers reduce to finite conjunctions and disjunctions, thereby yielding the Principle of the Excluded Middle in the form that $\sim(\forall x)A(x)$ is equivalent to $(\exists x)\sim A(x)$ and that $\sim(\exists x)A(x)$ is equivalent to $(\forall x)\sim A(x)$. "These equivalences", he continued,

> are commonly assumed, without further ado, to be valid in mathematics for infinitely many individuals as well. In this way, however, we abandon the ground of the finitary and enter the domain of transfinite inferences. If we were always to use for infinite sets a procedure admissible for finite sets, we would open the gates to error .... In analysis ... the theorems valid for finite sums and products can be translated into theorems valid for infinite sums and products only if the inference is secured, in the particular case, by convergence. Likewise, we must not treat the infinite logical sums and products $A(1)\&A(2)\&A(3)\&\cdots$ and $A(1)\vee A(2)\vee A(3)\vee\cdots$ in the same way as finite ones. My proof theory ... provides such a treatment. [24, 155].

In order to avoid introducing quantifiers directly, Hilbert introduced a new logical postulate, from which they could be defined. It relied on a new logical function τ which was a kind of "anti-choice" function acting on predicates. This new postulate was his Transfinite Axiom: $A(\tau A) \to A(a)$. It stated that if a predicate $A(x)$ is true for the particular object that the function τ assigns to $A(x)$, then $A(x)$ is true for any object whatever [24, 156]. The quantifier "for every x, $A(x)$" was then defined as $A(\tau A)$.

Thus Hilbert continued to avoid infinitely long expressions in his logic in the 1920s. Yet in 1931 he introduced an infinitary rule of inference.

A vestige of the Peirce–Schröder tradition survived in the philosopher C.I. Lewis. In *A Survey of Symbolic Logic* (1918), the logical notation remained largely that of Schröder. Moreover, Lewis defined the existential quantifier $\sum_x$ and the universal quantifier $\prod_x$ in terms of logical expressions that could be infinitely long: "We shall let $\sum_x \varphi x$ represent $\varphi x_1 + \varphi x_2 + \varphi x_3 + \cdots$ to as many terms as there are distinct values of x in φx. And $\prod_x \varphi x$ will represent $\varphi x_1 \times \varphi x_2 \times \varphi x_3 \times \cdots$ to as many terms as there are distinct values of x in φx .... The fact that there might be an infinite set of values of x in φx does not affect the theoretical adequacy of our definitions" [29, 234–235]. Lewis sought to escape the difficulties that he found in such infinitely long propositions by reducing them to finite ones: "We can assume that any law of the algebra [of logic] which holds *whatever finite* number of elements be involved holds for any number of elements whatever .... This also resolves our difficulty ... that the number of values of x in φx might not be even denumerable" [29, 236]. Thus Lewis was aware that his infinitary propositions might have uncountably many symbols. He made no attempt to justify his assertion that the laws holding in all finite Boolean algebras also hold in all infinite ones.

7. A DIFFERENT TRADITION: RUSSELL, WITTGENSTEIN, AND RAMSEY

Besides the fragments of infinitary logic that are found repeatedly in the Peirce–Schröder tradition, and in Hilbert's work, there is a second tradition in which other fragments occur. This is the tradition of *Principia Mathematica.* Russell's early sympathy with infinitely long formulas appears in a letter that he wrote to G.E. Moore in August 1900:

> Have you ever considered the meaning of any? I find it to be the fundamental problem of mathematical philosophy. E.g. "Any number is less by one than another number." Here *any number* cannot be a new concept, distinct from the particular numbers, for only these fulfil the above proposition. But can *any number* be an infinite disjunction? And if so, what is the ground for the proposition?

Shortly after this, Russell fervently adopted Peano's system, which was strictly finitary. In a letter of July 1905 to Russell, G.H. Hardy argued that the infinitary disjunction $x = a \vee x = b \vee x = c \vee \cdots$ was a legitimate "definition by extension" without there being a propositional function "definable in finite terms"; Russell disagreed. His interest in infinitely long formulas was only revived by that rather enigmatic book, Wittgenstein's *Tractatus Logico-Philosophicus* (1922).

There Wittgenstein argued that every proposition is a truth-function of atomic propositions [59, §5]. To do so, he introduced an operation N on a set A of propositions that made a new proposition out of the conjunction of 'not P' for each proposition P in A [59, §5.5, 5.502]. He regarded each proposition as built up from atomic propositions by a finite number of applications of truth-operations; in fact, the operation N suffices [59, §5.5]. Wittgenstein's idea was, in effect, to use a truth-function that operated on arguments whose number was variable (and perhaps infinite). Russell accepted Wittgenstein's idea and used it in the introduction to the second edition of *Principia Mathematica* in 1925.

The infinitary nature of Wittgenstein's proposal was explicitly recognized by Ramsey (1926): "Mr. Wittgenstein has perceived that ... there is no reason why the arguments to a truth-function should not be infinite in number. As no previous writer has considered truth-functions as capable of more than a finite number of arguments, this is a most important innovation" [39, 343]. Ramsey immediately expressed an important reservation: "Of course, if the arguments are infinite in number they cannot all be enumerated and written down separately; but there is no need for us to enumerate them if we can determine them in any other way, as we can by using propositional functions." Then he used a propositional function Px to define the set of propositions obtained by replacing x with all values a that make Pa true. He treated $(x)Px$, or "for every x, Px", as asserting the conjunction of all propositions of the form Px.

Ramsey's treatment of "all" as meaning infinite conjunction was met with silence from logicians, except Gödel. In his 1944 article "Russell's Mathematical Logic", Gödel alluded to

> the fiction that one can form propositions of infinite (and even non-denumerable) length, i.e., operate with truth-functions of infinitely many arguments, regardless of whether or not one can construct them. But what else is such an infinite truth-function but a special kind of an infinite extension (or structure) and even a more complicated one than a class, endowed in addition with a hypothetical meaning, which can be understood only by an infinite mind? [15, 142]

Gödel was very sceptical of infinitely long propositions and used them to argue against certain aspects of Russell's introduction to the second edition of *Principia Mathematica*:

> In *Principia* the notion of a propositional function of order n is neither taken as primitive nor defined in terms of the notion of a

> finite combination, but rather quantifiers referring to propositional functions of order n ... are defined as certain infinite conjunctions and disjunctions. But then one must ask: Why doesn't one define the integers by the infinite disjunction $x = 0 \vee x = 0+1 \vee x = 0+1+1 \vee \cdots$ ad infinitum, saving in this way all the trouble connected with the notion of inductiveness? [15, 146]

It is ironic that during the 1950s, when infinitely long propositions were at last legitimized and accepted, the use of just this kind of infinite disjunction to define the natural numbers within a logical system was taken very seriously. Thus Karp wrote in her book *Languages with Expressions of Infinite Length*:

> My interest in infinitary logic dates back to a February day in 1956 when I remarked to my thesis supervisor, Professor Leon Henkin, that a particularly vexing problem would be simple if only I could write a formula that would say $x = 0$ or $x = 1$ or $x = 2$ etc. To my surprise he replied, "Well, go ahead." The problem is now long-forgotten, but that reply has led to this monograph. [26, v]

By contrast, in Gödel's article, such an infinite disjunction was treated as the absurd conclusion that shows a premise to have been mistaken.

8. AGAINST TRADITION: ZERMELO, 1930–35

Zermelo, who had been influenced by Hilbert while at Göttingen around 1900, later deviated from Hilbert in regard to quantifiers. Whereas Hilbert wished to avoid quantifiers (as carriers of infinite conjunctions and disjunctions) by means of the Transfinite Axiom, Zermelo (1932) embraced such infinitely long formulas as a way of countering Skolem's countable models for set theory and Gödel's incompleteness theorems. "What is a 'mathematical theorem', a 'mathematical proof', a 'mathematical theory' ...?", he asked in [60]. He answered that, if the domain of individuals is fixed in advance for a particular theory, then logic can be based on set theory—in the following sense. Every mathematical theory consists of a "fundamental domain" of "individuals" with certain atomic relations between these individuals. These atomic relations generate further formulas by the use of "not", "and", "or". Here, he insisted, "and" as well as "or" can join any finite or infinite number of formulas, and thus can serve as universal and existential quantifiers [60, 6]. Such a system of formulas must be "well-founded" in regard to the atomic relations, and splits into levels just as does the cumulative type hierarchy of sets.

In 1935 Zermelo returned to this infinitary logic in an article, which ended as follows: "The foregoing considerations form only the beginning of an investigation that is not yet completed, which has as its aim the founding of an "infinitistic" and genuinely mathematical syllogistic and proof theory" [61, 146]. Zermelo saw mathematics as the "logic of the infinite", but unlike almost all researchers at that time, he encouraged the use of infinitely long sentences and also the use of infinitely long proofs [61, 144], not dismayed by the difficulty that our "finite understanding" may have with such proofs.

Zermelo's efforts to develop a strong infinitary logic vanished with hardly a trace. No one pursued his ideas on infinitary logic, although the legacy of his 1935 paper is the general notion of well-foundedness. It was quite ironic, given his critique of Skolem regarding cardinality, that Zermelo's infinitary logic, which was the propositional part of $L_{\infty\omega}$, was not strong enough even to define within it the relation of two sets having the same cardinality (i.e. Härtig's quantifier).[2]

9. HELMER AND CARNAP: 1938 AND 1943

In 1938 Helmer published in *Erkenntnis* a brief but intriguing article on a logic with infinitely long expressions. It was influenced by *Logische Syntax der Sprache* (1934), in which Carnap made a plea for tolerance in the use of languages. Helmer stated that all languages previously "the subject of syntactical investigation" had only expressions with a finite number of symbols. Apparently he did not know, e.g., Ramsey [39] or Zermelo [61] that would contradict his claim. There were, Helmer remarked, various kinds of infinite expression that could be introduced:

> One might, for instance, admit infinite logical products '$A_1 \cdot A_2 \cdot A_2 \cdot \ldots$' and infinite logical sums '$A_1 \vee A_2 \vee A_3 \vee \ldots$'. Or one might think of a language containing sentential functions or descriptive functions with infinitely many argument-places; for example, an infinite series $\sum a_n x^n$ might be construed literally as a function $f(x, a_0, a_1, a_2, \ldots)$ of infinitely many numerical variables. [16, 139]

Thus Helmer was aware of countably infinite conjunctions and disjunctions, although he did not credit them to any previous author. His use of sentential functions with infinitely many argument-places was reminiscent of Ramsey.

The particular infinitary language L that Helmer chose to develop was rather different. It included what he called "dual fractions", i.e. countably infinite strings of 0 and 1 that were intended to represent real numbers. He followed Carnap in giving a careful statement of the symbols of L and the formation rules for expressions of L; an expression would have length less than ω^2. Among the axioms for L were those for the real numbers, and among its rules of inference were one for complete induction and one for Dedekind's axiom of continuity [16, 140]. The rest of the article was devoted to arguing briefly that Gödel's incompleteness results would hold for this infinitary language L.

Helmer's article elicited almost no response. Beth gave it a slightly sceptical review in the *Journal of Symbolic Logic* [2], while Hermes gave it a five-line

[2] See [21, 1154]. On Zermelo's logical research during the 1930s, see [35, 120–136].

review in *Jahrbuch über die Fortschritte der Mathematik* (1942). Both reviews insinuated that expressions of infinite length were dubious.

Possibly influenced by Helmer's article, Carnap used infinitely long expressions in a different context in *Formalization of Logic* (1943). (Carnap was aware of Ramsey's arguments in favor of treating the universal and existential quantifiers as infinitely long conjunctions and disjunctions, respectively, since he had mentioned those arguments in his 1929 book [6, 14].) In 1943 Carnap's aim was to examine the adequacy of the usual calculi of propositional logic vis-à-vis non-standard interpretations (of the logical connectives), in which the axioms were true and the rules of inference preserved truth. He gave such non-standard interpretations, which he regarded as showing that the usual propositional logic was not fully formalized. He did so while well aware of the completeness theorem for propositional logic. A full formalization of propositional logic, he argued, required new syntactic concepts [8, 94–99].

Yet Carnap first considered the matter semantically. He noted that, if one follows the usual convention, a set Σ of sentences is called true (in a logic) if every sentence in Σ is true (in that logic). In propositional logic, he remarked, this means that the set of sentences $\{p, q\}$ is true if and only if the conjunction $p\&q$ is true. Thus he regarded the usual convention as construing the set in what he called "the conjunctive way". But, he added, there was a second way in which a set of sentences could be called true, and this he called "the disjunctive way": A set Σ of sentences is called true (in a logic) if some sentence in Σ is true (in that logic). Since he found both ways useful, he wanted to have both available. He introduced a notation for each way, calling a set of sentences construed in the first way a "conjuntive sentential class", or a "conjunctive", and in the second way a "disjunctive sentential class", or a "disjunctive". Conjunctives, disjunctives, and sentences were all called "junctives". Thereby he hoped to restore the symmetry missing in the foundations of syntax and semantics [8, 105].

Next, Carnap considered his junctives syntactically. They permitted the statement of new kinds of rules of inference, e.g. disjunctive rules of inference and rules of refutation. These permitted a "full formalization" of propositional logic [8, 113, 127].

To obtain a full formalization for first-order logic, Carnap used "transfinite junctives", i.e. countably infinite conjunctions (or disjunctions) of first-order formulas [8, 142]. In his full formalized system FC_1 he adopted a new rule of inference: Let $A(x)$ be a formula; the premise is the junctive consisting of the conjunction of all instances of $A(x)$, i.e. x is successively replaced by each individual constant (of which there is one for each individual, in other words, one for each natural number); the conclusion is $A(x)$. He also formulated an alternative rule of inference, where the premise was $(\exists x)A(x)$ and the conclusion was the junctive consisting of the disjunction of all instances of $A(x)$; for this rule, $A(x)$ was required to have x as its only free variable [8, 136, 145, 147]. Carnap was well aware that infinitely long expressions could be avoided by using rules of inference with infinitely many premises, as he had done in his *Logische Syntax der Sprache.*

10. CARNAP'S INFLUENCE ON ROBINSON: 1951

Around 1951, Robinson's work on logic shared several common features with that of Henkin and Tarski. First, all three of them used first-order logic as their basic logical system. (At that time the theory of types still had a claim to be the dominant, though waning, system of logic. Church's prominent textbook [13], for example, still treated first-order logic as a subsystem of ω-order logic.) Second, Robinson shared with Henkin and Tarski a strong desire to apply first-order logic to obtain results about algebra; this was the beginning of model theory in the modern sense. Third, all three used the Strong Completeness Theorem as their basic tool in this regard: A set Σ of sentences has a model if and only if Σ is consistent. Fourth, Henkin and Robinson used first-order logic with uncountably many symbols—abandoning the dogma, espoused by many philosophers, that languages may only have a countable number of symbols. They did so because many of their basic results depended on using uncountably many symbols.

Where Robinson differed from Henkin and Tarski (as of 1951, but by 1956 they all agreed again) was in accepting infinitely long formulas. In 1951 Robinson described his fundamental logical outlook:

> The formal language used ... is based on the restricted calculus of predicates [first-order logic] with certain extensions which prove natural in the course of the analysis .... In Formal Logic, [the author] accepts such constructs as infinite conjunctions and disjunctions, and believes that it is legitimate to ascribe definite truth values to such constructs in the usual way. However, ... the introduction of these concepts and constructs is not due to an unbridled desire for generalisation, but ... they are required for the formulation and proof of theorems which are in themselves quite orthodox. For instance, Archimedes' axiom can be formulated naturally in terms of an infinite disjunction. [40, 2]

Robinson took a first-order language with countably infinite conjunctions and disjunctions as "an extended language" [40, 22]. He gave a careful statement of the formulas of such extended languages, and noted that "languages involving infinite conjunctions and disjunctions are considered by Carnap in the [1943] book".

Robinson's primary use of languages with infinite conjunctions and disjunctions was to investigate the Archimedean axiom [40, 51–61]. He showed that in first-order logic there is no set of sentences characterizing Archimedean ordered fields (whereas there is a finite set of sentences characterizing ordered fields). By contrast, in his infinitary logic Robinson gave a finite set of sentences characterizing Archimedean ordered fields. He used his infinitary results to show the existence of non-Archimedean ordered fields F with various additional properties, such as the intermediate value property. Barwise later wrote that here Robinson "foreshadows important work to come by studying when such a [countable disjunction] is equivalent to one of its finite 'approximations'" [1, 103]. Henkin recalls [20] that by 1954, when he began his own work on infinitary logic, he had read Robinson's book and had rejected its infinitely long conjunctions and disjunctions because they were part of a deductive system.

11. RUSSIAN CONTRIBUTIONS: NOVIKOV AND BOCHVAR, 1939–1943

We turn back to work which preceded the efforts of Carnap (1943) and Robinson (1951) but which they did not know. In 1939, also apparently unaware of earlier work on infinitary logic, the Soviet logician P.S. Novikov formulated a propositional logic with countably infinite conjunctions and disjunctions. He extended the conjunctive/disjunctive normal form to this logic, and introduced a syntactic notion of a formula being "regular". His basic result was that a formula in his logic is regular if and only if it is provable. From this he deduced two corollaries: First, this logic is consistent. Second, for any countably infinite disjunction of finitely long formulas $F_1, F_2, \ldots$ in the logic, if that disjunction is provable, then there is some n such that the finite disjunction of F_1 up to F_n is provable [36]. (Thus an infinite propositional formula could be equivalent to a finite approximation, in contrast to the situation discussed by Robinson [40].)

In 1940 Church gave Novikov's article a negative review in the *Journal of Symbolic Logic*, describing the second corollary as "pointless". Church concluded: "In view of the non-effective character of the ... formulas ... it cannot be said that this calculus is a logic ... in the proper sense of the word" [10, 69-70].

Later that year Bochvar investigated the infinitary propositional logic of his compatriot Novikov, giving a proof that it was complete. Bochvar remarked that this calculus, which extended classical propositional logic, had uncountably many formulas. Intuitionism played a role in this calculus, as it had not done in the work discussed in previous sections. In the formation rules, an infinite sequence of formulas was required to be "intuitionistically given" in order to form its countable conjunction or countable disjunction [3, 66]. Likewise, such a sequence had to be "intuitionistically given" in order to use two infinitary rules of inference: (1) If, for every n, the formula $B \to A_n$ is provable, then the formula $B \to (A_1 \wedge A_2 \wedge A_3 \wedge \cdots)$ is provable; (2) if, for every n, the formula $A_n \to B$ is provable, then the formula $(A_1 \vee A_2 \vee A_3 \vee \cdots) \to B$ is provable [3, 69]. He then proceeded to derive various formulas in this calculus, including infinitary distributive laws [3, 72].

Next, Bochvar turned to completeness. After considering the possibility of an intuitionistic proof of completeness, he opted instead for a classical proof. To do so, he dropped the condition that a sequence of formulas be "intuitionistically given" from the formation rules for formulas and from the rules of inference. He interpreted his formulas set-theoretically in the two-element Boolean algebra, and gave a proof of soundness for his calculus [3, 79–82].

Bochvar then proved completeness: a formula is either satisfiable or refutable. He introduced the notion of the rank of a formula, defining a formula to be of 'first rank' if it is a countable conjunction of countable disjunctions of finite formulas. Next he established that every formula of first rank is either satisfiable or refutable. The method was based on Novikov's topological "Covering Theorem", which used a Baire space [3, 85–90]. Finally, Bochvar showed how to reduce a formula A of arbitrary rank to a formula B of first rank such that A is satisfiable if and only if B is satisfiable [3, 90–100].

Church was not much more positive toward Bochvar's paper than he had been to Novikov's, calling Bochvar's proofs "frankly non-constructive in character". Even for Bochvar's intuitionistic calculus, Church insisted that "there would not be an effective criterion by which it could always be determined whether an alleged proof were a valid proof in the sense of the calculus" [11, 119]. Curry also gave Bochvar a review that was somewhat sceptical, remarking explicitly that he had not checked all the proofs [14].

In a 1943 article, Novikov returned to his infinitary propositional calculus [37]. He assumed an intuitionistic metamathematics for his calculus, and spelled out in detail the syntactic results he had sketched in 1939. Church's review of it was still negative [12]. Novikov's contribution to infinitary logic was not appreciated in the West until 1980 when Barwise [1, 107–110] acknowledged the correctness of Novikov's results and gave a semantic proof for them.

12. CONCLUSION

This article has indicated how infinitely long formulas were used with various purposes prior to their definitive introduction in 1954–56. Is it surprising that infinitary logic was isolated from the mainstream of logic before that date? Not at all, given the foundational emphasis of mathematical logic during the previous decades. What is more surprising is that, beginning in 1954, the opposition to infinitary logic was so minimal. (One hint that it still existed is the fact that the seminal articles by Tarski (1958), Scott and Tarski (1958), and Henkin (1961) were not reviewed in the *Journal of Symbolic Logic* until 1965—in contrast to the speedy reviews given most articles there.)

The prehistory blends into the history when Robinson, speaking at the same 1957 Cornell institute where Tarski presented his infinitary logic, emphasized that, for field theory, the most natural way to introduce concepts going beyond first-order logic was to use a logic with infinitely long conjunctions and disjunctions [41, 326–327]. What we wish to emphasize is the richness of that prehistory.

Department of Mathematics
McMaster University
Hamilton, Ontario

ACKNOWLEDGEMENTS

I would like to thank J. Zucker for his comments on a draft of this paper.

REFERENCES

1. J. Barwise, "Infinitary Logics", *Modern Logic—A Survey*, ed. E. Agazzi (Dordrecht, 1980), 93–112.
2. E.W. Beth, Review of [16], *J. Sym. Logic* 4 (1939), 25.
3. D.A. Bochvar, "Über einen Aussagenkalkül ...", *Mat. Sbornik* n.s. 7 (1940), 65–100.
4. G. Boole, *The Mathematical Analysis of Logic* (London, 1847).
5. G. Boole, *An Investigation of the Laws of Thought* (London, 1854).
6. R. Carnap, *Abriss der Logistik* (Vienna, 1929).
7. R. Carnap, "Ein Gültigkeitskriterium ...", *Monatshefte für Math.* 41 (1935), 263–284.
8. R. Carnap, *Formalization of Logic* (Harvard Univ., 1943).
9. C.C. Chang, *Bull. AMS* 61 (1955), 325 (abstract).
10. A. Church, Review of [36], *J. Sym. Logic* 5 (1940), 69–70.
11. A. Church, Review of [3], *J. Sym. Logic* 5 (1940), 119.
12. A. Church, Review of [37], *J. Sym. Logic* 11 (1946), 129–131.
13. A. Church, *Introduction to Mathematical Logic* (Princeton Univ., 1956).
14. H.B. Curry, Review of [3], *Math. Reviews* 1 (1940), 321.
15. K. Gödel "Russell's Mathematical Logic", *The Philosophy of Bertrand Russell*, ed. P. A. Schilpp (New York: Tudor, 1944), 123–153.
16. O. Helmer, "Languages with Expressions of Infinite Length", *Erkenntnis* 7 (1938), 138–141.
17. L. Henkin,"The Completeness ...", *J. Sym. Logic* 14 (1949), 159–166.
18. L. Henkin, "The Representation Theorem for Cylindrical Algebras", *Mathematical Interpretation of Formal Systems* (North-Holland, 1955), 85–97.
19. L. Henkin, "Some Remarks on Infinitely Long Formulas", *Infinitistic Methods* (Pergamon, 1961), 167–183.
20. L. Henkin, E-mail of 10 August 1994 and 26 September 1994 to G.H. Moore.
21. H. Herre et al., "The Härtig Quantifier ...", *J. Sym. Logic* 56 (1991), 1153–1183.
22. D. Hilbert, "Über die Grundlagen ...", *Internat. Congress Math.* (1905), 174–185.
23. D. Hilbert, *Logische Principien des mathematischen Denkens*, lecture course, 1905.
24. D. Hilbert, "Die logischen Grundlagen ...", *Math. Ann.* 88 (1923), 151–165.
25. P. Jordan, "Zur Axiomatik der Verknüpfungsbereiche", *Abhand. Math. Sem. Hamburg. Univ.* 16 (1949), 54–70.
26. C. Karp, *Languages with Expressions of Infinite Length* (North-Holland, 1964).
27. M. Krasner, "Une généralisation de la notion de corps", *J. Math. Pures et Appliquées* (9) 17 (1938), 367–385.
28. K. Kuratowski, "Les types d'ordre définissables ...", *Fund. Math.* 29 (1937), 97–100.
29. C.I. Lewis, *A Survey of Symbolic Logic* (Berkeley: Univ. Calif., 1918).
30. L.H. Loomis, "On the Representation of σ-Complete Boolean Algebras", *Bull. AMS* 53 (1947), 757–760.

31. L. Löwenheim, Review of [44], *Archiv der Math. und Physik* (3) 17 (1911), 71–73.
32. L. Löwenheim, "Über Möglichkeiten ...", *Math. Ann.* 76 (1915), 447–470.
33. S. Mac Lane, "Addendum ...", *A Century of Mathematics in America, Part III*, ed. P. Duren (Amer. Math. Soc., 1989), 439–441.
34. A.I. Maltsev, "Untersuchungen aus dem Gebiete der mathematischen Logik", *Mat. Sbornik* n.s. 1 (1936), 323–336.
35. G.H. Moore, "Beyond First-Order Logic: ... Axiomatic Set Theory", *History and Phil. Logic* 1 (1980), 95–137.
36. P.S. Novikov, "Sur quelques théorèmes d'existence", *C.R. Acad. Sciences URSS* (Doklady) n.s. 23 (1939), 438–440.
37. P.S. Novikov, "On the Consistency ...", *Mat. Sbornik* n.s. 54 (1943), 231–260.
38. C.S. Peirce, "On the Algebra of Logic ...", *Amer. J. Math.* 7 (1885), 180–202.
39. F.P. Ramsey, "The Foundations of Mathematics", *Proc. London Math. Soc.* (2) 25 (1926), 338–384.
40. A. Robinson, *On the Metamathematics of Algebra* (North-Holland, 1951).
41. A. Robinson, "Applications to Field Theory", *Summaries ... Summer Institute of Symbolic Logic in 1957 at Cornell University* (1957), 326–331.
42. J.B. Rosser, "Gödel Theorems for Non-Constructive Logics", *J. Sym. Logic* 2 (1937), 129–137.
43. E. Schröder, *Vorlesungen über die Algebra der Logik*, Vol. 3 (Leipzig, 1895).
44. E. Schröder, *Abriss der Algebra der Logik*, part 2, ed. E. Müller (Leipzig, 1910).
45. D.S. Scott, *Bull. AMS* 61 (1955), 522–523 (abstract).
46. D.S. Scott and A. Tarski, "The Sentential Calculus ...", *Summaries ... Summer Institute of Symbolic Logic in 1957 at Cornell University* (1957), 83–89.
47. D.S. Scott and A. Tarski, "The Sentential Calculus with Infinitely Long Expressions", *Colloq. Math.* 6 (1958), 165–170.
48. T. Skolem, "Logisch-kombinatorische Untersuchungen ...", *Videnskapsselskapets skrifter, I. Matematisk-naturvidenskabelig klasse*, no. 4 (1920).
49. T. Skolem, "Einige Bemerkungen zur axiomatischen Begründung der Mengenlehre", Ibid., no. 6 (1923).
50. M. Stone, "The Theory of Representations ...", *Trans. AMS* 40 (1936), 37–111.
51. A. Tarski, Translation of 1933 "On the Concept of Truth in Languages of Deductive Sciences" in [56], 152–278.
52. A. Tarski, Translation of 1936 "On the Concept of Logical Consequence" in [56], 409–420.
53. A. Tarski, *Bull. AMS* 61 (1955), 523–524 (abstract).
54. A. Tarski, "Remarks on Predicate Logic with Infinitely Long Expressions", *Summaries of Talks presented at the Summer Institute of Symbolic Logic in 1957 at Cornell University* (1957), 160–163.
55. A. Tarski, "Remarks on Predicate Logic with Infinitely Long Expressions", *Colloq. Math.* 6 (1958), 171–176.
56. A. Tarski, *Logic, Semantics, Metamathematics* (Hackett, 1983), ed. J. Corcoran.
57. J. van Heijenoort, *From Frege to Gödel* (Harvard Univ., 1967).
58. R.L. Vaught, "Model Theory before 1945", *Proc. Tarski Symposium*, ed. L. Henkin (Amer. Math. Soc., 1974), 153–172.
59. L. Wittgenstein, *Tractatus Logico-Philosophicus* (London, 1922).
60. E. Zermelo, "Über mathematische Systeme und die Logik des Unendlichen", *Forschungen und Fortschritte* 8 (1932), 6–7.
61. E. Zermelo, "Grundlagen einer allgemeinen Theorie der mathematischen Satzsysteme", *Fund. Math.* 25 (1935), 136–146.

9
Ethics of Science and Technology

SHARON BEDER

ENGINEERS, ETHICS AND SUSTAINABLE DEVELOPMENT

ABSTRACT.

Most engineering codes of ethics worldwide exhort engineers to consciously put the public interest above all others. This seems to run counter to the market philosophy that the public interest will be achieved by individuals pursuing their own self-interest. It is this latter philosophy that is at the heart of sustainable development with its emphasis on economic valuation and economic instruments to achieve environmental protection. Sustainable development policies generally embody an economic determinism with respect to technological change. These policies avoid the issue of ethics and assume environmental and economic goals are compatible. Yet engineers today are grappling with the ethical dilemmas posed by everyday conflicts between the economic and environmental requirements of their work. In the past conflict between self-interest and public interest was seldom a problem for engineers, since engineering works were almost synonymous with human progress. Today environmental issues have created a divergence between self-interest, employer interest, professional interest and public interest. But how realistic is it to expect engineers to display higher ethical standards than those normally expected of the wider community? And can individual ethics play a significant role in influencing technologies that are collectively shaped by professional paradigms and philosophies?

Modern engineering codes of ethics require engineers to put the public interest before professional interests and business interests. In the first part of this paper I will show that professional interests and business interests have always been aligned and that codes of ethics have been developed to serve professional interests within this context. In the second part of the paper I will consider the ethical requirement for engineers to serve the public interest comparing this with the prevailing ethic of the market which stresses self-interest as the norm. In the third part of the paper I will consider, with the use of a case study, how the alignment of business interests and professional interests and the conflict between self-interest and public interest are exacerbated by environmental controversies and how the environment inevitably suffers as a result. Finally I will show that sustainable development does not provide a new supporting ethic for engineering codes of ethics that will protect the environment by raising public interest above professional interests, business interests and self-interest.

1. PROFESSIONAL INTERESTS, BUSINESS INTERESTS

The first engineering society to adopt a code of ethics was the Institution of Civil Engineers in England in 1910. The American Institute of Consulting Engineers used the British Code to derive their own which they formally adopted the following year. Other engineering societies quickly followed. Such

M.L. Dalla Chiara et al. (eds.), Structures and Norms in Science, 127–143.

codes were, and still are, a mixture of moral values and rules of business etiquette governing how engineers should relate to each other in their business dealings, a code of gentlemanly conduct rather than a code of ethics to protect the public welfare.

Codes of ethics serve several purposes. Firstly they are the hallmark of the professions. Engineers "lay claim to professional standing" to distinguish themselves from workers in general and to share in the prestige of the older more respected professions of medicine and law.[1] Edwin Layton[2] claimed that the emergence of a professional identity was also a reaction to the large authoritarian bureaucratic structures which engineers were increasingly finding themselves working in. The ideal of professionalism was based on esoteric knowledge and social service and Layton identifies three themes to the new ideology. Firstly, the engineer was identified as the agent of technological change and so essential to human progress. Secondly, the engineer was considered a "logical thinker free-of bias" and therefore able to lead and arbitrate between classes. Thirdly, the engineer was thought to be socially responsible for ensuring progress and the benevolence of technological change.[3]

The claim to professionalism and social responsibility also enabled engineers to assert their independence from business,[4] at least rhetorically if not in practice. Engineers had readily aligned themselves with business and capitalist values because it was business people and their capital which enabled them to build their great works.[5] Layton points out "Engineers accepted without question the structure, power, and basic ideological principles of business."[6] David Noble suggests that the modern engineer came into being to serve the purposes of the capitalist.

> From the outset, therefore, the engineer was at the service of capital and, not surprisingly, its laws were to him as natural as the laws of science. If some political economists drew a distinction between technology and capitalism, that distinction collapsed in the person of the engineer and in his work, engineering.[7]

Engineers, at least in the West, have therefore incorporated business values into their engineering activities. Zussman argues that "cost is itself a crite-

---

[1]Robert Zussman, *Mechanics of the Middle Class: Work and Politics Amomg American Engineers*, University of California Press, 1985, p. 10.

[2]Edwin Layton Jr, *The Revolt of the Engineers: Social Responsibility and the American Engineering Profession*, The Press of Cape Western Reserve University, Cleveland and London, 1971.

[3]Layton, *op.cit.*

[4]David Noble, *America By Design: Science. Technoloqy and the Rise of Corporate Capitalism*, Alfred A Knopf, New York, 1977; Layton, *op.cit.*

[5]David Noble, *The Forces of Production: A Social History of Industrial Automation*, Knopf, New York, 1984, p. 44; Zussman, *op.cit.*, 1985; Richard Ritti, *The Engineer in the Industrial Corporation*, Columbia University Press, 1971, pp. 54-5.

[6]Layton, *op.cit.*, p. 67.

[7]Noble 1977, *op.cit.*, p. 34.

rion of technical efficiency" which must be considered along with the physical properties of the materials. The purpose of technology, in a capitalist society, is determined by the market and engineering is seen as a means, not an end.[8] Ritti's study of an American systems design company also found that engineers placed great importance on having the opportunity to help their employing company increase its profits.[9] Whalley suggests that engineering employees "are socialised and selected" from the beginning to accept the legitimacy of both bureaucratic authority and the dominance of business values. These are secured by a career structure which rewards the trustworthy.[10]

Business interests and engineering interests have always been aligned and in the past there has been little conflict between engineering interests and the public interest since engineering works were perceived to be almost synonymous with human progress. However, as the community began to question whether new technology was always in their interest so engineering interests have been seen to occasionally conflict with the public interest and in recent years engineering codes of ethics have stressed the importance of prioritising the public interest. This element of the code of ethics can best be understood in terms of a social contact between the engineering profession and the community in which the engineering profession promises public service in return for professional status.

2. PUBLIC INTEREST, SELF INTEREST

Most modern engineering codes of ethics state that engineers should hold paramount the health and safety of the public or, in the words of the Australian Code of Ethics, engineers:

> shall at all times place their responsibility for the welfare, health and safety of the community before their responsibility to sectional or private interests...[11]

In a recently released discussion paper on "Dealing with Risk" the Institution of Engineers, Australia, argued that its code of ethics is part of a longstanding agreement the engineering profession has with the community; "its terms are that engineers will put the health, welfare and safety of the community before all other considerations; and the quid pro quo is, and should continue to be, that the community allows us to regulate the profession ourselves."

Traditionally morality has been defined by religion, laws and cultural conventions. In today's society many people are not religious and religious moral-

[8] Zussman, *op.cit.*
[9] Ritti, *op.cit.*, p. 54
[10] Peter Whalley, *The Social Production of Technical Work: The Case of British Engineers*, MacMillan, 1986; Zussman, op. cit., 1985.
[11] Institution of Engineers Australia, *Code of Ethics*, Canberra 1995.

ity seems to be confined to issues of life, death and sexuality; not reaching far into people's everyday working lives. Laws also, necessarily only have limited jurisdiction. They can cover blatant fraud and deception but are not able to force engineers to make "good" judgements.

Increasingly in a market economy, the morality of working life (apart from a work ethic that stems from Protestantism[12]) is based on cultural conventions which often fall within the category of ethical egoism or self interest. Ethical egoism is a minimalist form of moral reasoning which some would argue has nothing to do with morality. It reasons that each individual should look after themselves. Some argue that this form of reasoning involves considering other people's interests as well because in the long-term it serves one's own self-interest to do so. Adam Smith, demonstrated ethical egoism in his argument that social welfare is best served by individuals pursuing their own interests and companies pursuing maximum profits in a free market.

Engineering ethics normally go beyond ethical egoism, at least in principle. The ethical principle that engineers put the public interest before other interests seemingly works against their self-interest. However some philosphers, such as Hobbes and Rousseau, discuss ethics and morality in terms of a social contract that serves self-interest in the long-term. The terms of this contract are that if everyone follows the rules of morality rather than acting on personal self-interest, then everyone will be better off; society will be a better place to live in.

> Morality consists in the set of rules, governing how people are to treat one another, that rational people will agree to accept, of their mutual benefit, on the condition that others will follow these rules as well.[13]

However, as Rachels points out, there is a natural limit to the social contract. If people obey these rules so that they will be better off then, in cases where obeying the rules means they will be worse off, they won't do it. "We may not exact a sacrifice so profound that it negates the very point of the contract."[14]

This is particularly relevant to the engineering code of ethics because if it is a form of social contract which provides social status to engineers then it is not reasonable to expect engineers to obey any rules in the code of ethics which requires them to forfeit their status as engineers. If putting the public interest first requires them to risk their jobs and career then this will be seen as too great a sacrifice by most engineers. And because the individual status

[12]Max Weber, *The Protestant Ethic and the Spirit of Capitalism*, trans Talcott Parsons, Unwin University Books, London, 1967.

[13]James Rachels, *The Elements of Moral Philosophy*, 2nd ed., McGraw-Hill, New York, 1993, p. 143.

[14]Ibid., p. 151.

of engineers is so dependent on their employers, this social contract has little power for engineers.

Various studies have confirmed that codes of ethic and conduct have little power.[15] Engineers are essentially subordinate and their status derives from organisational mobility rather than technical expertise. As the Australian Institution of Engineers' Professional Practices Officer, Derek Baldwin, readily admits, "it takes a man or women of considerable strength and courage" to obey the code of ethics rather than his or her employer. And Michael Dack, a Director of the Institution, admits that the code of ethics has a "very weak moral power" compared to the employer who has the "power of economic life and death over an employee".[16]

3. ENVIRONMENTAL INTEREST — A CASE STUDY

For decades a combination of remnant religious beliefs, law and a culture of ethical egoism has served Western societies fairly well. Average standards of living have increased and nations have become wealthier and more powerful. But in recent times environmental issues have highlighted a conflict between self-interest, employer interest, professional interest and public interest. An Australian case-study will illustrate some of the issues involved.

In 1985 an engineering firm approached construction company Transfield with an idea for a car tunnel to cross Sydney Harbour. The Sydney Harbour Bridge suffered traffic congestion at peak hours and the Tunnel would provide an alternative route between North Sydney and the City Business District (CBD). Transfield joined with the Japanese firm Kumagai Gumi to form a consortium that then sold the idea of a toll-financed tunnel, which they would build, to the Department of Main Roads. The Minister for Main Roads was particularly keen on the idea.

Before approval could be given to the Tunnel an environmental impact statement (EIS) had to be prepared. An EIS, which is required by law in New South Wales (NSW) for major road proposals, is supposed to provide a justification for the project, a detailed assessment of the potential environmental effects of the project and consider other alternatives. The aim of an EIS is to ensure that development decisions take account of and where possible mitigate environmental impacts. Transfield-Kumagai hired engineering consultants Cameron McNamarra to prepare the EIS on their behalf.

---

[15] Kenneth Prandy, *Professional Employees: A Study of Scientists and Enqineers*, Faber & Faber Ltd, London, 1965; William Rothstein, Engineers and the Functionalist Model of Professions, in Robert Perrucci & Joel Gerstl, eds, *The Engineers and the Social System*, John Wiley & Sons, 1969; William Kornhauser, *Scientists in Industry: Conflict Accomodation*, University of California Press, 1962; Robert Perucci and Joel Gerstl, *Profession Without Community: Engineers in American Society*, Random House, New York. 1969.

[16] Sharon Beder, Engineers, Ethics and Etiquette, *New Scientist*, 25 September 1993, pp. 36–41.

In Australia there has been a progressive loss of faith in the environmental impact assessment process. Communities likely to be affected by proposed engineering projects such as freeways, chemical plants and waste facilities are often disappointed to find that EISs are not the independent, objective assessment of environmental impacts that they expected. Increasingly such documents are being viewed by the local residents as sales documents for the project and the engineers who prepare them as mouthpieces for the proponents. Subconsultants working on EISs have also become concerned that their findings are edited and selectively reported in the final document.

The Harbour Tunnel EIS, as in most cases, supported the project and argued that there would be no adverse environmental effects. However in this case the consultants were accused of breaching the Engineering Code of Ethics, by North Sydney Municipal Council and the Society for Social Responsibility in Engineering, for not putting the public interest first. These organisations made representations to the Institution of Engineers, Australia (IEAust) alleging that the consultants had overestimated the benefits and underestimated the environmental costs of the Tunnel project. John Gerofi, an engineer who conducted an inquiry into the tunnel proposal for the Council, stated that; "The inquiry can find no rational explanation as to why competent and respected consultants employing professional engineers and other qualified staff would have produced an EIS with so many questionable assumptions which favoured the project, and with so many deficiencies."[17]

The Institution never proceeded with an investigation of the engineering consultants who authored the Harbour Tunnel EIS. Bill Rourke, the Institution's chief executive at the time, said that it had not been given evidence that constituted a prima facie case against any individual member of the Institution. The North Sydney Council decided not to assemble a case against individuals. Gerofi later stated in a letter to *Engineers Australia*, that "the defamation laws, plus a lack of resources and a reluctance to accuse individuals will prevent all but the most blatant ethical transgressions from being raised" if the Institution continues to confine its attention to individuals.[18]

This case raises various questions about ethics, engineering and the environment. Does a favourable interpretation of the data consist of unethical conduct? Is it fair to single out individual engineers who are doing the job as their employers require them to and who are not acting very differently from other engineers in similar positions? Can the environment be protected by ethical conduct?

---

[17] Anon, Was there unethical behaviour in tunnel project, *Engineers Australia*, 14 July 1989, p. 10.

[18] J.P. Gerofi, Sydney Harbour Tunnel and the Code of Ethics, *Engineers Australia*, 20 October 1989, p. 4.

4. WHAT IS ETHICAL BEHAVIOUR IN THIS SITUATION?

An EIS, which includes scientific studies and economic analyses, purports to be an objective statement of environmental impacts. However the goal of an objective document is illusory if we assume that science itself is socially constructed. This is exacerbated by the circumstances of EIS preparation where large investments, careers and the viability of businesses are at stake. It is therefore inevitable that the values and goals of those preparing an EIS, and those who employ them, will shape its contents and conclusions through the way scientific data is collected, analysed, interpreted and presented.

In Australia environmental impact statements are done by or commissioned by the proponent of a project or development.[19] The engineer working on that impact statement is therefore, either directly or indirectly, employed by a party whose interests may differ in significant ways from the public interest. The immediate objective of their employer or client will be to get approval for the project to go ahead even though that project may have an adverse impact on the environment and on local residents.

Because the EIS is done rather late in the planning process the project proponent will almost certainly have committed considerable financial resources to a particular option at a particular site. From their point of view they will have balanced the community costs against their benefits and decided the project should go ahead. The EIS at this stage becomes another obstacle in a field of bureaucratic hurdles on the way to their end goal. But on top of this it is a public document that will be scrutinised by local residents, bureaucrats, politicians and environmentalists.

4.1. *Value Judgements in Scientific Studies*

Naturally, project proponents will want that document to emphasise the advantages of the project to the community and to downplay the disadvantages. To a large extent that environmental impact statement is like a public relations document for the project. Any expression of possible adverse environmental effect or even any mention of uncertainty will certainly be magnified by opponents of the project and used in argument against the project.[20]

Values and judgements enter at every stage of the preparation of an EIS beginning with definition of the problem. Both justification and the framing of alternatives will be shaped by the way the problem that the project is

[19] The reason for this is that it is assumed that the consultant needs to work closely with the proponent to understand the project and be able to recommend changes to the project as environmental impacts become evident.

[20] Sharon Beder, The Role of the Professional, in *Environmental Impact Statements: Selected Readings*, ed. by Sharon Beder, Environmental Education Project, Sydney University, 1990, pp. 45–48; Sharon Beder, Bias and credibility in environmental impact assessment, *Chain Reaction*, No 68, February 1993, pp. 28–30.

supposed to be solving is defined. For example, in the Sydney Harbour Tunnel EIS, the problem was said to be traffic congestion. Traffic built up and slowed down on the approaches to the Sydney Harbour Bridge, causing delays to people trying to cross the Harbour by motor vehicle. A second crossing was therefore justified on the grounds of traffic congestion, and alternatives framed, in terms of providing better road access across the Harbour.[21]

Opponents to the Tunnel did not perceive congestion to be a problem at all. Ted Mack, Mayor of North Sydney at the time the Tunnel was proposed, argued that congestion shaped a city by encouraging the movement of people and businesses to other parts of the metropolitan area so that new centres of activity were established. Ross Blunden, emeritus professor of traffic engineering, argued that congestion encouraged people to change their journey times or take public transport. Both men concluded that a second crossing, far from removing congestion, would merely attract more car traffic and that congestion on both crossings would be the eventual outcome.[22]

The scope of what is to be covered in the EIS is also a matter of judgement and the way it is decided varies from state to state. A narrow scope can make a project appear more desirable. With the Sydney Harbour Tunnel, one can see that, whilst the proponents argued that pollution would be reduced in the immediate vicinity of the Harbour because of the smoother flowing traffic, a broader scope would have ensured that the wider impacts of increased car usage encouraged by the Tunnel —greenhouse emissions, smog production and oil usage— were also taken into account.

The design of an EIS requires judgements of what types of impacts will be significant and the collection of data requires decisions about the time period and area over which samples are collected, the species to be studied and the quantities of individual specimens to be collected, and more generally the scale of study. Such decisions are not made only on the basis of what might be considered by a scientist to be appropriate, but will also be affected by considerations of cost, time availability, previous studies and perhaps even likely outcome. In the case of the Sydney Harbour Tunnel EIS many of these decisions were criticised.[23]

Similarly, methods of analysing data can vary in the sorts of results they produce and data they require and those preparing an EIS will choose the methods using many criteria, apart from the 'purely scientific'. Even where the method of analysis is uncontroversial, assumptions and judgements will need to be fed into the analysis. Most EISs require some form of forecasting of population numbers or other human activities and this requires assumptions

[21] Cameron McNamara, *Sydney Harbour Tunnel: Environmental Impact Statement*, Transfield-Kumagai Joint Venture, November 1986.

[22] *Sydney Harbour Tunnel, Environmental Impact Statement*, video produced by Sydney University Television Services, 1991.

[23] Ibid.; personal communication, Alan Jones, Australian Museum, 1990.

such as where people are likely to live and work and what their habits will be in the future.

Perhaps the most contentious analysis in an EIS is the cost-benefit analysis. A cost-benefit analysis (CBA) for a road project will require estimates of the value of time saved and may require estimates of the value of bushland or open space lost to the community. In the case of the Harbour Tunnel, a number of CBAs were done by various consultants. Those commissioned by proponents of the tunnel all showed that benefits outweighed costs and those commissioned by the opponents to the tunnel all showed that the costs outweighed the benefits.

In an EIS data collected and the results of analyses can be interpreted and presented in a number of ways. Even though real world engineering and environmental science is fraught with uncertainties an EIS can be carefully worded to avoid any impression that anything is uncertain.

The amount of judgement required in putting together an EIS raises the question of whether a person's ethical stance will affect the outcome. Here I am not talking about outright falsification or omission (which of course does happen occasionally). I am talking about subtle judgements within a range of legitimate and valid choices. Inevitably there will be a grey area between what is accepted as scientifically credible and outright deception which will depend on a person's viewpoint. (See the diagram on p. 136.)

Engineers have a *legal* obligation to avoid the outer area of falsehoods, omissions and deliberate distortions. So does ethical reasoning only apply to the grey area where they can get away with slight distortions? Or do they also have an ethical obligation, when operating within the range of scientific credibility, to make judgements in favour of the public interest and environmental protection? Is it reasonable to expect people to be conscious of the way they shape scientific studies and to endeavour not to be influenced by vested interests in that shaping?

From the standpoint of ethical egoism it makes sense for the engineer to prepare an EIS which favours the project. An engineer's career prospects are dependent on an employer's assessment of their loyalty and reliability. The employee/employer relationship is necessarily one based on trust.[24] The employer's judgement in this regard crosses organisational boundaries because job applicants need favourable references from previous employers. Engineers seldom have the sort of independent reputations that scientists sometimes build up through publications.

Even as self-employed consultants, engineers are dependent on the judgement of clients and that judgement is based on whether they are perceived to be able to deliver what is required by the client. Consultants with 'overdeveloped' consciences, who do not put the client's priorities first, are less likely to

[24] Whalley, op. cit., Zussman, op. cit.

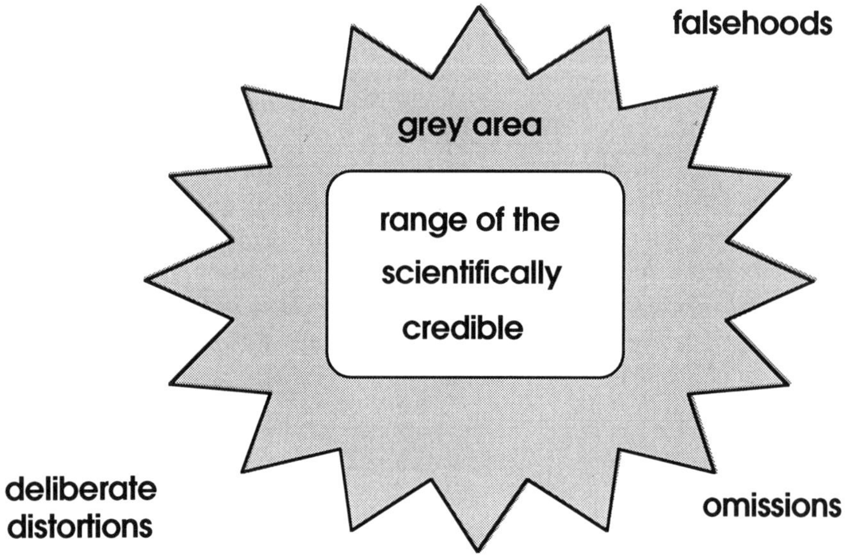

Fig. 1.

be given work in future. In many fields the number of potential clients is very limited and consultants with troubling tendencies toward social responsibility will soon be well known.

If engineers relied solely on ethical egoism for moral reasoning then there is no reason why they should look beyond their employer's interest in preparing an EIS unless they themselves would be adversely affected by the project. However, engineers are expected by the community to go beyond self-interest. At a minimum, the code of ethics should prevent an engineer from producing a favourable EIS for the client/employer if this will mean that public welfare is endangered.

However, if an engineer believes the project does not endanger the welfare, health and safety of the community then s/he may feel justified in ensuring that the EIS is supportive, whether or not this has involved consciously making favourable judgements in favour of the project within the range of scientifical credibility or the grey areas surrounding it. This is especially so since most engineering codes of ethics also include a tenet requiring them to apply their skill and knowledge in the interest of the client.

Despite debate within the Australian Institution of Engineers, the Code of Ethics deliberately leaves out specific mention of environment in its tenets leaving it up to the individual engineer to decide whether environmental protection is an essential ingredient of community welfare.

In discussions about their role in preparing EISs most engineering consultants refer to a different kind of moral reasoning, they argue that they have integrity (a virtue based ethic) and that therefore their EIS reports are not biased in favour of the proponent. Engineering codes of ethics also generally include tenets such as "engineers shall act with fairness, honesty and in good faith..." However, most engineers preparing an EIS would not consider making a judgement within the range of scientific credibility as being unfair or dishonest. This ethic may, however, prevent such an engineer from knowingly wandering into the grey area.

I would argue that most engineers are quite conscious of the way they manipulate an EIS to give a favourable outcome and that their claim of integrity has more to do with reputation than ethical behaviour. Engineering consultants that earn a reputation with the public for misleading and distorted EISs will be a liability to project proponents who want to gain public approval for their projects. On the other hand consultants who prepare an EIS that gives the opposition plenty of ammunition against a project or causes approval to be denied would also not get further work. Integrity for engineering consultants involves treading that fine line that enables a favourable EIS to be prepared without resorting to blatant bias; that is by remaining within or close to the area of scientific credibility.

4.2. *The focus on individual people and projects*

Despite the prevailing ethic of self-interest and the emphasis on individualism that modern capitalist societies convey, they depend on a sharing of values and collective activity. Engineering is one example of a collective activity that is built upon the work of past engineers and a consensus within the profession about what are appropriate technological solutions. The terms technological systems[25], technological traditions[26], technological paradigms[27], technological regimes[28] and technological trajectories used by scholars of technology all

[25]Thomas Hughes, *Networks of Power: Electrification in Western Society, 1880-1930*, John Hopkins University Press, Baltimore and London, 1983.

[26]Edward Constant, Communities and hierarchies: Structure in the practice of science and technology, in *The Nature of Technological Knowledge. Are Models of Scientific Change Relevant?*, ed. R. Laudan, D. Reidel Publishing Co, Holland, 1984.

[27]Giovanni Dosi, Technological paradigms and technological trajectories, *Research Policy*, no. 11, 1982, pp. 147-162; Sharon Beder, Pipelines and Paradigms: The Development of Sewerage Engineering, Australian Civil Engineering Transactions, vol. CE 35, no. 1, 1993, pp. 79–85.

[28]Richard Nelson and Sidney Winter, In search of useful theory of innovation, *Research Policy*, vol. 6, 1977, pp. 36–76.

refer to the way technological development is a social enterprise shaped by the context within which it occurs.

The design, choice and implementation of technology is seldom an individual endeavour. Most engineering projects rely upon teams of engineers working together and no single engineer feels individually responsible for the outcome, nor is any project dependent on the willingness and cooperation of any one engineer. As Cohen and Grace point out:

> "groupthink" typically is responsible for perceiving and articulating whatever justification there is for a project. A critical perspective is difficult for an individual to take, let alone sustain, in the course of "normal" engineering work under an accepted paradigm.[29]

Merton observed many years ago that engineers were unlikely to take full responsibility for their work and all its consequences because as employees they are "knit into a bureaucratic apparatus" with fixed spheres of competence and authority, and are "rewarded for viewing themselves as technical auxiliaries."[30]

Moreover the emphasis on individual ethics, which a professional code entails, tends to contain ethics within the microsphere of individual projects. In this case an EIS applies to a single project which on its own may have quite minor environmental impacts. It is often the cumulative impact of such projects that degrades the environment. Therefore an engineer working on a particular EIS may feel satisfied that the project will not significantly harm the environment but to what extent can such an engineer be expected to consider the cumulative impact of many such projects?

The Institution of Engineers, Australia clearly does not believe that this is possible. It has argued that EISs cannot adequately assess the cumulative impacts of projects and that "it was naive to expect the EIA [Environmental Impact Assessment] process to address longer-term sustainability issues."[31]

> "Imagine for one moment an EIA on the impact of a coal power station on the local environment," Dr Webster [IEAust chief executive] said. "An EIA would not consider the possibility that alternative energy sources could be developed elsewhere, and assess the nett environmental impact of the various options."[32]

What is more, an EIS is merely a tool for identifying environmental impacts and adverse environmental consequences do not mean a project will not be approved. These consequences are weighed against the wealth to be created

[29]Stephen Cohen and Damian Grace, Engineers and social responsibility: An obligation to do good, *IEEE Technology and Society Magazine*, Fall 1994, p. 13.

[30]R.K. Merton, The Machine, the Worker and the Engineer, *Science*, Jan 1947, p. 82.

[31]"IEAust rejects government proposals for environmental impact assessments", *Engineering Times*, June 1995, p. 1.

[32]Ibid.

by the project and in most nations priority is given to the wealth generating potential of the project.

Clearly it is unrealistic and perhaps even unreasonable to rely on the ethics of individual engineers applied to individual projects to protect the environment from the impact of engineering works, particularly within a market economy where the self-interest ethic of the market dominates. However, a growing recognition of environmental decline has prompted governments all over the world to turn to sustainable development as a way of managing the environment. But does sustainable development supply a social ethic to replace the individualistic ethic of the market which would be more supportive for engineers who act ethically?

5. THE ETHIC OF SUSTAINABLE DEVELOPMENT

The central ethical principle behind sustainable development is *intergenerational equity.* The Brundtland Commission defined sustainable development as: "development that meets the needs of the present without compromising the ability of future generations to meet their own needs."[33]

Intergenerational equity can be defended in both consequentialist and deontological terms. Firstly it can be considered in terms of ensuring the long term consequences of today's actions are beneficial. This utilitarian viewpoint fits the pragmatic concerns of some business interests. The environmental crisis threatens the sustainability of economic activity. Many activities such as agriculture, forestry, fishing, tourism and recreation are dependent on a healthy environment. Others are indirectly affected as it becomes more expensive to obtain resources and because pollution decreases the health of the work-force. Looking ahead to the future ensures the sustainability of business activities.

Intergenerational equity can also be considered a duty that current generations have to future generations or a right of future generations. However, if we examine the way that sustainable development is operationalised we see that it is done in a way that protects the market system and perpetuates individualism and self-interest above any ethic of equity.

David Pearce argues that if we are to ensure intergenerational equity then future generations need to be compensated for any environmental damage done by current generations and that this is best done by ensuring that damage is made up for by increased wealth and human-made assets. In other words natural capital (the environment) can be run down if human-made capital (money, equipment, infrastructure, knowledge etc) are increased.[34] In order to compensate future generations we need to value of the environment in

[33]World Commission on Environment and Development, *Our Common Future*, Australian edn, Oxford University Press, Melbourne, 1990, p. 85.

[34]Pearce admits that there are some environmental assets that cannot be replaced by human-made capital.

the same way as we value human-made assets; that is we need to give it a monetary price.[35]

Environmental economists, such as Pearce, also claim that environmental degradation has resulted from the failure of the market system to put any value on the environment. They argue that because environmental 'assets' are free or underpriced they tend to be overused or abused, resulting in environmental damage. Because they are not owned and do not have price tags then there is no incentive to protect them. This is a view shared by business people. The Business Council of Australia claims that the environmental problem;

> is that important environmental assets tend not to be priced in a market like other assets. These assets are common property - they belong to everybody, and to nobody. Without ownership rights there is not the incentive for any person or group to look after them properly... if the environment has a zero price to users it will eventually be used up.[36]

These views, which have been incorporated into sustainable development rhetoric and sustainable development policies, call for putting a price on the environment. However the whole process of pricing the environment to ensure that decisions take account of environmental degradation works against intergenerational equity and instead extends market logic and market morality into a wider sphere of operation.

There are two main ways of operationalising the idea of putting a price on the environment. The first is through cost-benefit analyses. The second is through the use of economic instruments.

5.1. *Cost-Benefit Analysis (CBA)*

CBA has traditionally been used by governments as part of their decision-making processes but in the past environmental costs and benefits have usually not been quantified and incorporated into the analyses but listed separately in a qualitative form for consideration. The sustainable development approach is to incorporate these environmental costs and benefits by pricing them and incorporating them into the calculations.

In a way CBA is the ultimate embodiment of consequentialist ethics in that it seeks to ensure that good consequences outweigh bad consequences and consequences are measured in money terms. In reality however CBA works against the ethic of equity and the measuring of consequences in financial terms fails to capture the consequences fully.

---

[35] David Pearce, ed. *Blueprint 2: Greening the World Economy*, Earthscan, London, 1991, p. 1–2.

[36] Business Council of Australia, *Achieving Sustainable Development: A Practical Framework*, BCA, 1991, p. 9.

CBA is about aggregated costs and benefits and does not deal with the issue of how they are distributed yet distribution of costs and benefits is of is of prime concern when considering equity. As long as the sum of benefits outweighs the sum of the costs, even if a small groups of people get the benefits and a whole community suffers the costs, the society as a whole is assumed to be better off by CBA. It is sometimes argued by economists that, if the total benefits outweigh the total costs, the winners could compensate the losers and still be better off; but this is only theoretical reasoning and seldom happens.

In a CBA, the value of future consequences is discounted (reduced) because it is assumed that costs and benefits in the future are not worth as much to people today. This is a direct result of using money as a measure. The logic behind discounting derives from the logic of money-that a person would prefer to receive money now than the same amount in the future. Daly and Cobb point out that the idea of discounting comes from the fact that money can be put in the bank to get interest and that people have a choice between putting their money in a bank or investing in the project in question. Economists forget this when they apply their models to things that don't grow like money in a bank, such as the environment.[37]

The use of a discount rate means that the further into the future that the costs or benefits are, the less they will be worth in the CBA calculations. Normal discount rates ensure that any costs or benefits more than 30 years in the future are almost valueless for the purposes of a CBA. Discounting therefore discriminates against future generations by saying that environmental damage in the long-term future can effectively be ignored.

CBA also rests on the assumption that environmental assets can be substituted by human-made assets and all that matters in the end is that the aggregated gains outweigh the aggregated losses. If a project generates more wealth than what it is calculated the environmental damage that is caused is worth, then the project should go ahead. The loss of environmental amenity is made up for by the wealth that is generated. As mentioned before CBA takes no account of who gains and who loses from the exhange of shared resources for profits; that is the substitution of common environmental amenity that is accessible to the whole community with private capital that belongs to the developers.

The method of calculating the value of the environment incorporates market values. The most popular method currently is to use contingent valuation, or willingness to pay surveys. People are asked how much they would be willing to pay to preserve a particular environment or their willingness to pay is inferred from their behaviour in the market such as the extra price they are willing to pay for real estate in non-polluted areas (hedonic pricing).

[37]Herman Daly and John Cobb, *For the Common Good: Redirecting the Economy toward Community, the Environment, and a Sustainable Future*, Beacon Press, Boston, 1989, pp. 153–4.

Using the market, whether an actual market or a contrived one, to value the environment tends to produce values that reflect and therefore maintain the prevailing distribution of income. Wealthier people are willing to and able to pay more for what they want so their preferences will have more weight in any survey. Moreover according to a CBA siting a dirty industry in an already dirty area will be less costly than siting it in a clean area where wealthier people can afford to live. This is because the decline in property values will be less as a result of the pollution.

The valuation of the environment in terms of the total of what each individual is willing to pay denies a separate concept of public interest. As Lenihan and Fletcher state "The welfare of society has meaning only as the summation of the welfare of its individual members".[38] Daly and Cobb also point out that the economic view of value is based on a reduction of human values to individualism and reduces the world to one in which "individuals all seek their own good and are indifferent to the success or failure of other individuals"[39]. Therefore valuation of the environment through CBA is a concept that embraces the values of ethical egoism and is in fact antithetical to an ethic of equity.

5.2. *Economic Instruments*

Another increasingly popular way of incorporating environmental values into decision-making is through the use of economic instruments. The idea is that prices of resources should reflect the true cost, including environmental costs involved in their extraction and manufacture. If this were the case then, the economists argue, people would use environmental resources more wisely.

Of course putting a monetary value on these costs suffers the same problems involved in cost-benefit analysis. However in practice, economic instruments seldom involve calculating the real value of environmental damage. Rather, in the case of price-based measures such as pollution charges, an extra amount is charged, chosen somewhat arbitrarily by the government, that is supposed to provide an incentive to change environmentally damaging behaviour. In the case of tradeable pollution rights, a level of emissions is chosen and rights to emit up to that level are traded between companies or auctioned off by the government. In both cases the idea is that by making economic adjustments, individuals and firms can continue to pursue their self interest and the environment will be protected at the same time.

---

[38] John Lenihan and William Fletcher, *Economics of the Environment*, Blackie, Glasgow and London, 1979, p. 4.

[39] Daly and Cobb, *op.cit.*, p. 159

Advocates claim that economic instruments provide a way that the power of the market can be harnessed to environmental goals.[40] They also serve a political purpose in that they reinforce the role of the 'free market' at a time when environmentalism most threatens it. Chant et al argue that "contrary to the popular view that a market system leads to the abuse of the environment" it is in fact the absence of a market which leads to environmental degradation.[41] Economic instruments attempt to make a virtue out of the profit motive and ethical egoism:

> The economists' preference for market solutions is an ideologically based one: Its first pillar comes squarely out of a philosophical tradition that grew from Adam Smith's notion that individual pursuit of self-interest would, in a regime of competitive markets, maximise the social good. That tradition is so firmly embedded in economics by now that most economists probably do not realize, unless they venture out into the world of noneconomists, that it is a proposition of moral philosophy... (Kelman 1983, p. 297)

6. CONCLUSIONS

A revolution in ethics is needed to displace the powerful ethical egoism that rationalises the market as the predominant decision-making tool in our society. It is unrealistic to expect engineers to manifest higher ethical conduct than is the norm throughout the community in which they live. Sustainable development, with its rhetoric of intergenerational equity, is in reality a way of endorsing market morality and is inadequate to the solution of modern environmental problems.

Department of Science and Technology Studies
University of Wollongong

[40] Robert Stavins, Harnessing market forces to protect the environment, *Environment* 31(1): 5–7, 28–35; T.H. Tietenberg, Using economic incentives to maintain our environment, *Challenge*, March/Apr.,1990, p. 42.

[41] John Chant, Donald McFetridge and Douglas Smith, The economics of the conserver society, In *Economics and the Environment: A Reconciliation*, edited by Walter Block. Canada: Fraser Institute, 1990, p. 62.

L. JONATHAN COHEN

ARE THERE ETHICAL REASONS FOR BEING, OR NOT BEING, A SCIENTIFIC REALIST

I

There are a variety of different kinds of normative questions that may be asked about science. We can ask, for example, whether laws should be conceived as stating no more than constant conjunctions? Or how should experimental controls be structured? Or is evidential support measurable? Or do theories function only as anti-realists claim that they do — viz. as instruments for the prediction of observable events? Or is there only one correct pattern of scientific finding, or, as some readers of Alasdair Crombie's *Styles of Scientific Thinking in the European Tradition* might suppose, are there several? These and their like are one familiar group of normative questions about one aspect of scientific enquiry. Let us call this the "epistemological" aspect, because it is this kind of question that needs to be answered in order to make progress towards resolving the larger issue that may be conveniently summarised in the question "What is scientific knowledge?"

A second group of normative questions may be said to concern the "ethical" aspect of scientific enquiry rather than the epistemological. What good can scientific discoveries achieve? What moral constraints should control the choice of options in experimental design? Ought the choice of research topic to be affected by the good, or bad, consequences that success in the research might have? Is there only one system of ethics with which science is compatible, or are there several?

Other groups of normative questions about scientific enquiry also exist. For example, questions may be asked about the social structure that is optimal for scientific progress — questions influentially raised by Francis Bacon in his *New Atlantis* and still exercising the minds of administrative and grant-giving organisations. Or questions may be asked about conventions in writing scientific papers, such as conventions about how names are ordered in the title of a jointly written paper. And there are also, corresponding to each of the above groups of normative questions, important groups of factual questions about science, which are not always easily separable, as in the history of scientific discovery, the history of scientists' value-systems, the history of conventions in scientific literature, and so on.

I am concerned here, however, only with the first two groups of normative questions and, in particular, with a problem about the relationship between those two groups of questions. I want to address the problem whether there are

M.L. Dalla Chiara et al. (eds.), Structures and Norms in Science, 145–153.

any connections between questions of the two kinds, such that some reasonable answers to a question of the second kind (the ethical kind) can support some particular answer to a question of the first kind (the epistemological kind). And I shall concentrate on only one possible question of such a type, viz. are there ethical reasons for being, or for not being, a scientific realist? Of course, some people may have an immediate intuition that such reasons cannot exist. But by them my purpose can be regarded as an attempt to explore the rational basis for such intuitions.

II

Let us consider first an argument for answering this question in the negative. The argument begins with the claim that, if some moral judgement is a member of a set of jointly sufficient and severally necessary premisses for the support of scientific realism, then any assertion of scientific realism must have a moral element. But — the argument continues — all moral judgements in science have a certain structural character: specifically, they all apply some general moral principle to the particular circumstances of science. And scientific realism does not have this structure: it makes no such application. Therefore, concludes the argument, scientific realism has no moral content and there are no ethical reasons for being a scientific realist.

Now here there seem to be two premisses that need substantiation. One is the premiss that all moral judgements in science have the specified structural character: the other that scientific realism does not have this character. Let us consider these two premisses in that order.

There are certainly no immediately obvious exceptions to the thesis that all moral objections in science involve the application of some quite general moral principle to the particular circumstances of science. Consider, for example, how policies for scientific research are standardly evaluated. What normally counts is their probable nett contribution to the achievement of human needs, wishes and benefits, even when this involves nothing but the satisfaction of curiosity. The contribution may vary in its degree of immediacy. "Experiments of light", as Francis Bacon called them, may themselves be justified by the leads they supply to "experiments of fruit". Admittedly animists, animal-lovers, or Gaia worshippers, may attach rather less importance to human welfare and evaluate other ends more highly. But, whatever the nature of the values that scientific activity seeks to promote, their relevance is not peculiar to science.

An analogous situation arises when the experiments needed to test a particular hypothesis cannot be carried out without causing what are thought to be morally unacceptable side-effects, such as through vivisectional experimentation in animal physiology or when misleading information is given to human subjects in certain kinds of psychological experiments. Scientific enquiry runs the risk of ethical involvement here because the processes, structures, etc.

that are experimented upon are associated with organisms that may suffer harm to their interests thereby. But there are many other ways — outside the ongoing process of scientific enquiry — in which such interests may suffer similar harm: it is not only in experiments that animals get hurt or people get deceived. So the ethical downside here is due to the situation's being an instance of a general category of situation; and there is nothing peculiar to scientific experiment that is responsible for this downside.

Another such type of situation arises from discreditable behaviour in the handling of information somewhere in the process whereby research findings are established and enter into the canonical literature. A person may make deliberately false statements in applying for a research grant; he may deceive a colleague in a research-team;he may plagiarise from another team or person, or misreport data to a journal, etc. etc. Now the room that exists in science for dishonesty of this kind is due to the fact that, at least since the 17th century and particularly under the inspiration of Bacon's *New Atlantis*, science has normally been thought of as a social activity with consequential rules, presumptions and expectations. Scientific progress is predicated typically of a community rather than of an individual: an individual's achievement is to contribute to this progress. But science is far from being the only type of rule-bound social activity, and correspondingly dishonesty in science owes the source of its immorality to a principle that governs all inter-personal dealings: namely, the principle that in normal circumstances one shouldn't tell lies to people or mislead them in any way.

Yet another such type of situation arises when a particular line of research is thought to be too dangerous to pursue because of the risk that its findings will inevitably be exploited and generate bad consequences. For example, if the technology of chemical weaponry is researched, some governments can perhaps not be trusted not to use this technology. Or, if the technology of the internal combustion engine has been improved, it may be that some governments cannot be trusted to maintain support for other measures designed to reduce global warming. And here too the ethically relevant features of the situation to be avoided — the unnecessary risk of death etc. — are not peculiar to scientific research. The core of the matter is that one shouldn't facilitate others' misdeeds.

In sum, if we argue inductively from familiar types of case like the four that have just been considered, it looks as though any moral judgement in science is an application of some general moral principle to derive a prescription or proscription in particular circumstances. Science, we may say is ethically heteronomous.

But now let us turn to the other premiss that is needed by the argument under consideration. Does scientific realism indeed have the logical structure that (as the above examples seem to show) is standard for moral evaluation of features of scientific activity? That is to say, does it rest on the application of some general moral principle to the issue about realism in the analysis of scientific theory? And the answer to this question seems to be a negative one. For, if we try to identify some general principle in terms of which scientific realism appears as a special case, or derivable consequence, that principle would presumably have to embody some more general form of realism. And any such principle, in vouching for the factuality of what is held to be the case, must inevitably contrast with any kind of moral principle that endorses some range of values or obligations and thereby endorses what ought to be the case rather than what is. The factual is, as it were, a paradigm case of the non-normative.

So, if I am right about the two premisses that I have been considering, it looks as though those premisses provide the necessary foundation for an argument that there cannot be any ethical reasons in favour of being a scientific realist.

III

The situation might be quite different, however, if scientific activity, or some features of it, were indepently valuable. For, if scientific enquiry, or some features of it, were good, or bad, in themselves, and not just as a consequence of a more general moral principle or principles, then the negative argument that I have been sketching would not get off the ground. Since it would not be necessary to endorse some general principle from which a value-judgement about certain scientific activity derived, the lack of such a general principle would not matter, and the possibility of an ethical reason for being a scientific realist would remain. That is to say, perhaps the inductive argument developed in the pevious section of this paper is invalid and there is indeed some degree or kind of ethical autonomy in science. Perhaps in appropriate circumstances, and within appropriate limits, and under an appropriate interpretation (such as a realist one might be) we can treat the process of scientific enquiry as an end in itself.

Let us briefly turn aside here to consider a possible analogy with art, since there is an old controversy in philosophy about whether art admits of being good in itself or only of being a means to good results. Thus a medieval altar-piece may have been painted, or a motet may have been composed, so as to facilitate the performance of a religious ritual. Or a statue may have been sculpted, or a poem written, so as to glorify a programme for social reform. In each case the altar-piece, motet, statue or poem may achieve its end just because it has the required features — the icon, tune, size, metre, etc. — that is needed in order to accomplish the religious or political purpose involved. But, though such examples may be cited to illustrate a heteronomous conception of ethical value, their existence does not suffice to establish that this kind of conception is the only admissible one. It makes perfectly good sense to claim at the same time that certain other paintings, songs, sculptures, poems, etc., have an intrinsic value because they conform so satisfactorily to accepted criteria of excellence within their several genres. Indeed three obviously different positions are possible here for anyone who thinks that aesthetic value exists at all. We can hold either that aesthetic merit is always derivable from service to some other purpose, or that it is always derivable from conformity to its own intrinsic criteria, or that sometimes the one holds good and sometimes the other.

Now, in any particular field of research, knowledge and the pursuit of knowledge may be treated analogously to painting, sculpture, music and literature, and their pursuit, within any particular aesthetic framework. Either this value always derives from their utility in the service of some otherwise justifiable purpose. Or their value always derives from their conformity to appropriate criteria intrinsic to science. Or their value derives sometimes from the one and sometimes from the other. Thus instrumentalist analyses of scientific theory may be construed as assuming that the value of the theory lies in the uses to which it may be put in the construction of inferences, predictions, retrodictions and other intellectual foundations for technology. But, according to Plato, because knowledge of the Forms culminates in knowledge of the Good, there is an ultimate state of knowledge that has to be good for its own sake: knowledge of the Good is not to be sought for the sake of realising some other value. Similarly a Spinozist conception of knowledge — so far as it concerns what it regards as the most perfect form of knowledge — takes the intellectual love of knowledge to be part of the infinite love with which God loves himself and thus takes scientific knowledge to be good in itself. Or — in more modern terms — we might suppose that it is just this that would be done by a religion which worshipped science.

So it at least makes sense to think of scientific enquiry as an activity that can, or even should, be sometimes pursued for its own sake and has an autonomous value. Moreover, when it *should* be so pursued a realist analysis of scientific theory seems more appropriate than an instrumentalist one. For an instrumentalist conception of scientific theory implies the pursuit of science for other purposes than the achievement of knowledge for its own sake. And, coversely, it would then follow that the autonomous ethics of scientific enquiry favours realism rather than instrumentalism in the analysis of fundamental theory.

IV

We need to ask, therefore, how strong are the arguments for saying that science — i.e. the systematic pursuit of knowledge — has, or has not, intrinsic value.

One plausible argument for a negative conclusion here might proceed from the premiss that, if a particular line of scientific research cannot reasonably be expected to help to do anybody else any good — i.e. cannot reasonably be expected to help to supply food, cure illness, or satisfy some other important purpose — then the only recognizable motive for pursuing it must be that of selfish curiosity or personal fame. And to use the requisite resources of skilled personnel and expensive equipment in pursuit of such a selfish goal is to reduce the quantity of resources and personnel available for research that might actually do somebody else some good. So, though it may sound allright to treat science, like art, as having an intrinsic value, in fact this is equivalent to pampering the supporters of an immoral policy.

However, the above argument is weaker than it may appear. It assumes that, if science *ought* only to be pursued for the benefits that it may convey to suffering humanity, then this is the motive that will *actually* determine individual or collective decisions of science policy. But, if there are many potential scientists who are not so charitably disposed, it is possible that recruitment into scientific research might be poorer in quality, and weaker in quantity, if scientific research were not attributed intrinsic value. Perhaps our communities need a complex system of values in which some activities of a philanthropic character may only get sufficiently promoted because they are encouraged by the value attributed to activities of a non-philanthropic character. Or, to put it more sharply, unless we reject instrumentalism we may fail to attract some of the best scientists into our laboratories.

But does the argument run the other way also? Are there some potential scientists who will work in the laboratories only if they can believe that their work may solve practical problems? Surely yes. So it looks as though the analysis of appropriate scientific motivation will not support a one-sided position on the question about realism and instrumentalism.

V

Let us turn therefore from motivation to methodology. Let us just assume that the scientific enquiries which get funded, and which competent people enthuse about and are honoured for undertaking, are worthwhile irrespective of whether they are pursued for their own sake or for their technologically applicable consequences. Then presumably any methods or practices or philosophical analyses that promote these enqiries are superior to any that do not.

For example, it is often necessary to draw comparisons between the degree of evidential support for one proposed solution of a problem and that for another. Philosophers who reject the possibility of such inductive reasoning can allow no room for rational technology. Why else should we think it reasonable to accept that one particular medicine is safer, or more likely to be successful, than another in the treastment of a particular disease, unless there is inductively relevant evidence for thinking this? Anyone who seriously and consistently rejected the possiblity of induction would be obstructing the path of worthwhile enquiry. And the same would be true for any other feature of correct methodology. So for present purposes we need to ask whether there is a case for treating realism as such a feature. That is, does realism describe an operation — not induction but some other operation — that, on a correct conception of scientific procedure, every scientific theorist should strive to pursue? Or does instrumentalism describe some such operation? Or, alternatively, do realism and instrumentalism just offer differing ontological options? Are they just opposing analyses of what scientific theories assert, rather than opposing accounts of correct methodology?

Let us consider these different versions of what the realism-instrumentalism issue is all about, and let us take the ontological version first. What needs to be remarked, and is often overlooked, is that ontological realism cannot avoid imputing a metaphor to any new scientific theory to which it is applied. The theory is interpreted as implying the "existence" — in nature — of unobservable, particles, forces, or whatever. But in any non-theoretical assertions of existence the normal criterion for what is to be said to exist in nature is observability. Analogously Darwin (in *The Origin of Species*, 1859) adopted "selection" as a metaphorical term for nature's non-purposive process of singling out individuals with a better chance of survival, despite the fact that in its normal use the term "selection" implies conscious purpose. Of course, metaphors die and pass into the language. This is certainly what happened to Darwin's use of the term "selection". The word has long since enjoyed a dictionary entry in its biological sense as well as in its pre-Darwinian sense. But it is not at all clear how "existence" or "reality" — as a term associated with the analysis of scientific theories — can achieve this status because there are no obvious criteria for its application other than the requirement that it have the very implications that are cited by instrumentalists as the content of their rival analysis: the experimentally detectable properties of the waves, particles, forces, etc. In other words ontological realism and ontological instrumentalism seem to collapse into each other. The ontological version of the realism-instrumentalism issue dissolves the conflict.

VI

The issue comes alive again, however, if we regard it as a methodological one, not an ontological one. The realist is then claimimg that the supposition of a particular type of empirically undetectable entity guides the choice of one type of empirically detectable entity rather than another as having the properties that the theory should aim to predict. The instrumentalist, on the other hand, claims that he has no need of such a hypothesis: he can make his choice without such guidance. The realist asserts that the scientist's theoretical assumptions suggest the experiments by which he can test his theory: the instrumentalist asserts that the theory is no more than a set of possible experiments. And if you think that realism, thus conceived, reflects an important feature of scientific methodology, viz, the importance of developing a good basis for ongoing imaginativeness and originality on relevant issues, then that is at least one good reason to be a realist.

Someone may object at this point that there is an *ignoratio elenchi* here. The paper began by asking whether there are any *ethical*, as distinct from methodolgical reasons for being, or not being a scientific realist, and it has now moved away from that topic to discuss whether there are any *methodological* reasons for being or not being one.

However, if you have a general moral duty to follow the best methodological principles available and one such principle is the principle P, then normally (i.e. unless there is a conflicting principle that has priority in your situation) you must have a derivative moral duty to follow P. More specifically, if realism is part of the best scientific methodology and if your professional ethics requires you to follow the best scientific methodology, then you have a professional duty to be a realist. There is a parallel here in jurisprudence. If you have a general moral duty to conform to the law of your community, and you have a legal duty to file your tax-return each year before October 1, then you have a derivative moral duty to file your tax-return each year before October 1. Being a virtuous, law-abiding citizen means that you will conform to all applicable laws, and if you do not so conform you are doing something that is morally as well as legally wrong.

It may seem, therefore, that we have tracked down here one genuinely ethical reason for always being a realist in one's attitude to scientific theory. But there are at least two arguments for not being overmuch impressed by this achievement.

The first is that the same line of reasoning may be pursued in relation to any other correct methodological principle. So, disappointingly, there is nothing special here about scientific realism.

Secondly, and perhaps more interestingly, the attribution of a moral duty here may seem important only because it exploits an *a priori* connection. If you have a moral principle in virtue of which you ought to do act A, and if A is done if and only if B is done, then you have a moral principle in virtue of which you ought to do act B. But the ethical obligation to follow a correct methodological principle here is not inherent in the principle itself even though *a priori* derivable from it.

VII

Let me now sum up. I began (I) by posing the question whether there are any connections between epistemological and ethical issues about science. I then (II) pointed out that, while moral judgements in science normally involve applying some general moral principle to particular circumstances, no such application takes place in the assertion of scientific realism. Next I considered (III) what it would be like to consider science as an end in itself, as art is sometimes regarded, and (IV) how strong is the case for taking this view of scientific motivation instead of an instrumentalist one. I then (V-VI) went on to consider two different versions of what is at stake in the controversy between instrumentalism and realism, namely the ontological version and the methodological one, and pointed out that a scientist's professional ethics, by requiring him to adopt the best available methodology transforms each methodological principle into an ethical obligation.

The Queen's College
Oxford

DAINIAN FAN

RIGHTS AND AUTONOMY OF SCIENTISTS

Cases in the People's Republic of China

1.

In 1949, the Chinese Communist Party (CCP) under the leadership of Mao Zedong seized the state power of mainland China, and the People's Republic of China was founded. As a radical Marxist, Mao Zedong set strict demands on scientists, such as: (1) natural scientists should exert themselves to develop natural sciences to catch up with the advanced level of the world quickly, and to give service to the industrial, agricultural construction and the national defense; philosophers and social scientists should research and interpret history, economy, politics, culture and international affairs with dialectical and historical materialism, to propagandize and justify the Party and Government's policies, to serve the socialist revolution and construction, (2) intellectuals including scientists should remold their ideology, abandon the bourgeois individualism, establish the proletarian communist world outlook, devote themselves to others without any thought of self, serve wholeheartedly the workers, peasants and soldiers, (3) scientists should obey the leadership of the Party, be the docile instruments of the Party, combat liberalism; in their scientific work, they should integrate theory with practice, their research should be brought in line with the state plan of construction. Although the first Constitution of the People's Republic of China (passed in 1954) stipulated that people had the freedom of speech, press, assembly and association, but was actually not guaranteed to implement. This kind of significant unbalance between rights and duties of Chinese scientists brought about disastrous effects. I would like to show these by the following cases:

(1) *The abolition of the China's Science Society.*

China's Science Society was the first modern, comprehensive, multi-disciplinary, autonomous, civil, scientific community which was founded in 1915 and since had been very active and influential in China. In 1950, China's Science Society was forced to merge into the All-China Federation of Specialized Societies of Natural Sciences under the leadership of the CCP. After that, the finance, personnel affairs, and publications of scientific communities all were under the control by the Communist Party and government.[1]

(2) *The ban of sociology.*

[1] Ren Hongjun, "A brief introduction of the history of China's Science Society", *Wenshi Ziliao Xuanji*. Vol. 15, 1961.

M.L. Dalla Chiara et al. (eds.), Structures and Norms in Science, 155–166.

According to the orthodox viewpoint of the Chinese Communist Party, philosophy and social sciences were superstructures and had class character; only Marxist philosophy and social sciences were scientific and represented the interests of the proletarian and the people; on the other hand, modern western sociology was bourgeois pseudo science which served the capitalists and imperialists. So, since 1952, the teaching and research of sociology were banned. In 1957, a group of sociologists who appealed to the CCP and government to recover the teaching and research of sociology were condemned to be the rightists—the targets of the proletarian dictatorship; their claim was condemned as the crime of the restoration of bourgeois pseudo-science.[2]

(3) *The criticism and suppression of Morgan's genetics.*

According to the orthodox viewpoint of the Chinese Communist Party, the natural sciences were not superstructures and had no class character, but the world outlook and methodology impregnated in the modern western theories of natural sciences still had class character, which were idealistic and metaphysical, and thereby should be criticized. Therefore, the Chinese natural scientists should learn from the Soviet advanced sciences which were guided by dialectical materialism. So, in the early 1950's, Morgan's genetics was criticized as idealistic, reactionary pseudo-science and seriously suppressed. On the other side, Lysenkoism was looked upon as advanced Soviet science which Chinese biologists and agronomists were forced to accept. In 1954, Bao Wenkui (1916–), a research professor of Sichuan Institute of Agricultural Sciences was forced to stop his research on the breeding of alloctoploid triticale and his experimental field was destroyed.[3] In 1955, a brilliant botanist Hu Hsien-hsiu (1894–1968) published a book entitled *A Short Course on Plant Taxonomy* in which Lysenko's view about biological species was criticized. As a result, professor Hu was openly criticized as opposing the Soviet Union, opposing Chinese Communist Party's leadership to science. His book was taken from circulation and all the copies of this book were destroyed by burning.[4]

In 1956, Mao Zedong and the leaders of Chinese Communist Party became aware that the suppression to Morgan's genetics was not adequate. In August 1956, at the Genetics Symposium held in Qingdao, the labels such as "idealistic", "reactionary" etc. on Morgan's genetics were canceled. Morgan School of genetics was allowed to contend with the Michurin-Lysenko school equally. But in 1958 and 1960, criticism of the Morgan's genetics had reached high tide twice. Early in 1966, at the beginning of the Cultural Revolution, a brilliant geneticist C.C. Tan (1990–) was condemned by red guards as a walking dog of American Imperialists and was forced to crawl in the camps with his

[2]See Han Mingyi (ed.), *History of Sociology in China.* (Tianjin, Tianjin People's Press, 1987), pp. 169–174; Yuan Yayu & Zhan Yizhi (ed.), *Sociology: History, Theory, Methods.* (Chengdu, Sichuan University Press 1989), pp. 65–72.

[3]Yu Youbin, "Bao Wenkui: Green objective", *Journal of Dialectics of Nature.* 1979, No. 3, pp. 85–93.

[4]Xue Pangao, "Hu Hsien-hsiu event", not formally published.

face smeared by black ink. Tan's wife could not endure such an insult, and committed suicide.[5]

(4) *Some scientists' claims of autonomy were condemned as anti-socialist.*

In response to Chinese Communist Party's call of "helping the Party to rectify its incorrect styles", on 9 June 1957, five scientists —members of a democratic party, the China's Democratic League— published a document entitled *Some Opinions about China's Scientific System.* They posed the call "to protect scientists", strove for the autonomy and freedom of scientific research and rights of equality for scientists, and put forward some suggestions about how to improve the scientific system with goodwill.[6] However, unexpectedly, these "Opinions" were severely reprimanded as "An Anti-Socialist Programme of Science" by the president of the Chinese Academy of Sciences (CAS) Guo Moruo (1892–1978) at the National People's Congress held in July 1957.[7] After repeated criticism and struggle, two of the five scientists, Zeng Zhaolun (1899–1967) and Qian Weichang (1912–) were condemned to be Bourgeois Rightists, their posts as members of the Academic Division, Chinese Academy of Sciences, deputy minister of High Education, vice-president of Qinghua University etc. were all dismissed. Three other scientists were also severely criticized. After this Anti-Rightists Campaign, Chinese scientists dared not criticize the Communist Party again and kept quiet out of fear.

(5) *The nationwide campaign to kill sparrows was initiated regardless of zoologists' objections.*

In 1956, under the leadership of Mao Zedong in person, "The National Programme for Agricultural Development (1956–1967)" (Draft) was formulated. It stated that, "to start in 1956, within 5, 7 or 12 years, basically wipe out mice, sparrows, flies and mosquitoes everywhere as possible." In October 1956, on the 2nd National Congress of the Chinese Zoological Society, some zoologists claimed that sparrows were not purely harmful birds. Their opinions were not accepted. In 1958, in many urban and rural areas of mainland China, a big campaign to wipe out sparrows was initiated. Only in Shanghai, according to statistics, nearly 700 thousand sparrows (certainly among them there was no lack of birds of other species) were killed, and thereby the ecology of the whole country was greatly damaged. After that, supported by the Leading Party group of the Chinese Academy of Sciences, some biologists posed their opinions again, performed research projects on the harms and benefits of sparrows, and got some conclusive evidence to show that the benefits of sparrows were larger than their harm. After that, until March 1960, Mao Zedong agreed and said: "Don't kill sparrows any more, instead we should

[5]According to an interview with a student of professor Tan.

[6]See: Wang Yuan, *Hua Luogeng.* (Beijing, Kaiming Press, 1994), pp. 218–227.

[7]Guo Moruo, "Denounce an anti-socialist programme on science", *People's Daily*, July 6, 1957.

kill bedbugs." Fortunately, in this case, some scientists who spoke boldly and frankly were not subject to attack.[8]

(6) *Biologists were forced to till unbelievable high-yield fields.*

In 1958, the great leap-forward movement was started in mainland China. Under the influence of calls such as "How brave the man is, how many crops will be produced in the field", "the lowly are most intelligent", proneness to boasting and exaggeration and practicing fraud was in vogue for a time. Peasant experts of "getting bumper crops" raised the target of production of wheat to be more than 150 tons per hectare, and that of sweet potato to be more than 1500 tons per hectare through practicing fraud such as combining the products of many hectares into one hectare. And one famous scientist of aerodynamics Qian Xaesen (1911–) argued that, according to his calculation, theoretically such high yields are feasible because the fields absorb enough solar energy. His argument became one reason why Mao Zedong believed such high targets of production. But, on the other hand, a plant physiologist Luo Zonglao (1898–1978) expressed his suspicion about such targets tactfully. He was immediately criticized. At this moment, some cadres of the Communist Party forced the biologists of the institutes of the Chinese Academy of Sciences to compete with the peasants to till the high-yield field. If the scientists failed in competition, the institutes would be canceled. Under such threats, the biologists of the CAS in Beijing raised the targets of production of wheat to be 375–450 tons per hetacre and that of sweet potato to be 3000–3750 tons per hectare in 1959, but the peasants raised the target of production of wheat to be 750 tons per hectare. Biologists took the measures such as close planting, applying great amount of fertilizers, deep plough, artificial ventilation and illumination etc., but at last the production of wheat was only less than 7 tons per hectare. Fortunately, at the end of 1959, the proneness of practicing fraud was exposed, so the competition between the biologists and peasants was ended up with nothing definite.[9]

(7) *Criticism of Ma Yinchu's demography which called for family planning.*

In 1955, the president of Beijing University, the famous economist Ma Yinchu (1882–1982) found that the rate of growth of Chinese population was not 2.2%, but 3.0% through investigation for three years. Developing in such a way, owing to the large population and the lack of cultivated land, even the food supply would be a big problem. Therefore, Ma advocated the control of population actively. In February 1958, on the fifth session of the 1st National People's Congress, he expressed his claim of birth control. However, this time Mao Zedong was intoxicated with pleasure by the high yields of agriculture boasted by the peasants, and he said: "it is good that we have

[8]Xue Pangao, "A historical retrospect on wiping out and then rehabilitating sparrows", *Journal of Dialectics of Nature*, 1994, No. 3.

[9]Xue Pangao, "A recollection on tilling the so-called high-yield field by the Academic Division of Biology, CAS, in 1958–1959", not formally published.

a large population."[10] Hence, the criticism of Ma Yinchu's demography was started. But, in November 1959, Ma Yinchu published his article entitled "My Philosophical Thought and Economic Theory" and expressed that he would "hold firmly to the truth."[11] This initiated further attack and criticism from all sides, and his posts as president of Beijing University, member of the Standing Committee of the National People's Congress were dismissed. In 1973, China's population was rapidly approaching 900 million. In 1974, Mao Zedong repeated what Ma Yinchu had said: "The population must be under control" but this time China's population had broken through 900 million. Now, the population of China has broken through 1.2 billion which really brings on tremendous difficulties to the development of China later on.

(8) *The establishment of Academic Divisions of the Chinese Academy of Sciences and the suspension of their activities.*

The Chinese Academy of Sciences was founded on the base of the Academia Sinica and the Peiping Academy of Sciences in the Republic of China and became the national scientific center of the People's Republic of China under the leadeership of the Chinese Communist Party.

In 1953, a Delegation of Chinese Academy of Sciences visited the Soviet Union and made a report to the Central Committee of the Chinese Communist Party and the State Council suggesting to establish a system of academicians like the Soviet Academy of Sciences. But leaders of the Chinese Communist Party worried that the system of academicians would give too much autonomy to Scientists. Therefore, they only agreed to establish the system of academic divisions. The Chinese Academy of Sciences asked more than 600 scientists to recommend the members of academic divisions. After that, the Committee of the Academic Affairs of the Chines Academy of Sciences drew up the preliminary name list of the members of academic divisions. Then, after soliciting opinions from the Party Groups of the Chinese Communist Party of the Ministries of High Education, Health, and Industries, the list of names was reported to the Central Department of Propaganda of the Chinese Commuist Party. From June 1954 to May 1955, the process was repeated many times and at last the list of names was examined and determined by the Central Department of Propaganda of the Party, and was issued by the Premier in the form of an order of the State Council. Members of academic divisions had no right to elect the President of Chinese Academy of Sciences and the directors of the academic divisions. Selection of the members of academic divisions was not completely according to academic standards. Some brilliant scientists or academicians of the Academia Sinica elected in 1948 such as Hu Hsian Hsiu were not appointed as members of academic divisions because they did not conform to the so-called political standards. Some scientists who were members of the

[10]Xin Ping, "The ins and outs of the criticism of Ma Yinchu", *Yanhuang Chunqiu.* 1994, No. 10.

[11]Ma Yinchui, *Selected Papers on Economy.* Vol. II, (Beijing, Beijing University Press, 1981), pp. 296–357.

Chinese Communist Party were appointed in order to guarantee the leadership of the Party even they were low in academic achievement. Especially, most members of the Philosophy–Social-Sciences Division were members of the Communist Party and Marxists. In 1958, 11 members of Academic Divisions were dismissed because they were condemned as Rightists. According to the regulation, the assembly of academic divisions should be held once every two years, but, actually, after three assemblies held in 1955, 1957 and 1960 respectively, it was suspended. On the third Assembly of Academic Divisions, many model workers and peasants were invited to attend, because the Party leaders distrusted the members of academic divisions. The task of the 4th meeting of the Philosophy-Social-Sciences Division held in 1963 which was organized by the Central Department of Propaganda of the CCP was to mobilize the philosophers and social scientists to take part in the struggle against the so-called modern revisionism represented by Khrushchev. During the Cultural Revolution started in 1966, Academic Divisions were formally canceled.[12]

(9) *The Chinese Academy of Sciences during the cultural revolution.*

In the middle of 1966, Mao Zedong and the extreme leftists launched the so-called Great Proletarian Cultural Revolution. All the cultural fields including natural sciences were regarded as super-structures which should be controlled by the proletarian dictatorship. In this period, intellectuals were looked upon as bourgeois elements, leading scientists as bourgeois reactionary authorities, directors of scientific organizations as capitalist roaders, and young brilliant scientists as successors of revisionists. They all were criticized and struggled. As a result of the ten-years great calamity, the research institutes of the Chinese Academy of Sciences were dismembered, most research work were stopped, most scientists were cruelly persecuted and among them, 106 scientists and technicians (0.041% of the total) were persecuted to death. Most scientific, technical and administrative personnel were sent to the May Seventh Cadre Schools in the rural area and did heavy physical labor for years. So, the Chinese Academy of Sciences were seriously damaged. The rights of scientists including the basic human rights were almost totally lost.[13]

The above cases indicate that, from 1949 to 1976, scientists in mainland China lost not only the autonomy, but also some basic human rights. Why would Chinese scientists abandon their rights? There were reasons in many aspects:

(i) The cultural tradition of China put much emphasis upon the nation, country, collective and obligations, but looked down on the individual and rights. Many Chinese scientists would willingly sacrifice themselves for the independence, unification, wealth and power of the nation. Since 1949, many

---

[12] Li Zhenzhen, "The preparation and establishment of the Academic Divisions of the Chinese Academy of Sciences", *Journal of Dialectics of Nature*. 1992, No. 4, pp. 40–50.

[13] *Chinese Academy of Sciences*. (Beijing, Contemporary China Press, 1994), pp. 141–157.

scientists had placed their hope for a glorious future of China in the leadership of the Chinese Communist Party.

(ii) With respect to the economy, after the Chinese Communist Party seized the state power in 1949, all the universities and colleges, and the institutes of scientific research were owned by the state; and the state appropriations became the unique financial source of the scientists, therefore they had no independent, autonomous economic foundation.

(iii) With respect to organizations, the Communist Party put all the "democratic parties", scientific communities, units of education and scientific research under the leadership and supervision of the Party, scientists lost their own autonomous communities in rigorous meaning.

(iv) With respect to ideology, the Chinese Communist Party strictly controlled the medium, educated scientists repeatedly with Marxism-Leninism and Mao Zedong Thought, asked them to abandon the individualism and liberalism learned from the West, thereby many scientists lost the confidence to insist on their own opinions.

(v) The Communist Party regarded scientists as bourgeois intellectuals, as dissidents of the proletarian dictatorship. So, after a series of political movements and campaigns, under the pressure of the proletarian dictatorship, many intellectuals including scientists with independent thoughts and skeptical spirits were tragically attacked and punished. Therefore, most scientists had no choice but subjected themselves to the leadership of the Party and abandoned their autonomy and basic rights.

In 1956 and the early 1960's, the Chinese Communist Party had made certain concessions to the intellectuals and scientists, such as admitting them as a part of the working class. But, these concessions were partial and transient. In the end, it led to the final tragic result of the Cultural Revolution.

The Chinese Communist Party had persistently forced the Chinese people and intellectuals to believe in the Marxism- Leninism, to believe in the Party and chairman Mao. The Chinese Communist Party totally negated the values of China's traditional culture and Western culture. But, finally, the practices of the Cultural Revolution which brought about tremendous harm to the Chinese people, impelled people to awake, thereby led to the crisis of confidence and a vacuum of values.

2.

Since late 1970's, Deng Xiaoping has been the actual leader of the mainland China. As a pragmatist, in order to get the support and confidence from the intellectuals and scientists, Deng reaffirmed: intellectuals were a part of working class; natural sciences had no class-character; scientists had the freedom to do scientific research and the freedom of academic contention. Deng even held that the generalizations of advanced experiences of the Western economic and administrative management were sciences which could be applicable in the socialist society; thereby turning on the green light for the introduction of the Western philosophy and social sciences.

Deng also emphasized that science and technology were the primary productive force and the key to accomplish China's modernization. Under his influence, the Chinese Academy of Sciences, institutes, universities and colleges were resumed and developed, the system of academic divisions of the CAS was also recovered and was renamed as the system of academicians in 1994 but the autonomy of academicians was not increased.

Thanks to the reform and open-door policy, the market economy was developed. With respect to employment, international academic communication and application for financial support, scientists had more opportunities for choice.

But, as a Marxist and communist, Deng had to emphasize the adherence to the four basic principles (socialism, proletarian dictatorship, Marxism-Leninism and the leadership of the Communist Party), the opposition to the bourgeois liberalization, thereby seriously limited the freedom of research in the fields of philosophy and social sciences. In the fields of natural sciences, during the process of the reform of science and technology system, the leaders of the Communist Party and government thought scientific and technological results were commodities, thereby asked the institutes of scientific research to get money through selling scientific and technological results. The budgets of the institutes were greatly decreased, so the research of basic sciences faced great economic threat.

Although Deng emphasized the reform of the economic system of the People's Republic of China, the reform of the political system and the democratization of the politics were not moving forward. The autonomy and basic rights of scientists still were not guaranteed. And during the process of promoting the market economy, some misconduct in science increased. The following cases are examples:

(1) Peer review about research on the function of a Chinese medicinal herb Sheng Mai San was disturbed.

In 1977, a young researcher Mr. Liu of the Institute of Microbiology, CAS wrote a paper entitled "The Effect of Sheng Mai San to the Metabolism of Nucleic Acids in the Cardiac Muscle" which presented the claim that Sheng Mai San can increase the rate of synthesis of DNA in the cardiac muscle. The Academic Committee of the Institute discussed this paper and held that there were big mistakes in the paper and the author was guilty of misconduct. But Mr. Liu thought that he was suppressed by the Committee and lodged complaint to some superiors in the Communist Party. In 1978, according to the instruction of a party leader, the Chinese Academy of Sciences investigated this event and asked the Chinese Medical Society to organize a peer review with about 50 related experts as participants. After sufficient discussion, the conclusion of the peer review was that, Mr. Liu 's design of his experiment was inconsistent with his goal, the method of the experiment was rough and the conclusion of the paper was non-reliable. But Liu paid no attention to these peer reviews and once again lodged the complaint to the party leaders and won over the support from a famous writer Ms. Yang and brought great pressure to bear on the Institute of Microbiology and the Chinese Academy of Sciences until the end of 1980.[14]

(2) The academic criticism of M.C. Niu's work was stopped by the party group of the Chinese Academy of Sciences. An American Chinese biologist M.C. Niu cooperated in China with a Chinese biologist Tung in 1970's. Their work was published in the *Scientia Sinica* (1973, 1975, 1977, 1980) and presented the basic claim that species-specific morphological characters are transmitted by injection of nucleic acids into carp eggs. In November 1980, an American biologist Eric H. Davidson wrote a letter in which he criticized Niu's work. In February 3, 1982, the party group of the Chinese Academy of Sciences issued a document forbidding two affiliated Journal *Scientia Sinica* and *Journal of Dialectics of Nature* to publish Davidson's letter. Yet the president of the Chinese Academy of Sciences, Lu Jiaxi (1915–) and the Chief Editor of *Scientia Sinica* Yan Jichi (1900–) and the acting Chief editor of the *Journal of Dialectics of Nature* all had agreed to publish this letter.

(3) So-called artificial inducement of the legume nitrogen- fixing bacteria (azotobacter) on the non-legumes.

In 1981, without the review by related scientists, a worker Mr. Wang released news to the media that he had successfully transferred the legume azotobacter on the roots of the non-legumes and vegetables with the results of increasing production. In 1982, the State Science and Technology Commission (SSTC) convened the peer-review three times and negated Wang's work. Some experts even repeated Mr. Wang's experiment and got negative results. However, Wang got support from the leaders of the State Council and the State Planning Commission. In 1986, *People's Daily* reported and boasted of Wang's work with the title "the overture of a green revolution". In view of this, 29 agricultural scientists lodged a protest.[15] However, in 1991, *Guanming Daily* reported that Wang's biological azotobacter got good effects in the field experiments thereby was rewarded with the golden medal on the 3rd National Exhibition of the New Technological Products which was convened by the SSTC and Guanzhou municipal government.[16] But, in 1994, *Wenhui Daily* reported that some scientists pointed out, in recent years, certain new types of biological fertilizers were boasted as new and original products with-

[14]See: Shi Xiyuan, "'That Kind of Man' — Commenting on Yang Mo's reportorial literature 'This Kind of Man'", *Journal of Dialectics of Nature*. 1980, No. 4; "Forum: Who is to assess scientific results and talents?", *Journal of Dialectics of Nature*. 1981, No. 1.

[15]Guo Xingxian, "So-called 'artificial azotobactor' and the policy of letting a hundred schools contending", *Journal of Dialectics of Nature*. 1986, No. 5.

[16]*Guanming Daily*. December 1, 1991.

out rigorous scientific appraisal and caused serious disastrous consequences.[17]

(4) The debate on the Three Gorges Dam Project.

After Deng Xiaoping expressed support for the Three Gorges Dam Project in 1982, the State Council endorsed "in principle" the project. Then, more than one hundred delegates of the Chinese People's Political Consultative Conference (CPPCC) and members of democratic parties (many of them were scientists, engineers and economists) and some cadres questioned the project's financial and technical feasibility, demanding a postponement and requesting a cautious attitude toward the project.

In 1986, after Deng said that the dam project had more benefits than shortcomings, but it needed more studies, the feasibility studies of the project restarted and the Three Gorges Project Demonstration Leading Group dominated by the Ministry of Water Resources and Electric Power (MWREP) was formed. All the 11 members of the leading group were supporters of the project. The opponents of the project such as Li Rui (former deputy minister of electric power), Huang Wanli (a famous expert of water conservancy and silt), Qian Jiaju (an economist and delegate of the CPPCC) and many others all were expelled from the process of demonstration.

In 1988, the Leading Group released a report with the conclusion that " the Three Gorges Project is indispensable for our four modernization; it is technically feasible and economically reasonable. It is better to build it than not, more beneficial to do it early than late." In spite of that, all the 412 experts involved in demonstration were picked out and appointed by the leading group, there were still 9 of them who did not endorse the report and expressed their opposition, such as the ecologist Hou Xueyu (1912–1991, an academician of the CAS), the famous expert of silt Fang Zhongdai, the expert of hydroelectricity Qin Xiudian etc. But their opposition was totally neglected by the leading group. Under premier Li Peng's patronage, the leading group dominated the major mass media and vigorously propagandized the project. In such a situation, Tian Fang and Lin Fatang, two retired officials of the State Planning Commission edited two books on the macro-decision making of the Three Gorges Project which collected scores papers written by the experts opposing the project and were published in 1987 and 1989 respectively. Meanwhile, some journalists headed by Dai Qing visited the opponents of the project and published a book entitled *Yangtze. Yangtze.* Although these second voices were much weaker than the official media, they still made considerable impact on the public opinion and to a certain extent precipitated the 272 National People's Congress delegates and 10 CPPCC delegates' petitions against the project in 1989. Under the pressure of the voices of opponents, vice premier Yao Yilin on behalf of Premier Li Peng announced at a press

[17] *Wenhui Daily.* August 28, 1994.

conference that before the end of the Eighth 5-Year Plan (1991–1995), there would be no plan to start the project in large scale.

However, the moratorium on the project was ended in 1990. In 1991, the State Council's Three Gorges Examination Committee adopted the Leading Group's feasibility study report. The Leading Group posed that the period of the Eighth 5-Year Plan would be the best opportune moment to start the project and since then, a propaganda movement for starting the project soon developed on a large scale. On the other hand, only a couple of journals dared to publish the papers written by the opponents. In 1992, the National People's Congress session endorsed the Three Gorges Project with about one third opposing, abstaining or refusing to vote. In the history of the National People's Congress, so many negative vote and abstention were really unprecedented. In spite of the resolution passed on the 1992 National People's Congress session had said: " In the works from now on, the various difficulties and problems which might emerge should be sufficiently explored and attended to, things should be done cautiously and treated seriously, the opinions from various quarters should be listened to", many suggestions to postpone the project were neglected. At last, in 1994, the Three Gorges Project was formally started. But, many problems, such as the deposit of silt and cobble, shipping, environment and ecological problem, relocation of the people, investment, etc. all have not been adequately settled. Many opponents are still worrying about the disastrous effects of the project.[18]

From above cases 2 & 4, we can see that, in spite of the fact that Deng Xiaoping reaffirmed the policy of let a hundred schools of thought contend, actually, the implementation of this policy was seriously interrupted. From cases 1 & 3, we can see that some persons were guilty of misconduct in sciences usually freed themselves from the peer review and sought support directly from the Party or government leaders and the media. In cases 3 & 4, although formally there were a series of peer reviews or demonstrations by experts, sometimes the experts were controlled by the organizer of the peer review. Recently, along with the development of the market economic system and the commoditification of the technological results, bribe or quasi-bribe has occurred even in the peer reviews. Therefore, the purity and sacredness of the hall of science were polluted by fraud.

3.

So, in conclusion, we must seek the balance between the rights and obligations of scientists. You cannot ask scientists to be honest, to serve the nation and the people, without guaranteeing their basic rights and autonomy. In 1982,

[18]See: Li Rui's articles on the three Gorges project, *Journal of Dialectics of Nature*. 1993, No. 2, No. 3, No. 5; Tian Fang, "Discussing the Three Gorges Project with frustration", *Journal of Dialectics of Nature*. 1993, No. 1; and Tian Fang & Lin Fatang (ed.), *A Third Macro Decision-making of the Three Gorges Project*. Opinion-solicitation version, 1993.

moral norms of science were worked out by Chinese scientists and technologists in Beijing and Shanghai.[19] After that, some publications entitled "Ethics of Science" were successively published.[20] These norms and publications stipulated the social responsibilities and obligations of scientists and the norms in scientific work they should follow, but all neglected the rights of scientists and the autonomy of science. According to the experiences of Chinese scientists for 45 years, in order to uphold the ethics and ethos of science, the basic rights and autonomy of scientists must be respected and protected, otherwise the consequences will be evil and serious. In recent years, a number of Chinese scientists and intellectuals repeatedly appealed in various ways to the authorities to implement the reform of political system and guarantee the basic rights of citizens and scientists but got no positive response. Maybe we should anticipate to achieve this goal in the post-Deng era.

Institute of Policy and Management
Chinese Academy of Sciences
Beijing

[19] "Moral norms of science for scientists and technologists in Beijing", "Moral norms of scientists and technologists in Shanghai", in *Science and Morals.* (Chengdu, Sichuan Press of Education, 1984), pp. 196–197.

[20] See: Wang Yushu (ed.), *Ethics of Science.* (Nanjing, Press of Nanjing Institute of Technology, 1988); Yang Tongjing and Zhang Yeqing, *Science and Morals.* (Taiyuan, Sanxi Press of Education, 1992); etc.

IMRE HRONSZKY

ON ETHICS AND CHANGING EXPERTISE IN TECHNOLOGY ASSESSMENT

Recent technological changes necessitate a critical assessment from an ethical point of view. The task of this paper will be to inquire into the manner how ethical considerations find their way into technology assessment (TA). The presentation moves in two directions. Partly it deals with the changing professional responsibility for doing TA as an expert activity, partly it criticises some attempts that apply ethics in assessment of technologies.

The presentation firstly enumerates some of the most important reasons why and how technology has become a changing focal point of ethics. Secondly, it makes some remarks on TA as an expert activity and overviews its history to show the changes it has undergone from a scientistic-positivistic form toward a constructivistic, discoursive-participative one, which is now beginning to be realised. Consequences of this change for the responsibilities the experts have to acquire will be drawn. The third part applies some of the earlier conclusions, among other things, to challenge the views of N. Luhmann who denies any importance of ethics for risk assessment. The presentation also remarks on the ethics code for engineers, set up by the VDI (Association of the German Engineers). Lastly, the paper criticises a model developed by a group of researchers in Darmstadt to include ethics for TA.

1. THE GROWING CHALLENGE OF THE FORWARDING TECHNOLOGISATION OF HUMAN ACTIVITIES FOR ETHICAL CONSIDERATIONS

The ethical interest in technological issues has dramatically increased during the 20th century. The responsibility dilemmas that arose from the tensions of the engineering profession in respect to its loyalty relations were the classical point. Later, ethical discussions also turned to the problems of the abuse of technologies. Mostly, the war experiences forced thinking to handle the problem of technologies alien to humanistic purposes. By now, ethical reflections on technological change are becoming comprehensive: any phase of the lifecycle of technologies is included and ethics refers to a high complexity of human relations the technological system entails and a high complexity of human relations, through which it is embedded in society.

Let me briefly summarize the most important elements of this deep transformation. Perhaps the most striking is the accelerated devastation of nature by the recent type global technological system, that accomodates to short

M.L. Dalla Chiara et al. (eds.), Structures and Norms in Science, 167–188.

range economic interests, threatening the earth with a global catastrophe, already in the not-too-far future. Secondly, there is a rapidly growing tendency to invent and use technologies with high catastrophic potential. Further, there is an acceleration and a (promise of) a qualitative jump in the reflective application of technologisation to human beings in medical and psychological practice. Fourthly, there is a forwarding system-technocratic technologisation process of social relations, realised through the rapid development of the application of new information technologies. It brings the threat with it that will lead to new social organisational dependencies and to new undesirable domination forms, making the information technological revolution a source of emancipatorical and opposite tendencies as well. One of the most important issues, perhaps less in the focus of concerns, is the fifth point, that the process of the technologisation of human activities, as a whole, rapidly changes not only the structure and dynamics of these activities, but, mostly, also their senses and role in human life. These changes may not only make technically more effective the working of, but also reificate, human interpersonal relations. On the other hand, there are partial but essential social agents, like economic ones, who are both ready and constrained by recent working mechanisms of the economic system to generalise their partial interests by trying, if needed, in a process of alienation, to change humans into their means to realise their interests in technological growth by urging to develop "false needs". (Technological inventions always showed the capacity in history of stimulating new needs. New technologies lead to new needs in an accelerated manner nowadays and a bifurcating critical attitude, including a strong criticism, is appropriate to reflect this issue from an ethical point of view.)

Sixthly, new technologies often contribute to producing new inequalities, like disproportional dissemination of harmful effects or changing chances to work, etc.[1] Seventhly, the long term effects of modern technology increasingly affect later generations. Eighthly, in contrast to earlier beliefs that technologies are "in themselves" neutral, only their "application context" and misuse can make them evil, technologies are now proven to be ambivalent by their nature. Hence, not only their breakdowns and misuses but also their successes are inextricably connected to non-desirable effects. The ambivalence of any technology is in its nature to produce harms through successes. This is inscribed, in historically varying forms, and differently for different technologies, in the social anthropological "genetic code" of any technological action.[2]

[1]In this respect the important observation by U. Beck on the "egalitarian nature" of catastrophes surely must be "dialectically" corrected. Catastrophes might be of an "egalitarian nature", but because of the recent strong chances of producing and raising inequalities through the non-intended effects of new technologies without catastrophic potential, the pluralisation of focus as far as relevance groups in relation to a new technology are to be found is surely needed. (U. Beck: *Risikogesellschaft*, Campus, Frankfurt, 1986)

[2]This ambivalence of technology lastly originates from the reductivity included in any purposeful action and the fact that this necessarily leads to non-intended effects in the complex environment, as its response, where the action takes place. These non-intended

(That is why one has to speak of "ambivalence" instead of neutrality of technology.) Ninthly, it proves false that technologies realise special partial social values through their application context only, but their designing and construction process is an inscription of intended action into tools and never can be fully abstract and general.[3] Technologies are socially constructed and are intended or non-intended social actors already in their designing and construction phase. Their basic social dynamics is a repeated definition process, in space and time, in interaction with — changing — relevant groups and institutions of users. Hence, their social role is always a concrete partial one. And lastly, partly as a consequence of this earlier point, modern technology is the lengthened arm of politics.

Please, notice that ethics has its competence in technological issues because it refers to human actions that are realised in a technological form, through technologies and artifacts are to be taken into consideration as "frozen actions", as well. More than this, they determinate clusters of further action-potentials. Technologies are not simply things and tools to be excempted from action analysis, by reduction to their natural-material content. Forms, into which action purposes are solidified (that determine their intended usage, including the "normal ones" in a special society) as well as the nature of the materials utilised for technological artifacts have impact on the possible actions utilizing technological artifacts. Society has to accomodate to the materials, introduced into societal application, by social organizational forms, as well.[4]

In conclusion: As part of the overall and justified need for ethical reasoning over every important human action one also has to reason over human action in its technological form. In the measure technologisation realises new power, the responsibility for technological change must also grow. The responsibility requirement must extend to the designing process and the possible non-intended impacts of a technology and partly must change its nature. One can say that the global crisis of the recent form of industrial society with its leading model of the "homo economicus", the acceleration of the (extensive) growth of human consumption, constrains us to face the acceleration of the long run negative effects of the system as a whole, as well. We are also urged to exploration of those factors that are responsible for these negative effects by the construction mode of the technological artifacts themselves. The recent

effects can accumulate, making the non-intended effects the decisive ones, in the long run. A further element of the problem is that action of technologies is always mediated by actions of human beings and will be included into their purposeful actions, sometimes leading this way to quite different results than the expected ones.

[3]Technologies can not be purely made with universalistic interests for "mankind in general".

[4]The dangereousity of nuclear technology can be an example that the material nature of an artifact has some definite impact on the social conditions of its utilisability. Any normal utilisation of nuclear sources prescribes extreme caution. It normally leads to centralised surveyance.

growing "passion for undertaking risks" in industry, by utilising the enormly growing potential of technoscience, which moves technology into a main author of developing new social senses of action and the fact that nature may be "overcharged" by new technologies in a new manner,[5] not only by the inherited, "polluting ones", require an overall criticism, based on comprehensive philosophical principles. This criticism must be concretised down to a form that includes the empirically detailing investigations and, where possible, is on experimental base. In other words this criticism initiates, and should be based on, a different type of engineering. Both technology assessment and the ethical questions belonging to the investigations on the human sense of technology are but some responses to this recognition. A comprehensive ethical approach to technology as social action is the appropriate frame for this critical and regulative effort. It will be a topic of this presentation to inquire what sorts of the aforementioned relations and their ethical reflex are taken into account by technology assessment as a special empirical investigation.

Finishing with the introductory remarks about why new technologies are challenges for TA and ethics, let me shortly touch on how the perception of the need for ethics has primarily occured. It is a small wonder that moral reflections concerning recent technological change are overwhelmingly impregnated by deep concern that can be referred to in an abbreviated form as having been frightened by the emerging new technological potentials in perspective of the obvious irresponsibility of dealing with technology recently. This concern led to the requirement by H. Jonas in his well-known formulation for a "new humility".[6] The widespread irresponsible handling of the present technological power is more than enough occasion for this concerned attitude. And there is the further, equally important problem of justice, that, actually, harms mostly affect different people, strata, countries and regions unequally.

This concernedness is fully justified and important for orienting toward a better action. It is nevertheless evident that concernedness is ambivalent by its nature: it may lead to very important partisanship and to biased attitude too, to dogmatic presumptions, to bias put on the cognitive situation. To stop the development because, allegedly, would be obviously unexceptable (at least for some ethical approaches), became a strong sense-giving feeling. It seems that many philosophers only rationalised their fear when committed to a sort of, say, "moralistic fallacy" when they have done nothing else but evaluated the course of the recent technological change on fundamentalistically presumpted

[5]In the overall frame of the existence of nature as irreversible process nature can only be "overcharged" in the sense that evolution potentials, favourable for mankind, will be destroyed in the natural environment surrounding it, also in this anthropocentric meaning, except somebody accepts a teleological interpretation of self-evolution of nature.

[6]Referring to the present situation by this term of Jonas does not include that I would like to argue for any type of "new humility" when accepting the call for making a new level of responsibility conscious. I am ready to acknowledge the practical historical importance of Jonas' call without mainly agreeing with its possible contential implications. (Compare with Jonas, H.: *Das Prinzip Verantwortung*, Fankfurt a. M., 1979, Insel Verl.)

ethical theories, instead of expressing a readiness for mutual learning in the practical situation. This mutual learning could be reasoned for by referring to the fact that the meaning of ethical principles can partly be learned through their application to individual cases, only, and also by acknowledging that even the principles of ethics might be challenged to transformation by the emergence of new human situations through the development of new technologies.[7] Not to accept that also ethical theories should be kept open to epistemological criticism based on new facts realised by technology leads to dogmatism. Dogmatic behaviour tries to realise a timeless normative or value constraint in technological controversies, mostly based on considerations about eternal virtues and vices. The main problem with this for me is not that an ethically based prohibition against economic forces is only a weak force in practice, as many correctly state (meanwhile often forgetting about the possibility of a deeper change of value attitudes that can constrain the so called "pure economic forces" and provide for an important role in the long run of ethical criticism in this process), but the epistemological issue. That is that a normative dogmatism leads to some undiscutable prohibition, trying to realise a dictatorship in the name of ethics. With this a normativistic approach, based on fundamentalistically accepted ethical theories, is nothing but the weak counterpart of the reification tendencies (that take technologies as if they were nothing but things having a naturalistic fate) to be found in the activities of those who develop technologies in practice, and is in the same manner unjustified. These two tendencies produce together a false arena (in which it seems as if dogmatic will fought against the quasi-objective calculations) and in which nothing can be solved by reason.

Many ethicians are inclined to argue against the recent process of technological development as a whole. I do think that no technology criticism can really be credible and successful, if it can not give a balanced evaluation and positively account for the recent technological change as a strong promise for human development, too. New technologies presently contribute to the development of new human capacities in a revolutionary manner even when they also lead to earlier unknown immense problems concerning both the desirability of many of their purposes and the side-effects they may realise. Then some cautionary requirements are already built in new technologies or proved to be superfluous. The dramatic picture, as the route-metaphor, that recent technological development is schizophrenic (owing it to its recent type of social construction) is perhaps more realistic than the tragic views, overwhelmingly found in ethical reflections on technology. A balanced effort to combine work on what can be called a "prospective ethic for fashioning the future" together with an "ethic of preservation" seems to be more appropri-

[7]It might already be obvious that I have in mind the responsibility for developing a clever discussion to be continuously done by the different actors in the arena of developing technologies and not an ethical judgement put on technology.

ate to the recent situation and more advisable then the recent basic trend in ethics for technology.[8]

Ethics can lead to commands, prohibitions set dogmatically "despite and against calculations". Some sort of ethical commitments prescribe this attitude. Other ethical commitments allow that they are confronted and negotiated by calculating them. A most promising approach for ethical discussion of technological issues can be based on showing that some ethics is inextricably webbed in any economic and technological calculation. The real practice of the so called "pure calculations", in contrast to its rationalistic reconstructions, is inextricably social value implemented and realises some normative attitude toward social issues. Because some sort of ethics is inextricably included into calculations (not only on the acceptability of technologies but ethics serves as a background in the calculations of their so called pure functioning too, realising some ethical commitment through cognitive methodological choices to be necessarily made), the rational discussion of these values too, is, theoretically, inescapable. Developing alternative calculations, based on different value systems, different normative preconditions can help to hold the ethical discussion of the technology issues within a comparable and rationally controllable field. Calculations realise value orientations and interests, practical orientations through and inside the calculations. Calculations enlighten the conditions of realisation and the measure of realisability of some normative attitude.[9] This mutual exposure to discutability requests that most ethically put norms will also be "calculated", that means they will be included into a whole system of calculations and this way rationally confronted with the other available ethical alternatives. This does not necessarily lead to any nihilism, just means that ethical considerations based on different values all may have chance to get accepted and (partial) justification can be given to a lot of alternatives. This also gives back my opinion that mostly consequentialist considerations have a chance today to get through and practically accepted even when their justification may raise immense problems.[10]

[8]The terms are from the article entitled On the Assessment of Genetic Technology: Reaching Ethical Judgements in the Light of Modern Technology in: Science and Engineering Ethics, v.1., n.1, Jan. 1995., p. 22–31, and may have a similar meaning to what I would like to give them.

[9]These normative attitudes may be reificated, by hiding the ethical constituent of the so-called objective results, they may be "black-boxed", or "naturalised". But this is nothing but giving the legitimatory force of being "natural" to some accepted values , and, of course, with this the force, the false appearance of "necessity" of the offered solution. Moral is inextricably present on every level in designing and realising of any technology.

[10]There are very serious ethical arguments that values as qualities can not be included into calculations, without loosing their specificity, and that value of life expressed in money, or even harms and benefits simply compared as quantities, or giving a measure for acceptable pollution is unethical as such. On the other hand, the real practice certainly requires a sort of practical compromise and it is arguable that ethics should find some way of giving some sort of critical justification to the real practice and not condemn it as such.

2. SOME PRELIMINARY REMARKS ON WHAT TA RECENTLY IS AND ON WHAT IT IS NOT.

TA shares many issues with the philosophy (and ethics) of technology but it is of different, empirical nature. In its original form TA is a special expert activity, an analytical and evaluative tool for policy making. To internalise possible consequences of technologies TA was born from an extension and specification of the general responsibility for social action. It was born to anticipate and evaluate the likely effects of new or changed technologies, with emphasis on their unintended ones, even when they are indirect or/and delayed. Let me give you a brief classification. In its narrowest realisation, TA takes the investigated technology as given and tries to explore its possible side effects, it is technology oriented. Several times, TA activities realise a different, a broader effort. Dealing then with technology choice, TA only takes the social goal as prefixed: the question is then about the comparative appropriateness of different possible technological solutions to a prefixed goal if the possible side effects of the different possible technological solutions also will be compared. From this type of activity one can go over to criticism of the social mechanisms and goals, orienting technological change. One can ask, for example, whether it is good if recent market mechanisms with their growth orientation determine the possible technological trends and hinder the development of alternatives. It is obvious that the flexibility of the reflection and with this the social critical potential (but not necessarily the the truth content of the reflection) also raises from the first type, to technology oriented TA, toward the broader one, the criticism of the social mechanisms that form the technologies. There is one more possibility to widen the object of investigation by going to the level of philosophical anthropology and ask about the overall characteristics of technology in human society. There is some sort of division of activities among philosophers and TA experts. Because this demarcation, which is misleadingly identified as the "natural" difference of a philosophical and an empirical research, is constructed, bearing in this case political implications as well, and not given by nature, the question may arise: what sort of demarcationism is advisable in our recent situation and what sort of cooperation, the abolition of demarcationism should be urged by developing research topics in which empirical and philosophical constituents are unified in a new manner.

A short warning should be made. This warning concerns the alleged epistemological purpose of TA activities as forecasting, as many still characterise the essence of TA by an alleged capacity required originally for burocratic planing purposes in a positivistically conceptualised TA. Please notice that TA today is not a forecasting effort to be measured by the exactness of its score. Instead, it tries to reach a so called orientation knowledge, with an unceasably hypothetical character. This will be realised by showing possible effects and evaluating them according to recent value systems. These are models of possible issues and their task is to indicate danger situations on the basis of recent values. From the point of view of political science TA attempts to provide evidential resource for (the different participants of) the technological controversy, within the technology policy arena. It is a recent recognition, in congruence with some new knowledge in the philosophy of science and "sociology of objectivity", that the different participants of the TA arena not only need different parts of a common possible pool of evidential resource, but that they regularly are inclined to different representations of the "same state of affairs", to be able to get their own values realised in the policy controversies over technology. This calls for the critical participation of the philosophy of science in these debates as well as for ethics to regulate the controversy.

There is a tension between the importance of TA and the frequent inexactness and hypothetical characteristics of the TA results. This inexactness and the paradoxicality of the "early warning" attitude, both inscribed into the epistemological "nature" of this activity, make the relation of TA to political authorities even more unbalanced than politics has to any other type of scientific advisorship. Politics can much more arbitrarily let serve itself with TA and move between scientism and decisionism even more undeterminatedly than it does in relation to other advisorships. Further, those social factors, like "credibility constraints", which work as social constraints for any cognitive expertise, may quite strongly effect the TA. This is because the experts have to seek for social acknowledgment among their reference groups too, not only among the TA public. In addition to this, any TA activity can only be made by a team as a microsociety, with all the sociological and psychological characteristics that define what sort of integrated knowledge will be arrived at as the valid one, according to the measures of the "microsociety of experts", which is institutionalised and reinstitutionalised in social negotiations.

TA is a problem-oriented research. One consequence of this is that it depends on the "employer(s)"' needs, and the money available. These characteristics may strongly effect not only the scopes of investigation but the results, too. As problem oriented research, TA is in permanent time and tool shortage. These constrain for further reductions of curiosity and may not be evident for all those who rest on the results. All these characteristics but show that the responsibility question at this expertise is more complex and more immediate than in many other types. To accept this one has to think of the delegation of confidence as basic condition to doing expertise.

Modern TA should serve as an evidential resource for the "parties" involved in the practice of technology politics and policy making. Remember that, in opposition to the usual economic forecasting (that is based on an overwhelming consensus concerning its basic assumptions or is conceptualised so by typical economics), TA was born from a communication divide. Inspite of all the weakening to be found in this divide by now it preserved a good amount of its original background. That is why, I shall later argue, TA is to be based on the acknowledgement that these parties need their own cognitive resource, and not only a "normatively neutralised" one, if TA wants to progress toward serving both a rational and democratic management of technological controversies. Therefore, the TA arena reasonably has to wear more openly its political nature, its mediating role by systematic comparison of possible technology policy consequences of committing to different, often conflicting social values. With this, that its cognitive activity is inextricably bounded to conflicting social values the most important reason is mentioned why TA is morally sensible: it is some sort of factually chequed "world view", which is naturally different for different parties. Its mere existence offends some interests, even in its most neutralised form.

This political nature of TA is overdetermined by a special situation. Even when it can be disappointing for many philosophers of science, inclined to work in a purified world, doing TA is not only a deep and inextricable involvement in social value controversy issues, as mentioned above, but more than that, mostly it is done in a playing field that is "tilted" from the very beginning, to use the term of Kr. Shrader-Frechette.[11] It is "tilted" in the sense that constrains to form biased opinions for one party (an offence of that type of value neutrality requirement that is to be shared by anybody who wants to engage in a rational conversation) are factually present (even when not omnipresent) in the practical field — before — any TA begins to work.[12] Hence TA, being a special approach to looking for the possible negative side-effects, contains some critical attitude, even in its most neutral form,[13] and in a fuller form, an advocatory attitude toward those who can loose with the introduction of a new technology or a new product. This advocatory attitude on behalf of those who are "on the other side", can be a structural precondition not only for letting all the opinions heard, a basic requirement for democracy, but also for providing a real epistemological framework for objectivity. (How commitment and partisanship may realise or exclude objectivity is another question impossible to be dealt with here.)

3. ON THE HISTORICAL EVOLUTION OF TECHNOLOGY ASSESSMENT

Two periods will be differentiated in the history of technology assessment, according to different social and state political perceptions of the problems

[11]Kr. Shrader-Frechette: An apologia for activism, in: Fr. Harré and P. Hartel (eds): *Ethics and Environmental Policy (Theory Meets Practice)*, The Univ. of Georgia Press, Athens and London, 1994, 178–194.

[12]And, more than that, these opinions are propagated as if they were objective. It is like this simply, because, mostly, the producers are already rather strongly committed to some technological solutions for their purposes before TA begins to work. That is why they are inclined to presume these solutions as "appropriate to everybody". This inclination simply belongs to their recent social role if they commit to it.

[13]The practically realised paradox that some sort of neutralised TA can be utilised as rhetorics against deeper approaches can not be dealt with in detail in this presentation.

of technological change. The classical conceptualisation from the late 1960s to the early '70s led to an institutionalisation of TA as a previously absent expertise, as a scientific advisorship to policy management, for political decision-makers. Presently, time is pregnant with a new, partly more complex, partly changed attitude toward doing TA.

It is important to remember that technology criticism became a mass movement in the USA in the middle of the 60s for several reasons. Institutionalised politics perceived this disturbance in the "Lebenswelt" (in the "lifeworld") as strong enough to challenge the political system to channel it by creating some new technology policies. Politics identified a lack of scientific knowledge for developing appropriate policy actions to restore the consensus. This way a restricted political perception of the problems led to the institutionalisation of TA. The evil was identified simply as the incorrect managing of the growth problem and the lack of knowledge for rationalisation of decision making. With this politicians met the attitude of scientists and engineers sharing their inclination to and bias for expert knowledge as the decisive missing link. What could have been seen in a different perspective as genuine political problem of mediating different interests, and the problem of the basic assumptions of modern industrial society was tried to fully immunise by this reduction. A technology assessment that concentrated on technology oriented "early warning" of the effects by experts was set. Technologies were identified to be cured, not the technological system with its underlying "homo economicus" assumption, which includes these technologies with their deficiencies. This evaluation of technologies was to be tied on, or even reduced to the authority of science (and engineers) allegedly leading to the objective, hence to neutral knowledge of the effects.[14] In a double process of "normative neutralisation" a demarcation was first realised to keep the "extremities" outside the problem definition, then the problem so formulated was given over to the authority and methods of scientific expertise to work out the valid knowledge about it.[15] One can say that a repeated "normative neutralisation" of cognition was made once again, this time not for science as a whole as in the 17th century, but for an expertise as a policy relevant advisorship.[16] Taking into account that the orientation of looking at the non-intended effects was born from some sort of basic social concern, this "normative neutralisation" provided for a new institutionalised "authority" to define the "real problem" and that was ready to keep the limits

[14]It is typical that most TA teams are set together from scientists and engineers (and economists) even today. Classical TA is inclining this way more to a technology than to a policy research and to a genuine mediating role for social conflicts induced by use or introduction of technologies.

[15]The institutionalisation of TA, the setting of the OTA, as it is well known, had a special condition of equally serving for both parties in the US congress. This puts a special, additional sort of neutrality for survival that worked until 1995.

[16]It is very likely that this "normative neutralisation", together with the above mentioned one, was the conditio sine qua non for getting institutionalised by inclusion into the state-mechanism for TA.

in criticism. Nevertheless, this authority, TA as institutionalised policy advisorship to the state, was heavily attacked by the industry, which identified it as an obstacle to competitivity. But the classical TA, through its (important but) moderate criticism, behaved loyally to the industrial society on the global level, by realising a special type of "not meddling with" (remember the famous words of the Royal Society) the basic assumptions of technological growth in industrial society.[17]

Within its limits classical TA worked out important scientific knowledge of consequences of setting new technologies. Because in an age of technoscience several times even the perception not only the identification of many possible problematic "side-effects" needs expertise, TA did a very important job. But it worked out this knowledge within the reductive framework mentioned above, and mystified it by the same step. This mystification was partly based on the neutralisation processes by classical TA and helped to discredit any radical criticism of the basic assumptions of the industrial society, as mentioned above. On the other hand, it was partly based on a scientistic interpretation of everyday life in relation to TA. TA, based on science, was expected to work out "the real problem behind people's fear". The determination of "the real problem" and of being "objective" (as equal to being non-biased) fall in the hand of experts, educated in science and engineering, as the last authorities. As the consequence, everything, resisting to their type of objectification was to be seen as "purely subjective". TA began to realise a double activity. It looked for that knowledge part of non-intended effects, that was to be identified by science and engineering and simultaneously followed and realised a political decisionism over the codified and authorized type of knowledge of the underlying events.

Concerning the ethos, classical TA committed itself to a reificationistic scheme, by redefining value impregnated questions into ones set by science. The possibility of the reduction of the original life world context of the problems was in the baggage which science willingly accepted from politics. In trying to reduce the reality of life world issues to their natural and engineering science content, actually to a special type of it, classical TA not only scientificated the field, but, by developing false consciousness, did a good deal of ideological job, as well. All this made the efforts of classical TA rather ambivalent. As the best follower of rationality approaches of its time, classical TA was science believer and produced a science of the non-intended effects for policy makers with burocratic intentions. Its ethos served with behavioural norms to explore the "objective content" of the (possible) controversies and nothing else. This behaviour included that classical TA experts concentrated

[17]Radical criticism soon pointed to this conservativism, characterising classical TA as setting a "technological superfix". (Compare the term "superfix" with the term "technological fix", introduced by A. Weinberg in 1972, and to which it is an answer.) I think the only chance for TA to get institutionalised was to make this, perhaps basic, opportunism, the nature of which was not recognised by its, several times exagerated, early practitioners.

on the refusal of obviously biased opinions (for their lacking of, perhaps, any scientific knowledge of the issues). They were proud of their standing on the basis of scientific facts (as ultimate realities, not being aware of their constructive participation in the production of the treasure of facts) and concentrated on the moral command of not speaking unscientifically about the possible effects of technologies. As another part of this attitude, accepting the full division of roles, classical TA institutionalised an ethos of not meddling with politics even in the form of policy suggestions. The classical TA ethos was a role ethos to realise objectivity as neutrality, and did not share anything with moral requirements, except those needed to realise the job as if it were a purely neutral, fact exploring cognitive activity for "enlightening" the politicians. Its moral was believed to be based on the interest of the "common as such" in a technocratically-burocratically conceived world of industrial society. Ethical duties resulted from this role understanding toward the public as the duty of their "enlightenment".

By now, an emerging type of TA goes to the opposite direction. It may be seen as a result of a 20-year long social learning process that first of all includes political learning (among others: how possible social conflicts over technologies should be dealt with, and that there are "parties" with equal right to get heard, even in technology political issues, and get access to the knowledge resource they need in the political arena). This political learning extends to a recognition and acceptance of the importance of public participation. The social learning process also includes an understanding that the "early warning" capacity is very restricted for structural reasons. It also entails some recognition that scientific knowledge and the functioning of any technology is of social constructive nature, which might help both to corrige, up to a possible measure, the structural limits of "early warning" by extending the TA investigations into an accompanying process, and to take into account partial value commitments. The new ideal model for TA is to be seen as processual, constructive and discursive-participative, as a conflict-oriented mutual social learning process, based on both scientific expertise and its public control through participation in technological controversies.[18]

Even role ethos is to be put differently for a participative, processual and social constructivistically understood TA activity. TA is impregnated by social values, just as any other "resource" in the political arena, only in a different form. If one does not want to make a contradiction of the knowledge to be worked out and the role ethos, it has to be openly committed and, perhaps, activistic. A continuous ethical reflection and self-reflection is indispensable to clear the repeated value choices in the whole process of investigation. (It is very difficult to realise this reflection and self-reflection on a purely neutral

[18]The real practice of the institutionalisation of TA recently moves in three directions. The recent dissolution of OTA in the USA directly shows how sensibly the desirability of any TA activity can hang on political changes. Traditional TA continues to be realised in many other countries and some begin with participatory TA.

base, quasi from outside.) It is also important that any commitment should be combined with full openness to criticism to preserve the discursive nature of the investigation as a precondition to the rationality of the process of technological controversy. This simply means that, while being committed to develop appropriate knowledge resource for some participants of the technological controversy, the expert (and the party served by him) has to preserve the openness by continuously engaging in a process of criticism developed by parties not sharing his (partisan) commitments. On the other side, an ethical commitment to some sort of partisanship may help to deeply work out a cognitive perspective and knowledge his/her commitment requests and possibly being neglected or denied by other parties.[19]

But it seems reasonable that a conditionless requirement of being continuously open for criticism does not include that an identical knowledge resource for everybody must be arrived at in the end. Instead, a clear understanding of the epistemologically unsolvable knowledge differences (resulting from all the cognitive-methodological decisions, impregnated by social values on the epistemic field as results of different value approaches) and, hence, the better understanding of what is to be solved through cognition and what through value negotiation, through a real political skill of making compromises, is only to be required. This way, a cognitive controversy can lead to a "rational dissensus" in the cognitive field. A technology assessor, through its partisanal cognitive activity, might help to clear the real cognitive and value orientation differences in the technological controversy, help to develop a clear view of the controversy and so that of the negotiation process, and contribute with this to the practical-political "closure" of the negotiation processes. They can help the "parties" to try to generalise their value and cognitive perspectives and experience the real practical limits of these generalisations which lead to a consensus of "second order" in cognition.

It seems that a "wholistic" engagement, by partial denying of a disjunctive division of labour between the expert and the politician, as the appropriate ethos for doing TA is to be required, too. If technology assessment means responsibility for non-intended effects (of technologies to be introduced), if it originates from a concerned attitude, then it is contradictory (as it is in the classical TA role, except in case somebody does believe in the invisible hand) that assessors do not embody any role-responsibility for the practical results, for the real fate of their advice.[20] To solve this contradiction, instead of real-

[19]Remember that any cognitive activity is only neutralised by a "normative neutralisation". More than that, the alleged neutralisation process can not be fully realised. For example, making a choice of shaping the inquiry can, through its possible differential impact on politics, ideology and juridical issues, provide for comparative advantage to some and disadvantage to other social groups just as any other methodological decision on the cognitive issue in practical fields.

[20]One can escape this conclusion if he/she insists on the basic belief and evaluates the situation so that just science can show what is "real" behind concernedness. In this case, the state of affairs of being concerned does not put any moral charge of responsibility for

ising a reduced, isolationistic responsibility to fulfilling the reduced cognitive role of "enlightening how things are", they should realise a more wholistic way and follow the practical fate of their cognitive results and preserve (or better to say: develop) some responsibility as technology assessors for how politicians should deal with the problems and what the practical results should be. A type of "lobbying" for possible political consequences of the knowledge of the issues they worked out should healthily counteract the possible and unavoidable arbitrariness, with which politicians handle their cognitive results. Equally important to lobbying is that this wholistic understanding of the role orients toward a special feed-back. This deals with the problem of what should be included into the cognitive activity of TA to be able to effect the politics.

The needed new type of expertise has different duties toward the public, the "lay people", too. A participatory type of TA expertise, aiming at democratic cognitive actions, should symmetrisize its position to a mutual criticism and forget about any paternalistically understood duty toward the public, even when a main bulk of knowledge is to be produced by expertise. The most important issue perhaps is that a non-scientistically, non-technocratically oriented TA expertise has to anchor its activity not only in the acknowledgement of the ontological reality (as it might be made through some "enlightenment"), but the ontological primacy of everyday life, of "Lebenswelt".[21] Being included on an existential basis into the practical issue at stake, can help more easily feel the inappropriateness for the practice, from their standpoint, of expert knowledge, that some expert approaches as such bring with them. This is the main advantage of existential involvement, I think, not that these people can provide the experts with the immediate knowledge of some fixed preconditions for their calculations, as it is referred to in some instrumentalistic reflections on the relation of expertise and existentially involved people in working on the cognitive problem. It is but superfluous to mention that this choice for the primacy of the Lebenswelt is first of all a practical one, is a "world view" and leads to political consequences.

The case of risk analysis clearly shows the need for an expertise (or expertises) developed for the interests of the "public(s)". Let exemplify this by reminding you that risk analysis as "science" identifies "objective risk", i.e. what risk "objectively is" as the multiplication of damage and probability. With this decision, put in the form of a definition, risk is made into a quantity having only an abstract, general quality. A critic can correctly remark that "risk in reality" is a real, concrete quality with quantities, a genuine

---

practical results of the TA activity but only calls the responsibility for investigating into the "objective state of affairs" of issues behind the concernedness. This controversy over the correct responsibility, too, can only lead to nothing else but a rational dissensus.

[21]This is the step that can help to critically accept the "non-scientific" approach of involved people, who are, by the way, sometimes, on the same scientific level as the experts themselves. To accept means to take their point of view to the base of investigation and to do criticism means to filter it through science, allowing for the possibility of the emerging need for enlarging or even basically changing the scientific approaches.

fact of (everyday) life. One can prettily discuss who is right about the "real" or "really objective" conception of risk but I think it does lead to nowhere, for there is not any criterion to make a neutral, "objective" evaluation of what "risk really is". More than that its existence would be a contradictio in adjecto. Risk definitions are "social warriors", actually. Defining risks, as scientists mostly do today, by characterising them as "objective risks," means (at least the, perhaps unconscious, rhetorical preparation of) a policy action in the robe of a cognitive activity. It is because everything else, not included into the category of "objective risk", must be seen as "subjective", and be handled in this way, based on psychological techniques, which help to accomodate to the "reality". To reduce risks to the natural science content of risks (and evaluations put on them as they simply were natural), taken it as their "objective" content, legitimates some sort of policy actions (and delegitimates others). It provides some sort of politicians and actors of the political arena, with the needed knowledge and legitimation base for some definite inclination. It also helps to include management of risks into the recently existing burocratic systems, through the needed type of formal generalisation. It is the important output, risks are somehow rationalised already. But, through the reductions realised in the conception of "objective risk", it serves the interests of those social institutions and parties that win by the definition.

Insisting on defining risk as a genuine element of Lebenswelt also means a value orientation and (at least the preparatory work for) a real policy action in the robe of cognitive activity. It means open participation as opposed to "objectivism", to the participation hidden behind inappropriate definitions of "objectivity". What is interesting about the recent issues is that the TA expertise uncovers either as a rather plain activity, reducing itself to those elements of the debates which are really common, shared by everybody and mostly have a form of natural scientific and engineering knowledge, or uncovers as a committed, perhaps as a partisan activity. In the latter case ethics has a lot to do with TA. Just that the cognitive activity called TA is not socially neutral means that the cui prodest question cannot be neglected within TA activities, either. Defining its own social role is a deeply moral issue for any TA assessor. To mention the case of risk assessment, efforts for bringing weighed values, which express many peoples' expectations in the formulas, changing utilitarianism for equity considerations as ethical basis, show how participatory efforts for embettering "exact calculations" needed for burocratic action, as something necessarily based on formal generalisations, can be realised. With this "calculatory step", the ethical content behind the formulas is changed. One further possibility is, perhaps to express by a calculus, the ratio of voluntariness and imposedness of risks, — to counteract the attitude that can not differentiate between voluntarily accepted and imposed risks, for they should be seen(!) as non-differing ones, if measured on the reductive basis of "objective risk" — and combine this step with the "objective measure" of risk.

Allow me just one more general remark. What to do, how to be reasonable in a pluralised world, having no supreme authority to turn to in ethical issues of technological development? I think the answer to be given has no specificity for technological controversies. It is that a clear understanding of the initial cognitive dissensus, beside the possible final consensus in the arena and a possible practical tolerance are only the ways to be pursued when the issues will be given over to political negotiations. If not any set of commonly acceptable general ethical norms can be used for achieving concrete deductions then it is reasonable that clever debaters try to accept a pluralised practice as justified one, as far as it is possible. (Of course, even tolerable limits may be very different for different participants of the controversies.)

4. SOME REMARKS ON RECENT ATTEMPTS TO INCLUDE ETHICS INTO TA

One could say that the faculties have begun to quarrel after 200 years once again. There is a group of philosophers who argue for including ethics in the designing activity of technological artifacts to prevent making mistakes (who argue for the ethical regulation of developing technologies instead of doing classical empirical TA that is unceasingly limited to a high measure to some sort of post festum activities) and there are TA specialists who doubt and refuse these claims. To solve the problems that arose from the social structural constraints of classical TA (that of concentrating on the effects of a given technology), summarises a TA specialist, Fr. Gloede, some develop ethics as a normativistic concept instead of analyses based on concrete, factually bound investigations, as TA does.[22] He makes with this an important point: intentions in the research and development phase necessarily go through the filtering of many different factors until a technology is realised and so the intended effects, if they can be forseen at all, might be quite different from the effects of a realised technology. With this he rightly argues against the claim of "more ethics" for research and development as it should be done for changing the focus of criticism from the application context to the designing one and not only for a balanced and very moderate change. It is questionable, however, whether the stand-point of his adversaries is simply an ethical normativism, as he judges. He also repeats the well-known point: there are very forceful regulation forms in economics and politics and any ethics as a normative regulator (simply juxtaposed to them?), fails against them. One can only wonder why he does not give more chance to the possibility of consumers' mass "disobedience", a possible mighty economic factor, which, as he correctly states, follows the longer way of changing value attitude. Otherwise, his objection is to be accepted: moral is a weak force — as long as it cannot

[22]In: Th. Petermann (ed): *Technikfolgen-Abschatzung als Technikforschung und Politikberatung*, Campus, Frankfurt/New York, 1991, 3O9 cc.

change into an economic and political one. Weak or strong, but this objection cannot deny that moral has an ontological place in human actions. To realise this role, I can only repeat what was said earlier, ethicians should get involved in the so called "pure" calculations and de facto show up and criticise the value orientations and moral commitments these calculations include, the moral commitments in economy and politics. (The fate of their action in the recent structure and dynamism of society is a different question.) Meanwhile, ethics is not allowed to forget about the principial fallibility of any of its theses, insofar they represent knowledge.

The moral attitude to risk issues depends on how somebody conceptualises risks in his/her basic belief about what risk for the present society is. It is easy now to differentiate between two basic attitudes. Let us call them paradigms. The one, that of U. Beck, identifies recent society as a potential "catastrophe society".[23] A special non-calculability and non-meliorability is connected to the potential catastrophic consequences of setting special types of new technologies. The practical consequence can be derived that a sort of cautionary action should be realised to minimize risks and a heuristics of fear is ethically advisable as the overall principle. N. Luhmann puts the issues diagonally differently.[24] He identifies the basic new feature in the present society as making action progressively decision mediated and hence contingent. Taken the importance of individual and social learning into the centre of his considerations, he, one could say, is basically pro-risks (as he identifies them). Risk is nothing but undertaking an uncertain action in the hope of a win. Some level of risk is simply needed for learning, and because risks are in principle calculable, a type of optimistic heuristics should be acquired as basic attitude toward risks. Beck puts an emphasis on risky action as something perhaps leading to catastrophes and on winners and loosers and claims to define risks by these characteristics. Luhmann neutralises the often fear and conceptualises risks as calculable issues of benefit and damage ratios. Interesting is that Beck and Luhmann speak about the same "state of affairs" differently, as the classical identification of a preparadigmatic period of cognition requires it. With this I do not want to say that the Kuhnian understanding of cognitive issues might be simply set to clear the possible future of the risk debates, but just the opposite. I want to make cautious. To overcome the opposition of the approaches, which partly, can be characterised as conflicting paradigms,

[23]Compare with the book referred already ftn. 1 and his subsequent books. Some irreversibility belongs to any sort of action. It seems important to work out a tenable concept of "catastrophe potential", which clearly differentiates between types of irreversibities and gives orientation knowledge of what should be seen as a "chatastrophic potential" for the different technologies and technological systems.

[24]Summary of Luhmann's view is in his article entitled Das Moral des Risiko und das Risiko der Moral, in: G. Bechmann (Hrsg): *Risiko und Gesellschaft*, Westdeutscher Verl., Opladen, 1993, otherwise in numerous books.

concerning their knowledge base, needs, first of all, an ontological solution on the level of societal existence.[25]

Luhmann rejects ethics for "system-sociological" reasons. According to his theory of progressing functional differentiation, moral, as a regulator, has no adequate functional place in modern society. "How should ethics help when the calculus can not, how does responsibility help when rationality doesn't?" — he asks.[26] "The problem is not in the responsibility for risks but in the probability of avoidable mistakes at the decision on risks" — he continues later. He systematically compares moral with calculation and with economic forces with some astonishing results, which just repeat some earlier mistakes from writers of the 18th and 19th century, and gives to them the justificatory force of his "system-sociological" theory: that something can be calculated or not and moral only comes on stage in the latter case, and that moral is a weak regulatory force, actually system alien in a functionally differentiated economy. Moral obviously exists for him only in dogmatic form, not as discursive partner in social process. In its alleged dogmatic functioning moral has a social psychological and ideological role: to put an end to the uncertainty of opinion-formation, a moral communication will be developed. Hence the evaluation: on complexity people should not react with ambiguity, he suggests. He has some problems with the preconditions of practical working of moral, too. From well-known psychological results, like the different evaluation of the same "objective risk" if it is self-imposed or from outside imposed, he concludes that basic conditions for moral commands, the consensus and the reciprocity requirements, do not empirically work. Hence no moral with generality claim can work in empirical reality of practice. While for him having a moral stand-point seems to be equal to dogmatism, somewhat different is his evaluation of ethics, the latter having a distance from its object, as he states. No allowance for a moral that stands on the basis of accepting a critical ethical surveyance, also realised as self-criticism, has a place in his system. When he requests the weighing of different risks in calculations, instead of ethical considerations, he is consequent to his system but as it is arguable these "calculations" inextricably include ethical premisses and hence always realise some ethical stand-point, even when the seeming naturalness of some premisses may cover this fact and gives calculations the appearance of pure descriptivity.

In this way Luhmann fully excludes moral and ethics from "calculations". In practice, engineers and economists, making the "calculations", and lawyers

[25]Luhmann correctly states that there is a basic ontological limit, set by society, to think about risk "rationally", as he defines rationality, just because of the differences concerning the relation of different groups to decision making. Those who identify risks, as he defines risk, are the decisionmakers and see in it only a challenge but those who define it as "damages", are those who are excluded from decisionmaking and this is a difference in existential relations, as he states it.

[26]p. 333. cc. in G. Bechmann op. cit.

together provide for a frame to guarantee the rationality in technological decisions. Because it is impossible and the least advisable both for effectivity of action and democratic political reasons that technological calculations should be regulated down to the minute details by law, one more, at least a"rest-place" is left for ethics: i.e. it has to fill the remaining regulative hole. Luhman, as far as I know, does not see this role, either.

Turning to the question whether and how ethical requirements for doing TA should explicitly be included into codes of ethics for engineers follows from the whole presentation that I am for developing these codes. The VDI-Guidelines provide a frame for ethics for engineers.[27] Its publication is important and there are many important points in it. Instead of enumerating these merits, I only have place here to describe the points with my reservations.

A general remark concerns the basic model of society. The Guidelines analyse the relation of technology to mankind only in a poor context, where one finds individuals and society — and nowhere social groups, regional clusters and so on. It begins with a narrow definition of TA as investigation of the state of a technology, instead of giving a broader one. Ethics is considered in relation to "general well-fare". There is no place in the Guidelines for a consideration that modern technology is the lengthened arm of politics, even when this is a very important point to be made, for example, in relation to the developing countries and their relation to technological development made in the name of "general well-fare."[28] Not relating the ethical issues to the dissected structure of the world (and the dissected structure in any country) gives a semblance as if industry would just work for the "general well-fare", if general ethical conditions were accepted.

Enumerating criteria to be set for a technology, the Guidelines give two distinct subsets of criteria. One of them is of technical and economic nature, the other one of ethical. Doing this demarcation, the Guideline is debt with accounting for the very important feedbacks among the subsets, between the so-called technical and ethical-social functionings, as they are called by the Guidelines. Just take functionability. Functionability is meaningful only in relation to special customers, with their special social values and expectations. To reduce functionability to pure engineering considerations might be appropriate for a rationalising ideology of engineering, but it covers then that

[27] VDI Vorentwurf, in: H. Lenk, G. Ropohl (eds): *Technik und Ethik,* Reclam, Stuttgart, 1987, 297 cc. I am using the text from this book, not the final version published in 1991. Any mistake originating from this step is mine.

[28] The considerations from political economy that genetic engineering probably increases the distance of developed and under-developed countries, or the differentiation of the competitive capacity between small and big farms, are well-known. Other political science considerations relate to the possible effects of information technology to political issues, especially to questions of democracy. There is no place to mention here either that some ethical criticism realises at the moment a voice of the "public opinion" against "the" industry, in important cases, such as the overall burning problem of nuclear waste incinerators among other issues.

the technical artifact does only function for relevance groups.[29] One of the important consequences is that reducing functionability to natural scientific criteria is not enough to account for any real case of a real artifact, and ethics should be put to evaluate the functionability of a technological artifact where the Guidelines is in silence.

Technological growth seems to be an axiomatically accepted value for the Guidelines, it is not even mentioned that some sort of "moderate ascetics" should be taken into consideration in engineering constructive work, as an alternative. To risk has only been given a quantitative characterisation. Instead, it would be helpful for an engineer to know that risk — at its basis — is a qualitative concept, (provided somebody accepts some considerations about the primacy of the lifeworld, as it was referred to above). Risk is characterised from a consequentialist ethics, only, even when some risks should be attacked on different ethical basis. Compare: "For an expected benefit new risks will be calculated, under conditions". Is it not ethical from a non purely consquentialist ethic's point of view to put some absolute limits to some risks? To put absolut limits to some technological actions might be non-acceptable for the authors, but perhaps it is arguable that engineers should know about a wide range of ethical approaches, if a Guideline wanted to give an orientation knowledge. Error toleration (Fehlerfreundlichkeit, Chr. Weizsacker and others) is also missing from the considerations of the Guidelines, as possible, ethically corroborated alternative for the recently sole, or at least overwhelming technology policy that accepts risks of whatever measure and quality of damage capacity, if they seem controllable on a "reasonable" level. One can find a utilitarian idea instead: big benefit makes big risk acceptable. There is no remark on distributive justice at all, either. Justice as basic value should perhaps be mentioned in any ethical codex for engineers just because some sort of distributive justice belongs to the vocabulary of any democratic society.

The discussion of the case of information/calculation inexactness is also missing. It is problematic, for more and more risky technologies suffer from the problem that extrapolative expert opinions prove rather inexact, and experts deeply disagree with each other, only on this base, as well. In this case, the producers' and the public's interest is regularly diagonally different. An ethical codex should raise the point that there is no neutral position for the engineer in this case and when the expert is partisanal for the "public", an ethical conservativism is to be suggested on moral considerations.

All the above said gives some ground for the suspect that an ethical codex, only based on ethical principles that take into account mankind as a whole, will give more an ethical basis for legitimation rhetorics for producers than a really

---

[29]Instead of beginning with a longer argumentation I only refer to the pioneering article of Bijker and Pynch on the "early byke-story" and the subsequent huge literature from then. The lesson is that functioning of any technological artifact is relevance group dependent, even when these might be huge ones and over historical periods. Compare: The Social Construction of Facts and Artifacts, *Social Studies of Science*, v.14.(1984), 399–441

acceptable common ground for ethically based discussion of the controversial issues of technology development, despite the non-deniable good will of the authors.

One last remark on the spirit of the Guidelines. It seems that its authors tried to provide engineers with a type of solidified knowledge of ethics. I think that any present Guideline must provide the engineers with some sort of feeling of uncertainty and insecurity, too. They should understand that the recently offered ethical guidelines are themselves under deep continuous criticism, and, therefore, are, for example, often nothing more than suggestions always to be criticised both on general level, and especially, in any concrete issue.

Let me turn to the last case to be dealt with in this presentation. A group of researchers from Darmstadt recently published a report on a remarkable project dealing with reaching ethical judgements in the light of modern technology, already referred to in my recent presentation.[30] They try do develop a comprehensive framework to put an ethical evaluative approach to high-tech issues to be able to arrive at judgements in concrete issues of technology oriented assessment, nevertheless, having the starting point at the comprehensive evaluation which involves judgement of the basic social problem that serves for technology choice. The task the group sets is very important in the present state of TA. But their model that combines the obligation of satisfying justified needs with responsibility for the consequences of realisation is to be criticised for several reasons. To define what should be accepted as obligation and responsibility, they simply refer to mankind, instead of society as a structured entity. More than that, the good for mankind is simply identified by Western values. It is small wonder that the concrete application of the model to the genetic engineering case of developing virusless sugarbeet is suspicious. The suspicion relates to two things. The one is concrete and is based on the argumentation mentioned already, that the application of genetic engineering for agriculture probably increases the distance between the rich and the poor countries, as, for example, medical applications of gen.-eng. also seem to do.[31] It is difficult to understand then, without any mediating argumentation, that the development of sugarbeet, which is probably very important for parts of Central-Europe, while sugar for the whole mankind, can get an ethical confirmation by referring to the increasing needs of mankind for more sugar. The other suspicious aspect is that the Darmstadt model claims generalisability. Does it refer to the issue with combining growing effectivity of new technolo-

[30] Cf. 8.

[31] Whether a technology can serve as a tool for more or less justice obviously depends on the social-economic conditions under which the technology may be set. But these structural conditions already work when the basic line of technological development is set, when technological choices are made, and so on. Different value orientations may be put into technology policies, into technological choices and lastly, into technologies themselves, but no technology, whithout committed value choices would serve special social values. Genetic engineering as "hope for the whole mankind" does not exist. As any technology genetic engineering, as well, can be directed toward different practical intentions.

gies with increase in production? If so, its application is problematic for a number of technologies, like energy production.

Having come to the end of the presentation there is a last word to say. This is to thank to the organizers that they included an ethics for science and technology section into the programme of the DLMPS conference. It may show that the importance of a unified investigation of technology issues from the viewpoints of ethics, philosophy and the methodology of science is acknowledged already.

Department of Theory of Science
Technical University of Budapest

MATTHIAS KAISER

UNPOPULAR AND PRECAUTIONARY SCIENCE: TOWARDS A NEW ETHICS OF SCIENCE

PERCEPTION OF THE PROBLEM

The ethics of science is a newly emerging field that currently is met with considerable interest by practising scientist from many disciplines. Many research foundations, as well as national or international professional associations express the view that great ethical challenges lie ahead, and need to be tackled competently[1]. This view originates not solely with the public concern over scientific honesty, scientific fraud and quality control of research. It is also largely influenced by the new types of ethical dilemmas that innovative areas like biotechnology or computer science create. Another cause for concern is the important role science is assigned in the attempt to change the global course towards sustainable development. It is by no means certain that science is ready to take on this important task. In fact, several critics doubt this strongly.

In spite of the growing interest in the ethics of science, professional philosophers of science have been amazingly reluctant to pursue this topic systematically[2]. In this paper I shall argue that science needs a revised and thoroughly reflected view on its implicit value stances. Philosophers do have an important task in contributing to this ethical and indeed methodological clarification. The ethics involved may be less of the grand-theory type; it is rather the bottom-up and down-to-earth type of practical ethics that may help us along the way. It will emerge in the following that the ethics of science is closely associated to questions of scientific methodology, as well as institutional and organisational matters of science policy.

I shall discuss two kind of problems, namely *unpopular science* and *precautionary science*, which I claim to be two sides of the same coin. I will illustrate the problems by reference to specific examples, and then try to summarize the results by some general characteristics. The analyses of the problems are taken to refer to general trends of modern science which I describe as inhibiting a larger and more comprehensive scientific responsibility towards public welfare

[1]There are many examples of national efforts to come to grips with these challenges. My own institution, NENT, may serve as one, another is the platform on science and ethics at the Rathenau Institute in the Netherlands. Also, the International Council of Scientific Unions has recently formed a working group on the ethical responsibility of scientists.

[2]There are, of course, notable exceptions. A recent and important contribution is Shrader-Frechette 1994.

M.L. Dalla Chiara et al. (eds.), Structures and Norms in Science, 189–206.

than currently realized. I shall conclude by outlining two tentative formulations of normative extensions of the celebrated Mertonian norms.

UNPOPULAR SCIENCE

Let me start by briefly discussing one of the classic examplars of scientific method which has figured prominently in Hempels textbook on the philosophy of science (Hempel, C.G., 1966), and has been cited in many other texts. The example is that of Ignaz Semmelweiss' discovery of the causal link between childbed-fever and "cadaveric matter". It is too wellknown to repeat the details of it. I assume that scholars of philosophy of science recall the main story in sufficient detail. It usually serves as an illustration of the "hypothetico-decuctive method" and of hypotheses-testing in general. The implicit assumption is that Semmelweiss produced a piece of good scientific knowledge in a reliable, inter-subjective and reproducable way.

However, the example also raises some important ethical queries which are not normally addressed. I shall start with the issue of unpopular scientific findings. In the next chapter I shall turn to the second theme implicit in Semmelweiss story.

Semmelweiss revisited, part I

It is wellknown that Semmelweiss' discovery was not immediately accepted by the medical community. In fact it took about 20 years before his contribution was widely recognized. Physicians had a hard time to swallow the fact that Semmelweiss' results blamed the practice of their own profession for the death of many young mothers. He blamed them in effect of being "murderers" as one contemporary Norwegian physician noted. Semmelweiss' insistence on the correctness of his results led to the isolation of Semmelweiss from his own peer group and contributed to his ensuing personal tragedy. Supposedly we have here a case of the production of good scientific knowledge and the lack of adequate scientific and public acceptance. In many ways it was the individual scientist, i.e. Semmelweiss, who paid the price for this, together with all those unfortunate women whose death could have been avoided had his results been accepted earlier. Truth turned out to be most unpopular, and it demanded great personal courage to defend it against the scientific peers for so long. Had it merely been an abstract theoretical finding with no immediate practical consequences, like the discovery of non-Euclidian geometry, or like Wegener's hypothesis of drifting continents, then one could at least understand why acceptance had been delayed for so long. In these cases, individual scientific contributions may be well ahead the theoretical and/or empirical development of their discipline, and acceptance has to await further supporting evidence. However, Semmelweiss' results had testable and immediate practical consequences of great import. They implied death or life for many patients.

Yet, no serious effort was undertaken to achieve an objective evaluation of Semmelweiss' results. Instead he was personally blamed for undermining the status of the profession.

The question arises to what extent it is the individual scientists' personal responsibility to stand up and defend his or her unpopular insights, even when such an attitude may have damaging effects on career and private life. Is it not the responsibility of the scientific community as a whole to make sure that scientific dispute and personal career are clearly separated? Does institutionalised science provide enough leeway to secure that unpopular findings may be voiced without negative repercussions for the individual scientist or the group of scientists involved?

One is tempted to reply that, of course, this episode was most unfortunate, but it is, after all, only indicative of the workings of medical science one and a half century ago. Present day science is supposedly very different, and the machinery of Big Science with its international journals and rigorous testing of claims would have guaranteed a different outcome today. But is this really so? Are we perhaps over-optimistic? I want to present some evidence that we may indeed be a bit too confident.

Unpopular science today: Traffic safety research as example

Let me provide an instructive example from the work of the NENT committee in Norway. The case is taken to be similar in several important respects to other reported incidents of unpopular science, as e.g. studies which result in the finding of some beneficiary effects of the common culprits of public health policy, namely nicotine and alcohol. Our example here is traffic safety. In Norway, research on traffic safety and economic aspects of public transport is usually conducted by one independent institute in Oslo. The researchers working there are all university trained, and come from various relevant disciplines like economy, psychology, engineering science etc. Aside from a small base fund, the institute is dependent on means mainly coming from the ministry for public transport and the traffic directorate. Thus it often has to evaluate measures which were put in place by the client who orders and pays the study. Sometimes considerable work has gone into the realisation of these measures, and large interests, e.g. professional or industrial interests, are tied to them. There often exists a clear perception of what constitutes a 'favourable' or a 'less favourable' outcome of these studies. For instance, the institute conducted a study on the measurable safety-effects of longer traffic education, some time after supplementary courses on traffic safety (with regard to driving at night and driving in wintertime) were made mandatory for all new driver-licenses. The expectation was that one would find a decrease in accident involvement for those who have undergone such additional training when compared to older data concerning those drivers who got their license with a much shorter training. However, the result was the opposite: drivers who had gone through the longer education were significantly more often involved in traffic accidents during the first months of driving than drivers with the old traffic education. In other words, longer traffic education turned out to reduce traffic safety. The results were surprising even to the researchers. They assumed that this might have something to do with the psychology involved. The long training might have induced over-confidence in own abilities in young drivers, and thus have led to more risky driving. There followed a long discussion at the institute about whether or not to publish these results. The conflict with the political bodies was predetermined, and they felt the clear threat that these results may backfire in terms of reduced funds and debate about their credibility. After long deliberation they finally decided to publish the results, but not inside Norway. They went to an international journal which accepted the study for publication. After a while the news sieved through to the Norwegian media. There was some mention of this study in the press, but neither the institute nor the ministry wanted apparently to engage in a public debate about the issue. It remained an anecdotal note in the press. The study was, however, picked up by research institutes in other countries. As far as I know, they all ended up with similar results, and apparently these studies raised similar concerns. I witnessed one such discussion in Australian media last year. Also there the researchers kept a surprisingly low profile, while public agencies and other 'experts' maintained that the study was of little value for the evaluation of traffic education, and had no influence on training programs. The Norwegian researchers reported to our committee that they found the more or less open mechanisms of repression less frightening, than their own "internalizing" of the client's expectations. They simply tried to avoid to engage in research where similar conflicts might be expected. It goes without saying that Norwegian traffic education remained largely uneffected by these results.

General characteristics

Other examples of unpopular sciences could certainly be provided here. They may be more numerous than we initially think, once we start to look out for them. They often have several of the following characteristics:

- The unpopular results turn up as a surprise or a by-product of other research.
- The unpopularity emerges as the tension between the socio-political implications of the study and the expressed or implicit policy goals of the public agency funding the research.
- Publication of the results is debated at length, and sometimes guided by the wish of the research team or the funding agency to achieve only limited visibility.

- There is little incentive to discuss the unpopular results in public media, and if they are discussed at all, the practical implications are played down.
- The researchers involved in the unpopular study fear being stigmatized as favoring extreme policy views, and of being punished indirectly by refusal of further funding.
- The researchers feel a lack of support from the larger scientific community which may note their study with interest, but does not supply sufficient support in conflicts with the funding agencies.
- Scientists in fields with a potential of unpopular results and great dependencies on politically constrained funding agencies may internalize expectations of their providers to such an extent that they avoid asking certain questions or pursuing certain hypotheses.

I put it to you that this situation is a challenge to the ethics of science, comparable to the Semmelweiss example. The scientific enterprise is designed around the central norms of scientific freedom and autonomy in order to avoid the systematic bias towards wishful thinking. The Mertonian norms of disinterestedness and organized scepticism presume mechanisms of communication and dissemination which are not hampered by political expediency. This is not to say that science should be blind concerning the socio-political dimensions of their findings. Rather the opposite: we need to recognize the important role science has for policy formation, and it is in general a good thing to let questions of relevance, including socio-political relevance, guide one's design of research. But precisely in order to maintain and strengthen its role as independent and, as far as this is possible, objective provider of options for and critic of policy, science must stick to and defend its privilege of occasional strong dissent. This privilege applies then not only to a strong and large community of unaminous scientists, but trickles down equally on each individual scientist and research group. Even though their findings may be at odds with those of the majority of their peers, and even though there may be reason to object to their findings, it is vitally important to the progress of knowledge that they are subjected to scientific scrutiny and debate. In times of increasing mutual dependencies and interchange between science, industry and politics it is more important than ever to make sure that science does not degenerate into a mere instrument of power and/or popular movements.

Asking for great personal courage to make one's research public and visible, should not be part of the scientific ethos. How serious this problem of unpopular science is, will in part depend on how widespread it is. It also depends on the importance of the subject, and the institutional variety in the field. In certain settings, for instance in small countries with only few professional possibilities in a field, and with few major funding agencies, the consequences may turn out to be very serious indeed. In the end it is the public interest which may suffer, e.g. when misguided policies are not corrected in time.

PRECAUTIONARY SCIENCE

Let me now turn to the other side of the same coin. I want to discuss not the lack of acceptance of good knowledge and unpopular science, but the dangers that lie in accepting knowledge claims too soon and building up an expert-culture of scientists advising policy and development. The dominant themes here are uncertainty and ignorance, and the paths to a sustainable development. As an introduction to the problems to be discussed let me go back to the Semmelweiss example.

Semmelweiss revisited, part II

It is accepted wisdom that washing one's hands in a solution of chlorinated lime as Semmelweiss did, destroys infectious material adhering to the hands and thus reduces the chances of transmitting infections to patients. Aside from killing harmful germs, the procedure is harmless. Yet, how could Semmelweiss have known this? Assume for the sake of the argument that Semmelweiss' results had an immediate impact on reform in public health care. Assume further that Semmelweiss had not used a solution of chlorinated lime, but had cleaned his hands in some toxic or carcinogenic fluid, not known to be toxic or carcinogenic at the time. Assume that this substance had the property of effectively killing off the germs, but had no other immediate health effect which one could observe. Let us suppose it did have a mortal low-dose long-term effect, perhaps only after several years when it finally produced cancer among health personell and patients. Semmelweiss' tests would have given the same results.

The tests performed by Semmelweiss could not provide any indication that this possibility of serious harm was excluded. In fact, had Semmelweiss' recipe been widely used immediately after his discovery, and the health effects had appeared years later, one would most probably have had a hard time to find a causal link between the two phenomena. Assumedly Semmelweiss conducted an exemplary piece of scientific research, and yet the results may very well have turned out to be disastrous for public health policy, if circumstances had been slightly different.

In terms of methodological appraisal, our variation of the historical example raises the very interesting problems of significance, complex phenomena, delayed cause-effect relationships, multiple causation, and perhaps several others as well.

In terms of the ethics of science, the above story highlights the problems of uncertainty and ignorance in science for policy which are closely related to the above methodological issues. It shows that some science does have immediate practical consequences without being able to exhaustively evaluate and describe the outcome. Note that this problem is not identical to the fallibility of science, but is additional to it. Fallibilism would only mean that the test was illusive in the sense that the reduction of childbed fever was due to some other factor than the washing of the hands. Uncertainty and ignorance is also different to the celebrated underdetermination of theory by data which leaves room for alternative theoretical explanations. Instead, uncertainty and ignorance relate to our inability to foresee developments taking place in complex environments. Given that only well-informed advice (to policy) is responsible advice, the existence of basic uncertainties and lack of knowledge in science raises the question when and under which circumstances the conscious release of scientific findings with obvious policy implications to policy makers is morally justified. In other words, what kind of consequences does the realisation of the policy dimensions of research have for the transmission of research results? If we feel that the responsibility resting on the scientists' shoulders means a double checking of the results for possible flaws and greater caution in the formulation of consequences, we must specify further what greater caution and double checking means in practice. In still other words, would we feel that Semmelweiss' propagating and recommending his results to the public was a moral act or would we tend to condemn him for carelessness?

These questions relate to what has been termed "culpable ignorance", discussed in Aristotle and Aquinas, and more recently taken up by Ian Hacking (Hacking 1986). Basically it asserts that an evaluation of our ignorance of a certain matter is an essential moral ingredient when acting in spite of this ignorance. There is a call for an intensified effort to find out more about the subject, or to act carefully, if at all, if we cannot overcome the ignorance. If we do not, then we may be to blame for the unexpected harms our action may cause. It is not so much the ignorance in itself that is blameworthy, but the acting in a state of ignorance, even if, by mere luck, no harm occurs. It follows an obligation to try to find out more about the issue, before turning to actions. And then one may still want to abstain from action, in fear that the ignorance may be so significant that possible harm cannot reasonably be excluded.

Uncertainty and ignorance in present day science

Let me now turn to present day science for illustration of my claim that this is indeed a pressing issue for the ethics of science.

Failures to assess the environment

Present day environmental policy is largely dependent on scientific research. National law in most industrialised countries, such as in the EU and the USA, provides for an important role of science in development projects that might harm the environment. Environmental law in these countries requires the production of so-called *Environmental Impact Assessments* (EIAs) before any larger industrial development, like e.g. the opening of a new oil-field in the

Norwegian Sea or the building of the Sydney Harbour Tunnel (cf. Sharon Beder's contribution to this volume), is allowed to proceed. EIAs are often rather big and comprehensive research efforts, resulting in lengthy documents, and following agreed-upon methodological guidelines. In some cases and in some countries, more detailed quantitative risk assessments or risk-cost-benefit assessments (RCGAs) are performed. It is the expressed aim of all these studies to assess the likely environmental effects of the planned development on a scientific basis. Consequently, a large number of scientists is involved in these activities, large funds are used, and important and consequential decisions are taken on the basis of these assessments[3]. Usually the laws requiring these assessments have been in place since the 1970's or the early 1980's. Consequently a larger number of EIA's has been performed since. The main business of these EIA's is what philosphers of science describe as *prediction*.

Predictions, of course, have the disturbing property of being correct or not being correct. If they are of the quantitative kind, they may turn out to be more or less correct. Therefore, the quality of EIA's can usually be assessed some time after the development has taken place. Given the widespread use of these EIA's and the importance they have for environmental policy, one would expect that EIA's were routinely evaluated and assessed on the basis of their success. One would expect some improving feedback on the methodology of EIA's given their testable consequences.

In reality very little effort has gone into the improvement of EIA's or the assessment of their reliability. The very first quantitative analysis of the accuracy of environmental impact predictions for any country was done in Australia by Ralph C. Buckley and published in AMBIO in 1991 (Buckley 1991). I am not aware of any later studies on a national basis either. The basis for the study were 800 to 1000 EIA's, containing many thousand predictions. However, adequate monitoring data were available only for 3% of them, leaving some 200 major and 175 subsiduary predictions. Excluding hierarchical sets of predictions and other considerations left a total of 68 actual predictions which were compared to actual impacts. The study showed that "our predictions are less than 50% accurate on average and over two orders of magnitude out on occasion" (ibid., p.161), with actual impacts ranging from $0.05x$ to $> 30x$ prediction. "Predictions where actual impacts proved more severe than expected are on average less accurate (33% ±6%) than those where they proved less severe (53% ±6%)" (ibid.), and the difference being significant at $p < 0.05$.

There are, however, some other studies comparing project-wise post-EIA audits of completed projects with the preceeding EIAs (Bassin 1986, Culhane et al. 1987, Culhane 1993). John Lemons and Donald Brown summarize their results as follows:

[3]Some critics, however, maintain that EIAs or RCBAs serve more as ex-post legitimization of decisions already taken instead of ex-ante preparation for decision-making. Whatever their true function, their important role with regard to large development projects either as guide to development or as provider of public acceptability cannot be denied.

> "(1) that most EISs [EIAs in our terminology] do not contain minimally satisfactory data or information on physiographic and biological impacts to enable postaudits to be conducted; (2) that predictive accuracy of EISs was low; (3) that the variance in forecast accuracy was significant and unsystematic; (4) that no noteworthy patterns, functions, or variables emerged that explained either accurate or inaccurate forecasts; and (5) that most EISs do not provide adequate scientific information to guide monitoring of critical impacts during project implementation or to use in ameliorating adverse impacts when necessary." (Lemons & Brown 1995, p. 29).

No one can claim that these results are a feather in the hat for science. In fact it seems to show that much of the trust put into science when regulating environmental policy is misplaced. In many cases, an educated guess or the considered judgement of an experienced practitioner might have done the same or even better job. This is, in fact, exactly what Brian Wynne argues in his study of the aftermath of the Tchernobyl accident for the sheep-farmers around Sellafield (cf. Wynne 1992). I shall come back to this point later.

Uncertainties and ignorance in the assessment of salmon fish farming.

Most practising scientist would readily admit that all science contains uncertainties. It is, however, one thing to assert this in general, and another to pinpoint these uncertainties in specific cases. I believe that the ethical relevance of uncertainty emerges most clearly only in the latter case. I shall therefore illustrate my point by reference to the assessment of salmon fish farming in Norway, a study performed by NENT two years ago.

During the past fifteen years Norway has propelled itself into the world's leading producer of Atlantic salmon (salmo salar). The technology which made this possible is quite recent, dating back to the early seventies. The development was followed by several assessments right from the start. In the beginning there was enthusiastic appraisal of the potentialities of salmon fish farming, which in turn stimulated investments. This was followed by partially incompetent regulations, misguided research policy, and an unexpected high growth rate in the industry. Money came rolling in. Then the development led to equally unexpected problems with fish diseases, the lowering of quality in the fish meat, and environmental problems related to waste from production and fish escapes. Public attitude turned against the industry, which now became the target of critical media reports. It led to bankrupcies and a major reorganisation of the industry, government regulation, training of professionals and the scientific research activities connected to it. The pattern of enthusiasm at the start, to fast growth, environmental problems and a problematic public image is not specific for this industry, but is rather typical of many such ventures.

The first thing to note with regard to the role of the scientific experts in this development, is their total failure to foresee any of the problems which turned out decisive since the mid-eighties. With hindsight, this seems surprising since many of the problems seem so obvious to many outsiders. For instance, some agriculturists felt that farming fish might be similar to farming chicken, and it could thus be expected that diseases may become the number one problem. Yet, agriculturists were no experts on fish or industrial development, and consequently their expertise had no weight. Therefore Norway failed to build up veterinary expertise on fish diseases until the diseases finally broke out on a large scale. Nobody expected that the industry was so vulnerable to sudden diseases which were not even known to the scientists. The resulting effect was the lack of primary measures to avoid these diseases, so a massive treatment with antibiotics started which at times outweighed the total human consumption of antibiotics of the Norwegian people.

The general problem behind this is, I think, best described as the systematic bias of expert cultures to focus on areas of own competence instead of the complexity of the system under consideration. It is a general source of partial ignorance which permeats scientific expert assessments, especially with regard to environmental issues. This is a kind of uncertainty which, as Wynne and Mayer say "encircle(s) the science rather than being encircled by it" (Wynne & Mayer 1993, 33).

Other uncertainties follow as part and parcel of scientific method. This is my second point. Consider e.g. the question of fish escapes: Salmon kept in large cages in the open sea tends to escape as soon as the possibilities arises. The possibility arises for instance in major storms, or during the transport of the cages, or simply by incompetent maintenance of the cages. This fact was also quite surprising to the experts involved in the early assessments. They did not think of the possibility that the salmon could escape. Well, happy the salmon that gets out, you may say. Some of the fish escapes into the open sea where it sometimes has the bad fortune of being caught by fishermen. Some other fish moves along the coast and then up some rivers or riverlets. There has been quite extensive research just to chart where the escaped salmon disappears to, and under what circumstances. This research is naturally beset with some major uncertainties. It is for instance quite a challenge first to set out tagged salmon into the open sea or a river (usually many thousands of them), and then to catch them again. Naturally only a certain percentage is caught again, if one is lucky up to some 3% or so which end up in the nets of the local fishermen. And then the results do not show a clear and uniform picture either. Sometimes the salmon stays at the release place or comes back to it. Sometimes the salmon follows the coastline or goes out into the open sea, and sometimes enters the rivers. The variation seems dependent on an impressive amount of parameters, like e.g. the age of the salmon (smolt), specifics of the particular salmon population, the season, local water quality, release at sea, from fish cage, or river, and so on. In effect after several years of intensive study, the scientists know very little about what escaped salmon does under varying circumstances, or why he does it. Yet, the evidence is that escaped salmon is all over the place, including the rivers. But what happens next? The rivers were once the domain of the wild salmon where spawning and breeding takes place. It is here that the salmon grows into the smolt that can tolerate salt water. Acid rain and pollution made many of these rivers inhabitable for the salmon which led to a dramatic and steady reduction of the number of salmon in the Norwegian rivers in the past decades. The wild salmon that remains in the few habitable rivers left is threatened by extinction because of human industry. Now farmed salmon enters these rivers, and another question arises. Farmed salmon is like any domesticised animal a product of several generations of breeding work with selected individuals. It lacks many of the wild salmon's natural resistance to environmental pressures, and is certainly genetically far removed from its wild relatives. There is also evidence that the genetic outfit of wild salmon is quite specific to the river where it grew up in and to where it returns. Yet there is also genetic variation inside a given river population, and scientists disagree about whether genetic variation is significantly greater between rivers and their specific populations, or between individuals of a population. But now the really great uncertainties enter: does the escaped salmon interbreed with the wild salmon; if so, do they produce progeny fit to survive, and what will the long term-effect of such crossings be for the river-populations? This is the point where scientists start either to speculate or to refer to theoretical models. Of course they interbreed, some scientists claim, why should they not. But since one lacks direct evidence of such offspring, many also assume that the progeny will not be fit to survive for long. The point is that there is no good method to get any data indicating the one or the other. It is difficult and uncertain enough to differentiate between wild and farmed salmon with the help of some morphological criteria, or traces of artifical fishfeed. Offspring of interbreeding could not be identified that way. One would perhaps need some genetic markers introduced into farmed salmon which would be released after several years, and then after some more years one would have to test a river population whether the marker was present in the genetic outfit. And in order to observe the overall effects on the whole population one would have to wait for several more years with repeated releases. There are simply many practical problems connected to this issue which make it hard to get any hard evidence. Scientists therefore make theoretical models. Yet all these models can do is vary some parameters and relate them to other parameters. What parameters to consider, how they vary in reality, and how they relate to other parameters, is a matter decided by abstract theory, e.g. population genetics. One ends up with an impressive number of theoretical possibilities, ranging from no difference at all to the total extinction of the wild salmon stock in a few generations. What is totally lost in these scenarios are the complexities of varying local conditions. What are e.g. the conditions for the salmon populations in some parts of South-West Norway where the risk of extinction emerges not only as possible result of interbreeding with farmed salmon, but as the sum of this factor together with risks from acid rain from England, pollution from cities, and near-by disease infected waters? Interference effects, cumulative effects, and deviations due to local conditions are lost in these models. They become part and parcel of methodological uncertainty.

In view of preserving biodiversity, the consequences of this are great. Do regulating agencies have to interfere in order to preserve the river populations and to act in a precautionary manner? Do we need to take active measures in order to reduce salmon escapes to a tolerable minimum? What measures? Governmental bodies turn to science when they want these questions answered. Scientists reply that there is as yet no hard evidence of serious threats to the environment or biodiversity, they will pursue this issue further, and come back to it when they know more.

A third point should also be raised in this connection. While the first one dealt with uncertainty that is best described as ignorance of parameters and causal relations which crop up as surprise factors for limited expert groups, and the second point dealt with uncertainties which are inherent to the scientific approach, the third point relates to uncertainties which science systematically leaves for others to formulate and consider. They are best described as value uncertainties.

Let us go back to fish farming. Fish farming in general raises the problem of the sustainable management of renewable resources, more specifically the management of ocean resources. In order to raise salmon, the fish must be fed. The fish feed is mostly made up of — well, other fish. More than half of total annual intake from the Norwegian fishing industry is utilised — in a second production cycle so to speak — as fishmeal that goes into the pellets fed to salmon. Herring and mackarel, themselves perfectly good nourishment for humans, when caught, end their lives more typically as fishmeal than as human consumption. Amazingly enough the end result pays better. As one expert put it: Fish farming is the transformation of cheap fish into expensive fish. Yet, we know by now that our fish populations in the oceans are threatened. Over-fishing has brought some species on the brink of extinction. We know at least that there just is not a lot more fish out there for us to catch. We are already now moving close to the limit of tolerance. Scientific estimates of what constitutes reasonable fishing quotas in our seas are beset with vast uncertainties. In spite of this, industrial development pushes towards more growth in fish farm production, including salmon production, which in turn means the production of more fish meal. This again means an increase of the catch or larger proportion for industrial use, unless other protein sources are utilised very soon.

Behind this issue lurks the question of global sustainable development. How are we to manage the fish resources in the oceans so as not to risk diminishing stocks through harvesting? And how are we to allocate the fish which is caught, given the global differences in food and protein supply and the expected population increase? To what extent can we tolerate recycling these resources for the production of luxury goods for the rich countries? And how does this balance with agricultural protein production? These are questions which involve very basic value issues. None of these questions can be answered without sufficient scientific information. Yet rarely any scientists addresses them in a systematic manner. I suspect that the reason for this is that they cannot be answered without explicitly incorporating values, an area which is not regarded as science proper. Another reason is that scientists may feel a closer alliance to, and indeed dependence of, national policies than to global development. The result is that these issues remain largely unaddressed by all.

General characteristics

There are a number of characteristics and tendencies in present day science which the example shows to be problematic. For all of them the problems of uncertainty and ignorance are central. They are in brief:

- Science is much more risk-aversive with regard to the truth of its claims, than with regard to the failure to state a truth. This is the wellknown asymmetry between statistical type I and type II errors. Kristin Shrader-Frechette has in her writings repeatedly pointed out the need to re-evaluate this bias (cf. e.g. Shrader-Frechette 1994). A contributed paper to this symposium by Lene Buhl-Mortensen from Bergen provided further arguments from marine biology in the same direction (cf. Conference abstracts). The value issue comes in when one realizes that this bias goes together with a favourable bias towards minimizing developer's risk versus the public risk[4]. The case can be made that a precautionary approach necessitates greater attention to type II risks. I shall not dwell on this particular point any longer, since it is discussed at length in the relevant literature (cf. also Gray 1990, Peterman & Gonigle 1992).
- Science has a tendency to disregard possible sources of knowledge which do not fit into its own format of empirical, quantifiable, and inter-subjective data acquisition. There is a clear hierarchy of knowledge sources inside science, and between different disciplines. Many philosophers have e.g. pointed out the alleged superiority of experimental quantitative data over observational field data. This may turn out to be decisive with regard to acceptance or rejection of theories dealing with very complex phenomena. Naomi Oreskes has in an important paper made this point with regard to the late success of the theory of continental drift (Oreskes 1988),

[4] I use here Kristin Shrader-Frechette's terms, which are a variation of Churchman's terms "producer risk" versus "consumer risk".

and I have added further observations to it with regard to the same example (Kaiser 1993). Knowledge sources from outside a narrowly defined field of science, as e.g. the experiences of agriculturists with intensive farming of domesticised animals, are often not even considered relevant.

- Science thrives on idealizations, abstractions, and neat simplifications (cf. e.g. Kaiser 1991, 1993, 1995a and b). They are the recipe to produce general phenomena and interesting theories. But as I pointed out in another paper, our scientific "theories are simply not designed to deal with complex singularities, but rather with simple generalities" (Kaiser 1995a, 195). The reason for this is the extent to which the phenomena that theories explain are a human construct. Philosophers like Nancy Cartwright, Ian Hacking, Jim Woodward and others have made similar points (cf. e.g. Woodward 1989, Woodward & Bogen 1988, Cartwright 1989, Hacking 1983). I claim that there is a fundamental epistemic difference between directing our interests towards general theories or, alternatively, directing them towards specific problem-situations. If our general high-level explanatory theories deal with phenomena which in turn are the result of elaborate abstraction and idealization processes, then the path from concrete data to theory may not be easily reversible. The epistemic difference between theoretical and applied science is accompanied by an ethically relevant difference of where the major responsibilities are and how they are to handle.
- A measure for the success of a scientific discipline is its degree of specialization. Thus science moves towards greater fragmentation among its practitioners. Boundaries arise which are hard to cross, and the more general or holistic an issue appears, the more difficult it is to place inside science. In an expert-driven society this leads to an increasing mismatch between expertise on the one hand and the themes of the issue for which expert advice is sought on the other. Often the issues are already framed in ways which reflect diverse interests. They are stated in language not adjusted to scientific representation and have to be 'translated' into scientific terms.
- Scientists are caught in institutional dependencies which inhibit alliances with less powerful social groups. The same mechanisms which dampen the production of unpopular science, account for the lack of critical reflection on the limits of their own activity and expertise. The danger for scientific autonomy lies in the dominant internalization of expectations other than public welfare and global sustainability.

My main point in this section is to argue that science for policy necessitates a moral obligation for the scientific community to address problems of uncertainty and ignorance in an explicit and transparent manner. In particular it needs to actively explore value-laden paths to sustainable development. One of the major principles of sustainable development is the so-called precautionary principle, endorsed by 178 nations in the Rio-Declaration (§15), and sometimes in other international agreements (like the Maastricht Treaty of the EU) or national law. Basically it says that the lack of scientific certainty should not be used as a reason to postpone measures to prevent environmental degradation. It has been noted by several people that this implies a shift in the legal burden of proof in environmental affairs, and means moving the science from end-of-pipe research to the possible sources for environmental trouble. Brian Wynne and Sue Mayer (1992) argue that a truly precautionary science would imply a much more radical reform of the scientific culture than most scientists are willing to recognize. What the more specific implications of this principle for science will be, has certainly to be worked out in greater detail. There is, as yet, no agreement on how a precautionary approach reflects on scientific methodology, model-building, and problem-approach. In particular, it is not clear how value-laden issues can be tackled in a responsible manner by scientific expertise. However, if the above example is in any significant sense typical for problems of our modern science, then I think it is decisive that science strives to contribute to and to realize a precautionary approach to our dealings with the world. That is why I think it appropriate to label this attempt "precautionary science".

CONCLUSION: A SUGGESTED EXTENSION OF MERTON'S NORMS OF SCIENCE

There are only few attempts to formulate the ethics of science. The most wellknown formulation goes back to Robert K. Merton (Merton 1942) and his four (or five) "KUDOS" norms:

kommunism
universalism
disinterestedness
organized scepticism
(originality).

The Mertonian norms have been criticised on various counts, but mainly because many people felt they do not adequatly describe scientific practice. Ian Mitroff, for instance, suggested therefore to supplement each norm with its counter-norm (Mitroff 1974). If we talk sociology, this may indeed be appropriate. But if our interests are of the normative kind, i.e. if we talk of the ethics of science, the Mertonian norms do have a very strong *prima facie* appeal.

I shall not discuss them in any detail here. I do, however, want to point out that the above discussion may be taken as evidence for the need to supplement these norms. While the Mertonian norms reflect demands on the individual scientists as member of the scientific community, they do not, I believe, adequately reflect the institutional responsibility of science towards either the individual scientist or society as a whole.

Unpopular science will become a real threat to the scientific spirit if the scientific community turns its back to those few who stand out with their unpopular knowledge claims. It is important to note that organized scepticism and disinterestedness may foster debate, scientific disagreement, or simply the judgment of irrelevancy. Yet this is very different from the solidarity that science ows to those whose research led them to pronounce unpopular views that do not fit into mainstream thinking. The more weight we put on our peer reviews, project fundings, and other measurable outputs as basis for scientific careers, the more we have to make certain to provide enough room for the occasional unpopular voice. I do not believe that the concept of scientific autonomy is strong enough to counterbalance the strong pressures scientists face today.

And similarly with precautionary science. If the apparent success of science for policy, i.e. its increasing importance, is not to hamper public welfare through unrecognised or unreflected uncertainties which in the long run undermine its credibility, then science has to institutionalize mechanisms to avoid systematic biases. What kind of appropriate mechanisms and innovative methods one should envisage, has to be explored in much greater detail. But some authors have already sketched some valuable suggestions. For instance, Funtowicz and Ravetz have developed the NUSAP scheme for expressing uncertainties in science for policy (Funtowicz & Ravetz 1990). This scheme has still many shortcomings, but it points in the right direction. Their ideas on post-normal science (Funtowicz & Ravetz 1993) contains the need for extended peer reviews, an idea that also Brian Wynne (Wynne 1992) seems to embrace. This is likewise a suggestion that precautionary science should take seriously and explore further. Shrader-Frechette, Buhl-Mortensen and others point to methodological remedies to avoid preference of statistical type II (not to reject a false null-hypotheses) over statistical type I (to reject a true null-hypothesis) errors. These measures have in common that they seek to reduce what has been termed "public-risk", instead of putting a primary weight on the avoidance of "developer-risk".

I suggest, therefore, that these views be expressed as explicit norms for the scientific enterprise. A very tentative formulation might be the following:

The Solidarity Norm: The scientific community demands of its members the expressed and active solidarity with those of their peers whose research produces results which are counter to, or deviant from mainstream science, or which are problematic because of their socio-political implications.

The Welfare Norm: Science serves public welfare as its ultimate utility through institutionalising precautionary mechanisms and approaches which recognise public risks as the primary causes of concern.

I conclude with the observation that though these norms may have been meaningful and helpful already in the days of Ignaz Semmelweiss, they are even more needed in our times where we have to counteract the destruction of our planet. If science is to be the ally in the shaping of a sustainable future, it has no choice but to embrace its ethical responsibilities with all their methodological and institutional consequences.

The National Committee for Research Ethics in Science and Technology
Oslo

REFERENCES

Bassin, N.J., 1986, Analysis of Selected Agencies' Post-Final Environmental Impact Statement Commitments. EMS-86-002. Environmental Management Support, Silver Springs, MD.

Bogen, J. & Woodward, J., 1988, "Saving the Phenomena", The Philosophical Review, vol. 97, pp. 303–352.

Buckley, R., 1991, "How accurate are environmental impact predictions?", Ambio, vol.20, no. 2–3, pp. 161–162.

Cartwright, N., 1989, Nature's Capacities and Their Measurement, Clarendon Press: Oxford

Culhane, P., 1993, "Post-EIS Environmental Auditing: A First Step to Making Rational Environmental Assessment a Reality", The Environmental Professional, vol. 15, pp. 66–75.

Culhane, P., H.P.Frieseman, & J.A.Beecher, 1987, Forecasts and Environmental Decision-making, Westview Press Social Impact Assessment Services, no. 14, Boulder, CO.

Funtowicz, S. & Ravetz, J., 1990, Uncertainty and Quality in Science for Policy, Kluwer Academic Publishers, Inc.: Dordrecht

Funtowicz, S. & Ravetz, J., 1993, "Science for the Post-Normal Age", Futures, September, pp. 739–755.

Gray, S., 1990, "Statistics and the Precautionary Principle", Marine Pollution Bulletin, vol. 21. no.4, pp. 174–176.

Hacking, I., 1983, Representing and Intervening, Cambridge University Press: Cambridge

Hacking, I., 1986, "Culpable Ignorance of Interference Effects", in: Douglas MacLean (ed.), Values at Risk, Rowman & Littlefield Publishers, Inc.: Savage, Maryland

Hempel, C.G., 1966, Philosophy of Natural Science, Prentice-Hall, Inc.: Englewood Cliffs, N.J.

Kaiser, M. 1991, "From Rocks to Graphs - The Shaping of Phenomena", Synthese, vol 89, pp. 111–133.

Kaiser, M., 1993, Aspekte des wissenschaftlichen Fortschritts, Verlag Peter Lang GmbH: Frankfurt a.M.

Kaiser, M. 1995a, "The Independence of Scientific Phenomena", in: Theories and Models in Scientific Processes, W. E.Herfel, W.Krajewski, I.Niiniluoto, R.Woicicki (eds.), Poznan Studies in the Philosophy of the Scienbces and the Humanities, vol. 44, Rodopi: Amsterdam, Atlanta, GA, pp.179–200.

Kaiser, M. 1995b, "Empirical versus Theoretical Progress", in: Realism and Anti-realism in the Philosophy of Science, R.S.Cohen, R.Hilpinen, Q.Renzong (eds.), Kluwer Academic Publishers: Dordrecht, pp. 171–193.

Lemons, J. & Brown, D.A., 1995, "The Role of Science in Sustainable Development and Environmental Protection Decisionmaking", in: Sustainable Development: Science, Ethics, and Public Policy, John Lemons and Donald A. Brown (eds.), Kluwer Academic Publishers: Dordrecht, Boston, London.

Merton, R.K., 1942, "The Normative Structure of Science", in: R.K. Merton, The Sociology of Science, Chicago 1973.

Mitroff, I.I., 1974, "Norms and Counter-Norms in a Select Group of the Apollo Moon Scientists: A Case Study of the Ambivalence of Scientists", American Sociological Review, vol. 39, pp. 579–595.

Oreskes, N., 1988, "The Rejection of Continental Drift", Historical Studies in the Physical and Biological Sciences, vol. 18, pp. 311–348.

Peterman, R.M. & M'Gonigle, M. 1992, "Statistical Power Analysis and the Precautionary Principle", Marine Pollution Bulletin, vol. 24, no. 5, pp. 231–234.

Shrader-Frechette, K. 1994, Ethics of Scientific Research, Rowman & Littlefield Publishers, Inc.: Lanham, Maryland & London, UK.

Woodward, J., 1989, "Data and Phenomena", Synthese, vol 79, pp. 393–472.

Wynne, B., 1992, "Uncertainty and Environmental Learning: Reconceiving Science and Policy in the Preventive Paradigm", Global Environmental Change, June, pp.111–127.

Wynne, B., & Mayer, S., 1993, "How Science Fails the Environment", New Scientist, 5 June, pp. 33–35.

NAOMI ORESKES

TESTING MODELS OF NATURAL SYSTEMS: CAN IT BE DONE?

1. INTRODUCTION

Testability is at the heart of virtually all philosophical conceptions of scientific theory. Verificationists, confirmationists, and falsificationists share the belief that scientific theories must be in some manner testable. The Kuhnian concept of a critical "anomaly" that undermines accepted standard concepts also implicitly invokes a notion of falsification—the anomaly stands out against a background of paradigmatic expectation, thus providing partial grounds for rejection of previously held views.[1] More recently, social constructionists have greatly challenged the determinative nature of empirical evidence, but recent constructionist accounts nevertheless allow an important role for experimental tests, even if placing the use and intepretation of those tests in a firmly social context.[2]

Scientists almost invariably agree that testability is the sina qua non of their enterprise. Roughly following Popper, many scientists would argue that testability is the unique demarcation criterion that separates scientific theories from other forms of human explanatory effort. An idea that cannot be tested may be stimulating, and therefore useful, but it cannot qualify as scientific knowledge, until new technology or ingenuity renders the concept testable. Like philosophers, scientists invoke various words to convey this belief—testability, refutability, reproducibility, replicability—but these words share a common essential element: the notion that a scientific theory must be rejectable on the basis of empirical evidence drawn from the phenomenological world.

The introduction and widespread use of numerical simulation models of complex natural systems seriously challenges conventional notions of testability in the sciences. In this paper, I shall argue that many numerical simulation models in the natural sciences cannot be tested, at least according to conventional notions of the meaning of testability. The non-testability of numerical models arises from at least four aspects of these models

[1] This point is also made by Richard Boyd, "On the Current Status of Scientific Realism," Erkenntis, 19(1983), pp. 45–90, reprinted in *The Philosophy of Science*, edited by Richard Boyd, Philip Casper, and J.D. Trout, (Cambridge, Mass: MIT Press, 1991), pp. 195-222, esp. p. 208.

[2] For example, Andrew Pickering, *Constructing Quarks.* (Chicago: The University of Chicago Press, 1984), 468 pp. The notion of experimental tests figures largely in the constructivist account, even if the conceptualization and interpretation of these tests is construed as socially determined.

M.L. Dalla Chiara et al. (eds.), Structures and Norms in Science, 207–217.

1. their complexity
2. the inaccessibility of their empirical input
3. a conflation of input and output; and
4. the inaccessibility of their predictions.

2. COMPLEXITY OF NUMERICAL MODELS

Pierre Duhem recognized nearly a century ago that the concept of theory testing involves a fundamental ambiguity: Any test of a theory is simultaneously a test of the various auxilliary hypotheses needed to draw out the observable consequences of that theory.[3] More recently, philosophers have emphasized that any test of a theory is also a test of the instrumentation and equipment required to enact the test.[4] How then do scientists presume to test their theories? Wholist objections notwithstanding, scientists hold to the conviction that it is possible to test scientific hypotheses by relying on instrumentation and auxiliary hypotheses that have been well confirmed in other, independent domains. For example, earth scientists routinely rely on instruments and knowledge borrowed from physics and chemistry. In this regard, tests of theories in earth science are almost always tests of the whole of science, as Quine would have it, but earth scientists rarely worry that their colleagues in physics and chemistry have got the world all wrong.[5] Rightly or wrongly, the confirmed nature of this knowledge is taken as a given. Assuming that the borrowed instrumentation has been properly constructed and the background assumptions adequately articulated, then scientists proceed as if it is possible, at least in principle, to test the components of a scientific theory. This view is at the heart of the notion of the controlled experiment: by the gradual and conscious adjustment of parameters, scientists attempt to isolate, and therefore test, component parts. In short, scientists rely on the conviction that it is possible to isolate the component parts of scientific theory. Epistemologically, complete isolation may be impossible, but in practice, science runs on

[3]Pierre Duhem, Physical Theory and Experiment, in *The Aim and Structure of Physical Theory*, foreword by Prince Louis de Broglie; translated from the French by Philip P. Wiener. (New York: Atheneum, 1974, c1954).

[4]As contemporary science relies on increasingly complex instrumentation (which correspondingly fewer people fully comprehend), this recognition becomes extremely important. See Peter Galison, *How Experiments End*, (Chicago: The University of Chicago Press, 1983); Ian Hacking, *Representing and Intervening*, (Cambridge: Cambridge University Press, 1987).

[5]W. V. O. Quine, "Two Dogmas of Empiricism," reprinted in *Can Theories be Refuted?, Essays on the Duhem-Quine Thesis*, Sandra G. Harding, editor, (Dordrecht: Reidel Publishing Co., 1976). For a recent philosophical critique of Quinian wholism see Yuri Balashov, "Duhem, Quine and the multiplicity of scientific tests," *Philosophy of Science* 61 (1994) 608–628. On the distinction between wholism and underdetermination, see Carl Hoefer and Alexander Rosenberg, "Empirical equivalence, underdetermination, and systems of the world," *Philosophy of Science* 61 (1994) 592–607.

the assumption that isolation is possible, at least to a reasonable degree.[6] But in the case of numerical simulation models, the possibility of isolation does not exist, even in principle, because the model is a complex, interdependent, whole. In order to see why this is so, it is helpful to look more closely at the structure of numerical models.

We can think of a numerical simulation model of a complex natural system as comprising three principal components: a conceptualization, a set of governing equations (and the numerical methods used to solve them), and data. The conceptualization is the representation of the system to be solved in the model, including such matters as initial and boundary conditions; in short, the piece of the world chosen to be modelled. The governing equations are the "laws" of nature invoked as relevant to the system being modelled. The data are the empirical input parameters needed to make the model representative of an actual system somewhere in space and time. The standard problem of verification is that if a model produces erroneous results, we have no precise way of knowing which part of the model is at fault. The conventional scientific response to this, as suggested above, is to attempt to isolate components by multiple, overlapping tests. But models are by their nature complex wholes. A model cannot be disaggregated: if you take the model apart, it no longer exists as a model. The model as a whole cannot be tested. But what of its component parts? Can you test model components? The answer is often no, because of the interdependence of model components.

The components of numerical models are always interpendent: the conceptualization determines the relevant governing equations, but the form of the governing equations also affects the model conceptualization. Likewise, the conceptualization tells us which data are relevant, and therefore guides data collection, but instrumental constraints on the availability of data may determine, at least in part, how we conceptualize the system. And, of course, the instrumental availabililty of data may have determined the form of the governing equations in the first place. A good example of this is Darcy's Law, which describes the relation between fluid flow and permeability of a porous medium. Like most physical laws, it was derived from "bench-scale" experiments: laboratory work whose aim was to examine the fundamental physical and chemical processes controlling natural systems. There is no way to use Darcy's Law to describe flow in a regional-scale natural system except in the context of a numerical model. But does Darcy's Law hold in large-scale, heterogeneous natural systems? In building a numerical model, one proceeds as if it does—there is no alternative but to make this assumption—but there is no way to test Darcy's Law in a large-scale system independently of the model in which it in contained. The choice of Darcy's Law as a governing equation pre-

[6]Scientists may be sophisticated enough to reject the idea of a single "crucial experiment," but they must retain a belief in the value of multiple controlled experiments. See Balashov, ref. cit. above.

supposes its relevance to , and accuracy for, the problem at hand.[7] A similar problem arises with respect to the numerical methods used to solve the governing equations. Can these methods be tested independently of the model in which they are being used? At first glance the answer would appear to be yes: A numerical solution can be compared with an analytical solution to demonstrate that the two match over a particular range of conditions under consideration. This process is often referred to as "verification" in the scientific literature[8]—elsewhere I have referred to it as "bench-marking."[9] Bench-marking is an important step in code development: the failure of a numerical solution to reproduce an analytic solution is certainly cause for concern. But in application, numerical solutions are invariably extended beyond the range of available analytical solutions; indeed, the raison d'être of numerical modelling is to go beyond the range of available analytical solutions. Thus, a test of a numerical solution based on comparison with an analytic solution is no test at all in the context of the actual model. In the model application, the numerical method remains untested.

3. INACCESSIBILITY OF EMPIRICAL INPUT

The governing equations and numerical methods embedded in a numerical model cannot be tested independently of the model itself. But what of the empirical input? Surely if any part of a numerical model could be independently checked, it should be the observable component; the model's "contact points with reality."[10] Yet, the observable data are often the most problematic aspect of numerical simulation models of natural systems. Although philosophers have long recognized that data cannot be wholly separated from the theories that bestow significance upon them, the tendency exists nevertheless to presume that data collection is relatively unproblematic—or at least uninteresting from a philosophical standpoint. Data collection may be difficult in practice: requiring wet feet at one extreme, or hugely expensive instrumentation at another, but these are not normally viewed as epistemological problems. But in models of natural systems, data input is an epistemological problem, because the heterogeneity of the natural world makes the empirical input fundamentally inaccessible.

Many numerical models of natural systems are based upon continuum theory, in which systems composed of discrete physical entities are treated as if

---

[7] While it is true that ressearchers are working on the question of how to scale measurements from the small to the large scale, in order to define an equivalent porous medium for which Darcy's Law would hold, at some point, the scale becomes sufficiently large that no field test can be conducted to measure the properties of the relevant blocks.

[8] For example, M.P. Anderson and W.M. Woessner, *Applied Groundwater Modeling: Simulation of Flow and Advective Transport*, (New York: Academic Press. 1992).

[9] Oreskes, N., Shrader-Frechette, K, and Belitz, K., 1994, "Verification, validation, and confirmation of numerical models in the earth sciences," *Science*, vol. 263, pp. 641–646.

[10] Ref. cit. Balashov, note 5 above.

they were a single entity: the continuum. An example of this is a sponge: at some scale the "sponge" does not exist, it is a composite of solid material and the wholes within it. The properties of the sponge are emergent: they exist at or above a particular scale. So it is for any continuum. Yet these emergent properties may be themselves heterogeneous. Modelling typically involves the assignment of a homogeneous or simply varying value for such emergent properties. Less commonly, modellers allow for regional heterogeneity but still assume homogeneity at the model block scale. But how does one know that the specified scale is appropriate? In most cases, scale is chosen for practical reasons, for example, the number of model cells that available computers can accomodate. One might test the accuracy of the numerics by refining the model grid, but it is rarely the case that modellers would return to the field to collect empirical data appropriate to the new scale. Furthermore, eventually one would reach a sufficiently fine scale—i.e., sub-continuum—that the equations being used would no longer be operational.

If one had adequate financial, technical, and temporal resources to undertake exhaustive sampling on the scale of model elements, then one would run the risk of modifying the continuum properties that one is trying to measure. For example, the insertion of closely spaced drill holes into a porous medium may change the hydraulic properties of that medium. More commonly, the property of interest simply cannot be measured at the model scale: Global Circulation Models rely on input of average temperatures for model cells, but temperature cannot be measured on a model-cell scale. The required empirical data are not accessible at the chosen scale, because they do not exist at that scale.

The problem of detemining average model-cell temperature from various point measurements highlights another issue: the scaling-up of non-additive properties. Typically, the scale of model elements is much greater than the scale on which input parameters are measured—in some cases orders of magnitude greater. The important point is this: many of these properties are non-additive, and there are no laws available to tell us how to scale-up from the measured properties to the required input on the model scale. The relation between the empirical measurements and the larger-scale model parameters is always uncertain (Figure 1).

Finally, we must recognize that no matter what the chosen scale, the use of continuum theory necessarily entails a loss of information at the scale lower than the averaging scale. Finer-scale structure and process are always lost from consideration, a loss whose significance cannot be judged independently of the problem at hand.

4. CONFLATION OF INPUT AND OUTPUT

One could argue that the issues I have raised regard the input to a numerical model, but that scientists would understand testing as the evaluation of

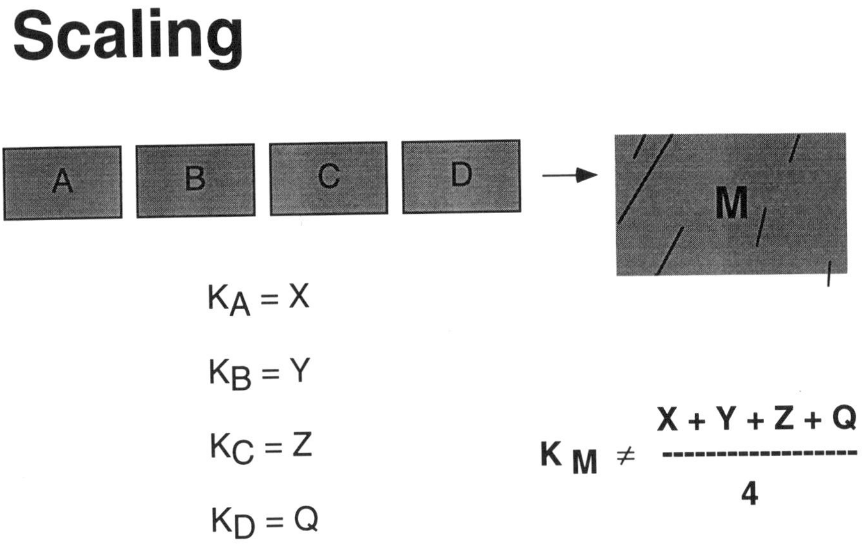

Fig. 1. The measurements of permeability on a regional scale illustrates the problem of scaling up of non-additive properties: we can measure permeability on samples of any convenient size, but we know that permeability on a large scale is almost always greater than on the bench-scale, because of the contribution of fractures (or other large scale features, such as karst cavities) to regional permeability. How much greater, however, is uncertain.

the output from a model. This presupposes that a clear distinction can be made between input (i.e., independent variables) and output (i.e, dependent variables). In practice, this distinction is blurred in many forms of scientific endeavor, but it is especially blurred in numerical modelling of natural systems.

In the earth sciences, modellers are commonly faced with what is known as the "inverse problem": The outcome is known, and the task is to find that input which will produce the known results. For example, a hydrologist knows the distribution of hydraulic head over a given geographic area, and solves for a distribution of permeability that can produce the observed hydraulic head.

Or a seismologist knows the arrival times of P– and S– waves generated by earthquakes around the globe, and solves for a structure of the Earth that can produce the observed arrival times. In these cases, the models produced are invariably non-unique: there is always more than one model configuration that can reproduce the available data. The output from such a model is a set of values for inaccessible properties of the Earth. This output cannot be tested because it cannot be compared with empirical observations, because such observations cannot be made. In other cases, the modeller may have access to a mix of dependent and independent variables. In this case, observational data that could (in principle) be used as a test of the model is built into that model. The distinction between theory and prediction is fundamentally blurred, and any "test" of such "predictions" may amount to little more than a circularity.[11]

This ambiguity is recognized by many modellers, and is evidenced by the neologisms some of them have developed to describe the results of their work. Philosophers and scientists have traditionally spoken of theory and experiment, or theory and observation, or hypothesis and data, but in recent years modellers have begun to speak of "pseudo-data," "numerical experiments," and "numerical data." For example, the neologism "pseudo-data" is used by Hartley and Prinn (1993) to denote model output in a three-dimensional chemical transport model that predicts atmospheric concentrations of trace gases.[12] The pseudo-data are the model predictions of atmosperic chlorinated fluorocarbons (CFCs) for five years into the future, based on past and present industry emissions of these gases. The model output is of the same form as the model input—i.e., the data; numbers representing atmospheric gas concentration—but these numbers are emphatically not data in the sense of deriving from the phenomenological world. Rather, they are derived from model itself. Hartley and Prinn show that their pseudo-data can be useful in directing the modeller towards areas of physical world in which additional observational data would be most effectual in improving model sensitivity. The risk, however, is that over time the prefix might be omitted, and the pseudo-data mistaken for the real thing.

5. INACCESSIBILTY OF MODEL PREDICTIONS

In the example cited above, to test the model against its predictions requires five years. In many cases, the predictions of geological and geophysical models are made on geographical or chronological scales that are simply not accesible.

[11]Observations may be theory laden, but many if not most philosophers still use the terms theory and experiment, or theory and observation, as if theories have definite empirical predictions that can be related to empirical or experimental tests.

[12]Dana Hartley and Ronald Prinn, "Feasibility of determining surface of emissions of trace gases using an inverse method in a three-dimensional chemical transport model," *Journal of Geophysical Research* 98 D3 (1993): 5183–5197.

If the predictions of a model are not accessible, then it becomes impossible for the model to be modified or rejected on the basis of faulty predictions. Here we can distinguish between two cases: models whose predictions are accessible in principle, but may not be accessed for practical reasons, and models whose predictions not accessible, even in principle.

Model predictions accesssible in principle, but not in practice

In the case of global climate models, predictions are generally made over time frames of years to tens of years. Such predictions are clearly accessible in principle; with a moderate amount of social organization and a funding mechanism in place, the predictions of global climate models can be compared with empirical evidence in the years to come. However, in practice, the builders of such models have revealed few if any plans for such comparisons. One reason is that the time frame involved, albeit shorter than a human life span, is longer than the time-span of most research careers, and far longer than most normal research projects. Forty years is not a long time in human history, but it is a long time in patterns of contemporary scientific research. In the United States, research grants from the National Science Foundation typically last two or three years; five years is considered a “long-term” study. A second reason, which is perhaps more important, is that policy decisions are being asked for right now. One cannot wait forty years to test a model if the information required of it is required immediately. Risking circularity, some scientists argue that we cannot wait to confirm the results of Global Circulation Models, because the models predict that forty years from now, it may be too late to effect the necessary change.

When a model predicts negative changes unless remediative action is taken forthwith, a serious dilemma develops: we cannot know in advance if the model predictions are correct, but if we wait to find out, it may be too late to use that information. It may also be unethical to wait. For example, if a geochemical and hydrological transport model were to suggest that a proposed toxic waste site were unsafe, then it would be unethical to ignore this information. The result is that many confirmable predictions are being put forward as basis for sound decision-making in advance of any confirmation actually taking place. Because the predictions are confirmable in principle, they may be treated as if they had in fact been confirmed.

There is an additional point that may be relevant when models are used as the basis of public policy. Even if a mechanism were to be emplaced to test the predictions of a model, the researchers who are build the model would not be the ones to test it. Although this is not a particularly unusual circumstance in the history of science—there are many examples of theories being tested after the deaths of their proponents—it may change the intellectual climate of research if one knows in advance that there is no way for one’s model to be tested in one’s own lifetime. Put another way, the attitude of an engineer building the Titanic might vary depending upon whether or not she planned to sail on it.

Model predictions inaccesssible

The builders of Global Circulation Models may not be active as scientists over the relevant time-frame, but their students will be. However, some models make predictions on time-frames over which science may not even any longer exist. I am thinking here particularly of models of nuclear waste repositories, in which predictions of repository behavoir may be drawn over several millenia. For example, transport models of proposed nuclear waste repository sites must predict repository behavoir for the next thousand to ten thousand years. Models of this kind raise a fundamental question with regard to what we mean by prediction. One could argue that, like global climate models, these models are testable in principle, even if in practice such tests will be awkward or even impossible to perform. The difference here might appear to me merely a matter of degree. I would like to argue otherwise.

Although individual scientific research projects in the United States and Europe normally last over three to five year time frames, there are many research programs that have been continuous for a number of decades; a professor's students may continue the same research direction, and so might their students after that. Certainly, one can imagine creating an institutional structure that would facilitate the confirmation of model predictions over the decadal time-frame. But if we jump to the millenial time frame, the situation is radically different. We have no reasonable grounds for belief that our culture will still be functioning 10,000 years from now, much less our scientific and governmental institutions. If the history of science is any guide, the cultural institution we call "science" may well not exist 10,000 years hence. Even if it does, it will almost certainly be in a form that will be unrecognizable to us. Conversely, and more important, our work, and our models, will most likely be unrecognizable as science to our successors. Therefore, even if an intellectual tradition were to exist that identified itself as the successor to our science, it is unlikely that its practitioners would be able to make much sense of the predictions we left for them to confirm or deny. In short, there is no way, now, for us to set up a mechanism to confirm the predictions of models 10,000 years in the future. And, as I have already suggested for the case of global climate models, even if we could set up such confirmatory tests, by the time we got the results, they would be superfluous. Decisions must be made now on the basis of predictions about a future to which we have no access. Our models are a surrogate for access to the future. How good a surrogate they are is an open question.

6. COMPLEXITY IS STRENGTH AND WEAKNESS OF NUMERICAL MODELS

In Physical Theory and Experiment, Duhem suggested that some very simple physical theories derived directly from empirical observations, such as early observational physiology, might be refutable. Implicitly, he suggested that the more complex a theory and the instrumentation upon which its empirical support relies, the greater the difficulty of refutation. The heterogenity of natural systems ensures that any model that attempts to represent a natural system must contain a large quantity of data. But this raises an insuperable difficulty: on the face of it, the more data we have, the better the model, because the more likely it is to capture the heterogeneity of the natural world. But the more data, the more complex the model becomes, and the philosophical problems of wholism correspondingly worsened. The same may be said for the governing equations and model conceptualization. The more complex the model, the worse the problem of testing becomes. One might suppose, therefore, that scientists would strive to simplify their models, but in fact the opposite may be the case. As Shackley and co-workers (1995) have emphasized in the context of Global Circulation Models for predicting climate change, because the natural systems being modelled are understood as highly complex phenomena, increased complexity of the model is interpreted as evidence of closer approximation to reality.[13] The oft-cited virtues of simplicity and parsimony in scientific theories are in this case superceded by the value of complexity, which is interpreted as evidence of the increased sophistication, and therefore presumed realism, of the model. The greater the scope of the model—the more ambitious the model—the more this is the case.

Complexity is thus the strength and weakness of numerical models. The ability of contemporary computers to deal with complex numerical solutions and large data sets permits scientists to examine processes that are not susceptible to reductionist analysis. Modelling therefore gives credibilty to synthetic work (double entendre intended). But complexity is also the weakness of numerical models: the more complex the model, the greater the interdependence of its parts, and the more remote the possibility of isolation. The problem of wholism in numerical modelling may not be different in kind from the problem of wholism in scientific theories in general, but the quantity of data involved, the complexity of the equations and their solutions, and the temporal dimension of their ambitions suggest that the problem may be much greater in degree.[14]

[13] Simon Shackley, Brian Wynne, Stuart Parkinson, and Peter Young, "Uncertainty, Complexity and Concepts of 'Good Science' in Climate Change Modelling: Are GCMs the Best Tools?", manuscript submitted to *Climatic Change*.

[14] Another example of this comes from mathematics. Some mathematicians are using computers to generate mathematical proofs, which are themselves so complex as to require computers to check them. The implications for infinite regress are obvious. See John Horgan, "The death of proof," *Scientific American*, October 1993.

7. ARE NUMERICAL MODELS SCIENCE?

I began this paper by stating that testability was at the heart of virtually all philosophical conceptions of scientific knowledge. Therefore, an obvious implication is that if numerical models are not testable, then they are not science. Insofar as the answers to the problems that models address are needed today, one could argue that modelling is not so much science as engineeering. Perhaps much of contemporary science is engineering, according to this criterion. But as a historian and practitioner of science, I have difficulty with that conclusion: sociologically, numerical modelling is science: it is done by scientists, it is funded by science-funding agencies, and the results of this work are published in scientific journals.

Perhaps the problem is the conflation of scientific theory with scientific knowledge. A scientific theory may be untested, in which case it cannot properly be considered scientific knowledge. Numerical models may be viewed as complex, untested, hypotheses. An untested hypothesis is by no means useless: it can be a heuristic tool, useful for guiding further study, and it can help to focus the direction of future data collection. But if one confuses scientific models for scientific knowledge, then problems may arise. Scientific knowledge implies reliability; the predictions of a well-confirmed theory are expected, *ceteris paribus*, to come true. But if models are untested, then the reliability of their predictions is an open question. At best, they provide insight into possible outcomes given various courses of action. At worst, they can be used in the public policy arena to justify faulty and potentially dangerous decisions.

Numerical models are machine-assisted thought-experiments. To the degree that modelling forces scientists to articulate assumptions in an exact and overt way, then models are a step in the direction of greater self-consciousness and greater insight on the part of scientists. The process of articulating quantified input parameters and conditions can lay bare assumptions that might otherwise be inarticulated. The difficulty is that this same quantification may make the model extremely precise, while leaving its accuracy unknown.

Department of Earth Sciences and Department of History
Darthmouth College
Hanover

11
Foundations of Physical Sciences

G.C. GHIRARDI

MACROSCOPIC REALITY AND THE DYNAMICAL REDUCTION PROGRAM

1. INTRODUCTION.

A quite natural question which all scientists who are concerned about the meaning and the value of science have to face is whether one can elaborate a worldview which can accommodate our knowledge about natural phenomena. As is well known, this desideratum does not raise particular problems within classical physics for various reasons which have been lucidly pointed out by J.S. Bell [1]: *Of course it is true that also in classical mechanics any isolation of a particular system from the world as a whole involves approximation. But at least one can* **envisage** *an accurate theory, of the universe, to which the restricted account* **is** *an approximation,* ... and moreover ... *even a human observer is no trouble (in principle) in classical theory — he can be included in the system (in a schematic way) by postulating a "psycho-physical parallelism" — i.e., supposing his experience to be correlated with some function of the coordinates.*

The situation is quite different in standard quantum mechanics (SQM), due to the specific peculiarities of the formalism. In particular, the linear nature of the Hilbert space description of physical systems gives rise to well known difficulties with macroscopic objects. The theory allows linear superpositions of macroscopically different states which render problematic to attribute definite properties to the systems. In particular, as is well known, this situation occurs in measurement-like processes in which, after the system-apparatus interaction is over, one has, in general, a linear superposition of macroscopically distinguishable apparatus states.

Various solutions to this puzzling situation have been proposed; for our purposes it turns out not to be relevant to discuss their specific features and/or to comment on their pros and cons. What has to be pointed out is that now everybody agrees that one needs a reinterpretation or a modification of the formalism (such as breaking the von Neumann chain, introducing hidden variables, limiting measurability or modifying the evolution law) which does not appreciably alter quantum predictions for microsystems but implies or makes legitimate to state that macro-objects have definite macroproperties.

All these attempts attribute to positions a privileged role in the description of the macroscopic world. This is quite natural since the definiteness, the particularity of the world of our experience, derives from our perceiving physical

M.L. Dalla Chiara et al. (eds.), Structures and Norms in Science, 221–240.

objects in definite places and this is also why the prescriptions for establishing the psycho-physical correspondence usually involve positions.

This paper is devoted to show how, within the context of the recently introduced nonrelativistic models [2,3] of spontaneous dynamical reduction, one can give a consistent description of the universe overcoming all the above-mentioned difficulties.

2. THE DYNAMICAL REDUCTION PROGRAM.

The program we are interested in discussing in this paper is based on the consideration, within a genuinely Hilbert space description of physical processes, of nonlinear and stochastic modifications of Schrödinger's dynamics which imply, without entailing any violation of established experimental facts, wave packet reduction with fixed pointer positions in measurement processes and, more generally, forbid the persistence of linear superpositions of macroscopically distinguishable states.

2.1. *Quantum Mechanics with Spontaneous Localization.*

We begin by discussing a slightly modified version (in which mass has a privileged role) of the first and simplest model of this kind [2], Quantum Mechanics with Spontaneous Localization (QMSL). The model is based on the assumption that, besides the standard evolution, physical systems are subjected to spontaneous localizations occurring at random times and affecting their elementary constituents. Such processes, which we will call "hittings", are formally described in the following way.

We consider a system of N particles. We suppose that when the i-th particle of the system suffers a localization the wave function changes according to

$$\Psi(\mathbf{r}_1,\ldots,\mathbf{r}_N) \rightarrow \Psi_{\mathbf{x}}(\mathbf{r}_1,\ldots,\mathbf{r}_N) = \Phi_{\mathbf{x}}(\mathbf{r}_1,\ldots,\mathbf{r}_N)/||\Phi_{\mathbf{x}}||\,,$$

$$\Phi_{\mathbf{x}}(\mathbf{r}_1,\ldots,\mathbf{r}_N) = (\alpha/\pi)^{3/4} e^{-\frac{\alpha}{2}(\mathbf{r}_i-\mathbf{x})^2}\Psi(\mathbf{r}_1,\ldots,\mathbf{r}_N). \tag{2.1}$$

The probability density of the process occurring at point $\mathbf{x}$ is given by $||\Phi_{\mathbf{x}}||^2$. For what concerns the temporal features of the processes we assume that the hittings for the various particles occur independently and at randomly distributed times with a mean frequency λ_m which depends on their mass. We choose $\lambda_m = (m/m_0)\lambda$, where m is the mass of the particle, m_0 is the nucleon mass and λ is of the order of 10^{-16} sec^{-1}. The localization parameter $1/\sqrt{\alpha}$ is assumed to take the value 10^{-5}cm.

Let a macroscopic system be in a state $|\Psi\rangle = |\Psi_1\rangle + |\Psi_2\rangle$ which is a superposition of two states $|\Psi_1\rangle$ and $|\Psi_2\rangle$ in which a given number of particles are well localized within two different (and far apart with respect to the characteristic distance $1/\sqrt{\alpha}$) spatial regions A and B, respectively. When one of these particles suffers a hitting that localizes it in the position corresponding to the state $|\Psi_1\rangle$ ($|\Psi_2\rangle$), the other term of the superposition is exponentially suppressed. Therefore, the macroscopic system jumps either to the state $|\Psi_1\rangle$ or to $|\Psi_2\rangle$ every time one of the particles differently located in the two states suffers a hitting. There follows that the rate at which one of the two terms of the superposition is suppressed increases proportionally to the number and the masses of the particles which are displaced.

The QMSL mechanism does not respect the symmetry properties of the wave function in the case of identical constituents. Its generalization satisfying such a requirement, the Continuous Spontaneous Localization Model (CSL), has been presented and discussed in various papers [3].

2.2. *The CSL Model.*

The model is based on a linear stochastic evolution equation for the statevector. The evolution does not preserve the norm but only the average value of the squared norm. The equation is:

$$\frac{d\,|\Psi_w(t)\rangle}{dt} = [-\frac{i}{\hbar}H + \sum_i A_i w_i(t) - \gamma \sum_i A_i^2]\,|\Psi_w(t)\rangle\,. \tag{2.2}$$

In eq. (2.2), the quantities $w_i(t)$ are c-number Gaussian stochastic processes satisfying

$$\langle\langle w_i(t)\rangle\rangle = 0$$

$$\langle\langle w_i(t) w_j(t')\rangle\rangle = \gamma \delta_{ij} \delta(t - t') \tag{2.3}$$

while the quantities A_i are commuting self-adjoint operators. Let us assume, for the moment, that these operators have a purely discrete spectrum and let us denote by P_σ the projection operators on their common eigenmanifolds.

The physical meaning of the model is made precise by the following prescription: if a homogeneous ensemble (pure case) at the initial time $t = 0$ is associated to the statevector $|\Psi, 0\rangle$, then the ensemble at time t is the union of homogeneous ensembles associated with the normalized vectors $|\Psi_w(t)\rangle \,/\, \||\Psi_w(t)\rangle\|$, where $|\Psi_w(t)\rangle$ is the solution of eq.(2.2) with the assigned initial conditions and for the specific stochastic process w which has occurred in the interval $(0, t)$. The probability density for such a subensemble is the "cooked one", i.e. it is given by:

$$P_{Cook}[w] = P_{Raw}\, \||\Psi_w(t)\rangle\|^2 \tag{2.4}$$

where we have denoted by $P_{Raw}[w]$ the "raw" probability density associated, according to eq. (2.3), to the Gaussian process w in the interval $(0, t)$, i.e.:

$$P_{Raw}[w] = \frac{1}{N} e^{-\frac{1}{2\gamma}\sum_i \int_0^t d\tau w_i^2(\tau)}, \tag{2.5}$$

N being a normalization factor.

One can prove that the map from the initial ensemble to the final ensemble obeys the forward time translation semigroup composition law. It is also easy to prove that the evolution, at the ensemble level, is governed by the dynamical equation for the statistical operator

$$\frac{d\rho(t)}{dt} = \frac{i}{\hbar}[\rho(t), H] + \gamma \sum_i A_i \rho(t) A_i - \frac{\gamma}{2} \sum_i \{A_i^2, \rho(t)\}, \qquad (2.6)$$

from which one immediately sees that, if one disregards the Hamiltonian evolution, the off-diagonal elements $P_\sigma \rho(t) P_\tau$ $(\sigma \neq \tau)$ of the statistical operator are exponentially damped.

For our concerns, the relevant feature of the dynamical process (2.2) with the prescription (2.4) is that it drives [3] the statevector of each individual member of the ensemble into one of the common eigenmanifolds of the operators A_i, with the appropriate probability.

This concludes the sketchy exposition of the general structure of the CSL model. Obviously, to give a physical content to the theory one must choose the so-called preferred basis, i.e. the eigenmanifolds on which reduction takes place or, equivalently, the set of commuting operators A_i. The specific form that has been shown to possess all the desired features has been presented and discussed in refs. [3]. Recently, various considerations have led to a slightly modified version making the mass to play a specific role [4,5]. In this version one identifies the discrete index i and the operators A_i of the above formulae with the continuous index $\mathbf{r}$ and the operator

$$M(\mathbf{r}) = \sum_k m^{(k)} N^{(k)}(\mathbf{r}), \qquad (2.7)$$

where $m^{(k)}$ is the mass of the particles of type k and $N^{(k)}(\mathbf{r})$ are the number density operators giving the average density of particles of type k in a volume of about 10^{-15}cm^3 around the point $\mathbf{r}$:

$$N^{(k)}(\mathbf{r}) = \left[\frac{\alpha}{2\pi}\right]^{\frac{3}{2}} \sum_s \int d\mathbf{q}\, e^{-\frac{\alpha}{2}(\mathbf{q}-\mathbf{r})^2} a^\dagger_{(k)}(\mathbf{q}, s) a_{(k)}(\mathbf{q}, s). \qquad (2.8)$$

Here $a^\dagger_{(k)}(\mathbf{q}, s)$ and $a_{(k)}(\mathbf{q}, s)$ are the creation and annihilation operators of a particle at point $\mathbf{q}$ with spin component s, satisfying the canonical commutation or anticommutation relations. Correspondingly, one has a continuous family of stochastic Gaussian processes satisfying:

$$\langle\langle w(\mathbf{r}, t)\rangle\rangle = 0,$$
$$\langle\langle w(\mathbf{r}, t) w(\mathbf{r}', t')\rangle\rangle = \frac{\gamma}{m_0^2} \delta(\mathbf{r} - \mathbf{r}') \delta(t - t') \qquad (2.9)$$

In the above formula m_0 is the nucleon mass, the parameter α is assumed to take the same value (10^{10}cm^{-2}) as in the case of QMSL and γ is related to the frequency $\lambda = 10^{-16}\,\mathrm{sec}^{-1}$ of that model according to $\gamma = \lambda(4\pi/\alpha)^{3/2}$. With the above choices the theory exhibits the same features of the standard CSL model of refs. [3] for what concerns the localization of nucleons. Obviously, also other massive particles suffer localizations but with a rate depending on their masses. The nonhamiltonian terms in the dynamical equation lead to the objectification of the position variables of massive particles.

It goes without saying that, due to the extremely low frequency of the localizations (for simplicity we use here the language of the QMSL model) of the elementary constituents of physical systems, all microscopic systems evolve in (practically) perfect agreement with the predictions of SQM. On the contrary, due to the amplification mechanism discussed above, macroscopic systems are strongly affected: a linear superposition of macroscopically different states like those corresponding to different locations of a macroscopic pointer cannot last for more than about 10^{-7}–10^{-6} seconds.

2.3. *Some Historical Remarks.*

Before concluding this Section we consider it useful to look at the proposed modifications of SQM from the point of view taken by Einstein in one of his last writings on quantum theory [6]: *Reply to Criticisms.* Contrary to what is usually referred as his firm position about the Copenhagen interpretation of the theory, in the paper under consideration he states clearly that one could accept the peculiar features which, according to this interpretation, characterize the microscopic processes, provided one could save the definiteness, the particularity of our experience with macro-objects. He remarks that in the case of the decay of a radioactive atom, the orthodox quantum theorist would claim that *within the framework of our theoretical total construction* it is not reasonable even *to posit the existence of a definite point of time for the transformation of a single atom.* He plainly declares that such a position *is certainly by no means absurd.* However he also stresses that Schrödinger taught us that the question about the time of decay can be straightforwardly transformed (by simply coupling the atom to an appropriate apparatus) into the question of the presence of a macroscopic registration mark on a paper strip. And he states: *the location of the mark on the registration-strip is a fact which belongs entirely within the sphere of macroscopic concepts* ... and ... *there is hardly likely to be anyone who would be inclined* to accept *that the existence of that location is essentially dependent upon the carrying out of an observation* ... In fact: *in the macroscopic sphere it simply is considered certain that one must adhere to the program of a realistic description in space and time; whereas in the sphere of microscopic situations one is more readily inclined to give up, or at least to modify, this program.* Few pages later he comes back to this point and he stresses that his realistic attitude does not derive from a philosophical prejudice: *the "real" in physics is to be taken as a type of program, to which we are, however, not forced to cling a priori. No one is likely to be inclined to attempt to give up this program within the realm of the "macroscopic" (location of the mark on the paper strip "real").* And then he draws his conclusion: *But the "macroscopic" and the "microscopic" are so inter-related that it appears impracticable to give up this program in the "microscopic" alone.*

We have chosen to devote so much space to this quotation to call attention to the fact that, with reference to the previous analysis, the crucial problem of our best theory is to succeed in accounting for the smooth merging of the cloudy microworld with the definite macroworld; stated differently, to make precise where and how acts occur which, in J.A. Wheeler's words, can be unambiguously identified as [7] *irreversible acts of amplification.* Quite appropriately, in our opinion, J. A. Wheeler maintained that locating the shifty split cannot be related to acts of empirical assertions. On the contrary, the occurrence of the irreversible acts of amplification must be part of the physical description and one should not ascribe to them any exceptional position above the rest.

The Dynamical Reduction Program represents an attempt to answer the puzzling questions we have discussed by identifying in a precise way within a strictly quantum context (i.e., by assuming completeness of the Hilbert space description of the states of individual physical systems) the shifty split between micro and macro, reversible and irreversible. The very existence of precise dynamical reduction models shows that, contrary to Einstein's expectations, the program of building up a consistent and unified theory allowing the microsystems to be wavy and nevertheless implying the definiteness of the macroscopic world is viable. It is important to stress that the emergence of such definiteness is implied and precisely described by the formalism so that, to account for it, there is no need to invoke (as many other attempts do) the practical impossibility of revealing interference effects at the macroscopic level. We leave to the reader the evaluation of the potential relevance of the approach and to identify its advantages and drawbacks with respect to the many interesting recent proposals. At any rate, it seems to us that having the explicit proof that a theory exhibiting such features is possible has, by itself, a certain relevance for the investigations about the foundations of quantum mechanics.

3. HOW MACROSCOPIC REALITY IS DESCRIBED.

A scientific theory is a conceptual scheme dealing with formal elements which are supposed to embody some information about something existing and to account for physical processes *which take place* "out there". Among these processes there are experiments performed by observers and also what have been called measurement-like events which [8] *we are obliged to admit are going on more or less all the time, more or less everywhere.*

To be something more than a simple list of facts and to allow one to go beyond a purely instrumentalistic position, i.e. *to explain* rather than simply *to describe*, a theory must be supplemented with some sort of interpretation. We are interested in interpretations which allow one to make claims about properties (related to specific variables) possessed by individual physical systems. With reference to this crucial point, which is the core of the debate about the foundations of quantum mechanics, we start by making precise the statement that such properties are objectively possessed or accessible.

Definition: we will claim that a property corresponding to a value (a range of values) of a certain variable in a given theory is *objectively possessed or accessible* when, according to the predictions of that theory, there is an experiment (or a physical process) yielding reliable information about the variable that would, if performed (or taking place), give an *outcome* corresponding to the claimed value. Thus, the crucial feature characterizing *accessibility* (as far as statements about individual physical systems are concerned) is the matching of the claims and the *outcomes* of physical processes testing the claims.

3.1. *Introductory Remarks.*

At this point we are ready to tackle our central problem, i.e. the one of giving an acceptable and appropriate description of the macroscopic world. Before doing that we would like to recall once more the lucid position [9] of J.S. Bell about what one has to require from a theoretical model in order that it can be considered as satisfactory. He points out that a theory is a mathematical framework related to empirical facts. And he states: *you must identify some things as being really there, as distinct from the many mathematical concepts that you can easily devise — like the projection of the side of a triangle to infinity and so on. We must decide that some things are really there and that you are going to take them seriously.* This requires a neat distinction between what he calls the kinematical and the dynamical aspects of the theory. The kinematics should list the possibilities one is envisaging, the dynamics should give the probabilities to be attached to the possibilities. And he puts forward strict requirements for the theoretician: *I would want the kinematics of your theory tell me what it is you are talking about before you tell me what about it.* We are perfectly aware that this position could be considered as quite naive from a philosophical standpoint, but it expresses in very simple terms our firm convictions about reality and the reasons which have motivated our interest in the foundations of science. Having stated this we can now go on with our analysis.

3.2. *A Short Digression about Macroscopic Similarity.*

Before tackling the problem of property attribution at the macroscopic level it is useful to point out the inappropriateness of the Hilbert space topology to describe the similarity or difference of macroscopic situations. Suppose a physical system S is an almost rigid body and let us consider the following three states: $|\varphi^A\rangle$, $|\varphi^B\rangle$ and $|\tilde{\varphi}^A\rangle$. The state $|\varphi^A\rangle$ corresponds to a definite internal state of S and to its centre of mass being well localized around A, the state $|\varphi^B\rangle$ is obtained by simply displacing $|\varphi^A\rangle$ so that it is well localized in a far region B, the state $|\tilde{\varphi}^A\rangle$ differs from $|\varphi^A\rangle$ simply by the fact that one or a microscopic number of its "constituents" are in states which are orthogonal to the corresponding ones in $|\varphi^A\rangle$.

It is obvious that, under any reasonable assumption about similarity or difference of the states of the macrosystem, $|\tilde{\varphi}^A\rangle$ must be considered very close (identical) to $|\varphi^A\rangle$ while $|\varphi^B\rangle$ must be considered very different from $|\varphi^A\rangle$. On the other hand, according to the Hilbert space topology

$$\|(|\varphi^A\rangle - |\tilde{\varphi}^A\rangle)\| = \|(|\varphi^A\rangle - |\varphi^B\rangle)\| = \sqrt{2}. \tag{3.1}$$

This shows with striking evidence that the "natural" topology of SQM is totally inadequate for the description of the macroscopic world. In Subsection 3.6, taking advantage of its dynamical features, we will show how CSL suggests a natural topology which allows one to overcome the now mentioned difficulty.

3.3. *The Mass Density Distribution.*

Let S be a physical system which constitutes our universe, and let $H^{(S)}$ be its associated Hilbert space. Let us make an attempt to describe the physical reality of *what exists out there* by giving the status of objectively possessed properties to the mean values of the average (over the characteristic volume of CSL) mass density operator (2.7):

$$\mathcal{M}(\mathbf{r},t) = \langle\Psi(t)|\, M(\mathbf{r})\, |\Psi(t)\rangle\,. \tag{3.2}$$

If one assumes, as one can consistently do within a non relativistic quantum framework, that the system S contains a fixed and finite number of particles, Eq. (3.2) defines, for a given t, a mapping of the unit sphere of $H^{(S)}$ into the space of positive and bounded functions of $\mathbf{r}$.

A question naturally arises: within the conceptual framework we are interested in does $\mathcal{M}(\mathbf{r},t)$ represent an accessible property of physical systems? Before answering this question a comparison with the case of SQM is appropriate.

3.4. *Mass Density and SQM.*

The map (3.2) is obviously many to one. To better focus on this point as well as for future purposes it turns out to be useful to compare two statevectors $|\Psi^{\oplus}\rangle$ and $|\Psi^{\otimes}\rangle$ defined as follows. Let us consider a very large number N of particles and two space regions A and B with spherical shape and radius

R. The state $|\Psi^{\oplus}\rangle$ is the linear superposition, with equal amplitudes, of two states $|\Psi_N^A\rangle$ and $|\Psi_N^B\rangle$ in which the N particles are well localized with respect to the characteristic length (10^{-5}cm) of the model and uniformly distributed in regions A and B, respectively, in such a way that the density turns out to be of the order of 1 gr/cm^3. On the other hand, $|\Psi^{\otimes}\rangle$ is the tensor product of two states $\left|\Phi_{N/2}^A\right\rangle$ and $\left|\Phi_{N/2}^B\right\rangle$ corresponding to $N/2$ particles being uniformly distributed in region A and $N/2$ in region B, respectively:

$$|\Psi^{\oplus}\rangle = \frac{1}{\sqrt{2}}\left[|\Psi_N^A\rangle + |\Psi_N^B\rangle\right], \quad |\Psi^{\otimes}\rangle = \left|\Phi_{N/2}^A\right\rangle \otimes \left|\Phi_{N/2}^B\right\rangle. \tag{3.3}$$

It is easy to see that the two considered states give rise to the same function $\mathcal{M}(\mathbf{r})$.

Let us now investigate the status of the quantity $\mathcal{M}(\mathbf{r})$ within SQM. To this purpose let us suppose that one can use quantum mechanics to describe the gravitational interaction between massive bodies and let us consider the following gedanken experiment: a test mass is sent through the middle point of the line joining the centres of regions A and B with its momentum orthogonal to it. In the case of the state $|\Psi^{\otimes}\rangle$ quantum mechanics predicts that the test particle will not be deflected. Thus the outcome of the considered test agrees with the value for the mass density in regions A and B given by eq.(3.2); we can legitimately state that such a value is objectively possessed. On the other hand, in the case of $|\Psi^{\oplus}\rangle$, it is quite obvious that $\mathcal{M}(\mathbf{r})$ cannot be considered as an "accessible" mass density function. In fact, if the same test is performed when the state is $|\Psi_N^A\rangle$ ($|\Psi_N^B\rangle$) quantum mechanics predicts an upward (downward) deviation of the test particle. Then, the linear nature of the evolution implies that if one would be able to prepare the state $|\Psi^{\oplus}\rangle$ the final state would be

$$|\phi\rangle = \frac{1}{\sqrt{2}}\left[|\Psi_N^A\rangle \otimes |\phi^{UP}\rangle + |\Psi_N^B\rangle \otimes |\phi^{DOWN}\rangle\right], \tag{3.4}$$

with obvious meaning of the symbols. If one includes the test particle into the "universe" and considers the mass density operator in regions corresponding to the wave packets $|\phi^{UP}\rangle$ and $|\phi^{DOWN}\rangle$, one discovers that nowhere in the universe is there a density corresponding to the density of the test particle. In a sense, if one would insist in giving a meaning to the density function he would be led to conclude that the particle has been split by the interaction into two pieces of half its density so that the test does not give a definite outcome. According to our criterion this shows that the mass density distribution $\mathcal{M}(\mathbf{r})$ in regions A and B, in the case of the state $|\Psi^{\oplus}\rangle$, is unaccessible.

3.5. *CSL and Objectively Possessed Properties.*

Coming back to CSL, it should be obvious that the situation is radically different from the one of SQM since such a theory does not allow the persistence for more than a split second of states like $|\Psi^{\oplus}\rangle$ which are strictly analogous to those giving rise to the measurement problem. To analyze this point it is useful to introduce a precise mathematical criterion emboding, for the mass density function, the accessibility requirement put forward at the beginning of this Section. To this purpose, let us consider the mass density variance at point $\mathbf{r}$ at time t defined by the following map from the unit sphere of $H^{(S)}$ into $\Re^3$:

$$\mathcal{V}(\mathbf{r},t) = \langle\Psi(t)|\left[M(\mathbf{r}) - \langle\Psi(t)|\,M(\mathbf{r})\,|\Psi(t)\rangle\right]^2|\Psi(t)\rangle\,, \tag{3.5}$$

and let us evaluate the ratio

$$\mathcal{R}^2(\mathbf{r},t) = \mathcal{V}(\mathbf{r},t)/\mathcal{M}^2(\mathbf{r},t). \tag{3.6}$$

We then claim that at time t the mass density at point $\mathbf{r}$ is accessible if:

$$\mathcal{R}(\mathbf{r},t) << 1. \tag{3.7}$$

This criterion is clearly reminiscent of the probabilistic interpretation of standard quantum mechanics. Actually, within such a theory, Eq.(3.7) corresponds to the fact that the spread of the operator $M(\mathbf{r})$ is much smaller than its mean value. Even though we take a completely different attitude with respect to the mean value of $M(\mathbf{r})$, it turns out to be useful to adopt the above criterion also within the new context. In fact, as we will discuss in what follows, when one has a space region such that for all its points relation (3.7) holds, all reliable tests aimed to ascertain the mass density value give outcomes corresponding to $\mathcal{M}(\mathbf{r})$.

With reference to the previous example we stress that in the case of $|\Psi^{\otimes}\rangle$ all points within regions A and B are such that criterion (3.7) is very well satisfied. In the case of $|\Psi^{\oplus}\rangle$ for the same points one would have:

$$\mathcal{M}(\mathbf{r}) \cong \frac{d}{2}m_0,$$

$$\mathcal{V}(\mathbf{r}) \cong \frac{d^2}{4}m_0^2, \tag{3.8}$$

where d represents the average density in a volume of 10^{-15}cm^3 around $\mathbf{r}$. There follows

$$\mathcal{R}(\mathbf{r}) \cong 1, \tag{3.9}$$

so that, as expected, the average mass density in regions A and B is nonaccessible.

Obviously for a microsystem, exception made for the case in which its wavefunction is appreciably different from zero only within a region of linear extension smaller than 10^{-5}cm, its average mass density turns out to be almost always nonaccessible since CSL does not forbid superpositions of arbitrarily far apart microscopic states. This fact does not constitute a difficulty for the proposed interpretation, on the contrary, it embodies the fact that nature has compelled us to allow [10] *electrons* (in general microsystems) *to enjoy the cloudiness of waves, while* our experience of the world requires *tables and chairs, and ourselves, and black marks on photographs, to be rather definitely in one place rather than another, and to be described in "classical terms"*. The possible nonaccessibility of the mass density distribution does not entail that it is not real. Actually even regions in which the mass density does not represent an objectively possessed property according to our mathematical criterion may play an important physical role[1]. We point out that within a theory like CSL which implies the definiteness of the macroscopic world, the request of being able to speak of objectively possessed macroproperties and the desire, we will discuss in the next subsection, to identify a topology which is appropriate for describing the similarity and the difference of macroscopic situations, lead to consider "two levels of reality". The attribution of a real status to $\mathcal{M}(\mathbf{r})$ even when it is not accessible does not give rise to the difficulties one meets when one tries to interpret the wavefunction as giving the charge or the mass density of the elementary constituents of nature. Actually this corresponds to plainly accept the well-known fact that [8] *however far the wavefunction has extended, the reaction of a detector ... remains spotty.*

[1]A typical example is represented by an intergalactic cloud with a very low mass density and wavefunctions for its particles with a large spread. According to our strict criterion, the mass density of the cloud is unaccessible, while, if we adopt the physically meaningful criterion put forward at the beginning of this Section and to test it we resort to the consideration of its gravitational effects, e.g., on the trajectory of a far asteroid, we would be led to state that our claims about the mass density are objectively true since they agree with the "physical outcome". Here, a comment is appropriate. The theory contains a specific dynamical mechanism and specific parameters which allow a precise mathematical formulation of the accessibility criterion. Obviously, when one is interested in a particular class of physical processes the requirement that the outcomes match the claims about properties could be satisfied even though the strict mathematical one does not. This does not raise any problem. It finds its counterpart in the fact that to analyze a specific physical situation one can use a coarser graining than the one corresponding to the fundamental characteristic length of the theory. Accordingly, if we adapt the graining to the problem of interest, we can assert that the mass of the cloud is accessible. Obviously, while *for practical purposes* one can change the graining according to the problem he is interested in, adopting a graining which is finer than the fundamental one of the theory makes no sense. In fact, if one takes the theory seriously as describing the laws governing the evolution of the universe, mass density values would almost never be objectively possessed when such a graining is chosen.

3.6. *The Appropriate Topology for the CSL Model.*

We come back to the problem of giving a precise mathematical criterion which allowes one to consider "similar" or "indistinguishable" various macroscopic situations. Let us denote by $\mathcal{U}^{(S)}$ the unit sphere in $H^{(S)}$ and let us consider the non linear map $\mathcal{M}$ associating to the element $|\varphi\rangle$ of $\mathcal{U}^{(S)}$ the element $\mathcal{M}(\mathbf{r},|\varphi\rangle) = \langle\varphi| M(\mathbf{r}) |\varphi\rangle$ of $L^{(2)}$, the Hilbert space of the real square integrable functions on $\Re^{(3)}$.

On $\mathcal{U}^{(S)}$ we define a topology by introducing a mapping $\Delta\colon \mathcal{U}^{(S)} \otimes \mathcal{U}^{(S)} \to \Re^{+}$ according to:

$$\Delta(|\varphi\rangle, |\psi\rangle) = \{\int dr[\mathcal{M}(\mathbf{r},|\varphi\rangle) - \mathcal{M}(\mathbf{r},|\psi\rangle)]^2\}^{1/2}. \tag{3.10}$$

Such a mapping is not a distance since it may happen that $\Delta(|\varphi\rangle, |\psi\rangle) = 0$ even though $|\varphi\rangle \neq |\psi\rangle$. However Δ meets all other properties of a distance:

$$\Delta(|\varphi\rangle, |\psi\rangle) = \Delta(|\psi\rangle, |\varphi\rangle) \geq 0,$$

$$\Delta(|\varphi\rangle, |\psi\rangle) \leq \Delta(|\varphi\rangle, |\chi\rangle) + \Delta(|\chi\rangle, |\psi\rangle), \tag{3.11}$$

as one can easily prove.

From now on we will limit our considerations to the proper subset $\mathcal{A}^{(S)}$ of $\mathcal{U}^{(S)}$ of those states which are allowed by the CSL dynamics[2]. For any element $|\varphi\rangle$ of $\mathcal{A}^{(S)}$ we consider the set of states of $\mathcal{A}^{(S)}$ for which $\Delta(|\varphi\rangle, |\psi\rangle)\tilde{<}\varepsilon$. Here the quantity ε has the dimensions of a mass and is chosen of the order of $10^9 m_0$, with m_0 the nucleon mass. From the properties of the map Δ it follows that:

i . $\{\Delta(|\varphi\rangle, |\psi\rangle)\tilde{<}\varepsilon$ and $\Delta(|\varphi\rangle, |\chi\rangle)\tilde{<}\varepsilon\}$ implies $\Delta(|\chi\rangle, |\psi\rangle)\tilde{<}\varepsilon$,

ii. $\{\Delta(|\varphi\rangle, |\psi\rangle) >> \varepsilon$ and $\Delta(|\varphi\rangle, |\chi\rangle)\tilde{<}\varepsilon\}$ implies $\Delta(|\chi\rangle, |\psi\rangle) >> \varepsilon$.

We have introduced the parameter ε in such a way that it turns out to be sensible (see Section 4) to consider similar to each other states whose "distance" Δ is smaller than (or of the order of) ε. More specifically, when

$$\Delta(|\varphi\rangle, |\psi\rangle)\tilde{<}\varepsilon \tag{3.12}$$

we will say that $|\varphi\rangle$ and $|\psi\rangle$ are "macroscopically undistinguishable".

We now discuss the "distorted" (with respect to the Hilbert space one) topology associated to the "distance" Δ. First of all we stress that the two states $|\varphi^A\rangle$ and $|\tilde{\varphi}^A\rangle$ which are maximally distant in the Hilbert space topology, turn out to be undistinguishable, i.e. to satisfy condition (3.12), in the

[2]For a more precise identification of such a set see ref. [5].

new topology. This is an example showing how such a topology takes more appropriately into account the fact that, under any sensible assumption, the "universes" associated to the considered states are very similar.

Obviously, one problem arises. Criterion (3.12) leads to identify states which may be quite different from a physical point of view, even at the macroscopic level. To clarify this statement we take into account two states $|\varphi\rangle$ and $|\psi\rangle$ corresponding to an almost rigid body located, at time $t = 0$, in the same position but with macroscopically different momenta, let us say $P = 0$ and $P = P^*$, respectively. Even though the two states are physically quite different, their distance at $t = 0$ is equal to zero. However, if one waits up to the time in which the state $|\psi\rangle$ has "moved away" from $|\varphi\rangle$, the "distance" $\Delta(|\varphi(t)\rangle , |\psi(t)\rangle)$ becomes large and the two states are no longer undistinguishable.

3.7. *A Worldview Based on CSL.*

As we have pointed out at the beginning of this Section, we are interested in describing something "existing and evolving out there" in such a way that the resulting physical picture is compatible with our experience of the world.

For the reader who has followed the above analysis, it should be clear that, within CSL, physical systems are fully described by the statevector which evolves according to equations emboding genuine elements of chance. However, even though [11] *there is nothing in this theory but the wavefunction*, due to the fact that as we have extensively discussed the dynamics itself tends (on the appropriate scale) to make objectively possessed just the values of the mass density distribution, it turns out to be correct to identify the "basic kinematical elements" in Bell's sense with such a quantity. This is what we have done in the previous Subsections by putting forward explicit criteria allowing one to recognize its being accessible and by showing the appropriateness of resorting to the mass density function to define the idea of similarity and difference between macroscopic situations.

In a sense we could legitimately state that the elements of the theory *to be taken seriously* are the values of the mass density function. All other properties which emerge as a consequence of the dynamical evolution derive from the basic feature of the theory of making this function *accessible*. For example in the case analyzed at the end of the previous subsection the macroscopic momentum of the system can, after a certain time, be related to an objectively possessed property just as a consequence of the objectification of the mass density.

With reference to the position we have just taken we feel the necessity of mentioning two other possible attitudes about the formal elements which have *to be taken seriously*. One has been mentioned by David Albert in his comments to our talk and coincides with the one taken by Bell himself in his last years. It amounts simply to consider as the only physically meaningful quantity of the theory the modulus square of the wavefunction in the $3N$-dimensional configuration space: such a quantity should be interpreted as representing the density of stuff of which the world is made[3]. The second one has been discussed in a recent preprint [12] by P. Pearle. He denotes, as we did in ref.(5), as *objective reality* the properties we have preferred to call *accessible* in this paper. But, besides defining objective reality, this author introduces two other concepts that he considers useful: projective reality and subjective reality. Moreover he takes into account all these three "levels" of reality both for macro and for microsystems. Projective reality is described by the mean values of all self-adjoint operators of the physical system under consideration. On the other hand, since any attempt to get a relativistic generalization of the dynamical reduction models implies that the mean values of local observables of microsystems depend, in general, on the space-like surface one is considering to evaluate them[4] such mean values can be different for different observers, a fact justifying the use of the expression "subjective reality" to describe them. The basic reason to take into account all these levels of reality is that them all have precise dynamical consequences for the system and may play an important role in determining the properties which will become accessible at subsequent times.

Let us discuss these two proposals. Concerning the first one, we perfectly agree (this firm request has guided our investigations from the very beginning) that the statevector of an individual physical system contains the most detailed information one can have about it. But we cannot avoid stressing various important facts. Here we are primarily concerned in accounting for the world as we perceive it, a problem on which J.S. Bell has made some cute remarks: *when Von Bortkevitch collected statistics on the kicking of soldiers to death by horses, in the Prussian Army, in different years, he found a Poisson distribution. Now, you don't go into the world looking for the Poisson distribu-*

[3]We feel the desire to mention that we had a series of discussions with J.S. Bell on this point and to call attention on his firm position about the matter. On october 3, 1989, answering to repeated questions I had raised in previous correspondence, he wrote: *As regards Ψ and the density of stuff, I think it is important that this density is in the 3-N dimensional configuration space. So I have not thought of relating it to ordinary matter or charge density in 3-space. Even for one particle I think one would have problems with the latter. So I am inclined to the view that you mention after "... as it is sufficient for an objective interpretation ... " And it has to be stressed that the "stuff" is in 3-N space — or whatever corresponds in field theory!* The reader can imagine how reluctant we are in taking a position differing to some extent from the one suggested by this great scientist and deep thinker. However, we would like to mention that, in a sense, the above letter strengthens our position. J.S. Bell is perfectly right in pointing out that "for one particle one would have problems" in passing to real space. In fact we have repeatedly stressed that the microscopic mass density is almost never accessible. But, at the macroscopic level, such density is dynamically objectified in real 3-space, and accounts for our definite perceptions.

[4]A detailed discussion of this feature and a precise analysis of the reasons for which it does not give rise to any conceptual difficulty has been presented in refs. [13]. The nice fact is that, why such a dependence is unavoidable at the microscopic level, once more, no ambiguity arises when macrosystems are considered.

tion, you go out looking for soldiers and horses and kicks. We could rephrase this sentence by stating that you do not go out in the $3N$-dimensional configuration space looking for a density of stuff, you go out in 3-dimensional space and look for masses in definite places. But there are other reasons to consider the mass density distribution as the fundamental element of the theory: if the theory is taken seriously such a quantity represents precisely the only property that nature makes accessible, just as particle positions represent precisely the only real and objective properties of physical systems within the de Broglie-Bohm theory[5]. The mass density is a quantity whose meaning is quite clear and which is directly related to our experience. Stressing its prominent role has not to be taken as a specific ontological choice, but simply as the appropriate way to call attention on the basic physical features of the theory which make the "status of the world" consistent with our perceptions: the dynamics running all natural processes strives to make definite the positions of massive objects and this mechanism embodies the new features of nature which the theory attempts to grasp.

There are also some obvious and trivial practical reasons to assign a prominent role to the mass density function. For instance, looking at the wavefunction in the $3N$-dimensional space does not yield any immediate information about the fact that the system you are dealing with is, e.g., a moving pointer. On the contrary, the changes of the mass density give precisely the relevant information. Obviously, the same argument could be used to claim the unappropriateness of the phase-space description of the state of a classical system. But one should not forget that within a quantum context the situation is much more peculiar since incompatible properties (which exist also for a macro-object) cannot become simultaneously accessible. Among all conceivable properties, the theory makes precisely accessible the mass density distribution, all the others (if they emerge) are a consequence of this fundamental dynamical feature of the theory.

Finally, as we have discussed in Subsections 3.2 and 3.3, if one is interested in the relevant problem of giving a precise formal status to the vague idea of similarity at the macroscopic level one needs a topology. As we have shown the appropriate topology cannot be defined simply in terms of the wavefunction or of the density of stuff in the $3N$-dimensional space, it requires in an essential way the consideration of the mass density distribution, an accessible and objectively possessed feature of our universe.

[5]Obviously there are important differences. The pilot wave theory puts on the same footing the positions of micro and macroscopic systems. The wave function is an additional and independent fundamental element of the theory. But, leaving aside the technical points, we think that, if one takes into account that our primary interest is in the macroscopic level of reality, one can legitimately claim that just as [14] *the moral* (of the pilot wave theory) *is that in physics the only observations we must consider are position observations*, in a completely analogous way one can assert that the moral of CSL is that the only observations we must consider are those ascertaining the value of the accessible mass density distributions, typically those corresponding to the positions of macroscopic objects.

We come now to the second attitude mentioned above. Concerning this we stress that the consideration of unaccessible properties (mean values of observables referring to microsystems or to nonobjectively possessed properties) is uninteresting for the description of reality as we perceive it. It is true that such quantities play some role in determining the properties which will become accessible in the future. But taking them into account seems to us redundant. In fact, as it is well known, the assignment of the mean values of all bounded self-adjoint operators is equivalent to assigning the wavefunction itself. Thus, since the quantities under discussion do not have a direct physical significance, the information they give is simply a substitute of the knowledge of the statevector. This being the case it would be more natural to take the position suggested by Bell. Analogous remarks, with the due changes, apply to subjective reality. For lack of space we cannot analyze here this interesting aspect which would require a detailed discussion of the features of "possible" relativistic reduction models. We limit ourselves to mention that any such model must unavoidably make accessible appropriate properties at the macroscopic level. These properties are those which should play, within such theoretical schemes, the role we have given here to the mass density function in accounting for physical reality.

4. ACCESSIBLE REALITY AND DEFINITE PERCEPTIONS.

The most characteristic and appealing feature of CSL and of the interpretation of it we have proposed consists in the fact that it allows one to give a satisfactory account of reality, to take a realistic view about the world, to talk about it as if it is really there even when it is not observed. However, one cannot avoid raising the problem of including also conscious observers into the picture, for [1]: *what is interesting if not experienced?* Thus one is led to consider the problem of the psycho-physical parallelism within the considered theory.

The previous analysis should have already given clear indications about the lines to be followed to reach this goal. In this Section we will first of all make clear how to relate the "different states of the universe" as characterized by the formalism according to the lines of the previous sections to the specificity of conscious perceptions. Secondly we will show how the theory itself supports the proposed correspondence. To get this we will perform a very sketchy analysis, from the point of view of CSL, of the current ideas about the physical processes leading to perceptions.

4.1. *External World and Internal Perceptions.*

As we have seen the CSL dynamics, at the macroscopic level, makes objectively possessed the values taken by the mass density function, and consequently suggests to give to this function the status of the basic element accounting for reality, as we perceive it. According to this picture, "what is out there" is identified with a precise and accessible mass distribution in real space. The reality of a massive object in front of us corresponds to the fact that in the region it occupies the value of the mass density which characterizes it is an objectively possessed property.

On the other hand it is a fundamental feature of our perceptions that they correspond to objects having precise locations and extensions. The problem of establishing a map between reality and perceptions is then naturally solved by correlating our perceptions to mass density distributions[6].

After these simple remarks we can pass to discuss how one can account, within CSL, for the emergence of perceptions.

4.2. *Describing the Perceptive Process.*

To make clear how CSL is able to "describe" the occurrence of definite perceptions of conscious beings, it turns out to be quite appropriate to start by discussing a criticism to this theory which has been put forward in some recent papers [15]. The idea is quite simple but raises a problem which deserves a detailed investigation. One considers the following process: a neutral microsystem with spin is sent through a Stern-Gerlach apparatus. The spin state is such that the system ends up in the superposition of being deflected, with equal probabilities, upwards or downwards, respectively. The two "potential trajectories" cross a fluorescent screen in two macroscopically far apart regions A and B, respectively. The particle-screen interaction is such as to lead to the excitation of a small number, e.g. of about ten atoms, which subsequently undergo a transition to the ground state accompanied by the emission of photons.

The argument of ref. [15] goes as follows. Since only few atoms of the screen are excited during the process, since their excitations involve displacements which are much smaller than the characteristic localization length of the model, since photons are not localized, there is no way for the CSL mechanism to yield the suppression of one of the two superposed states. Thus, the superposition of states corresponding to ten photons emerging from the different space regions A and B will persists for extremely long times. On the other hand, since the visual perception threshold is quite low (about 7 photons) there is no doubt that the naked eye of a human observer is sufficient to detect whether the luminous spot on the screen is at A or at B. This raises an interesting question: in the situation considered how can it happen that a definite perception about the location of the spot on the screen emerges? Are we compelled to accept that, at least in some circumstances, also within CSL

[6]Obviously our perceptions are much richer than those (corresponding to position and shape) we have listed here. In the next Subsection we will make clear how also more complex perceptions (such as colour etc.) can be naturally included in the picture we are presenting.

the conscious observer and its perceptions play a (peculiar) role analogous to the one attached to them in Wigner's views?

That this is not the case has been discussed in great detail in a recent paper [16]. If one takes, as one must, the above remark seriously one is compelled to consider the actual systems which enter into play and to analyse the implications of the CSL dynamics for them. We simply sketch here the argument of ref. [16]. One takes into account what we know about the transmission of nervous signals from, let us say, the retina and the higher visual cortex. Such a transmission requires, among other changes, the displacement of ions along the axons involved in the process. A very rough estimate of the mass associated to these ions and of the displacements they have to perform to flow through the ion channels which open at Ranvier's nodes to transmit the electric pulse can be given. This estimate makes perfectly plausible that the conditions which are sufficient, according to CSL, for the suppression of one of the two states (nervous signals) of the superposition within the visual perception time (which is of the order of 10^{-2} secs), are satisfied[7] .

We do not want to be misunderstood; this analysis does by no means amount to attribute a special role to the conscious observer or to his perception. The observer's brain is simply the only system among those which are present in which a superposition of two states involving different locations of a large number of particles occurs. As such it is the only place where the reduction can and actually must take place according to the theory. If, in place of the eye of a human observer, one puts in front of the photon beams a spark chamber or any device leading to the displacement of a macroscopic pointer or producing ink spots on a computer output, reduction will take place. In the considered example, the human nervous system is simply a physical system, a specific assembly of particles, which makes the same function as any one of these devices, if no such device is interacting with the photons before the human being does[8] .

[7]The reader will now easily understand the appropriateness of the choice we have made in section 3 for the parameter ε, marking the transition between similar and different macroscopic situations. According to CSL the suppression of one of the two states in a superposition is governed by the expression $\exp[-(\lambda t/2)\sum_i (n_i - n_i')^2]$, n_i and n_i' representing the nucleon occupation number of the "elementary cells — of volume $10^{-15}\mathrm{cm}^3$ — in the two states appearing in the superposition. In the worst case $n_i - n_i'$ equals one for the only cells which have different occupation numbers, so that, taking into account that $\lambda = 10^{-16} sec^{-1}$ and that the typical perception times are of the order of 10^{-3}-10^{-2} sec, one sees that within such times, superpositions of "macroscopically distinguishable" situations are dynamically suppressed.

[8]We consider appropriate a specification. The above analysis could be taken as indicating that we adopt a very naive and oversimplified attitude about the deep problem of the brain-mind correspondence. We do not claim and we do not pretend that CSL yields a physicalistic explanation of consciousness. We simply point out that, for what we know about the purely physical aspects of the perceptual process, the conditions guaranteeing that superpositions of different perceptions cannot occur are satisfied.

This example should be sufficient by itself to clarify our problem. Our perceptions are triggered by our sensory apparatuses. In many cases, such as for the hearing or tactual perceptions, the stimulus itself cannot be ambiguous, i.e. it cannot correspond to superpositions of different perceptions, since it requires macroscopic displacements (of the hearing membrane or of the skin). In other cases, as the one discussed above, the nervous signal can be triggered by a microscopic system, which can very well be in a superposition of states capable to induce different perceptions. But in all cases the nervous transmission of the signal involves a "macroscopic" (on the appropriate scale) displacement of mass in the brain. And, as discussed above, the CSL dynamics suppresses in a time of the order of the perception time superpositions characterized by different (in the precise mathematical sense we have given to this term in subsection 3.6) macroscopic mass distributions. In a sense the analysis shows that the problem of the psycho-physical correspondence admits a simple solution quite similar to the one which is usually assumed to hold for classical theories. Reality and perceptions involve the same fundamental quantity: the values of the mass density function. On one side "macroscopic reality" is, quite in general, related to accessible mass density distributions. On the other, definite perceptions are related to accessible macroscopic mass density distributions within the brain.

We do not want to spend further time in elaborating on this point. Our aim here is not that of performing a detailed technical analysis of any conceivable situation but simply of making plausible that the problem of the psycho-physical parallelism admits, within CSL, a solution which is quite analogous to the one of the unproblematic classical case. Thus CSL can be claimed, according to J.S. Bell's definition [1], to be an "exact theory" in the precise and limited sense that *it neither needs nor is embarrassed by an observer.*

Dipartimento di Fisica Teorica
Universita degli Studi di Trieste

REFERENCES

1. J.S. Bell, in: *Quantum Gravity* 2, C. Isham, R. Penrose and D. Sciama eds., Claredon Press, Oxford (1981).
2. G.C. Ghirardi, A. Rimini and T. Weber, Phys. Rev. **D 34**, 470 (1986).
3. P. Pearle, Phys. Rev **A 39**, 2277 (1989); G.C. Ghirardi, P.Pearle and A. Rimini, Phys. Rev. **A 42**, 78 (1990); G.C. Ghirardi and A. Rimini, in : *Sixty-Two Years of Uncertainty*, A. Miller ed., Plenum Press, New York and London (1990).
4. P. Pearle, and E. Squires, *Phys. Rev. Lett.* **73**, 1 (1994).
5. G.C. Ghirardi, R. Grassi, and F. Benatti, *Found. Phys.* **25**, 5 (1995).
6. A. Einstein, in: *Albert Einstein: Philosopher Scientist,* P.A. Schlipp. ed., Tudor Pub. Co., New York (1949).
7. J.A. Wheeler, in: *Quantum Theory and Measurement,* J. A. Wheeler & W. H. Zurek eds., Princeton University Press, Princeton (1983).
8. J.S. Bell, in: *Sixty-two Years of Uncertainity,* A.I. Miller ed., Plenum Press, New York (1990).

9. J.S. Bell, in: *Themes in Contemporary Physics II,* S. Deser and R.J. Finkelstein eds., World Scientific, Singapore (1989).
10. J.S. Bell, in: *Proceedings of the Nobel Symposium 65: Possible Worlds in Arts and Sciences,* Stockholm (1986).
11. J.S. Bell, in: *Schrödinger. Centenary celebration of a polymath,* C.W. Kilmister ed., Cambridge University Press (1987).
12. P. Pearle, *Tales and Tails and Stuff and Nonsense*, preprint, 1995.
13. G.C. Ghirardi, R. Grassi and P. Pearle, Found. Phys., **20**, 1271 (1990); in: *Symposium on the Foundations of Modern Physics 1990*, P. Lahti and P. Mittelstaedt eds., World Scientific, Singapore (1990).
14. J.S. Bell, Found. Phys., **12**, 989 (1982).
15. D.Z. Albert and B. Loewer, in: *PSA 1990*, vol. 1, A. Fine, M. Forbes and L. Wessels eds., Philosophy of Science Association, East Lansing, Michigan (1990). D.Z. Albert and L. Vaidman, Phys. Lett. **A139**, 1 (21989); D. Z. Albert in: *Sixty-two Years of Uncertainity,* A.I. Miller ed., Plenum Press, New York (1990); D.Z. Albert, *Quantum Mechanics and Experience,* Harvard University Press, Cambridge, MA (1992).
16. F. Aicardi, A. Borsellino, G.C. Ghirardi and R. Grassi, Found. Phys. Lett., **4**, 109 (1991).

12
Foundations of Biological Sciences

RICHARD M. BURIAN

ON CONFLICTS BETWEEN GENETIC AND DEVELOPMENTAL VIEWPOINTS — AND THEIR ATTEMPTED RESOLUTION IN MOLECULAR BIOLOGY

> Embryologists and geneticists never used to see eye to eye. As I have indicated in this book the two disciplines have now become united in a new subject formed by the fusion of developmental genetics with molecular biology [Peter Lawrence, *The Making of a Fly: The Genetics of Animal Design* (London: Blackwell Scientific Publications, 1992), p. 195].

This paper concerns the apparent rapprochement, now underway, between genetics and developmental biology.[1] For there to be a rapprochement, of course, there must have been long-standing disagreements. The disagreements between embryology and genetics bear all the stigmata of a deep discordance between research traditions built on conflicting assumptions and practices. They have been discussed by many historians of science and shown to rest on conceptual and institutional differences, on differences in the experimental and field practices of geneticists and embryologists and on the distinctive behaviors of the biological materials they traditionally employed.[2] Suffice to say that the disagreements were based in part on the absolute inability of geneticists to show how genes could account for the *Bauplan* of an organism and on their failure to give any weight to such phenomena as cytoplasmic gradients in the egg, polarities in the egg and the early embryo, cell death in organogenesis, and so on.[3] These complaints against genetics were just and

[1]I have benefitted greatly from the penultimate draft of "Resynthesizing Evolutionary and Developmental Biology" by Scott Gilbert, John Opitz and Rudy Raff 1996, and from discussions with Prof. Gilbert. I am grateful to him for sharing this work with me, for permission to utilize it here, and for permission to reproduce figures 3 and 7 below. Thanks also to Dr. Opitz and participants at the ICLMPS session in Florence for constructive comments and to Prof. Walter Gehring for supplying the photographs reproduced in Figures Five and Six below.

[2]See, e.g., Allen 1986, Burian MS 1985, 1993, Harwood 1984, Keller 1995, chap. 1, Maienschein 1986, Sander 1986 and Sapp 1983, 1986, and 1987.

[3]Cf. Sander 1994 for a brief account of some little-known early work along these lines and Morgan (1934, 1935) for one of the first widely-known articulations of an in-principle means of combining cytoplasmic gradients with rigid nuclear and genetic determination of all potentialities of the organism. I do not know any good histories of such concepts as that of a morphogenetic field. Some useful materials are included in Gilbert (ed.) 1991; Horder *et al.*(eds.) 1986, especially Wolpert's chapter; see also De Robertis *et al.* 1991 and Opitz 1985.

M.L. Dalla Chiara et al. (eds.), Structures and Norms in Science, 243–264.

remain so, although that fact does not reveal whether they are biologically or philosophically important.

The rapprochement of interest here results from the application of powerful new molecular techniques to basic biological problems and from fundamental conceptual revisions now underway in both genetics and developmental biology. The conceptual changes, which are strongly influenced by recent findings employing molecular techniques, depend in part on paying increased attention to the many ways in which genes are activated. Very briefly, one tendency to which we will pay attention treats development in terms of the activation of genes by complex higher-level systems rather than treating it as a passive consequence of gene action. I describe below a few illustrative molecular findings, discuss conflicting interpretations of the putative reconciliation, highlight some conceptual issues, and begin an examination of some ways in which molecular findings require changes in basic concepts and patterns of explanation in both genetics and developmental biology. The discussion leads me to suggest that the hoped-for reconciliation between genetics and developmental biology will likely undermine the prior conceptual foundations of each discipline.

In particular, I suspect that the reconciliation between developmental biology and genetics will not rest on a theoretical unification of the prior tenets of these disciplines regarding the control of organismal development. Rather, the reconciliation will require partial abandonment of a number of key commitments on both sides. Such as it is, the ongoing reconciliation rests on a detailed understanding of the bits and pieces of machinery out of which organisms are built and how they work. For the present, at least, we understand development better by working from the molecular bottom up rather than from the organismic (or some theoretical) top down. But, contentiously, the molecular bottom is not properly construed either as the master genes for which geneticists hoped, nor as the gene-independent gradients, pre-patterns, and morphogenetic fields for which many embryologists hoped. Rather, the molecular bottom involves bounded, sometimes very short-lived, interactive regulatory networks built by extraordinarily close interactions between genes and proteins. These networks, in turn, have extraordinarily complex and diverse relations to the many diverse pieces of molecular machinery that produce the phenotypic structures whose development the networks regulate. Many of the regulatory networks operate within specific geographies, most of which are not yet laid down in the earliest embryos; some of them are open to "resetting" only during brief critical periods. The control systems both require and build polarities and gradients that define the boundaries within which further control systems operate. Thus geographically and temporally bounded domains, gradients, and control systems all play an inescapable role in ontogeny. Accordingly, aspects of both traditional embryology and traditional genetics are represented in the detailed understanding of developmental processes. What remains obscure and is hotly contested is how these aspects can be put together coherently.

We are very far from understanding the presumptive global controls by means of which the bodies of organisms are integrated during the course of development. We don't understand how the many interlocking processes are coordinated so as to allow the various structures to fit together and achieve functional coordination among body parts. But it is already clear that many highly modular local control mechanisms go into the making of an integrated organism. The resulting picture is thus vastly different in detail and in character than was anticipated by classical genetics or by classical embryology — or, for that matter, by molecular genetics before the 1980s. At the same time we know, for example, that very strongly conserved regulatory devices contribute to the control of such processes as the making of body segments and of eyes. In the specific case of eyes, to be discussed below, the same control system regulates the beginning steps for making such radically different eyes as the compound eye of a fruitfly and the eye of a mouse or a human being. This example, like many others, raises and exemplifies enormous conceptual problems, among them the interplay of the universal and the specific in control of development. Indeed, it serves as a useful vehicle for posing a number of the conceptual problems raised by recent molecular work on development. To illustrate these problems, I will discuss the difficulty in making sense of the distinction between homology and analogy when, as I just hinted is the case for eyes, homologous processes controlled by homologous genes serve to regulate the construction of such seemingly non-homologous, but analogous, structures as dipteran and mammalian eyes.

I maintain that the integration of organisms is not readily understood within a unified account of the sorts that many geneticists and developmental biologists have hoped to forge. Organisms are, in many respects, like complex machines built from a diversity of different, hierarchically interacting, partly independent multi-level units or modules on different scales. The boundaries of important units in development only sometimes correspond with our prior intuitions and with obvious phenotypic markers. The development of many modules is under partly autonomous control, modulated by extremely diverse interactions with neighboring units during ontogeny. I shall put a bit of biological flesh on this skeletal picture below in an attempt to set forth an account of some of the obstacles to forging a unified and satisfactory theory of development on any familiar theoretical foundation. But let us turn to a description of some recent findings in molecular genetics and get to work.

SOME RECENT FINDINGS IN MOLECULAR DEVELOPMENTAL GENETICS

It would be very easy to get lost in the details of recent molecular work bearing on the genetics of development. Any attempt to summarize this work must be incredibly unrepresentative of a large, complex, and flourishing enterprise, one that employs extremely powerful tools to examine the details of extremely complicated and diverse systems. Nonetheless, the following slight sketch is meant to convey the spirit of one main line of findings. It discusses some work that is likely to require fundamental revisions of concepts in developmental biology, evolutionary biology, and genetics. The study of work like that discussed below may help us to rethink — and perhaps revise — the patterns of explanation we employ for many biological phenomena.

Major surprises, with wholly unexpected ramifications, stem from the discovery, sequencing, and functional analysis of such nucleic acid sequences as those of so-called homeoboxes, first characterized in 1982.[4] Homeoboxes are sequences, typically of 180 nucleotides (often plus introns), encoding a sequence of 60 amino acids. These form a so-called homeodomain in the corresponding protein. Homeodomains are transcription factors that regulate interactions between the proteins that contain them and DNA; they ensure that the proteins occupy specific sites on DNA. By occupying the specific sites, the proteins in question alter the likelihood that nearby stretches of DNA will be transcribed.[5]

Homeoboxes have an enormously broad phylogenetic distribution; they are found in all metazoans and are known in a number of fungi and plants. So far as is known, they occur exclusively in genes regulating cell fate or producing proteins that mark segment boundaries during development; these often produce patterns on which body parts are laid down. The proteins they encode play numerous roles in processes regulating the expression of concerted groups of genes and/or the formation of structural units in the bodies of ani-

---

[4] For brief accounts of the discovery of homeoboxes in late 1982, see Lawrence 1992 and Gehring 1994. All good recent textbooks of molecular genetics, developmental biology, or evolution already contain an account of homeoboxes, but detailed knowledge is moving very fast in the primary literature.

[5] Sixty amino acids are the canonical number in homeodomains. A few shorter homeodomains are known and there are both upstream and downstream extensions in some cases. See Bürglin 1994. The homeodomain "was the first of the DNA binding motifs to be identified in *Drosophila*... The sequence encodes four main α-helices linked by turns. The third of these helices fits into the major groove of the DNA, 'the recognition helix', and several amino acids on one face of this helix contact the bases of the DNA. The structure has been studied and this has helped identify those amino acids that make particular contacts, not all of which lie inside the third helix. When sequences from different genes are compared, these contacting amino acids are rather variable and may well determine which binding sites the protein seeks out. If one of these amino acids is altered, so are the *in vitro* binding properties of the domain" (Lawrence 1992, p. 110). Current results suggest that there are generally three helices, although the third often has a kink, thus producing what was, and sometimes is, counted as a fourth helix (see Bürglin 1994, p. 28).

mals. The amino acid sequences in homeodomains are enormously conserved in evolution — for example 59 of the 60 amino acids in a homeobox of the toad *Xenopus* are the same as those in the homeobox of the *Antennapedia* gene in Drosophila (Lewin 1990, p. 770).[6]

Homeoboxes were originally discovered in Drosophila; their name stems from the fact that the original homeoboxes were found in genes in which homeotic mutations occur. Homeotic mutations are those that replace one body part with another. One of the most famous of these, the Bithorax mutation first discovered by Calvin Bridges and T. H. Morgan in 1915 and worked on extensively since 1948, primarily by Ed Lewis, replaces the balancers (halteres) of Drosophila with wings, yielding a four-winged fly. By now, over 100 drosophila genes are known to contain homeoboxes, and every one of those genes appears to play some role in regulating cell or segment fate. Proteins containing homeodomains are often part of extraordinarily intricate networks of genes and proteins that regulate the fate of a phenotypically identifiable unit. At the same time, it is important to know that the boundaries defined by the proteins do not necessarily correspond to the familiar phenotypic units and that alterations in one protein in a given cascade can drastically alter the boundaries laid down by another (see Figure One).[7] The best understood example, far richer in detail than can be hinted at here, illustrates this point nicely. It concerns the genes in the bithorax complex of Drosophila which form part of the apparatus that regulate the division of the embryo into segments and determine the fates of those segments.

Homeotic mutants provide important clues to the segmentation of a body. In Drosophila and other segmented organisms, it is possible for the phenotype of one segment to be replaced with that of another segment — witness the bithorax mutant fly (**di**pteran) with four wings (**quadri**pteran?) or mutations in mice which result in cervical vertebrae occupying a place that should be

[6]The pairwise sequence similarity of 21 homeodomains from a total of 346 homeodomain amino acid sequences analyzed in Bürglin 1994 ranges from 20 to 67 percent. The 21 homeodomains were chosen to represent distinct classes; thus they differ far more than randomly chosen homeodomains. The 21 domains chosen come from drosophila, mouse, a brachiopod, caenorhabditis (a roundworm), a yeast, and arabidopsis (a plant). Among all 346 homeodomains, seven of the 60 sites in the homeodomain are occupied by only one or two amino acids, six have 3–5 amino acids, and thirteen have 6–9 amino acids. The substitutions in these sites are also conservative in that they are typically function-preserving. (For example, hydrophobic amino acids are replaced by hydrophobic amino acids.) Thus 26 sites are especially conservative and can serve as diagnostic tools for recognizing homeodomains. Parallel claims apply for the corresponding nucleotides of the DNA, although the nucleotide substitution rate is, of course, higher than that of amino acids.

[7]For example, the genes that define the boundaries between what are now called parasegments in the early drosophila embryo are shifted one-half segment away from the segments of the adult fly. This became clear only when it became possible to visualize the boundaries in question by making images showing the distribution of the proteins produced by the homeobox-containing genes, such as *engrailed, fushi tarazu*, and *paired*, which demarcate parasegment boundaries starting at about the third hour after the egg is laid.

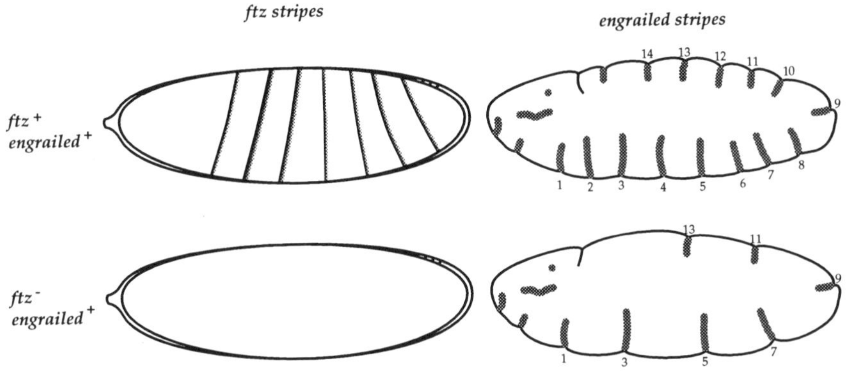

Fig. 1. Striped spatial patterns of the products of the *fushi tarazu* (left) and *engrailed* (right) genes at about 3–4 hours after a Drosophila egg is laid. Top row, both genes wild type. Bottom row, no *fushi tarazu* gene product. Note how the pattern of engrailed gene product stripes is entirely changed in the absence of this protein. Since the half life of the fushi tarazu product is about five minutes, there is a brief period when the interactions of the two gene products are critical to establish the correct boundaries for further events. This interaction is typical of a large number of other interactions affecting segment and other geographical boundaries within the embryo. Redrawn by David Rudnick after Figure 4.7 in Lawrence 1992.

occupied by thoracic vertebrae.[8] But there are also severe restrictions on such replacements — e.g., in a large class of cases posterior parts can be replaced by anterior parts but not vice versa. In the case of bithorax, there is an exact correspondence between the order in which certain contiguous genes are turned on and the determination of the body segments proceeding from the anterior end of the embryo toward its posterior end. I can only give a primitive first approximation account of this story, worked out largely thanks to the ingenious persistence of Ed Lewis, recently followed by many others.

The three genes in the bithorax series are located contiguously on the third chromosome. These are turned on sequentially in both time and space along an anterior to posterior axis in the embryo. Thanks to a series of alternative splices, a series of ten distinct products are made in this way (see Figure Two). This process depends heavily on protein-DNA interactions. Formation of the normal sequence of body segments requires that the homeoboxes of the bithorax genes each specify the exact sequences of amino acids required for the protein-DNA interactions that occur in normal development. When something goes wrong with this process, one does not obtain the normal sequence of

[8]John Opitz (personal communication) reports that it has not yet been demonstrated that such mutations, which are quite rare in humans, are associated with genes containing homeoboxes.

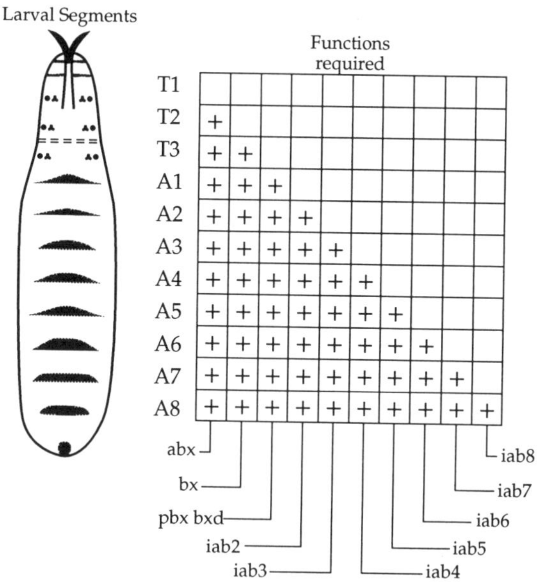

Fig. 2. Ed Lewis's proposal, now largely confirmed, for the manner in which the genes of the bithorax complex specify distinct fates for the thoracic and abdominal segments of drosophila and other insects. Successive genes are activated sequentially both in space and in time. The function of an additional gene product in the series is required to specify the identity of each succeeding segment. Redrawn by David Rudnick after Lewin 1990, Fig. 38.15.

thoracic segments. Equally interesting, part of the apparatus that regulates whether or not these genes are turned on is a feedback loop requiring that a protein boundary be established at the midpoint of the preceding segment, or, rather, of the preceding parasegment, in the series.[9] And one of the proteins required to be present at that boundary in order to turn on the next gene in the series is the product of the preceding regulatory gene in the series. This

[9]Thus, phenotypic segments in arthropods are divided into anterior and posterior half-segments. It is with respect to these boundaries that the *para*segments are shifted one half segment anterior to phenotypic segments. Parasegments appear to be the natural units for determining segment fate in the early embryo. Parasegments are laid down first; the boundaries within them, which define segment boundaries are laid down in a subsequent step. (See, e.g., Lawrence, 1992, pp. 91 ff., Gilbert 1988, pp. 638 ff., or virtually any recent standard text.) The details of the bithorax system with respect to boundary formation and parasegment fate determination are very important, but cannot be adequately summarized here. A useful summary is found in Lawrence 1992, pp. 111 ff. In fact, the three gene products need to be produced in the correct spatial sequence to correctly posteriorize *para*segments, not segments; interactive effects of the gene products characterize and determine some parasegments, rather than simply the presence of one, two, or three, of the gene products.

has the consequence that in the absence of the appropriate anterior function, it is often not possible to produce either the anterior *or the adjacent posterior* part and that it is never possible to place the posterior part in front of the anterior one. For this reason, although, in the bithorax mutation, balancers (halteres), a more posterior structure, have been replaced by wings, no case is known of halteres, the posterior part, being placed in front of wings.[10]

In general, before a homeotic mutation can have its characteristic effect, there must be defined boundaries that determine the geographical limits of its action. But there are many systems that yield homeotic changes and many relevant boundaries, determined by additional control systems. For example, the specification of whether a part of a segment is proximal or distal is independent of the determination of whether that segment yields a leg, an antenna or a wing. Thus in homeotic mutants in which parts of an antenna are homeotically transformed into parts of a leg, the transformed proximal parts of the antenna are always proximal parts of a leg, and the transformed distal parts of the antenna are always distal parts of the leg.

A further finding sharpens the connection with issues about homology. There is now very strong molecular evidence, nearly universally accepted, that the three bithorax genes, together with five 'antennapedia' genes, were initially produced by gene duplication followed by differentiation of function. (The same is true of many other groups of genes involved in determining segments.) That is, these two groups of contiguous genes match the classic nineteenth century definition of Richard Owen for being serial homologues; an initial gene reproduced first by "vegetative repetition" (gene duplication), followed by subsequent differentiation (see Gilbert 1980). The genes of the control system of which we have been talking manifest serial homology at the genetic level. Their relationship is reminiscent of that of the vertebrae in Owen's diagram of the archetypical vertebrate; like the differentiated vertebrae of the various vertebrates, the current genes result from differentiation of the iterated ancestral gene. And, in precise parallel with this, it is very likely that an annelid ancestor yielded other insects by the specification of different functions per segment (see Figure Three). The different segment identities depend on the activation of serially homologous genes which, themselves, have evolved by vegetative repetition followed by differentiation while in partial, though indirect, interaction with one another.

There are yet more interesting surprises about the segment-determining genes which, taken together, imply that these genes are also phylogenetically homologous (that is, exhibit what Owen called special homology) to genes that carry out similar segmentation functions in nearly all metazoans, from

[10]The production of wings can be blocked or preempted in various ways. Accordingly, it is possible to have halteres in the absence of wings, but not in the absence of the expression of the genes required for determination of a segment as a wing-bearing segment.

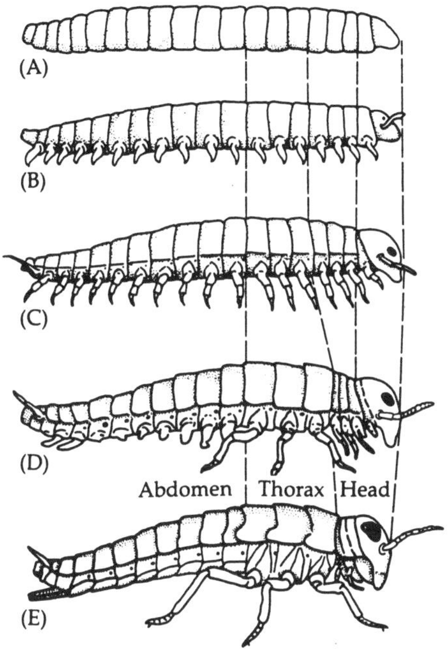

Fig. 3. Diagram of insect evolution, as hypothesized by Raff and Kaufman 1983, based on the addition of homeotic gene functions. The insects depicted are (A) an annelid, (B) an onycophoran, (C) a myriapod, (D) an apterygote, and (E) a pterygote. Reproduced, with permission, from Gilbert 1988, Fig. 29, p. 654, based on Raff and Kaufman 1983.

flatworms to mammals.[11] Because of space limitations, I cannot go into great detail. But, an examination of Figure Four will help to summarize some salient points. There is a precise sequential correspondence between four series of genes containing homeoboxes in mice and the *antennapedia* and *bithorax* series in drosophila. If one were to line up the genes in each group by examining only the degree of sequence similarity in their homeoboxes, one would obtain exactly the same linear ordering that, in fact, obtains. Furthermore, just as in Drosophila, the mouse genes are expressed in the mouse embryo sequentially, with the leftmost (3′) genes in the figure being expressed anteriorly; as one proceeds to the right, each gene is expressed more posteriorly. Yet further, the

[11] Very early in the cloning of homeoboxes, it was found that vertebrates have homeoboxes. In the late 1980s, it became clear that these occur in genes that are molecularly similar to the genes that control early embryonic development in Drosophila, including those of the bithorax complex. It also become clear that certain genes and gene clusters controlling segmentation in Drosophila are contiguous on a single chromosome and that their molecular mates in mice and other animals occur in the same serial order as in Drosophila.

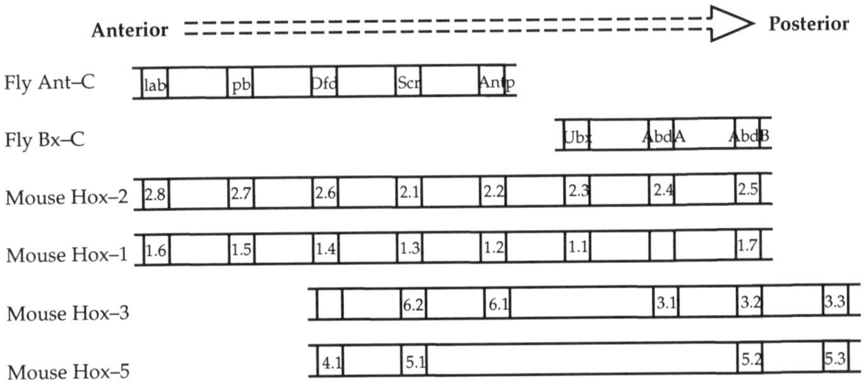

Fig. 4. Correspondence between the drosophila *antennapedia* and *bithorax* gene complexes and the four clusters of homeotic *Hox* genes of mice. In both organisms, the genes are expressed sequentially in time, starting at the left side of the diagram (3′ end of the DNA molecule) and proceeding to the right (5′ end). This is also the order in which the genes are expressed in the anterior to posterior axis of the organism. The genes of the *Hox* clusters are aligned with the fly genes according to the sequence similarity of the homeodomains of the genes in the respective clusters. Redrawn by David Rudnick after Lewin 1990, Fig. 38.19. For a more recent version of this figure, see De Robertis 1994, p. 14. There one finds the new designations of the *Hox* genes and additional information regarding parallels between the genes and their control systems.

genes in mice are "turned on" sequentially in time and specify segment identity in just the manner that they do in drosophila. Yet further, the enhancers of the mouse genes serve as enhancers of the drosophila genes and vice versa. (For one example, see Malicki *et al.* and McGinnis *et al.*) These genes exhibit homology of function and arrangement; the sequence similarities we have mentioned are not accidental but rest on a deep homology of some as-yet ill-understood sort. (For further details and references see Lawrence 1992, Gehring 1994, and Gilbert *et al.* 1996.) The recent evidence (summarized by Gilbert *et al.* 1996) suggests that this sequential system is virtually universal in metazoa. Assuming that all this stands up to further scrutiny, they summarize the results dramatically with the somewhat simplified formula: "Every animal has similar genes, has them in the same chromosomal order, and uses them to specify the same relative positions along the anterior-posterior axis" (p. 364). The underlying idea here goes back not to Cuvier, but to Geoffroy St.-Hilaire; the specification of the sequence of segments fundamental to the structure of animals, the basis for specifying their architecture or *Bauplan*, is not different among the four *embranchements*; it is the same in *all* animals.

A final example will help make the conceptual issues stand out more sharply, both because it is so dramatic and because it breaks one of the general rules about homeotic genes. The example, based on work done in Walter Gehring's laboratory in Basel, concerns the machinery for making eyes. One of the most exciting new findings (Halder, Callaerts, and Gehring 1995) concerns three genes, clearly homologous, denominated *ey* [eyeless] in Drosophila, *Sey* [small eye] in mice, and *Aniridia* [no iris] in humans. In each of these animals, the homologous gene appears to serve as a switch to turn on the entire cascade of phenomena required for formation of eyes. Homologues are also present in ribbon worms, ascidians, cephalopods — probably in nearly all metazoa. The sequence similarities of homeodomains of the mouse and drosophila proteins is 90% and that of an associated transcription factor affecting protein-DNA interactions and DNA transcription (called a paired domain) is 94%.[12] Remarkably, genetically engineered ectopic (ubiquitous) expression of *ey* in Drosophila imaginal disks induces the formation of eyes independent of locational information, i.e., independent of whether the gene is expressed in antenna, leg, wing, or haltere-forming tissue (Figure Five). This breaks the rule, suggested earlier, that homeotic mutants require additional positional information to yield one body part rather than another; in this respect *ey* behaves differently than the other homeotic genes about which I know. (On the other hand, the eye is anterior to any of the other organs in question; perhaps something in the cascade initiated by the eye-forming switch simply blocks or reverses the action of posteriorizing genes).

Yet more remarkably, the mouse protein, expressed in Drosophila via genetic engineering, yields Drosophila eyes exactly as does the Drosophila gene product (Figure Six). Four other Drosophila genes are known which, when appropriately mutated, yield eyeless Drosophila, but all of them are expressed posterior to the *ey* gene in the cascade of phenomena leading to eye formation. Indeed, it is estimated that at least 2500 genes, and probably more, are activated in the cascade of events that follows expression of the *ey* gene in Drosophila. To the authors, these results suggest rather strongly that *ey* is something like a master gene for turning on the eye-making machinery. Such an interpretation is explicitly advanced by Gehring 1994 (pp. 7, 9) for homeotic genes, and applied by Quiring *et al.* 1994 and Halder *et al.* 1994 to the *ey* gene. The last two papers suggest that the ancestor of the *ey* gene controlled a module for forming primitive eyespots very early in metazoan evolution and that the eye-making cascade evolved from this phylogenetically primitive starting point.

[12]The paired domain, incidentally, is a member of a family of transcription factors distinct from the homeodomain; the combination of both domains on a single protein "illustrate[s] evolutionary tinkering at the molecular level" (Quiring *et al.*, 785). In this case, the tinkering with modular control systems has yielded a novel control system.

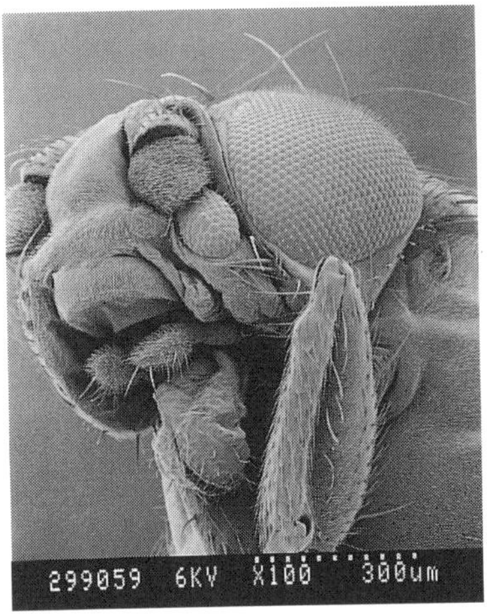

Fig. 5. Scanning electron micrographs of ectopic eyes in Drosophila; photographs kindly supplied by Walter Gehring. In A the small eye near the center of the photograph is an ectopic eye formed on an antennal disc. It occupies the space normally occupied by an antenna. B is a close-up of the ectopic eye shown in A. Such eyes are histologically normal or near normal and are connected to nerves which lead to the brain. The receptors are photosensitive and signals are transmitted by the nerves that innervate them. It is not known, however, whether the eyes are functional and whether the signals reach the correct domains of the brain. These photos correspond to figures 3A and C of Halder *et al.*, 1995.

CONFLICTING INTERPRETATIONS

How should we conceive of the eyeless gene in Drosophila? Gehring and his colleagues, who showed that *ey* [the Drosophila gene] is homologous with mouse and human genes and that the mouse gene can set off the Drosophila eye-making cascade, interpret *ey* and its homologues as "a master control

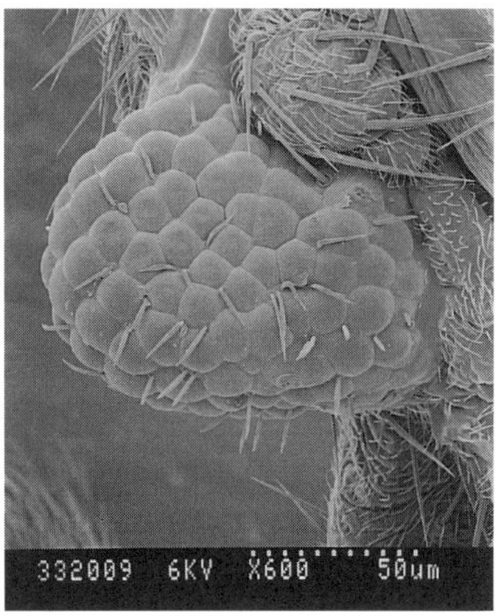

Fig. 6. Scanning electron micrograph kindly supplied by Walter Gehring. This is a close-up of an ectopic eye grow on a drosophila leg. This eye was induced by expression of the product of the mouse *Pax*-6 gene; the gene was activated by the same genetic engineering procedures as were employed to produce the eyes shown in the previous figure. The only significant difference is that the product of the one mouse gene has replaced that of the corresponding drosophila gene. The eyes produced in this way are virtually indistinguishable from those produced by activation of the *ey* gene as shown in Figure 5 above. This photograph serves as Figure 5 of Halder *et al.*, 1995.

gene for eye morphogenesis[13] that is shared by vertebrates and invertebrates"

[13] At p. 1791, the following rationale is used to support this characterization of the role of *ey* in Drosophila: it counts as a master gene "because it can induce ectopic eye structures in at least the imaginal discs of the head and thoracic segments." See also Quiring *et al.*, p. 788.

(Halder *et al.*, p. 1792; see also Quiring *et al.*). The label 'control gene' designates genes that set off the cascade of events that changes the identity of a body part and/or produces a specific organ. Control genes, including those yielding homeotic mutants, can be detected by the mutant phenotypes they cause. A master control gene is characterized as the first of the genes in the relevant cascade of control genes (Halder *et al.*, 1788). They claim (with all the necessary cautions) that *ey* and its homologues are master control genes.

But the phrase 'master control gene' is controversial — and for good reasons. Gilbert, Opitz, and Raff, for example, would reject any talk of master genes, even in such dramatic cases as this. Their interpretation stems from developmental biology rather than genetics. They employ a modified notion of morphogenetic fields that are laid down by interaction among genes and proteins within bounded domains. These fields are discrete, modular units that respond differently to the activation of a given control gene. They would therefore reject the more traditionally genetic interpretation of the Gehring group. The interpretative differences are significant, for they enter into the construction and evaluation of hypotheses regarding a wide range of developmental, evolutionary, and genetic phenomena and reflect underlying differences about the ontology in terms of which to interpret organisms.

To illustrate the fundamental character of the issues at stake, consider the claim put forward quite naturally by the Gehring group, that "because [the same master gene] is involved in the genetic control of eye morphogenesis in both mammals and insects, the traditional view [reference to Plate 1924] that the vertebrate eye and the compound eye of insects evolved independently has to be reconsidered" (Quiring *et al.*, 785).[14] In other words, Gehring and his colleagues believe that their startling findings show drosophila and mouse eyes to be homologous structures. This revisionist claim apparently rests on the view that structures either *are* or *are not* homologous as wholes, so that finding a genuine homology deep within the controls for the manufacture of corresponding structures proves that they are homologous. This commitment follows naturally from the view that there is a master gene responsible for making an entire structure and from the finding that the master gene, slightly altered but still functional is shared between two organisms. I will argue that both the implicit claim about homology and the explicit characterization of *ey* as a master gene should be rejected.

The objection to master genes can be made in part from within the genetic tradition. The phrase 'master gene' *could* be replaced with one that is far less committal — after all, why not call the first gene in such a sequence a *switch*

[14]See also p. 788: "If the corresponding genes in flatworms and nemerteans are also involved in eye morphogenesis, the concept that the eyes of invertebrates have evolved completely independently from the vertebrate eye has to be reexamined. Also, the hypothesis that the eye of cephalopods has evolved by convergence with the vertebrate eye is challenged by our recent finding ... of *Pax*-6- [the mouse homologue of *ey* — RB] related sequences in the squid *Loligo vulgaris*."

gene, say, rather than a master gene? 'Switch gene' is a standard genetic term. As is well known, it is commonly used, for example, for genes that control wing pattern in butterflies. Homeotic mutants are those in which a body part (or a portion thereof) has switched identity. Gehring himself (1994, p. 6) mentions as an alternative term "homeotic selector gene." Genes that set off a cascade that alters the identity of a body part or a fundamental pattern or morphology are thus rather naturally characterized as switch genes.

But is this just trivial semantics? Is there a significant difference between speaking of master genes and speaking of switch genes? I claim that the terminological difference is *not* innocuous. The concept of a master gene carries strong heuristic (not to mention ideological) overtones with it. In particular, it supports claims that lie at the heart of the old divisions between genetics on the one hand, and embryology or developmental biology on the other. It supports the metaphor of the genome as a program that contains the entire organism *in potentia*. This is a commitment at the heart of the human genome program, one that the concept of a switch gene need not support, especially when one recognizes that sometimes the switch in question may be thrown by environmental rather than genetic or other internal changes (cf. seasonal dimorphism in butterflies). We face here all of the familiar issues about the degree to which the so-called information in the genome depends on its cellular and organismic, even its environmental, context. Thus it is not surprising that many analysts of science and scientists, including many developmental biologists, resist the terminology of master genes.[15] I share their resistance and wish to examine further the issue of the role that should be ascribed to genes in even such an extreme case as that of the putative master gene for eye formation.

HOMOLOGY

To help drive home the biological importance of these conceptual issues, consider the problem of homology. Grant that the Gehring group is right, as it surely is, that the ancestor of the *ey* gene activated a module for forming primitive eyespots very early in metazoan evolution and that the eye-making cascades of different lineages evolved from this phylogenetically primitive starting point. How are we to take account of the fact that eyes — and the later parts of the eye-making cascade — evolved so differently in different lineages? Don't the differences in material composition, structure, and morphology between mouse eyes and drosophila eyes resolve the issue in favor of analogy rather than homology? Don't the enormous differences between the genes and gene products recruited in forming mouse and drosophila eyes — and the structural differences in the ways those products are organized force the traditional answer — *analogy!* — on us? But, then, how are we to account for the genetic

[15]See for example Keller, 1995 and Nijhout 1990.

homology that we just granted? This is a piece of the problem of understanding how to define and apply the concept of homology properly. The topic is too large to pursue at length here; the most that can be hoped for is to show that there is a conceptually coherent way of answering these questions in favor of *analogy* rather than *homology*.

For brevity, I shall borrow the approach developed by Roth 1984, with an addition from the work of Gilbert *et al.*, 1996. Roth grants that homology requires reproductive community, so that "*genetic relationship* provides the thread which unites all forms of biological homology into one concept, with one definition" (1984, 17). She proposes that homology requires the sharing of a common developmental pathway, and points out that this proposal encompasses both serial and special homology. She also emphasizes that *features of structures, rather than the structures themselves, are homologous* and that *the extent of shared differentiated pathways is a matter of degree.* Thus,

> homology is not an all-or-none phenomenon: it is important to recognize *degrees* of homology. The hierarchical nature of phylogenetic homology is widely recognized: as forelimbs, bird wing and bat wing are homologues; as wings they are not — the reptilian common ancestor had forelimbs, not wings. By analogy with phylogenetic relationship, in which taxa can be closely or distantly related, structures can be biologically homologous to varying degrees — from nearly identical, strongly homologous structures, on down to the very weakest degree of homology, manifested by structures which simply derive from the same germ layer. The points at which developmental paths diverge determine the strength of the homology (pp. 18–19).

Furthermore, homology must be measured in a way that is appropriate to the structures being examined; different measures of homology are appropriate for genes than are appropriate for limbs or eyes or for behaviors.

> By stressing the genetic basis of homologous relationships I do not mean to imply that one could measure degree of homology in morphological structures most effectively by counting the number of identical codons involved in their developmental process. ... While I attribute homology to the genealogy of genes (whereas I would reject a suggestion of homology if structures are created by unrelated sets of genes), the *measure* of homology in morphological structures resides in part at the level of descriptive embryology (p. 18).

Although Roth's official view is that *features* not *structures* are homologous, she does not consistently enforce this position. Her account would be greatly improved by so doing. On the resultant account of homology, the fact that certain genes are intimately involved in, or even trigger the entire cascade for the formation of, say, the eyes of two different organisms cannot settle the degree of homology between those eyes, for the degree of homology that a comparison will yield will depend on the features compared:

> Tying the definition [of homology] to developmental pathways [indeed, more strongly, tying the *degree* of homology to the extent of differentiated developmental pathway held in common — RB] establishes a link between the concept, and the element of continuity (genetic reproduction) at its basis. However, applying it at the level of embryology rather than at the level of genes, in addition to making it more practical, keeps the definition at the organismal level, where the concept was applied originally and where it is most biologically meaningful (p. 19).

An addition from the work of Gilbert *et al.*, 1996, helps clarify the point. They make a very useful distinction between what they call *process* homology and *structural* homology. The findings regarding segment formation and specification discussed above show that basic processes underlying segmentation and specification of segment identity in insects and mammals are homologous (and are initiated by genes whose sequence similarity is due to homology) even though the relevant features of the structures produced on the various segments — e.g., to speak only of insects, wings and halteres — are not homologous. Putting it more carefully, since homology *is* a matter of degree, the *processes* underlying segment specification and organ formation can be strongly homologous even when the relevant aspects or characteristics of the structures under investigation cannot be counted as homologous. Accordingly, the findings regarding homeotic mutants during the last ten to fifteen years may be summarized as follows: there is now strong support for the immensely surprising result that both segmentation and the production of eyes in vertebrates and invertebrates are initiated by a cascade of homologous processes even though the features of the structures that corresponding segments bear and the eyes of these organisms are, at best, analogous. Thus, in spite of the strong homology between the series of process control systems in invertebrates and vertebrates that initiate eye formation, nearly any suitable characteristic of the vertebrate and invertebrate structures on corresponding segments should count as analogous rather than homologous. The great surprise at stake is the deep and unsuspected homology of major aspects of the processes and genes underlying the formation of the structures in question, not the homology of the resultant structures. Biologists will be coping with the ramifications of this surprise for a long time to come.

EXPLANATORY APPARATUS

I take one of the morals of my story to be that we do not yet know what is required for a deep improvement of the explanatory apparatus with which to explain many aspects of organismic development. It is clear that many incredibly intricate multi-leveled domains, mechanisms, processes, structures, and so on enter into development. Furthermore, many of these are formed (pardon the pun) "on the fly" — that is, they are not laid out in advance but arise in interactions between genes and proteins that come to form a rapidly-shifting tartan of boundaries between domains and define something like morphological fields in the midst of the ongoing processes of cell-type specification, tissue formation, organogenesis, etc. At any stage of development some of the relevant modules that enter into normal development preexist, others are formed in the course of events, and others will or will not be formed according to the status and condition of interacting units and modules at key moments in the processes in question. The shuffling of process modules and alterations of their interactions is a form of evolutionary tinkering. Recognition of this topic opens up a wide range of interesting investigations, which is at its very beginning.

I mention two final examples, both of xenoplastic transplantation, to make the point stand out. Over sixty years ago, Spemann and Schotté (1932) showed that reciprocal transplants between presumptive flank ectoderm in early frog and newt gastrulae, placed in the ectodermal region destined to become the mouth in the other animal's gastrula, yielded parts related to the mouth. But the frog material in the newt yielded tadpole suckers, whereas the newt material in the frog yielded newt balancers (Figure Seven). Harrison (1933) reports that Spemann described this results as follows: The ectoderm says to the inducer, "you tell me to make a mouth; all right, I'll do so, but I can't make your kind of mouth; I can make my own and I'll do that."[16] This result has been widely reproduced. In contrast, the second is still contested. But it is remarkable in seeming to recover potentialities lost about ten million years ago. Some fifteen years ago, Kollar and Fisher (1980) reported that they had activated the machinery for making non-mammalian teeth, presumably the teeth of a proximal ancestor of chickens, in chick epithelium transplanted into mice. If this result is really reliable, it shows that there are deeply buried processes which, under ill-understood circumstances, can be activated and recruit the determined tissue into quite different pathways. I interpret these distinct results as showing not only that morphological fields are changing entities, but that their very existence (if we take them to be causally efficacious) is at best relative to processes and events that we do not understand; we have no reason to believe that there is a tooth-forming morphological field in chickens, yet, under rather bizarre circumstances, chicken cells can produce or respond to something of the sort.

I am sceptical (though my scepticism is not an argument) whether, at this point, we have anything resembling an adequate way of describing the multi-leveled causally relevant units for most of the developmental processes of interest. But the Spemann and Schotté and Kollar and Fisher results illustrate very nicely the fact that the controls of such processes as making

[16]Scott Gilbert kindly called this example to my attention. It is described in Gilbert 1988, 561–562; the quotation from Spemann, reported in Gilbert, is in Harrison 1933, p. 318.

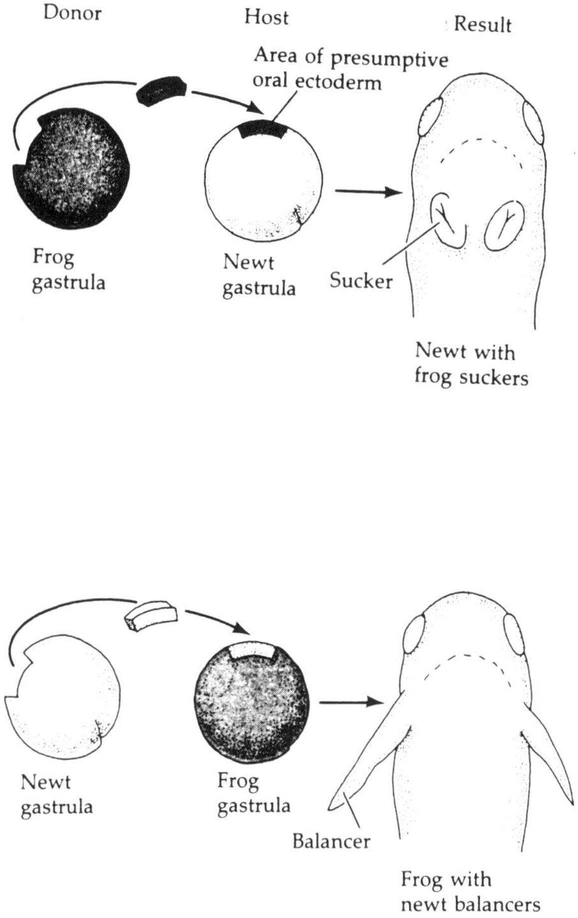

Fig. 7. Reciprocal transplantation between newt and frog gastrulae. In each case indifferent ectoderm is transplanted into the area of presumptive oral ectoderm of the foreign host. The result is a product that would normally belong near the mouth of the *donor* not the host species. (The figure shows newt larvae with tadpole suckers and tadpoles with newt balancers). Such findings show the genetic specificity of the donor material and its responsiveness to the control system (morphological field?) that determines the fate of the region into which it is transplanted. Reproduced, with permission, from Gilbert 1988, Fig. 6, p. 561. This figure was drawn after Hamburgh, 1970, based on experiments first reported in Spemann and Schotté 1932.

mouths or teeth are preserved across species barriers, while the details of the product are controlled at or close to the level of the species. On grounds such as these, I am as sceptical of transformations of concepts taken from developmental biology, such as the concept of morphogenetic fields, as I am of restricting ourselves to narrowly genetic concepts. Until we know how to delimit adequately the boundaries of the causally relevant units in a developmental network, to determine when they are brought into existence or made causally efficacious, and to detail the mechanisms that yield different results when different materials from different sources are placed within the boundaries of a given unit, we will not be able to assess adequately the value of any attempted integrative explanatory apparatus for the phenomena of development. And since, to repeat, the interactions in question involve multi-leveled processes and sometimes evanescent modules, made by bootstrapping "on the fly" and effective only in interaction with other such modules, we are far from understanding what, exactly, it is that we are talking about.

This sceptical stance is meant as a spur, not a barrier, to research. The challenge is to articulate the strengths and limitations of various approaches attempting to integrate genetics with our current understanding of developmental phenomena and to enforce contact between them as best we can. More especially, the deep challenge is to work out how to demarcate the relevant boundaries of the many interactive and dynamic processes required to cause the formation of particular structures so that one can balance the need for identifying the relevant units in a causal analysis with the dynamic and multi-leveled character of the processes under investigation. This is a task that will require fascinating conceptual, as well as empirical, research — a task I challenge the reader to take up.

Center for the Study of Science in Society
Virginia Polytechnic Institute and State University
Blacksburg

REFERENCES

Allen, Garland E., 1986. "T. H. Morgan and the split between embryology and genetics, 1910–1935." In T. J. Horder, J. A. Witkowski, and C. C. Wylie, eds., A History of Embryology (Cambridge: Cambridge University Press), pp. 113–146.

Bürglin, Thomas R., 1994. "A comprehensive classification of the homeobox genes." In D. Deboule, ed., Guidebook to the Homeobox Genes (Oxford: Oxford University Press), pp. 27–71.

Burian, Richard M., MS 1985. "Lillie's paradox — or some hazards of cellular geography."

Burian, Richard M., 1993. "How the choice of experimental organism matters: Reflections on an epistemological aspect of biological practice." Journal of the History of Biology 26: 351–367.

De Robertis, Eddy M., Elaine A. Morita, and Ken W. Y. Cho, 1991. "Gradient fields and homeobox genes," Development 112: 669–678.

De Robertis, Eddy M. 1994. "The homeobox in cell differentiation and evolution." In D. Deboule, ed., Guidebook to the Homeobox Genes (Oxford: Oxford University Press), pp. 13–23.

Gehring, Walter J., 1994. "A history of the homeobox." In D. Deboule, ed., Guidebook to the Homeobox Genes (Oxford: Oxford University Press), pp. 3–10.

Gilbert, Scott F., 1980. "Owen's vertebral archetype and evolutionary genetics: A Platonic appreciation," Perspectives in Biology and Medicine 23: 475–488.

Gilbert, Scott F., 1988. Developmental Biology, 2nd ed. (Sunderland, MA: Sinauer).

Gilbert, Scott F. (ed.), 1991. A Conceptual History of Modern Embryology (New York: Plenum).

Gilbert, Scott F., John M. Opitz, and Rudolf A. Raff, 1996. "Resynthesizing evolutionary and developmental biology", Developmental Biology 173: 357–372.

Halder, G., P. Callaerts, and W. J. Gehring, 1995. "Induction of ectopic eyes by targeted expression of the *eyeless* gene in Drosophila," Science 267: 1788–1792.

Hamburgh, M. 1970. Theories of Differentiation (New York: Elsevier).

Harrison, Ross G., 1933. "Some difficulties of the determination problem," American Naturalist 67: 306- 321.

Harwood, Jonathan, 1984. "The reception of Morgan's chromosome theory in Germany: Inter-war debate over cytoplasmic inheritance," Medizinhistorisches Journal 19: 3–32.

Horder, T. J., J. A. Witkowski, and C. C. Wylie, eds., A History of Embryology (Cambridge: Cambridge University Press).

Keller, Evelyn Fox, 1993. Refiguring Life: Metaphors of Twentieth Century Biology (New York: Columbia University Press).

Kollar, E.J. and C. Fisher, 1980. "Tooth induction in chick epithelium: Expression of quiescent genes for enamel synthesis," Science 207: 993–995.

Lawrence, Peter A., 1992. The Making of a Fly: The Genetics of Animal Design (London: Blackwell Scientific Publications).

Lewin, Benjamin, 1990. Genes IV (Oxford: Oxford University Press).

Maienschein, Jane, 1986. "Preformation or new formation — or neither or both?" In T. J. Horder, J. A. Witkowski, and C. C. Wylie, eds., A History of Embryology (Cambridge: Cambridge University Press), pp. 77–108.

Malicki, J., L. C. Cianetti, C. Peschle, and W. McGinnis, 1992. "Human HOX4B regulatory element provides head-specific expression in *Drosophila* embryos," Nature 358: 345–347.

McGinnis, N., M. Kuziora, and W. McGinnis, 1990. "Human *Hox 4.2* and *Drosophila Deformed* encode similar regulatory specificities in *Drosophila* embryos and larvae," Cell 63: 69–976.

Morgan, Thomas Hunt, 1934. Embryology and Genetics (New York: Columbia University Press).

Morgan, Thomas Hunt, 1935. "The relation of genetics to embryology," [Nobel Lecture, reprinted in] Scientific Monthly 41: 5–18.

Nijhout, H. Frederik, 1990. "Metaphors and the role of genes in development." BioEssays 12: 441–446.

Opitz, John M., 1985. "The developmental field concept," American Journal of Medical Genetics 21: 1–11.

Plate, Ludwig, 1924. Allgemeine Zoologie und Abstammungslehre (Jena: Fischer Verlag).

Quiring, Rebecca, Uwe Walldorf, Urs Kloter, and Walter J. Gehring, 1994. "Homology of the *eyeless* gene of *Drosophila* to the *small eye* gene in mice and *Aniridia* in humans," Science 265: 785–789.

Raff, Rudolf A. and Thomas C. Kaufman, 1983. Embryos, Genes, and Evolution: The Developmental-Genetic Basis of Evolutionary Change (New York: Macmillan).

Roth, V. Louise, 1984. "On Homology," Biological Journal of the Linnean Society 22: 13–29.

Sander, Klaus, 1986. "The role of genes in ontogenesis — evolving concepts from 1883–1983 as perceived by an insect embryologist." In T. J. Horder, J. A. Witkowski, and C. C. Wylie, eds., A History of Embryology (Cambridge: Cambridge University Press), pp. 363–395.

Sander, Klaus, 1995. "Of gradients and genes: Developmental concepts of Theodor Boveri and his students," Roux's Archive for Developmental Biology 203: 295–297.

Sapp, Jan, 1983. "The struggle for authority in the field of heredity, 1900–1932: New perspectives on the rise of genetics," Journal of the History of Biology 16: 311- 342.

Sapp, Jan, 1986. "Inside the cell: Genetic methodology and the case of the cytoplasm." In J. A. Schuster and R. R. Yeo eds., The Politics and Rhetoric of Scientific Method (Dordrecht: D. Reidel) pp. 167–202.

Sapp, Jan, 1987. Beyond the Gene: Cytoplasmic Inheritance and the Struggle for Authority in Genetics (New York: Oxford University Press).

Spemann, Hans, and Oscar Schotté, 1932. "Über xenoplastische Transplantation als Mittel zur Analyse der embryonalen Induction," Naturwissenschaften 20: 463–467.

Wolpert, Lewis, 1986. "Gradients, position, and pattern: a history." In T. J. Horder, J. A. Witkowski, and C. C. Wylie, eds., A History of Embryology (Cambridge: Cambridge University Press), pp. 347–362.

JEAN GAYON

THE "PARAMOUNT POWER OF SELECTION": FROM DARWIN TO KAUFFMAN

1. INTRODUCTION: "EXPLANATORY POWER" AND "POWER" OF NATURAL SELECTION

For approximately two decades now, the Darwinian interpretation of evolution has now been challenged in many ways. Modern criticisms make it difficult, even for the staunchest Darwinians, not to take a distance from Darwin's bold phrases on the "power" of natural selection. Let me remind you of some famous declarations of Darwin on the subject:

> "It may be said that natural selection is daily and hourly scrutinising, throughout the world, every variation, even the slightest; rejecting that which is bad, preserving and adding up all that is good; silently and insensibly working, whenever and wherever opportunity offers, at the improvement of each organic being in relation to its organic and inorganic conditions of life".[1]
>
> "What limit can be put to this power, acting during long ages and rigidly scrutinising the whole constitution, structure and habits of each creature,—favouring the good and rejecting the bad? I can see no limit to this power, in slowly and beautifully adapting each form to the most complex relations of life. The theory of natural selection, even if we looked no further than this, seems to me to be in itself probable".[2]
>
> "The long-continued accumulation of beneficial variations will infallibly have led to structures as diversified, as beautifully adapted for various purposes and as excellently co-ordinated, as we see in the animals and plants around us. Hence I have spoken of selection as the paramount power, whether applied by man to the formation of domestic breeds, or by nature to the production of species".[3]

---

[1] Charles Darwin, *On the Origin of Species*, London: Murray, 1859, (Facsimile: Cambridge: Harvard University Press, 1964), p. 84.

[2] *Ibid.*, p. 469.

[3] Charles Darwin, *The Variation of Animals and Plants Under Domestication*, 2nd ed., 1875, quoted in *The Works of Charles Darwin*, New York, AMS Press, vol. 8, p. 426; see also p. 236.

M.L. Dalla Chiara et al. (eds.), Structures and Norms in Science, 265–282.

The purpose of this paper is to provide some landmarks in order to better assess the implications of assertions, old and new, about the "power" of natural selection.

One important distinction must be made from the outset. There are in fact two possible meanings for the expression "power of selection", but only one can be found in Darwin himself. The "power of selection" has been often understood, particularly in contemporary literature, as "the explanatory power" of the principle of natural selection, or in other words the answer to the question: —what classes of phenomena does it account for?[4] In that sense, extinction, divergence, or patterns of classification, as far as they can be presented as consequences of natural selection, belong to its explanatory power. However, Darwin himself did not speak of the "power" of selection in that sense. Rather, he habitually restricted it to the explanation of only one class of phenomena, the adaptive modification of species. It is only in this precise context that he qualified selection as a "paramount power", or simply (in *The Origin of Species*) as a "power", or else as an "agent". In Darwin's spontaneous philosophical vocabulary, the *power* of selection, either artificial or natural, consisted in the immediate result of its *action*. In other explanatory contexts, he preferred to say that natural selection "entails", "leads to", "explains"[5], "induces" or even "causes"[6], for instance, extinction, divergence or affinities.

We will see that this distinction between two senses of "power" is more important than it looks at first sight. In particular, it can shed light on contemporary challenges to Darwinism. Most criticisms addressed to the Darwinian paradigm in recent times, especially in the context of macroevolutionary debates, have dealt with the explanatory power, or theoretical responsibility[7], of natural selection, or its claim to account for many classes of facts other than adaptive modification. Moreover, as strange as it may appear, attacks against what Darwin what called the "paramount power of selection" —its

[4]In a similar sense, Jonathan Hodge speaks of the "responsibility" of the principle of natural selection ("Natural Selection as a Causal, Empirical, and Probabilistic Theory", in *The Probabilistic Revolution*, 1987, L. Krüger, G. Gigerenzer, M.S. Morgan, Eds., Vol. 2., Cambridge, M.I.T. Press, pp. 233–270).

[5]"Whether natural selection has really thus *acted* in nature, *in modifying and adapting* the various forms of life to their several conditions and stations, must be judged of by the general tenour and balance of evidence given in the following chapters. But we already see how it *entails* extinction... Natural selection, also *leads to* divergence of character..." (Charles Darwin, On the Origin of Species, London: Murray, 1859 [Facsimile: Cambridge: Harvard University Press, 1964], pp. 127–128. Emphasis added). "Natural selection, as has just been remarked, *leads to* divergence of character ant to much extinction of the less improved and intermediate forms of life. On these principles, I believe, the nature of the affinities of all organic beings may be *explained*". (*Ibid.*, p. 128. Emphasis added.)

[6]"We shall then see how Natural Selection almost inevitably *causes* much extinction of the less improved forms of life, and *induces* what I have called Divergence of Character". (*Ibid.*, p. 5. Emphasis added.)

[7]On this term, see above, n. 4.

ability to modify and adapt species—, have been much rarer in contemporary literature. Attacks of this latter sort were in fact commoner in the early history of Darwinism, when, for example, Lamarckian or mutationist theories of evolution contested natural selection as a sufficient or even conceivable explanation for the genesis of adaptations. However, a strong exception in recent literature can be found in the work of Stuart Kauffman. In his recent book on *The Origins of Order*[8], the problem of the "power" (and correlatively the "limits") of selection is explicitly central from the beginning to the end.

These preliminary remarks justify the direction of my argument. The first step will be to clarify Darwin's assertions about the power of selection. I will examine this notion in relation with other philosophical categories used by Darwin for characterizing the status of Natural Selection. I will then turn to contemporary evolutionary biology. I will first provide a recent example of the independence of the meanings I have previously distinguished for the "power" of natural selection. I will then consider two clear cases in which more or less genuine Darwinians have tried to face the difficulties raised by Darwin's argument on the "paramount power" of selection in modifying and adapting species. One will be taken from classical population genetics. The other will concern Kauffman's theoretical reflections on self-organisation and selection.

2. DARWIN AND THE "POWER" OF SELECTION

2.1. *Natural Selection as Hypothesis, Theory and Power*

Darwin had relatively clear ideas about the philosophical status of natural selection. These ideas were not explicitly formulated in *The Origin of Species*, but in *The Variation of Animals and Plants under Domestication* (1868). There, he used three important words for qualifying the status of natural selection: "hypothesis", "theory", and "power". Let us begin with "hypothesis" and "theory". This distinction is made in the introduction of *The Variation*:

> "In scientific investigations, it is permitted to invent any *hypothesis*, and if it explains various large and independent classes of facts it rises to the rank of a *well-grounded theory*. (...)
> The *principle* of natural selection may be looked upon as a *mere hypothesis*, but *rendered in some degree probable* by what we positively know of the variability of organic beings in a state of nature, —by what we positively know of the struggle for existence, and the consequent and almost inevitable preservation of favourable variations, —and from the analogical formation of domestic races.
> *Now this hypothesis may be tested*, and this seems to me the only fair and legitimate manner of considering the whole question, —*by trying*

[8]Stuart A. Kauffman, *The Origins of Order: Self-Organization and Selection*, New York and Oxford, Oxford University Press, 1993.

> *whether it explains several large and independent classes of facts*, such as the geological succession of organic beings, their distribution in past and present times, and their mutual affinities and homologies. If the principle of natural selection *does explain* these and other large bodies of facts, it ought to be received" (emphasis added)[9].

These sentences make the structure of the argument of *The Origin of Species* explicit. Considering natural selection first as "a mere hypothesis", Darwin says that it is "rendered (...) probable" by certain classes of empirical data. This assertion refers to the first five chapters of The Origin, where natural selection appears to be a conclusion drawn from a set of highly probable universal premises (about the rate of reproduction of organisms, about the limitations of resources, and about variation and heredity). Note also the allusion to "the analogical formation of domestic races". Here, Darwin does not refer to a premise leading to the natural selection hypothesis, but to an experimental analogical model, the main function of which is to convince the reader of the efficacy of selection in modifying species. I will return later to this aspect, which is important for the interpretation of selection as a "power".

As for the "well-grounded theory", it is a clear reference to the second half of *The Origin* (chapters 7 to 12 in the 1st edition), where natural selection plays the role of a principle in the old Newtonian sense, that is to say a proposition able to unify and explain various classes of independent facts. Thus, for instance, Darwin invokes natural selection for the explanation of animal instincts, extinction, divergence, geographical distribution of species, affinities and homologies. Hence a second strategy of justification of natural selection, through consequences of the hypothesis.

To sum up, we find in Darwin two levels of justification for natural selection. The "mere hypothesis" is concerned with direct evidence for the existence of natural selection, and with the concepts required for thinking coherently about it. The "well-grounded theory" is a reconstruction of the whole theoretical structure of natural history: in this grand theory, natural selection functions as a remote principle for many classes of facts. As noted by many historians and philosophers, the whole argument is in good agreement with the traditional strategy of confirmation of hypotheses in physical sciences since Newton, or, in other words, the ideal of the *vera causa*[10]

---

[9]Charles Darwin, *The Variation of Animals and Plants Under Domestication*, 2nd ed., 1875, quoted in *The Works of Charles Darwin*, New York, AMS Press, 1972, vol. 7, p. 9.

[10]David Hull, *Darwin and his Critics*, Cambridge, Massachusetts, Harvard University Press, 1973; V.C. Kavaloski, *The* vera causa *principle: a historico-philosophical study of a metatheoretical concept from Newton through Darwin.* University of Chicago, Ph.dissertation, 1974; Michael Ruse, "Darwin's debt to philosophy: an examination of the influence of the philosophical ideas of John F.W. Herschel and William Whewell on the development of Charles Darwin's theory of evolution", *Studies in History and Philosophy of Science*, 6: 159–181; M.J.S. Hodge, "Natural Selection as a Causal, Empirical, and Probabilistic Theory", in L. Krüger, G. Gigerenzer and M.S. Morgan (eds.), *The Probabilistic Revolution*, Cambridge, Massachusetts: The MIT Press, 1987, vol. 2, pp. 233-270.

Let us now come to the third concept Darwin utilized for characterizing the status of natural selection —selection as a "power", or, as said in *The Variation...*, a "paramount power". I have already given the key quotation. Here it is again:

> "The long-continued accumulation of beneficial variations will infallibly have led to structures as diversified, as beautifully adapted for various purposes and as excellently co-ordinated, as we see in the animals and plants around us. Hence I have spoken of selection as the paramount power, whether applied by man to the formation of domestic breeds, or by nature to the production of species"[11].

This text states that selection is the major force orienting change in species. What is the relation between this thesis and the "hypothesis" vs. "theory" distinction? The "paramount power" thesis bears on a special aspect of the theory: the explanation of adaptive change. Darwin's claim is that selection, if not the sole force orienting evolutionary change, is able to overcome any other force (such as: random variation, correlation of organs, or the effect of use and disuse). Although this thesis was absolutely crucial for Darwin, it must not be confounded with the representation of selection as unifying the whole field of the natural history of life. Modification of species is one thing; extinction, geographical distribution of species, divergence and diversity (as reflected in classification) are other things. The "paramount power" thesis is concerned only with adaptive modification of species.

At this point, it can be useful to represent the relations between "hypothesis", "theory" and "power" in a diagram. In most of its features, this diagram is classical. Please note that adaptations appear as a consequence of the hypothesis of natural selection. In a certain sense, this way of speaking is consistent with Darwin's own declarations. This can be seen easily for instance in the quotation above: "The long-continued accumulation of beneficial variations will infallibly have led to structures...as beautifully adapted as...". But, as we will see shortly, in other respects this presentation, as far as one does not take into account the "power" aspect of selection, is misleading. Indeed, the "paramount power" thesis might well be already there in the very formulation of the hypothesis.

---

[11]Charles Darwin, *The Variation of Animals and Plants Under Domestication*, 2nd ed., 1875, quoted in *The Works of Charles Darwin*, New York, AMS Press, vol. 8, p. 426; see also p. 236. In *The Origin of Species*, there is a similar sentence, but it concerns only the case of artificial selection: "Over all these causes of change I am convinced that the accumulative action of Selection, whether applied methodically and more quickly, or unconsciously and more slowly, but more efficiently, is by far the predominant Power" (1st ed., 1859 [Facsimile: Cambridge: Harvard University Press, 1964], p. 43).

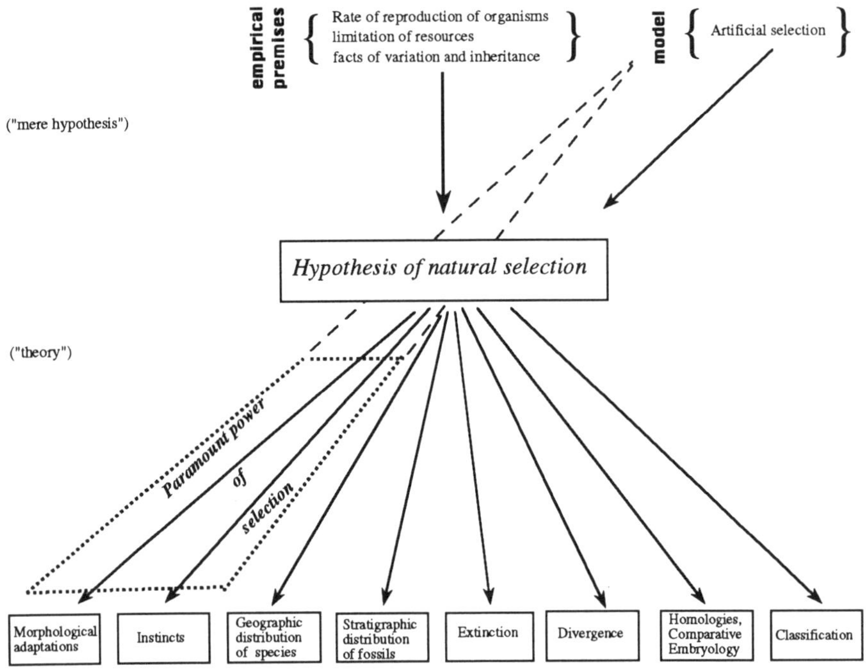

Fig. 1.

2.2. *Further observations on the "power" of selection in Darwin's discourse*

We must now answer two further questions. First, why is selection said to be a *paramount* power for the modification of species? Second, what is the function of Darwin's use of the word "power" in the general economy of his big "argument"?

Why is selection a *paramount* power for the adaptive modification of species? The argument consisted in an examination of the relation between variation and selection. Darwin admitted his complete ignorance of the laws of variation, but he believed that variation obeyed some laws, which were the precondition for any event of selection. Thus, he never said that selection could produce any arbitrary thing. He admitted for instance that there are strong correlation constraints restricting the range of possible variation. Other factors involved in the production of variation were, for example: direct or indirect action of changing conditions of life, use and disuse, reversion[12]. What he claimed was that, if there is variation of some sort among the members of species, then

[12] See *On the Origin of Species*, ch. I & V; *The Variation of Animals and Plants Under Domestication*, ch. XXII–XXVI.

this causal factor of change will necessarily be in a subordinate position in comparison with selection. Whatever may be the origin of variations, selection will both use them and overcome them:

> "I have spoken of selection as the paramount power, yet its action absolutely depends on what we in our ignorance call spontaneous or accidental variability. (...) The variations of each creature are determined by fixed and immutable laws; but these laws bear no relation to the living structure which is slowly built up through the power of selection, whether this be natural or artificial selection. (...) Although variability is indispensably necessary, yet, when we look at some highly complex and excellently adapted organisms, variability sinks to a quite subordinate position in comparison with selection"[13].

Thus, for Darwin, provided that there is variation, selection will necessarily modify species and increase their adaptation. Hence the bold declaration of the concluding chapter of *The Origin*, already quoted: "I can see no limit to this power..."[14]

Let us come to our second question: what is the function of Darwin's systematic characterization of natural selection as a "power" in the context of adaptive modification? Many historians and philosophers have pointed out the archaic character of Darwin's statements on natural selection as an "agent" or "power". Assuredly, there is some truth in this remark. However the interesting point, again, is that it is only when he speaks of modification and adaptation of species that Darwin uses such a vocabulary. Why does he not merely say that natural selection explains, or entails, or causes adaptive modification, just as it explains, entails, or causes extinction, divergence and affinities[15]? The reason for this is that, for Darwin, adaptive modification is a little more than a consequence of natural selection, even if an immediate consequence. Actually, Darwin tends to incorporate adaptive modification in the very formulation of the hypothesis of natural selection. Let us consider again this hypothesis. As classically presented[16], the hypothesis comes as the conclusion of an argument of the following sort:

1. In every species, all organisms tend to increase in geometrical ratio.
2. Many more individuals are born than can possibly survive.
   Hence:
3. Individuals compete for survival and reproduction.
4. For a number of characters, there is heritable variation.
   Hence:
5. (hypothesis of natural selection)

---

[13] Charles Darwin, *The Variation...*, *loc. cit.*, vol. 8, p. 236.

[14] See above n. 2.

[15] For quotations, see above n. 5 & 6.

[16] See for instance Julian Huxley, *Evolution: the Modern Synthesis*, London, Allen & Unwin, 1942, chap. I, § I.

Now what is the precise content of this hypothesis? In fact one finds two versions of it in Darwin's text. Let us call them the weak and the strong version. The weak or sober version is the following:

> "Each slight variation, if useful, is preserved"[17].

And here is an example of the stronger version:

> "Every slight modification, which in any way favoured the individuals of any of the species, by better adapting them to their altered conditions, would tend to be preserved"[18].

Most often, Darwin introduces the second version shortly after the first one[19]. In his own vocabulary, the first version describes the "process" of natural selection[20], the second one describes its "action" (adaptive modification). Darwin would never say that extinction, divergence or affinities describe the way selection "acts" in nature. And whenever Darwin evokes the "power" of selection, it is always in relation with its immediate action consisting in "modifying and adapting". Of course, this power becomes the greater as selection accumulates successive variations. But this is only a question of degree. Furthermore, one must observe the parallel with artificial selection: both natural and artificial selection are described in terms of their "power" and "action": both "modify" and "adapt"[21], and both are "predominant" (or "paramount") in that respect[22]. This confirms that Darwin's reflections on the "power" of selection definitively bear upon an intrinsic philosophical feature of the "hypothesis" of natural selection.

I will not pursue this terminological enquiry further. What I have tried to show is this: one should not confuse the explanatory capacity of the hypothesis of natural selection with what Darwin calls the "power" of natural selection. Of course, in a sense, adaptations can be viewed as "explained" by natural selection. But in the deductive hierarchy of Darwin's argument, adaptations do not stand on the same level as extinction, divergence, affinities and other explananda. For this reason, the classical diagram on which I commented earlier is not satisfactory. Under the light of artificial selection (the analogical model), the "power" of selection thesis percolates through the "hypothesis" *and* the "theory" of natural selection. But it does not diffuse in all regions of the theory. Hence the unusual "triangle" I have introduced in my diagram.

If we now turn to various challenges and criticisms addressed to Darwinism in the course of 20th Century evolutionary biology, this analysis has interesting consequences. One can indeed conceive of two very different kinds of criticism. One kind consists in denying that natural selection explains a vast

---

[17]Charles Darwin, *On the Origin of Species*, London: Murray, 1859, (Facsimile: Cambridge: Harvard University Press, 1964), chap. III, p. 61.

[18]*Ibid.*, chap. IV, p. 82.

[19]See for instance *Ibid.*, chap. III, p. 61; chap. IV, pp. 81–82, & 127.

[20]E.g. *Ibid.*, chap. IV, p. 102.

[21]*Ibid.*, chap. III, p. 61.

[22]See above n. 12.

array of "independent classes of facts" other than adaptations —such as: extinction, divergence, geographical distribution of species, fossil gaps, laws of comparative embryology, phylogenetic patterns, etc. This kind of criticism has been particularly intense in the field of macroevolutionary studies for the past two decades, but attacks coming from morphologists or embryologists would also fit this schema. A very different kind of criticism comes from those who have challenged Darwin's conviction that natural selection has the power of increasing adaptation. This kind of criticism implies a challenge to what Darwin named the proper "power" of selection — or its immediate and apparently self-evident "action". The following section will be mainly devoted to the question of possible intrinsic limits to the adaptive power of selection.

3. LIMITS TO THE EXPLANATORY POWER OF NATURAL SELECTION AND LIMITS TO THE "POWER" OF SELECTION: TWO DIFFERENT ISSUES

I now turn to contemporary evolutionary biology. In previous papers[23], I have claimed that the questions about the identity of Darwinism, past and present, could be clarified by admitting what I called a "decoupling conjecture". This conjecture consists in saying that there have been various regimes of Darwinian discourse, corresponding to the three philosophical roles assigned to natural selection in Darwin's own writings: those of "hypothesis", "theory", and "power". In the present section, I will use the two latter categories in order to distinguish two very different kinds of challenges to Darwinian schemes in recent evolutionary debates. It is one thing to deny that natural selection entails various large classes of facts other than adaptation, such as extinction and divergence. It is another thing to question the very idea that natural selection is able to improve organisms by better adapting them. The first kind of criticism aims at what Darwin called the "well-grounded theory" of natural selection (its explanatory capacity), the second one is concerned with what he named the "power" of selection. I do not want here to give an exhaustive account of these two kinds of criticisms, but to illustrate them by means of clear-cut cases.

3.1. *Challenging the "theory" of natural selection (its explanatory capacity)*

Recent macroevolutionary debates provide extensive evidence of criticisms of the grand Darwinian argument. One can think here, of course, of the con-

[23] Jean Gayon, "What does "Darwinism" mean?", *Ludus vitalis*, 2 (1994), pp. 105–118; "La biologie darwinienne de l'évolution est-elle réductionniste?", *Revue philosophique de Louvain*, 93 (1995), pp. 111–139; "Neo-Darwinism", in *Concepts, Theories and Rationality in the Biological Sciences — The Second Pittsburgh-Konstanz Colloquium in the Philosophy of Science*, ed. by Gereon Wolters and James G. Lennox in coll. with Peter McLaughlin, Universitätsverlag Konstanz/University of Pittsburgh, 1995, pp. 1–25.

troversy over punctuated equilibria, but this case is a rather complex one. I will therefore analyze a much simpler case, that of David Raup's reflections on extinction of species, in his recent book on the subject[24]. Raup challenges the classical Darwinian interpretation of extinction. In the classical Darwinian view, perfectly explicit in Darwin's Origin of Species, extinction is interpreted as a consequence of natural selection. As natural selection transforms species, some of them happen to be fitter in the competition with other species. The less favored forms decrease in number and finally go extinct[25]. This is the argument that permits Darwin to say that "the extinction of species... almost inevitably follows on the principle of natural selection"[26]. Raup does not deny that species go extinct in this way. But he says that a considerable number of extinctions, particularly mass extinctions, might well have been caused by non-biotic factors, of major physical changes affecting the earth, such as big climatic changes, impacts of comets, etc. Species which happened not to be equipped for such brutal perturbation in the physical environment went extinct for reasons which were independent of their relative success in the ecological theater.

Now the issue is: is this attitude non-Darwinian, and in what sense? In the conclusion of his book, Raup himself raises the question whether his interpretation of extinction challenges "Darwin's natural selection". He clearly answers "no", specifying that natural selection remains the only possible explanation for adaptations. But he adds that natural selection alone could not have produced mass extinction events, and the diversification of living beings which probably followed. Thus, Raup does not contest Darwin's idea of natural selection as the "paramount power" or cause of modification of species; but he denies that selection provides a sufficient explanation for at least one major class of facts other than adaptations. Thus, if there is any significant departure from Darwinism here, it is at the level of what Darwin named the "well-grounded theory of natural selection".

I have chosen Raup for the simplicity of the example. Raup does not question either the hypothesis of natural selection nor its power in modifying the adaptive features of organisms. His argument bears on natural selection as a sufficient principle for a general theory of the history of life. And even in this perspective, only one traditional Darwinian deduction is contested, that concerning extinction.

From this example, it is relatively easy to locate various other challenges arising within the field of macroevolutionary studies. Divergence of species and patterns of classification, for instance, have been major targets for criticism, particularly by the defenders of punctuated equilibria. More generally, serious doubts have been raised about the possibility of explaining patterns of

[24] David Raup, *Extinction. Bad Genes or Bad Luck?*, New York, Norton and Co., 1991.

[25] Charles Darwin, *On the Origin of Species*, London: Murray, 1859, (Facsimile: Cambridge: Harvard University Press, 1964), pp. 109–111.

[26] *Ibid.*, p. 475.

distribution in higher taxa or large scale ecological phenomena, solely in the language of natural selection. These various kinds of criticisms can be easily visualized on the diagram above. They boil down to suppressing one or the other of the "boxes" in the lower part.

3.2. *Challenging the (adaptive) power of selection*

I now come to the last part of my analysis. It will bear mainly on Stuart Kauffman's recent reflections on self-organization and selection. Reading Kauffman's recent book on *The Origins of Order*, I have been struck by his explicit concentration on the question of the "power" of selection. The first part of the book is indeed explicitly devoted to the examination of "the power and limits of selection when acting on complex systems exhibiting spontaneous order"[27]. However, before analysing the precise meaning of such adeclaration, let us briefly examine some previous circumstances in which evolutionary biology faced the question of the power of selection.

There are at least two cases in which the modifying and adapting power of selection has been seriously questioned in post-Darwinian biology. The first case corresponds to a period running from approximately 1870 to 1920. In this period, the main challenge natural selection had to face was heredity. Was the Darwinian idea of modification through gradual accumulation of small individual differences compatible with existing theories and experimental evidence about heredity? Galton, who was the first to try to provide a quantitative description of hereditary phenomena, thought that this was not possible; the mutationists, and most of the early Mendelians, held similar views, saying that selection was unable to create or modify anything, and that it was limited to picking out big changes that had arisen from a sudden process of "mutation". It is only when it appeared that continuous variation could be accounted for by Mendelism that biologists again considered it plausible that natural selection could gradually modify and adapt species.

The other major case in which the power of selection came to be seriously questioned is the heroic period of theoretical population genetics. Population genetics consisted of a systematic exploration of the consequences of Mendelian genetics at the population level. Although a highly mathematized discipline, its fundamental concepts were all distinctively biological: the concepts of mutation, recombination, migration, selection, population size and structure — all refer to biological phenomena. The proper description and understanding of these phenomena belong to various laboratory and field disciplines, such as cytogenetics, biogeography, ecology, demography, ethology. However, in population genetics, these phenomena were described in the ho-

[27]Stuart A. Kauffman, *The Origins of Order: Self-Organization and Selection*, New York and Oxford, Oxford University Press, 1993, p. xv. In his preface, Kauffman says also that a major theme of his book is to "understand the extent to which selection can achieve systems able to adapt" (*Ibid.*, p. vii).

mogeneous language of their measurable effects on gene frequencies. Within the new framework, many qualitatively different factors of evolution were describable as forces — forces which determined commensurable effects in evolution. This methodology entailed a new vision of the hypothesis of natural selection. In this new methodological context, selection could no longer be seen as a principle with exceptional privileges. In fact, selection has no privilege whatsoever in the equations of population genetics; it is a force among others, and there is absolutely no a priori nor conceptual necessity that this force should be the major evolutionary factor. For instance, if the population is small, or if there is a high rate of mutation in comparison with the selective pressure, it is perfectly possible that a population be driven mainly by forces other than selection. Of course, as soon as one applies the models to particular situations, selection will stand as a good candidate for the status of a major or even predominant force. But this is precisely the point: theoretical population genetics made it possible to formalize the various conditions under which selection could or could not be a preponderant factor in biological evolution.

Both these stories are well known[28]. My purpose here was only to indicate that Darwin's ideas about the power of selection have been repeatedly and seriously discussed since the publication of *The Origin of Species.* What must be stressed from a philosophical point of view is the following point: the question of the modifying and adaptive power of selection has always been relative to a certain theoretical and experimental context; it has never been a self-evident issue. For more than a century, biologists have looked for limits to the power of selection: they have found them in theories of heredity and population structure. And these limits have been incorporated into the very theory of selection itself. This point had to be stressed before turning to Stuart Kauffman's examination of the power and limits of natural selection.

It is quite out of the question for me to set forth, or even summarize, the ideas developed by Stuart Kauffman in his book on *The Origins of Order*. It is a luxuriant work, endowed with a strong sense of paradox, which makes it somewhat difficult to decide whether the author has brought decisive arguments against the Darwinian view of evolution or whether he has genuinely deepened it and taken it a step forward[29]. I will do no more than classify some theses defended by Kauffman which have a direct bearing on the question of the "power" of selection.

Let us first point out the bold question addressed by Kauffman to the Darwinian interpretation of adaptive modification. Early in the book, Kauffman

[28]For a detailed historical account, see William B. Provine, *The Origins of Theoretical Population Genetics*, Chicago, Chicago University Press, 1971; Jean Gayon, *Darwin et l'après-Darwin: une histoire de* l'hypothèse *de sélection naturelle*, Paris, Kimé, 1992.

[29]David J. Depew and Bruce H. Weber provide interesting information and judgment on this point in their *Darwinism Evolving: Systems Dynamics and the Genealogy of Natural Selection*, Cambridge (MA) and London (England), Bradford & MIT Press, 1995, chap. 16.

evokes what he names "the unchallenged core of evolutionary theory", and makes a clear and orthodox reference to Darwin:

> "A curious, logically unnecessary, but powerfully influential feature of Darwin's thinking was that the variation within one species which paved the way for emergence of well-marked varieties constituting two species was of an indefinite range. The idea that variations could occur in virtually any direction, an idea which dominates Darwin's work despite attention to correlations among traits under selection, has had important consequences. It follows that selection could discriminate which new variants will be found in later generations. Here is one root of our current idea that selection is the sole source of order in the biological world"[30].

Kauffman's challenge to Darwinism flows from these declarations. He questions the very idea that living systems are able to adapt through the accumulation of advantageous variations. Even if there is variation and if individuals differ in fitness, it is not obvious that selection can achieve adaptive modification of organisms. "Darwin told us that adaptive evolution occurs by gradual accumulation of useful variants but failed to tell us what kinds of systems can evolve successfully by random variation and selection for fitter variants"[31]. This is a leitmotiv of the book. One can hardly imagine a more straightforward formulation of the question of the "power" of selection.

Now, the perspective within which Kauffman raises this question is not indifferent: it is complexity. Whereas traditional evolutionary biology used to deal with particular traits, or limited groups of traits of organisms, Kauffman questions the power of selection from the point of view of organisms as complex systems, that is to say entities composed of many parts with many interactions between them. Hence the general question: is there anything in complex systems as such which could limit the adaptive and formative power of selection?

A word must be said about Kauffman's method. In most cases, it consists in speculating on formal properties of systems consisting of N parts, these N components having on the average K interconnections of some sort between them. Kauffman calls this the "NK model". By introducing a certain number of specific parameters, this basic idea can be applied to many biological, or non biological problems. In the case of evolution, it will lead, for instance, to the formulation of the following kind of question: —what will happen to a system of many genes with many epistatic relations between them, if submitted to a process of random mutation and selection? In particular, what will be the effect on the mutation-selection process of increasing the degree of complexity, by modifying both the number of parts and the average number of intercon-

[30]Stuart A. Kauffman, *The Origins of Order: Self-Organisation and Selection*, New York and Oxford, Oxford University Press, 1993, p. 6.

[31]*Ibid.*, p. 29. See also, p. v, xiv

nections? Similar questions can be raised about genetic regulatory networks, coevolving systems, and many other complex entities. In all these cases, the purpose is to evaluate the effects of increasing complexity on the evolvability and stability of the system, if exposed to some kind of mutation-selection process. In order to do this, Kauffman makes extensive use of computer calculation and simulation. In some cases, the computer is used for exploring numerical solutions to problems inaccessible to ordinary algebraic calculus. In other cases, the method consists in studying the dynamic properties of such and such a complex system by observing the behavior of an analogical system such as a Boolean network.

I do not want to go further into the detail about the methodology. I will concentrate on Kauffman's conclusions, focusing solely on the power of selection problem. These conclusions, although they are most often inextricably mingled with one another, fit into three different categories.

Kauffman's first assertion is that a number of the properties of complex systems arise as a function of their degree of intrinsic complexity, whatever the selective pressures are applied to them. These properties, which he calls "generic properties", cannot be viewed as the specific result of selection. In other words, they will arise in spite of selection, in virtue of "spontaneous sources of order". As a concrete example, Kauffman often cites the relation between the number of cell types and the number of genes in organisms. Most often however, the examples point to more formal self-organized properties. The general spirit of this first kind of limitation to the power of selection is that selection is not the "sole source of order". Complexity by itself restricts the field of possible solutions: in relation to the degree of complexity, there exist properties which will arise independently of, or in spite of, selection, even here selection is the proximate cause of change. This kind of limitation is in a certain sense exterior to the proper action of natural selection: just as selection cannot contradict the law of gravitation, selection must also coexist with regularities associated with complex organization. I am not sure that this comparison would resist a careful epistemological analysis, but there is no doubt that Kauffman has something like that in mind[32].

Beside this first kind of limitation, Kauffman explores another type, more closely related to the process of selection itself. A large part of the book on The Origins of Order tries hard to build models of selective evolution under the hypothesis that selection acts on a system of many genes with many interactions between them (epistasy in some models, regulatory loops in others). Using a generalized notion of fitness landscape, Kauffman tries to show that beyond a certain degree of complexity, there are serious limits to the ability of systems to evolve toward a higher fitness or even to maintain themselves at

[32]This first class of limits to the power of selection has been very carefully analyzed by Richard M. Burian and Robert C. Richardson, in "Form and order in evolutionary biology/ Stuart Kauffman's transformation of theoretical biology", *PSA*, 2 (1990), pp. 267–287. This article provides a deep analysis of the philosophical implications of Kauffman's enterprise.

a given level of fitness. Various kinds of "complexity catastrophes" can arise, as a function of the organizational constraints imposed on the system. Again, I will not enter here into the details of the models, but I would like to draw a parallel with classical theoretical population genetics. In the twenties and thirties, Fisher and Wright established that population structure imposed crucial limits on the capacity of selection to control the evolution of a Mendelian population. However their models were restricted to a very small number of loci, most often one. Kauffman's program is comparable in a certain sense: instead of looking for limits imposed by the population structure, he looks for them in the structure of the genome. And just as Wright believed that some kinds of population structure were more favorable than others to the action of natural selection, Kauffman tries to show that some kinds of connectivity within the genome are more favorable than others to the action of selection.

Finally, one finds in Kauffman a third category of proposals regarding the power of selection, which he calls "the bold hypothesis"[33]. The bold hypothesis does not place a limit on the power of natural selection; rather, it is a deepening of it. In fact, in a sense, Kauffman proposes more for the direct action of selection than Darwin did. The idea comes itself as an answer to the issue of the origin of adaptability:

> "We must ask that which Darwin did not broach: What kinds of integrated dynamical systems harbor the ability to adapt?"[34].

Kauffman's ultimate answer to this question is now well known. He supports the thesis that the most favorable situation is a regime intermediate between "ordered" and "chaotic". In other words, selection will be better able to improve the adaptation of organisms within a lineage as the system submitted to its action will be in this poised regime, "between solid and liquid". But what produces this miraculous state? Here comes the "bold hypothesis", repeated a number of times throughout the book. Let me quote a few passages stating the bold hypothesis in various ways:

> "[we are led] to the hypothesis that *the target which selection achieves* is complex systems poised in the complex regime on the boundary between order and chaos" (emphasis added)[35].
>
> "We examine the attractive hypothesis that networks poised at the edge of chaos can perform the most complex tasks. Furthermore, *we consider whether selection can achieve such poised systems*. If both answers are yes, as they begin to appear to be, then we may have succeeded in discovering the characteristic kind of complex systems which selection achieves in order to optimize both evolvability and fitness" (emphasis added).[36]

---

[33]For the characteristic formulas, see pp. xvi, 30, 175, 183, 209, 218, 221, 232, 280, 645.
[34]*Ibid.*, p. 209.
[35]*Ibid.*, p. 30.
[36]*Ibid.*, p. 175

> "Parallel-processing systems lying in this interface region between order and chaos may be those best able to adapt and evolve. Further, *natural selection may be the force which pulls complex adaptive systems into this boundary region*. If so, we begin to have a powerful tool with which to examine the collaborative interaction between self-organization and chaos" (emphasis added)[37].
>
> "Living systems exist in the solid regime near the edge of chaos, and *natural selection achieves and sustains such a poised state*" (emphasis added)[38].
>
> "We expect *that selection can mould the entities it acts on to improve the characters of the landscapes these entities explore*... In short, the capacity to evolve is itself subject to evolution and may have its lawful properties" (emphasis added)[39].

These quotations are puzzling. Here is an author who claims many times that there are strong limits, external and internal, to the power of selection. Nevertheless, the same author comes to support the view that, ultimately, selection has the power of "tuning" complex systems in such a way that they optimize the action of selection itself. In other words, selection is able to evolve adaptability itself. David Depew and Bruce Weber, In their recent book entitled *Darwinism Evolving*, say that in exploring such an idea, Stuart Kaufmann has indeed attempted to provide a theoretical basis for Dobzhansky's "intuitive vision of adaptations for adaptability"[40]. In that sense, the discourse about the limits of natural selection turns into an expansion, or better, a deepening of its power. Note that this is not Darwin's sense of the "power of selection" as I explained it above. Darwin's sens concerned the ability of selection to "adapt" and "form" the organisms within a lineage. Kauffman's question concerns the ability of selection to mold evolving systems so that they have the property of being subject to what Darwin called the power of selection[41].

Therefore, to summarize, on the one hand, Kauffman increases significantly the external and internal limits imposed to the power of selection in Darwin's sense; on the other hand, he increases this power, by extending it to the generation of properties such as "evolvability" and "adaptability".

4. CONCLUSION

What lessons can we draw from this story? I will formulate three.

First, I think I have made clear that historians and philosophers of biology should clearly distinguish questions related to the explanatory capacity of

---

[37] *Ibid.*, p. 218.
[38] *Ibid.*, p. 232
[39] *Ibid.*, p. 645.
[40] David J. Depew and Bruce H. Weber, *op. cit.*, p. 443.
[41] I am indebted to Richard M. Burian for this nice formulation.

natural selection and issues related to what Darwin called its "power". This distinction is particularly useful for anyone who wants to clarify contemporary criticisms of Darwinian evolutionary biology. It is one thing to contest the ability of natural selection to "entail" (and therefore explain) such and such a class of phenomena , it is another thing to question the intimate "action" of natural selection.

Second, one can be puzzled by the continuing importance of the question of the power of selection in the history of evolutionary biology. In the time of Darwin himself, his manner of speaking of natural selection with this kind of philosophical vocabulary was criticized as an archaism, and there was certainly some truth in this. However, in the course of time, this archaic philosophical vocabulary may have been useful. Darwin's confidence in the power of selection has indeed stimulated many people to seek for its limits. From biometry to population genetics, and today to the dynamics of complex systems, the "power of selection" issue has led to a repeated deepening of the concept of natural selection.

Finally, a philosopher cannot avoid asking about the strange use Darwin made of the term of "power". The distinction we find in Darwin's writings between the "power" and the consequences of natural selection can be captured as follows. In a sense, adaptation is one consequence of natural selection among others; in another sense, adaptation is the immediate result of the proper action of natural selection, and is a little more than a mere consequence, although it is indeed a consequence. This subtle distinction means that Darwin's notion of power is not merely epistemological. The explanatory capacity of natural selection is an epistomological issue: it is related to the deductive structure of a theory. But what Darwin called the power of natural selection requires one to think of it in terms of its action or causal efficacy in a defined material space. This is a material issue. Again, one must stress the fact that Darwin described both artificial and natural selection as a power. The model of artificial selection was indeed immensely important because it was the sole evidence Darwin had of the efficacy of a selective process. If Darwin had been interested only in the deductive structure of his "theory", he would not have needed to depend on this analogical model.

Département de philosophie
Université de Bourgogne, Dijon

and

Institut d'histoire et de philosophie des sciences et des techniques
Paris

ACKNOWLEDGEMENTS

I want to thank R.M. Burian and M. Grene for their linguistic corrections and numerous fruitful comments. Let also E. Fox Keller and J. Hodge be thanked for their fruitful comments on an earlier and oral version of this paper; these comments are more or less incorporated in the ultimate conclusion of this text.

C.N. VAN DER WEELE

ENVIRONMENTAL CAUSES IN ONTOGENY AND AN ETHICS OF ATTENTION

Doing science involves making many choices. Since those choices have consequences outside science, we need a moral philosophy of science that takes into account such consequences.

This is the general thesis of this paper. The argument, however, will not be so general, because I defend and illustrate the general idea with the help of a specific case, which is the study of embryological development in animals. The paper is based on my conceptual work on developmental biology during the last few years (Van der Weele 1995, 1993a,b), which has focused on the place of environmental influences in this field. I begin by giving examples of environmental influences in animal embryological development. Second, I sketch how a genetic approach deals with those influences and third, how some alternative approaches differ from the genetic approach—and resemble it as well. After that I go into the unavoidability of scientific choices in the study of development. Finally, I shall associate 'ethics of attention' with a consideration of consequences of such choices.

1. ENVIRONMENTAL INFLUENCES IN ONTOGENY

How can the environment be involved in development? Let me distinguish three different ways, and give examples of each of them in turn:

- a Different normal environments may make a specific difference for developmental outcomes.
- b "Background conditions" do not make a specific difference for the phenotypic outcome, though they are vitally important.
- c Abnormal conditions may disturb development.

1.1. *Making a difference*

Environmental influences often make a difference for developmental outcome. In many cases, the sex of an animal is determined by environmental factors, such as temperature or food. Crocodiles and other reptiles are probably the most well-known example here. Temperature during some part of embryological development determines whether a crocodile becomes male or female (For overview and hypotheses concerning mechanisms, see Janzen and Paukstis 1991). In many other reptiles, too, this phenomenon occurs. Let me mention a few more examples of environmental sex determination.

M.L. Dalla Chiara et al. (eds.), Structures and Norms in Science, 283–300.

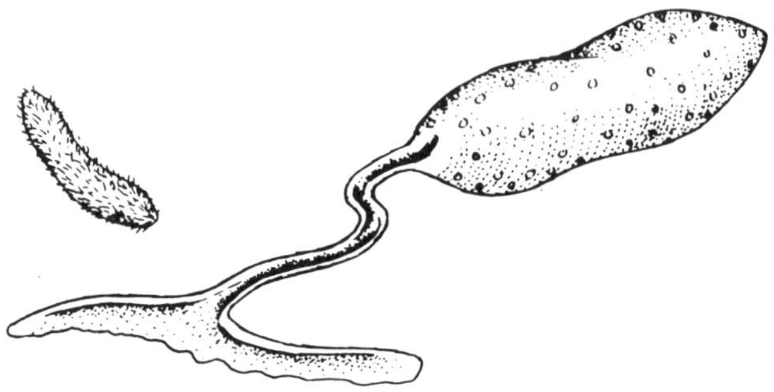

Fig. 1. *Bonellia viridis* Right: Female. Left: male, greatly magnified compared with the female.

- In 1912, the marine echiurid *Bonellia viridis* was one of the first organisms for which environmental sex determination was described. Larvae are free swimming, and sex develops only after settling. If a larva lands on a rock, it develops into a female with a body of about 10 cm and a long proboscis, which can become more than a meter in length. This proboscis is used for feeding, but it also has a function in sex determination. If a larva lands on the proboscis of a female, it migrates into the female and develops into a tiny (1 to 3 mm) parasitic male. The male stays inside the female throughout its life and fertilizes her eggs (Bull 1983:110; Gilbert 1994:783).
- *Mermithids* are a family of roundworms which are free living as adults, but parasitic on insects in the larval stage. In this larval period, they feed on the host, which is killed when the worms leave it. The sex of the worms is determined by the amount of food they get, as can be inferred from:

  a Crowding effects. When there are few worms per host (1–5), they become almost always female. The percentage of males increases with the number of worms per insect; when there are more than 16 worms per host, nearly all of them become male. Several studies have demonstrated that the decisive factor is the amount of host available to the developing worm.

  b Host nutrition. The sex ratio is much more female-biased when the host is well-fed than when it is starved (Bull 1983:111–115).
- Individuals of the snail *Crepidula fornicata* pile up to form a mound. Young individuals, on top of the mound, are always male. After a while, they enter a period of lability, and after that they can become male or female, depending on the underlying individual: if that is female, it

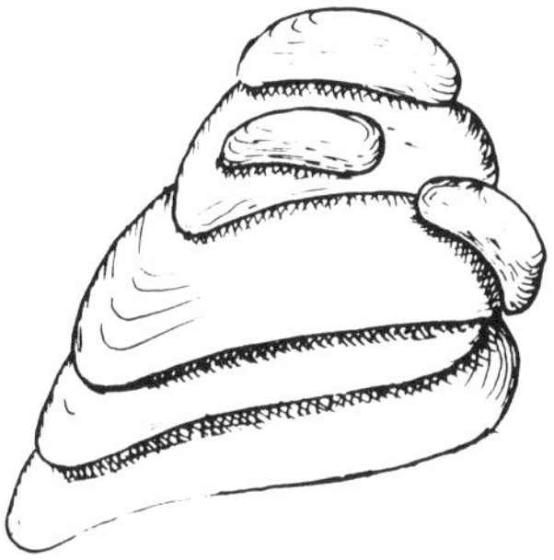

Fig. 2. Cluster of *Crepidula* snails.

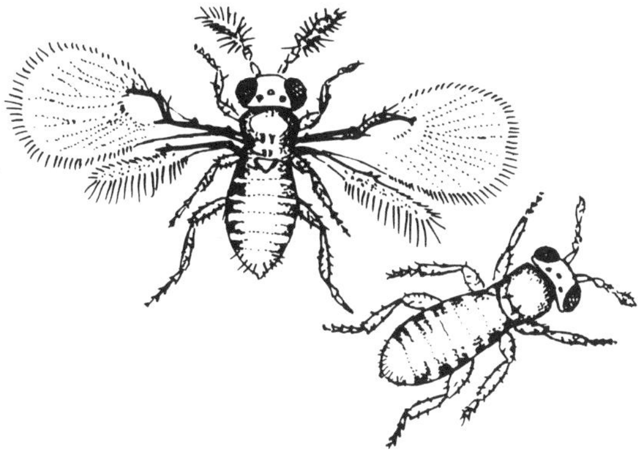

Fig. 3. The two forms of *Trichogramma semblidis* (after Gottlieb 1992).

becomes male again. When the number of males grows, some of them will become female (Gilbert 1994:784; see also Gould 1985:56–63).

So far for sex determination. Other traits than sex can also be involved. For example,

- The tiny parasitic wasp *Trichogramma semblidis* sometimes lays its eggs in a butterfly host and sometimes in an fly. The outcomes of devel opment in both hosts are strikingly different: an animal with wings in the one (butterfly) case, and an animal without wings in the other (Gottlieb 1992:153).
- Many butterflies show seasonal polyphenisms, that is, they have more than one generation a year, and the animals of different seasons are different, due to environmental differences, mostly in temperature and

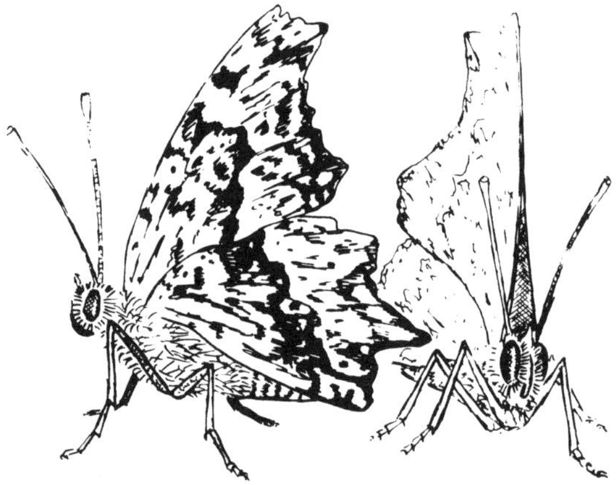

Fig. 4. The comma butterfly, *Polygonia c-album*; dark and light morph.

daylength. The map butterfly is an example. Its spring form and summer form differ so strikingly that Linneaus classified them as different species. Another example is the comma butterfly, *Polygonia c-album*, which has a light and a dark morph. Dark adults hibernate and give rise to light as well as dark offspring. The light morph reproduces much sooner; it gives rise to an other generation (of dark butterflies) in the same year, while the dark one goes into diapause in winter, only reproducing the next year. Which morph develops depends on the photoperiod. Not only daylength matters, but also the change in daylength. Increasing light yields the light morph. So the early offspring of this first generation is light, the later offspring is dark (Nylin 1989).

- Rotifers or wheel animalcules are very small animals that live in water. There are many different species, and some of these species eat other ones. Rotifers of the genus *Asplanchna* for example, eat rotifers of the species *Brachionus calyciflorus*. Now the presence of *Asplanchna* influences the development of *Brachionus*. *Asplanchna* releases a chemical into the water which causes *Brachionus* to develop a pair of long spines, which it does not possess in the absence of *Asplanchna*. The spines protect *Brachionus* from predation by *Asplanchna* (Gilbert 1966).

1.2. *"Background conditions"*

The second category of examples contains what may be called 'background conditions'. Apart from causing specific differences, the environment is always a vital condition. Developing organisms can only develop in certain ranges of

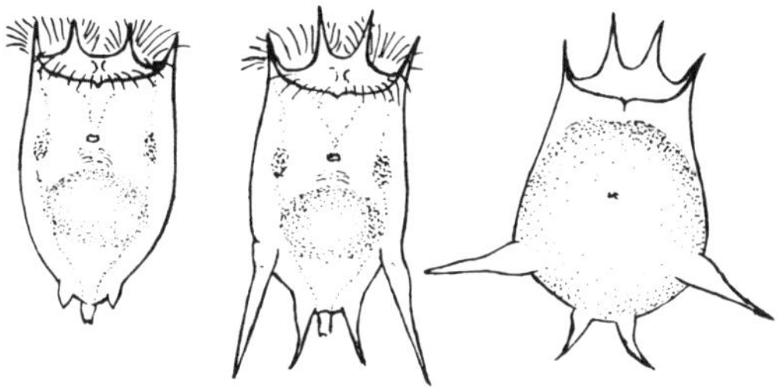

Fig. 5. *Brachionus calyciflorus*. Left: a young individual with three short pairs of spines (two anterior pairs and one posterior); no postolateral spines are present. The middle and right figure show an individual with long, *Asplanchna*-induced, postolateral spines (after Gilbert 1966).

circumstances. Variations in such conditions do not make a specific qualitative difference for developmental outcomes, say between male and female or black and brown. But they make a difference between life and death, or health and disease.

The environments that animals need are often very specific. For example, many larvae need specific kinds of food to survive. Think of the caterpillars of many butterflies, that can sometimes feed only on one kind of plant. Or think of the larvae of the mantid-like insect *Mantispa uhleri*. They must find a female spider to overwinter and after that the eggs of the spider are their obligatory food (Redborg and Macleod 1983). This category of vital background conditions may in some respects not be the most interesting one for the scientific study of development. But for some purposes such knowledge is important. In many cases of species decline, the reasons are unclear. For example, many species of amphibians are in dramatic decline, for reasons often known, such as habitat destruction, but often unknown. For two species of frogs it has recently been found that stratospheric ozone depletion is the main cause of decline. The species live high up in the mountains. They lay their eggs in shallow water where embryos are exposed to sunlight and thus to increasingly damaging doses of radiation due to decrease of the stratospheric ozone shield (Blaustein and Wake 1995). Clearly, the ozone layer is just a background condition for the development of frogs. But such knowledge is important nevertheless because it is at least a necessary condition for responsibility and action.

In fact, this example is already a bridge to the next category of environmental influences.

1.3. *Abnormal conditions may disturb development*

That environmental conditions can be a source of disturbed development is clear, but information on such effects has long been lacking. One important reason is that toxicological research has focused primarily on effects on survival and reproduction in adult animals. Recently, however, effects on offspring are beginning to be studied, and they turn out to be alarming. Many man-made chemicals that have been released into the environment have the potential to disrupt the endocrine systems of animals, including humans. The effects of DES give an idea of the kind of influence of these chemicals. DES is the synthetic estrogen that for many years was given therapeutically to countless women during pregnancy. Daughters born to mothers who took DES suffer increased rates of vaginal cancer, various genital tract abnormalities, abnormal pregnancies and some changes in immune response (see e.g. various papers in Colborn and Clement 1992).

Related things happen in animals. Several pesticides and other chemicals that are abundant in the environment have hormone-like effects. For example, in gulls and terns, DDT and PCB's are broken down in the body of the mother bird into estrogen-like components, which turn male birds into intersex sterile animals. Such effects have also been found in other animals, such as crocodiles and fishes (Fry and Toone 1981, Stone 1992, 1994, Mestel 1994, Culotta 1995, various papers in Colborn and Clement 1992). Many such chemicals with very bad effects on development are probably extremely common in the environment now.

The relevant disciplines here are toxicology and developmental biology. In toxicology, attention to environmental effects on development is now increasing. In developmental biology, on the other hand, a genetic approach is increasingly dominant and is rather guiding attention away from these phenomena. Let me say a little more about this genetic approach and how it tends to deal with the environment. After that I will sketch how a genetic approach is criticized and/or widened.

2. DIFFERENT APPROACHES

2.1. *How does a genetic approach deal with the environment*

In a genetic approach, the basic assumption is that genes control development. Development is pictured as a hierarchical process in which genes higher in the hierarchy control the activity of other genes, according to the instructions of a genetic program (see e.g. Raff and Kaufman 1991, Wolpert 1994). The idea of a genetic program is that the sequence of developmental events is laid down in the genome; the underlying idea is that something that is so precise and complicated as development must be coordinated by a central controlling agency, which regulates the process as a whole. The genome is then the place to look.

Characteristically, the environment is considered to be just a background condition, which is vitally important but scientifically not so interesting. It should mainly be kept constant, in order to find out what difference genetic mutations make. Of course an embryo needs oxygen and water, but these things do not make a specific difference, is the typical view. As the examples have shown, this cannot be maintained for many cases, such as environmental sex determination, where the environment does make a specific difference. But even in such cases, if they are recognized at all, the causal importance of the environment tends to be marginalized. The environment is then seen as a 'mere switch' (see Waddington 1942, Gilbert 1991a,b); it switches development into a track which is in itself genetically determined. Since switching is not a function that requires much information, the causally interesting search is for the 'underlying' genetics. The odd thing, however, is that a gene that works as a switch is called a master gene (see, e.g. Raff and Kaufman 1991, Wolpert 1991), which controls genes lower in the hierarchy of developmental decisions. So there is a remarkable asymmetry here. Genes are clearly privileged causes in the eyes of many biologists. I will not go into all the reasons of this privileging, but let me mention one assumption about the environment that certainly encourages this way of looking, and which stems from evolutionary theory. Neo-Darwinian theory has a strong picture of genetic variation, on the one hand, and the environment as selecting force on the other hand. The environment is important as a selecting force, not as a cause of development. The standard associations of development with genes, and of environment with selection, discourage a consideration of the direct role of the environment as a causal influence in ontogeny.

The distinction between a transformational and variational model of change, as it has been worked out for analytical purposes by both Lewontin and Sober (Lewontin 1983, see also Levins and Lewontin 1985; Sober 1985), reinforces this discouragement. The distinction has been introduced in order to explain what was new in Darwin's account of evolution, compared with older views. In the older views, of Lamarck and others, evolutionary change originated in ontogenetic change in organisms. Darwinian evolutionary change on the other hand takes place through the selection of random variation, change within individual organisms is not the point. Thus, evolution became a population phenomenon. Both Lewontin and Sober see this individual-population distinction as inherently coupled to a second distinction, between internal and external causation, suggesting that transformational change (of individual organisms) is necessarily caused from within, while the environment plays only a role, as a selecting force, in the variational model of evolutionary change, where it selects among static varieties. But though these associations are historically accurate, there is no inherent reason for them (see also Oyama 1988, 1992).

2.2. *Two critical approaches: constructionism and structuralism*

A gene-centric approach to development is not the only one around. I will now sketch two critical approaches, which have in common that they picture developmental causality as a complex network of relations instead of a set of one way genetic instructions. They both reject the idea of a genetic program of development, but their alternative is different.

The first alternative approach is 'constructionist', after one of its central metaphors, that pictures development as construction. Susan Oyama's work is representative. In her book *The ontogeny of information* (Oyama 1985; see also Oyama 1989, 1994; for related approaches see e.g. Gray 1992, Griffiths and Gray 1994, Johnston and Gottlieb 1990, Nijhout 1990) she explicitly criticizes the genetic program metaphor and emphasizes the decenteredness of causation. Development is a process in which control continually shifts from one interactant to another. These interactions need not be laid down in advance or governed by a privileged contributor. In other words, there need not be a representation of the process in the genome. The outcome of development is highly reliable, but this is not due to the actions of a program, but to all the contingent interactants being reliably present. Developmental environments, for example, which are often assumed to be a source of great fluctuation which cannot be counted on, are in fact typically highly predictable and reliable. Think of wombs, which afford very constant environments for many developing organisms. Or think of seeds. Seeds of a particular plant may be found in extremely heterogeneous environments, but they germinate only under certain conditions, so the environments for young developing plants are much more homogeneous. Not only the internal conditions are essential for development, but the external ones as well. Indeed, a constructionist approach takes the unit of development to be an organism-environment system. The gene is not a causally privileged unit, not even the organism is a privileged unit, and neither is the environment a privileged unit; all components in interaction determine developmental processes, and they can and should not be taken apart.

The second alternative approach, represented here by the work of Brian Goodwin (see Goodwin 1994, also Goodwin 1984, 1985, 1988, 1990), is 'structuralist'. Unlike Oyama, Goodwin does emphasize the organism as the most important unit in biology. He, too, criticizes the genetic program idea, and holds that genes should not be made so extremely special. But while the constructionist alternative emphasizes the importance of heterogeneous kinds of interactants, Goodwin favours a central explanatory device in developmental biology. It consists not of a program but of fields of developmental forces. In the ideal case, these fields are described by field equations, which include all the relevant forces. Genes 'merely' set parameter values for such field equations, and the same applies to the environment. Parameter settings are not really explanatory; for Goodwin, real explanation comes from a description

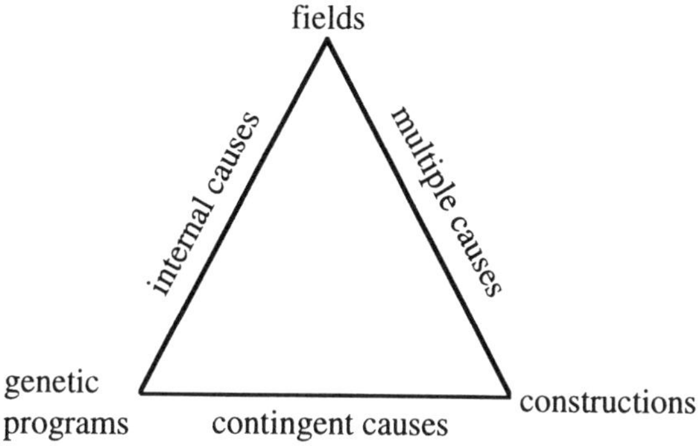

Fig. 6. Similarities and differences between the three approaches. Explanation in the text.

of fields. Only in this way will a complete understanding of potential and realized forms of organisms be possible. This implies that processes in time are not really central here; laws of nature that determine fields are the scientific goal. Further, Goodwin stresses that this field-approach makes organisms the central units of biology.

2.3. *Comparison: similarities and differences*

The field approach as well as the constructionist approach are proposed as alternatives for genetic programmes. Yet a comparison of the three views shows similarities as well as differences. Pairwise similarities are shown in the diagram. The corners of the triangle show different positions. The zoom in differently: on genes, organisms (through fields), and organism-environment systems. Concerning the causal understanding of development, there are similarities and differences between all the three.

1. Inside-outside. A genetic approach fits in with a long tradition that understands development as coming from within. Constructionists want to break with this tradition. They see organism-environment systems as the units of evolution and development; the boundaries of the organism are not causally fundamental. Goodwin's structuralism does not address this issue explicitly, but implicitly joins the internalism of the tradition.
2. Contingency. Constructionists also have something in common with geneticists: their causes are contingent in principle; no general laws are involved in their explanations. The metaphor of conversation used in both these perspectives likewise suggests (by association rather than by logical force) the importance of time, history, contingency. It is absent

from Goodwin's writings. Historical contingency is secondary for him, necessity defines the scientific content of biology.

3. Relational causation. Both Goodwin and Oyama criticize a causal overemphasis on genes, and stress the relational and systemic nature of causation. The development of organisms is a dynamic process in which no discrete causes and effects can in fact be distinguished.

2.4. *Epigenetics*

The three approaches discussed so far are not just constructs for analytical purposes; they describe real differences. But of course, science is not static. For example, a predominantly gene-centric approach is evolving — significantly though not uniformly — into what is more appropriately called an epigenetic perspective, which bears similarity with the constructionist outlook. The term 'epigenetic' deserves some elaboration. From Aristotle onward, epigenesis has been defined in opposition to preformation. The precise meaning of these terms has been changing during history, but by and large preformationist theories hold that development is the unfolding of structures already present in the egg, while epigenesis involves the progressive formation of new structures during development. The present fate of these positions is summarized by Gould, who writes that "modern genetics is about as midway as it could be between the extreme formulations of former times" (Gould 1977:18). Others, too, have observed that neither position has 'won out'. In one sense, epigenesis has triumphed because embryonic structures are not preformed in the egg. In another sense, preformation is right because DNA is seen as the preformed basis for development. This feeling of synthesis is well expressed by the word 'epigenetics', which was coined by Waddington in the 1950's, and which blends 'epigenesis' and 'genetics'. Epigenetics has come to be seen as the causal analysis of development defined as the mechanisms by which genes express their phenotypic effects (Hall 1992:89).

On the one hand, that 'epigenesis' turned into 'epigenetics' is illustrative of the central place that genes now have in developmental biology. On the other hand, epigenetics goes beyond pure genetics in that it studies how genes are activated and repressed, in short, how they are regulated. The study of gene regulation began with Jacob and Monod's model for the regulation of the lactose-operon in *E. coli.* Jacob and Monod in 1961 emphasized the relevance of their results for "the fundamental problem of chemical embryology [which] is to understand why tissue cells do not express, all the time, all the potencies inherent in their genome" (Jacob and Monod 1961, quoted by Gilbert 1994:47). Mechanisms of gene regulation show how cytoplasmatic factors are involved in embryology. So, after a period of narrow gene-centrism, the cytoplasm has got a real place in developmental biology. The idea is no longer that instructions flow only in one direction: epigenetics talks not only about gene action but

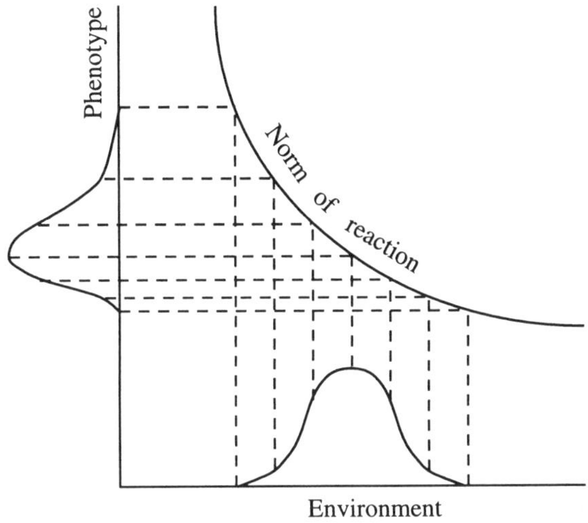

Fig. 7. Norm of reaction, which converts the environmental distribution on the horizontal axis to the phenotype distribution on the vertical axis (after Griffiths et al. 1993).

also about the regulation of genes. Indeed, this is how the field is defined. To what picture of development this ultimately leads depends on how epigenetics will develop further. Up to now, there is still much talk about genetic programming and a heavy focus on genes regulating other genes. But models are starting to emerge about heterogeneous factors that influence and regulate genes in a great network of interactions. When the idea of genetic programming is dropped, the constructionist and epigenetic perspective might become merged positions in the end.

2.5. *Study of environmental influences*

In principle, the study of the environment can easily get a place in an epigenetic as well as a constructionist view of development. One useful concept is 'reaction norm', which is the relation, given a certain genotype, between a range of environmental conditions and the outcomes of development under those conditions. The concept is visualized in the diagram above.

When the environmental factor on the x-axis makes a difference for development, the norm of reaction is different from a flat line. The phenotypic characteristic is then 'phenotypically plastic' under different environments. Two main cases can be distinguished: a continuous effect, and a discrete effect (e.g. sex, discrete wing patterns). The discrete case is not always called a reaction norm; it is more often called a polyphenism. Further, this distinction between polyphenism and reaction norm is not always made in the same way, but I will not go into that. Both effects are forms of 'phenotypic plasticity', which refers to the phenomenon of environmentally induced phenotypic variation in general.

Once reaction norms or polyphenisms have been described, a next possible step is to study in molecular detail how the environmental factor in question influences gene expression. This is a quite feasible move within the domain of epigenetics as well as in a constructionist approach.

Though this move is feasible, and though it occurs (see for example Niimi et al. 1993), it does not characteristically occur. The epigenetic picture in practice is often internalistic and much of the emphasis is still on genes, that is to say, the idea that genes are regulated by the products of other genes is still dominating the field and guiding much of the research. Thus, the cytoplasm has become included in the work terrain of genes, but the idea of genetic programs and the central emphasis on genes are still powerful.

The reasons may have to do with internalistic presuppositions which are dropped only gradually. So for the near future we might have to look at constructionism, which explicitly challenges the internalistic assumption about development, for a widespread initiation of the study of environmental influences in development. But in fact, constructionist authors typically do not advocate specific study of environmental influences. Internalist orientation cannot be the reason here, because it is explicitly criticized. Instead, there is a reluctance to privilege any cause. I will deal with this issue in the next part of this paper, which concerns the unavoidability of restricting choices.

3. THE UNAVOIDABILITY OF CHOICES

3.1. *The ideal of completeness*

When we look at the various approaches to development, we can distinguish between their general outlook (which the triangle earlier in the paper dealt with) and their specific research priorities.

	General picture:	Prom. research questions
(epi-)genetics	genetic programmes/ gene regulation	regulatory genes
structuralism	Developmental fields	role of non-genetic factors e.g. calcium
constructionism	organism-environment-systems	-

In principle, the environment could be studied within any of these approaches, and particularly, of course, in the constructionist one. But it happens little, and even constructionists do not advocate it, because they do not systematically want to isolate ('privilege', 'reify' (see Gray 1992)) incomplete aspects of developmental processes. The background ideal is complete explanation. Let us look closer at that ideal, which is much alive in biology.

All the approaches to development I discuss here aim at complete explanation. Beginning with the the genetic picture of development: it is often said that a complete insight in the regulatory genes associated with development will generate an essentially complete account of the process. The critical pictures reject this completeness-claim of the genetic approach, holding that completeness is to be reached by taking in more than genes. Their central goal is to offer a more complex, more complete, and more true vision of development. In order to create a framework that can illuminate how incomplete the genetic approach is, constructionist and structuralist writings also deal with issues of causation in general terms (see Van der Weele 1995).

What about this ideal of completeness from a philosophical point of view? The ideal of completeness can also be found in the philosophy of scientific explanation. Salmon, for example, defends the notion of an ideal explanatory text (Salmon 1989), which, for a given phenomenon, would spell out all the nomic and causal connections and which would be connected with an ideal theory. He admits that in no case the ideal text is known (in most, if not all, cases it can be expected to be 'brutally large'); nevertheless in his eyes it should be a firm yardstick for explanations. That is to say, the ideal explanatory text should offer the framework in which more pragmatic considerations can find their place. Yet it is not hard to see that when ideal texts are not known, neither can they be a yardstick for anything. The abstract and ideal views of theories that abound in this type of the philosophy of science are not really helpful for concrete science. For practical purposes we can be brief about this completeness-ideal: in practice, scientific theories are nowhere near ideally complete. For biology, this is shown by Robert Brandon's attempt (Brandon 1990) to specify an ideal explanation in evolutionary theory. Even the best and most complete explanations in evolutionary biology rest on many theoretical and practical restricting decisions, and even so contain large knowledge gaps. This situation is representative. Not only are knowledge gaps everywhere around, but it is also unavoidable that theories are built on many restricting choices. Because of their restrictions, the usefulness of theories must be evaluated in relation to their purpose. Richard Levins has argued, for example, that mathematical theories in biology cannot simultaneously maximize generality, realism and precision, and that, more generally spoken, there can never be a single, all purpose model (Levins 1968, 1993). Therefore, in order to find yardsticks it makes more sense to focus on the specific choices that are made than on the ideal of completeness.

3.2. *Choices in developmental biology*

In approaches to developmental biology, choices are abundantly made. In the first place, the general pictures rest on choices, through the images and definitions of development. They draw attention in specific directions, with some questions near at hand, and some never arising. For example, defining development as the unfolding of a genetic program, or even as a process of gene regulation, most naturally focuses attention on genes. A field of physical forces on the other hand draws attention to physical and chemical aspects of pattern formation.

So already the general pictures involve restricting choices. Further, a scientific approach is more than a general picture of nature. It also addresses research problems, and here again are choices to be made. Some problems are addressed while others are left untouched, and some but not all causal links that are feasible within an approach are pursued. Such restrictedness is not at all a vice. Causal relations are like logical relations in this respect. Both kind of relations can be of widely different kinds, but never of all kinds at once. Further, they can be established between two or more items, but never between all potentially interesting items at once. Thus, within a relational view of causation one may choose to focus on reaction norms, or on the regulation of some particular gene, or a mathematical model for certain aspects of pattern formation, or whatever. Never can the whole interactive process of development of whatever organism be studied 'completely'.

Thus, restricted questions are a necessity, not a problem. And genetic, epigenetic and structuralist approaches do clearly make them. The structuralist approach, for example, despite its ideals of completeness and comprehensiveness, is really addressing specific problems, for example concerning the role of calcium in pattern formation. The constructionist approach on the other hand more or less manages to avoid these further restrictions because of the idea that no type of causal influence should be isolated, privileged, and/or reified. More specifically, constructionists do not wish to favour the study of environmental influences for fear of falling into the trap of the gene-environment dichotomy. This dichotomy, which creates the illusion that there are two alternative causes of development, genes and environment, is central in the long and depressing history of the nature-nurture debate. I agree that the trap is there and that the way to solve nature-nurture problems is to deny the causal importance of the internal-external distinction. Instead, organism-environment systems are the relevant causal units for development and I also agree that overall views of causation should be interactive. In principle, no part of the process is isolated. In practice, however, you cannot do everything at once. When you refuse to ask restricted questions, you are in great danger to become stuck in broad and general proposals, which in themselves do not yield empirical data.

In sum, though research on environmental influences in animal development is a real possibility, for various reasons the environment has not yet received a serious place in developmental theory and practice. I finally address the question whether a larger place for environmental causes of development is desirable.

4. ETHICS OF ATTENTION

The necessity of choices in science may give rise to a next question: are there reasons to prefer some choices and regret other ones? One thing is at least clear as a beginning of an answer: it is certainly not to be preferred if one type of questions, say concerning genes, is taken to exhaust the domain of potential scientifically interesting questions on development. This is an argument for scientific pluralism. I take it, without going further into that point here, that in order to save science from onesidedness which is presented as the whole picture, pluralism is a necessity.

But we can go on and ask: apart from reasons that have to do with pluralism, are there reasons to specifically encourage more attention to environmental influences in developmental biology? As I see it, the answer is yes. Here I come to the 'ethics of attention' announced at the beginning of the paper, which focuses on scientific choices and their consequences. If scientific choices had no consequences, they would not matter much. But science is not only a free floating system of views and perspectives on nature. It also does things; its choices make a difference outside science. More in particular, science is used in many social contexts to solve old problems and define new ones; in other words, factual or causal information prestructures problem definitions, including moral problem definitions.

Taking such consequences into account inspires an ethical approach to science. Note that it is a special kind of ethical approach. It does not focus, for example, on researchers' motives or interests. Even if you have no particular motive at all, you can never avoid to make restricting causal choices, you always generate specific and incomplete knowledge, and unavoidably these choices have consequences, there is no way out of that (Garfinkel 1981 argues this point extensively). Insight into those consequences can be a basis for moral evaluations of science. This may be called an 'ethics of attention'.

What does this mean for developmental biology? As I emphasized before, choices on two levels can be distinguished: the level of overall approaches and the level of specific research questions. Different overall approaches to development give different starting points to answer questions about what it means and takes to live and flourish and what can possibly go wrong. This is not merely an academic affair. On the one hand, when genes are thought to afford the whole causal picture of development, this supports the idea that genome projects and genetic studies yield the most important information about the development of organisms. This idea is now taking up increasing amounts of money and attention. On the other hand, a developmental biology that is more ecologically oriented and that includes the environment could point to the existence of environmental problems, concerning pollution, food, climate and so on, and perhaps contribute to solutions. For this latter purpose, however, a general picture is not enough; environmental influences cannot get a real place in thinking about development if they are not specifically studied. In other words, choices on the level of research questions are also important. In the present situation I think there are moral reasons to specifically encourage the study of environmental influences in development. The environment deserves this specific attention not because it is special in some ontological sense, nor because it stands in opposition to genes, but because man-made environmental changes cause developmental problems on a massive scale, which have been neglected from a scientific point of view. As a consequence, there are large gaps in our knowledge about relevant developmental environments. Knowledge is a necessary precondition for responsibility and action, and it is certainly due to a lack of scientific interest and knowledge that all kinds of harmful chemicals have been able to become so abundant in the environment.

So far for developmental biology. Finally, as I announced at the beginning, this message about developmental biology is at the same time meant as an argument for a more general view. In many other scientific fields, too, choices have morally relevant consequences which deserve study from the perspective of an ethics of attention. What this paper gives is of course a far cry from a fully worked out ethical approach to science. I think we should work hard to develop such an approach.

Department of Philosophy
Utrecht University

ACKNOWLEDGEMENTS

I thank Bas Kooijman for the animal drawings. The investigations for this paper were supported by the Foundation for Research in the field of Philosophy and Theology which is subsidized by the Netherlands Organisation for Scientific Research. The research was done at the Free University of Amsterdam. For useful suggestions and comments I thank Wim van der Steen, Arno Wouters, Diedel Kornet, Richard Burian and Jane Maienschein.

REFERENCES

Blaustein, A.R. and D.B. Wake (1995). The puzzle of declining amphibian populations. *Scientific American*, April: 56–61

Brandon, R.N. (1990). *Adaptation and Environment*. Princeton, Princeton U.P.

Bull, J.J. (1983). *Evolution of Sex Determining Mechanisms*. Menlo Park, Benjamin/Cummings.

Colborn, T. and C. Clement (1992). *Chemically-Induced Alterations in Sexual and Functional Development: The Wildlife/Human Connection*. Princeton, Princeton Scient. Publ.

Culotta, E. (1995). St. Louis Meeting Showcases 'Creature features'. *Science* **267**: 330–331.

Fry, M.D. and C.K. Toone (1981). DDT-induced feminization of gull embryos. *Science* **213**: 922–924.

Garfinkel, A. (1981). *Forms of Explanation*. New Haven, Yale U.P.

Gilbert, J.J. (1966). Rotifer ecology and embryological induction. *Science* **151**: 1234–1237.

Gilbert, S.F. (1991a). Induction and the origins of developmental genetics. In: S.F. Gilbert (ed.): *A Conceptual History of Modern Embryology*. New York, Plenum Press: 181–206.

Gilbert, S.F. (1991b). The role of embryonic induction in creating self. In: A.I. Tauber (ed.): *Organism and the Origins of Self*. Dordrecht, Kluwer: 341–360.

Gilbert, S.F. (1994). *Developmental Biology* (fourth edition). Sunderland, Sinauer.

Goodwin, B.C. (1984). A relational or field theory of reproduction and its evolutionary implications. In: M.W. Ho and P.T. Saunders (eds.): *Beyond Neo-Darwinism*. Academic Press: 219–241.

Goodwin, B.C. (1985). What are the causes of morphogenesis. *BioEssays* **3** (1): 32–36.

Goodwin, B.C. (1988). Problems and prospects in morphogenesis. *Experientia* **44**: 633–637.

Goodwin, B.C. (1990). Structuralism in biology. *Sci. Progress* **74**: 227–244.

Goodwin, B.C. (1994). *How the Leopard Changed its Spots*. London, Weidenfeld and Nicolson.

Gottlieb, G. (1992). *Individual Development and Evolution*. New York & Oxford, Oxford U.P.

Gould, S.J. (1977). *Ontogeny and Phylogeny*. Cambridge (Mass.), Harvard U.P.

Gould, S.J. (1985). *The Flamingo's Smile*. London, Penguin.

Gray, R. (1992). Death of the gene: developmental systems strike back. In: P. Griffiths (ed.): *Trees of Life: Essays on the Philosophy of Biology*. Dordrecht, Kluwer: 165–209.

Griffiths, A.J.F., J.H. Miller, D.T. Suzuki, R.C. Lewontin, W.C. Gelbart (1993). *An Introduction to Genetic Analysis*. New York, Freeman.

Griffiths, P.E. and R.D. Gray (1994). Developmental systems and evolutionary explanation. *Journal of Philosophy* **91**: 277–304.

Hall, B.K. (1992). *Evolutionary Developmental Biology*. London, Chapman & Hall.

Jacob, F. and J. Monod (1961). Genetic regulatory mechanisms in the synthesis of proteins. *J. Mol. Biol.* **3**: 318–356.

Janzen, F.J. and G.L. Paukstis (1991). Environmental sex determination in reptiles: Ecology, evolution, and experimental design. *The Quarterly Review of Biology* **66** (2): 149–179.

Johnston, T.D. and G. Gottlieb (1990). Neophenogenesis: A developmental theory of phenotypic evolution. *J. Theor. Biol.* **147**: 471–495.

Levins, R. (1968). *Evolution in Changing Environments*. Princeton, Princeton U.P.

Levins, R. (1993). A response to Orzack and Sober: formal analysis and the fluidity of science. *The Quarterly Review of Biology* **68** (4): 547–555.

Levins, R. and R. Lewontin (1985). *The Dialectical Biologist*. Cambridge (Mass), Harvard U.P.

Lewontin, R.C. (1983). Darwin's revolution. *New York Rev. Books* (June 16): 21–27.

Mestel, R., (1994). Gulls and terns all at sea about sex. In *New Scientist* 3 September 1994:9.

Niimi, T., O. Yamashita, T. Yaginuma (1993). A cold-inducible *Bombyx* gene encoding a protein similar to mammalian sorbitol dehydrogenase; yolk nuclei-dependent gene expression in diapause eggs. *Eur. J. Biochem.* **213**: 1125–1131.

Nijhout, H.F. (1990). Metaphors and the role of genes in development. *BioEssays* **12** (9): 441–446.

Nylin, S. (1989). Effects of changing photoperiods in the life cycle regulation of the comma butterfly, *Polygonia c-album* (Nymphalidae). *Ecological Entomology* **14**: 209–218.

Oyama, S. (1985). *The Ontogeny of Information*. Cambridge (Mass.), Cambridge U.P.

Oyama, S. (1988). Stasis, development and heredity. In: M.-W. Ho and S.W. Fox (eds.): *Evolutionary Processes and Metaphors*. Chichester, John Wiley & Sons. 255–274.

Oyama, S. (1989). Dogma: do we need the concept of genetic programming in order to have an evolutionary perspective? In: M.R. Gunnar and E. Thelen (eds.): *Systems and Development*. Hillsdale, N.J., Lawrence Erlbaum Associates: 1–34.

Oyama, S. (1992). Ontogeny and phylogeny; A case of metarecapitulation?. In: P. Griffiths (ed.): *Trees of Life*. The Netherlands, Kluwer Ac. Publ.: 211–239.

Oyama, S. (1994). Rethinking development. In: P.K. Bock (ed.): *Handbook of Psychological Anthropology*. Westport, Greenwood: 185–196.

Raff, R.A. and T.C. Kaufman (1991 (1983)). *Embryos, Genes and Evolution*. Bloomington, Indiana U.P.

Redborg, K.E. and E.G. Macleod (1983). *The Developmental Ecology of Mantispa uhleri Banks (Neuroptera: Mantispidae)*. Urbana & Chicago, Univ. of Illinois Press.

Salmon, W.C. (1989). Four decades of scientific explanation. In: P. Kitcher and W.C. Salmon (eds.): *Scientific Explanation*. Minneapolis, Univ. Minnesota Press: 3–219.

Sober, E. (1985). *The Nature of Selection*. Cambridge (Mass.), MIT Press.

Stone, R. (1992). Swimming against the PCB Tide. *Science* **255**: 798–799.

Stone, R. (1994). Environmental estrogens stir debate. *Science* **265**: 308–310.

Van der Weele, C.N. (1993a). Explaining embryological development: Should integration be the goal? *Biology and Philosophy* **8**: 385–397.

Van der Weele, C.N. (1993b). Metaphors and the privileging of causes. *Acta Biotheoretica* **41**: 315–327.

Van der Weele, C.N. (1995). *Images of Development; Environmental Causes in Ontogeny*. Thesis, Amsterdam.

Waddington, C.H. (1942). Canalization of development and the inheritance of acquired characteristics. *Nature* **150**: 563–565.

Wolpert, L. (1991). *The Triumph of the Embryo*. Oxford, Oxford U.P.

13
Foundations of Cognitive Science and AI

FRANK JACKSON

MENTAL CAUSATION WITHOUT THE LANGUAGE OF THOUGHT*

Here is a familiar kind of story. Harry believes that there is coffee to the right and beer to the left. His belief that there is coffee to the right explains, when combined with certain desires, his moving right; his belief that there is beer to the left explains, when combined with certain other desires, his moving left. The beliefs play different causal explanatory roles. Accordingly, we need to acknowledge individual beliefs conceived of as distinct, presumably internal, states of subjects that are thus able to play distinct causal roles—or so it seems.

I will call any doctrine of this general kind, individualism about belief. It is a pity that the word 'individualism' is already used for another doctrine in the philosophy of mind, but I hope that explicitly noting the possible source of confusion is the best way of guarding against it.

The language of thought hypothesis gives us a way of fleshing out individualism about belief. According to it, beliefs are, or are relations to, stored sentences of mentalese, and storings of different sentences of mentalese will be distinct states. We then account for the difference between explaining behaviour in terms of a subject's believing that P and their believing that Q in terms of the different causal powers of the distinct structures that are the storings of the relevant sentences in the head. Likewise, we can account for interconnections between what a subject believes in terms of causal interconnections between distinct stored structures.

This paper is concerned with what to say should it turn out that individualism is false, that nothing like the language of thought hypothesis is true. How then, if at all, can we account for the difference between causal explanations of behaviour in terms of subjects' believing that P and their believing that Q? This paper sketches an answer to this large question. I will not address the related and important question of what to say about explanations of subjects' believing that P in virtue of believing that Q—the interconnection issue—should individualism turn out to be false, but for the record I believe that a similar treatment applies.

*I am much indebted to philosophical exchanges, some supporting and some dissenting, with Michael Devitt, Georges Rey, Karen Neander, and most especially David Braddon-Mitchell, Philip Pettit, and David Lewis.

M.L. Dalla Chiara et al. (eds.), Structures and Norms in Science, 303–318.

1. BELIEF, MAPS AND HOLOGRAMS

It will be helpful to have before us a reasonably concrete example of an alternative to the language of thought. What I am about to describe is, I take it, what many supporters of F.P. Ramsey's idea that belief is a map by which we steer—David Armstrong, Robert Stalnaker and David Lewis to name but a few—have had in mind.

It is, or should be, common ground that belief states are states of heads that represent things as being one way or another, and crucial to their being representational states is that there is a correspondence between the relevant possible head states and various possible world states. Otherwise they would not count as representational states. Now one way such a correspondence can obtain is by virtue of more piecemeal correspondences obtaining. Some part of a belief state might play a distinguished role in settling some part of how it represents things as being: there is, as it might be, a part of a subject's belief state that is specially responsible for how she takes things to be in South America. By disturbing that part of her belief state we could change what she believes about South America while leaving everything else much the same.

The part of the belief state that has special responsibility for South America need not be localised. It might be spread through the state. Think of the part of a book responsible for what the book says about South America. The part might be made up of sentences and words scattered throughout the book. It is still the case that there is a part, a scattered part, that has special responsibility for South America.

However the correspondence between what does the representing and what it represents might be radically holistic. Holograms provide an example. (Or so I understand—it is not crucial that what I am about to say be the truth of the matter, only that it might be.)

Holograms are 'laser photographs'. When light from the laser is projected through the negative, the well known, three dimensional, coloured array is produced. The negative can be thought of as representing things as being the way the coloured array depicts them. However no part of the negative has special responsibility for some part of the array, and so for some part of how things are represented as being. Each part contains information about the whole array. In consequence, what happens if you damage a part (local or scattered) is a loss of detail, a blurring, of the three dimensional array, not a loss in any particular part of the array.

This is not an example of representation without structure. We could have systems of representation that involved arbitrary correspondences between a finite number of representing states and a finite number of represented possibilities. The past use of flags to represent diseases in sailing ships may be an actual example. Or close to one: the choice of a yellow flag for yellow fever was not totally arbitrary, obviously. The correspondences would be arbitrary in that there would be no right way to go on when presented with new cases.

We would have a correspondence between state S_1 and possibility[1] P_1, S_2 and $P_2, \ldots$, and S_n and P_n, but there would be no answer to what S_{n+1} represented flowing from this information. Hologramic representation, however, is not like this. There is a right way to go on. The same goes for maps and diagrams, of course. In all these cases structural connections between what does the representing correspond to structural connections between what is represented. This is how a new map or hologram negative can be created to represent a totally new situation.

The fact that maps and hologram negatives are structured representations means that they possess the desirable properties of systematicity and productivity. It is a commonplace that a map (or globe) which can represent that more of the southern hemisphere than of the northern hemisphere is covered by water can also represent (falsely) that more of the northern hemisphere than of the southern hemisphere is covered by water. Likewise, maps can represent indefinitely many new possibilities: before the discovery of Australia, people speculated in maps as well as in words about the great southern continent.

It seems to me that our heads might represent in the way maps or holograms represent. It might, for all we now know, be the case that we find our way around the world by virtue of having an internal structure that represents how things are in broadly the way that maps or negatives of holograms do. If this is the case, then when Jones believes that snow is white, there is no individual structure in her head that has assigned to it the job of representing that snow is white. Rather, inside Jones there is a structure that constitutes, as we might put it following David Lewis and intentional systems' theory, a system of belief that represents, among a great many things, that snow is white: Jones believes that snow is white by having a system of belief according to which snow is white.[2] We can still talk of individual beliefs in the sense of aspects of the way heads represent things as being which can be captured in individual sentences, but there will be no underlying individual states that these belief sentences report. We may say that there are individual belief contents but no individual beliefs. A similar point is sometimes made in terms of maps. A map of the Earth might represent that the taller mountains are mostly near the deeper oceans, but there is no part of the map that says just that in the way that there may be a sentence that says just that—for instance, the sentence you heard a moment ago.[3]

The same goes for hologramic representation except that it is more radically holistic than map representation. In the case of maps, we can match

[1] Possibility, not possible world as the state will not represent how things are in full detail. The metaphysics of possibilities (and possible worlds) is an important issue but one we can fairly set aside here. There had better be some way representing representational content.

[2] See, e.g., Lewis (1986).

[3] The maps we use often contain words, and to the extent that they do they represent in the way language represents. In what follows it is perhaps best to think of 'wordless' maps.

up changes to parts of what does the representing to changes in parts of how things are represented as being. Rubbing out one part of a map typically has especially drastic effects on what the map says about some particular part of the world, whereas damaging a part of a hologram negative blurs the whole hologram.

We can now sharpen our initial question. If when Harry believes that there is coffee to the right, he has a head according to which there is coffee to the right, but there is nothing in his head which is the belief that there is coffee to the right, how do we make sense of causal explanations that advert to his believing that there is coffee to the right? Even if we reject explanations of behaviour by individual beliefs, we had better allow for explanations by individual belief contents. Our question is whether and how the radical anti-individualist can do this. But before we address it directly, it will help to have a more detailed account of the view of belief that reflection on maps and holograms suggests.

2. RADICAL HOLISM

I will call the following view of belief radical holism.[4] It is not an analysis of belief in the traditional sense, for it rests on the empirical, contingent hypothesis that heads represent the way things are in somewhat the way hologram negatives and maps do. But it is a piece of analysis in the sense that it delivers a truth-making story. The view tells us how a certain possibly true account of how heads represent the way things are would make true certain claims about belief.

The first tenet of radical holism is that there are no individual beliefs. There are only systems of belief. We can say that Jones' belief that snow is white is true but her belief that all swans are white is false. But what makes such a claim true is her having a system of belief according to which snow is white and a system according to which all swans are white, and its being the case that the first claim is true and the second is false.[5] In the same way a map might represent that more of the southern hemisphere than of the northern hemisphere is covered by water, and that the largest island is in the northern hemisphere. The first claim would be true and the second false. But for both head and map, it is false that there are two distinct parts, one of which represents truly and the other falsely.

[4]Not to be confused with holistic doctrines about how some system of representation—belief, language or whatever—gets to represent what it does represent.

[5]As David Lewis and Robert Stalnaker have emphasised, it is no part of the system's view that there is at most one system for each head. Thus, we need to distinguish two ways we might give an account of when S believes that P: we might say that S believes that P if every system of belief in S's head is one according to which P, or we might say that S believes that P if some system of belief in S's head is one according to which P. See the discussion in Lewis (1986), p. 31. But we will not be concerned with this complication in what follows.

Secondly, radical holism takes as fundamental not just the notion of a system of belief but that of a rich system of belief; a system that says a lot; a system that represents how things are in considerable detail, including as a tiny part that, say, snow is white. This is not an accidental feature of radical holism. It is a consequence of the way radical holism approaches the problem of content.

In the case of language of thought views a strategy worth serious attention is that of giving the content of belief in terms of causal links, or patterns of co-variation, or something along these general lines, between parts of individual beliefs and parts of the world. The belief that snow is white, for example, might be held to be about snow by virtue of its 'snow' part having the right kind of connection to snow. And of course this story can be supplemented by including as contributors to content inter-connections between mentalese sentences and their parts, as in conceptual role semantics. Strategies of this kind are not available to radical holists. They must instead assign content at least in large part behaviourally. It will be the actual and potential way that the relevant head structure steers Jones around a putative world that is crucial for the content of that head structure and so the content of her system of belief. Behavioural approaches to content fit nicely with recent emphases on informational approaches to content. The head would have a hard time moving the body through the world unless it recorded information about the world, unless, that is, its states were apt to co-vary systematically with its surroundings. Also, those who wish to can allow a role for evolutionary history in the determination of content provided they retain a major place for behaviour.

Now behavioural approaches to content inevitably deliver rich content. There is no behaviour that the belief that there is a mine near that tree as such points to. It is instead a rich system of belief to the effect, say, that there is a mine near the tree, that the mine is likely to be triggered by going near it, that moving one's legs in such and such a way will or will not bring one near the tree, that there is not a bigger mine that can only be avoided by going close to the tree, that triggering mines tends to cause death, and so on and so forth, along with the appropriate desires, that points to behaviour.

The problem is not merely that the behaviour subjects are disposed to manifest depends on a big part of how they take things to be. The way a particle moves under the influence of massive bodies depends on the totality of the component gravitational forces acting on it. Nevertheless, for each force there is a distinctive contribution specifiable in terms of the motion the particle would undergo if that were the one and only force acting on it—a contribution that is explicitly represented in the diagrams for calculating resultant forces. Again, the action of a poison can be blocked by an antidote. Nevertheless, there is a result that taking a poison typically points towards: the existence of antidotes simply means that typically pointing towards is not the same as ensuring. However, there is no such thing as the distinctive behaviour associated with too small an account of how an agent takes things to be or wants them to be. Jones' belief that there is a mine over there has no direction whatever in and of itself.

Thirdly, radical holism holds that there is no interesting distinction between core and derived beliefs. The problem of clutter means that language of thought theories must distinguish core from derived beliefs. If we allowed an internal sentence for everything Jones believes on an inclusive sense of 'belief', we could not fit the sentences into a head. Accordingly, language of thought theories only treat a small set of beliefs, the core beliefs, in terms of encoded sentences, and offer a separate treatment for beliefs which, it is argued, are in some sense derivative on the core beliefs.[6] For radical holism, however, there is no fundamentally interesting distinction between core and derived belief. Although there are natural ways of carving up passages of prose at their representational joints, there are no natural ways of carving up maps or hologram negatives at their representational joints. For instance

> Jones lives in the north of England. The north of England is colder than the south of England

represents inter alia that Jones lives in the colder half of England. But we extract this content by cutting across the more natural way of itemising what the passage says. By way of contrast, consider a map of the world that represents the largest island as being in the hemisphere with the most water, the hemisphere with the most water as being the southern hemisphere, and the largest island as being the smallest continent. It equally represents the largest island as being in the southern hemisphere, the smallest continent as being in the hemisphere with the most water, and the largest island as being the smallest continent. There is no sense in which one way of itemising its content carves nearer to the joints than the other, or any other. At most there is a reporting-relative distinction. Some things a map represents are more easily reported on in language by someone using the map than are others. A quick look enables us to say that a map of the World represents the southern hemisphere as having more water than the northern hemisphere; it takes more study to report that it represents the largest island as being the smallest continent.

Fourthly, the fact that there is a reporting-relative distinction to be drawn means that radical holism can distinguish a more inclusive from a less inclusive sense in which we may believe that P. According to radical holism what makes it true that S believes that P is that S has a system of belief according to which P. This is the inclusive sense. But there is, I take, a pressure in our ordinary talk to tie belief that P to the ability to produce a relevant sentence on request—or at least there is when we focus on belief in humans rather than belief in animals. Radical holism can allow that there is a more exclusive sense of 'belief' in which an ability to produce the right kind of sentence is called

[6] See, e.g., Field (1981) and Fodor (1987), ch. 1.

for. The underlying representational reality is that there is a rich system that takes things to be thus and so, but different aspects of this rich content may vary in, as we might put it, how near they are to the linguistic surface. And anyone who wishes to is free to restrict their use of the word 'belief' to aspects that are near to that surface.[7]

Finally, we should note that any view based on the idea that the way belief represents is like the way maps or holograms represent cannot be accused of representational incompetence. It is sometimes suggested that maps and the like are good for representing geographical facts but not for representing the way house prices vary over time, or more generally the kinds of facts we typically use numbers to capture; and nor are they good for representing rather abstract, theoretical beliefs.[8] However, we can be confident that nothing escapes the representational powers of a map-like structure.

This is because everything perfectly represents itself. The plans of a house may get one or another detail about the house wrong but the house itself gets it exactly right, down to where each and every nail is located. Likewise, the world itself represents the way things are perfectly. It gets everything there is to get right exactly right down to the last sub-atomic particle. But the world itself is a huge four dimensional map-like structure. So we know that there is nothing a map-like thing cannot get right.

3. CAUSAL EXPLANATION BY APPEAL TO DISPOSITIONS

The story I will tell about how radical holism accounts for the evident difference between causally explaining behaviour by reference to believing that P and by reference to believing that Q arises from a certain view about explanation by appeal to dispositional properties. So this section of the paper is about explanation by appeal to dispositional properties. It is, of necessity, short and somewhat dogmatic.

Properties stand in causal relation. The film *The Way We Were* is about how the way its protagonists were in their youth led to how they became in middle age. But the way its protagonists were is a matter of the properties they had back then, and what they became in middle age is a matter of the properties they acquired in middle age. The film makers simply took it for granted that properties can be causes. And so do we all. The remark that when the US sneezes the rest of the world gets a cold is nothing other than

[7] And so can reinstate the core/derivative distinction by recasting it as one relating to differences in how easy it is to recover the relevant sentences.

[8] For instance Stich (1990), p. 109, asks —in a highly sceptical tone— how the metaphor of beliefs as maps of extramental reality can handle examples like 'I believe that if Robert Kennedy had not been assassinated, he would have become President of the United States' and 'I believe that we do not now have a good theory of consciousness'. However, the way he words his criticism suggests that he may be thinking of the map analogy as illuminating the topic of individual beliefs, whereas in fact it is put forward (here anyway) as a way of thinking about systems of belief.

a claim that how things are in the US causally affects how the rest of the world is. I am of course aware that many insist that it is general causal laws that relate properties, and that singular causal statements connect events, thought of as quite distinct from properties. But surely general causal laws deliver generalisations about what does or would cause what. It thus escapes me how general causal laws could be about properties without singular causal statements also being about properties. Likewise, I find strange the objection that properties pertain to causal explanation and not singular causation. To explain causally is to tell about what caused what, and so if explanation involves properties, as surely it does, that is support for properties being causes.

There remains the question of how to integrate the property view with the familiar events' framework. Perhaps the best strategy is to think of events as property-like and view the term 'event' as reminding us that it is properties qua instances that cause.[9] In any case I will talk of properties and state types as causes.[10]

Nevertheless, causal explanation by appeal to dispositional properties is not causal explanation that cites dispositional properties as causes. Although causes are properties, not all properties are causes. In particular, dispositions do no causing. Their categorical bases do all the causing.[11]

There are two reasons for insisting that the causing is done by the categorical bases. Consider a fragile glass that shatters on being dropped because it is fragile, and not (say) because of some peculiarity in the way it is dropped. Suppose that it is a certain kind of bonding B between the glass's molecules which is responsible for the glass being such that if dropped it breaks. Then the dispositional property of being fragile is the second order property of having some first order property or other, bonding B as we are supposing, that is responsible for the glass being such as to break when dropped. And the first order property, bonding B, is the categorical basis of the fragility. But then it is bonding B together with the dropping that causes the breaking; there is nothing left for the second order property, the disposition itself, to do. All the causal work is done by bonding B together with the dropping. To admit the fragility also as a cause of the breaking would be to admit a curious and ontologically extravagant kind of overdetermination. Or take an even simpler example, the way a thermostat consisting of a bi-metal strip turns a refrigerator on and off. What causes the refrigerator to turn on and off is the bending and straightening of the strip, and this in turn is caused on any

---

[9] I am here indebted to Graham MacDonald's dissent in discussion of the discussion after Jackson (1995).

[10] I here skate over a large debate. For further references and a bit more argument for the view I (unoriginally) favour, see Jackson (1995). For a recent statement of the other side see Donald Davidson (1993).

[11] Or at least they do when they are there to do the causing. We will not be concerned with 'bare' dispositions. Perhaps they are possible, perhaps not. It does not matter for anything I say here.

particular occasion by the differential expansion of the two metal strips: there is no causing left to be done by the relevant dispositional properties.[12]

The second reason for insisting that it is a disposition's categorical basis that does any causing relates to what is right about Hume's thesis that causal connections are contingent. There are necessarily true statements asserting causal connections. Thus, if a certain chemical structure is highly poisonous and thereby gets to be correctly described as 'the killer structure', the statement 'The killer structure typically kills' is necessarily true (modulo worlds where there is no killer structure).[13] However, what Hume had in mind is untouched by such examples. Hume's thesis is that what a state causes, or would cause were such and such to happen, is not an essential property of that state. His claim is that for any state it is true that that very state might not have caused what it did in fact cause, and might not have had the causal powers it in fact has.[14] The killer structure does not kill in some possible worlds—namely, in those worlds where the laws of nature or the internal nature of humans differ from how they actually are in such a way that that very structure is harmless to humans. In these worlds it is wrong to call the property in question 'the killer structure' but it is the same structure nevertheless.[15]

However, to allow that the second order state, the dispositional property of fragility itself as opposed to its categorical basis, causes breaking on dropping, would be to allow that there are properties that have causal powers essentially. If fragility does or would do the causing in one world, then it does or would do the causing in any world. In every world the fact of being fragile is intimately connected with the fact of breaking on dropping, and if that intimate connection counts as fragility causing or having the power to cause breaking in any world, it will so count in every world.

This argument from Hume only directly impinges on the claim that dispositions cause their own manifestations, and someone who thought that they had an answer to the first argument, the redundancy argument, against dispositions as causes might be tempted to allow dispositions as causes of things other than their own manifestations. Thus, some philosophers of colour identify yellow with a disposition to reflect light differentially, not with the disposition to look yellow. Part of their reason seems to be that they want looking yel-

[12]I intend these remarks to be in part stipulative of how I will use the terms 'disposition' and 'dispositional property', though I take it that my usage is reasonably standard. The substantive claim is that the second order properties for which I am reserving the terms 'disposition' and 'dispositional property' are not causes. The thesis that dispositional properties, and functional properties in general, are not causes has been much discussed recently in connection with the question of the causal efficacy of content, see e.g. Block (1990), and Jackson and Pettit (1988). By 'a second order property' I mean one whose specification requires mention of properties, not a property of a property like being relational or much discussed.

[13]Or see the discussion in Donald Davidson (1967).

[14]David Hume (Treatise).

[15]On some reasonable way of counting structures.

low to be a typical response to yellow, and believe that looking yellow could not be a response to the very disposition to look yellow. But putative cases of dispositions causing are where what is putatively caused is downstream from the dispositions' manifestations. The yellow object manifests its disposition to reflect light differentially and then the differentially reflected light causes the object to look yellow, as it might be. But if the disposition does not cause its manifestation, how can what the manifestation causes count as something caused by the disposition? If dispositions do not cause their own manifestations, it is hard to see how they could cause anything.

However it would be wrong to infer that we cannot give causal explanations in terms of dispositions. Richard Feynman explained the Challenger disaster in terms of the fragility at low temperatures of an O-ring. It would have been ridiculous for a member of the congressional committee of enquiry to respond 'Please Mr Feynman. In Philosophy 201 we were taught that dispositions cannot be causes. You must have made a mistake.' What then are we doing when we explain in terms of dispositions? It is tempting to give a simple answer. When we explain in terms of a dispositional property, we are citing the disposition's basis, which we may or may not know, as the cause. Thus, Feynman was right if and only if the basis of the O-ring's fragility caused the Challenger disaster.

There are, however, clear counter-examples to this simple answer. Here is one due to David Lewis. I will simplify the science but in a way which does not damage the philosophical moral. Opacity, electrical conductivity and thermal conductivity in metals have the same categorical basis. The way free electrons permeate metals that are opaque, are good thermal conductors and are good electrical conductors is responsible for the input-output behaviours distinctive of these three dispositional properties. Nevertheless, an explanation in terms of a metal's opacity is clearly not the same as one in terms of its electrical conductivity; nor is it the same as one in terms of its thermal conductivity. The transmission of a telephone call down a wire is explained by its electrical conductivity, not by the wire's opacity. Good results from cooking in copper-based saucepans are explained by copper's high thermal conductivity, not by its opacity or electrical conductivity.

Peter Menzies holds that we should account for the difference in these causal explanations by viewing the different dispositions as different causes. Thus he holds that we should abandon the doctrine that dispositions are not causes; for if they are different causes, they are causes. I admit the prima facie force of this argument but reply by insisting that it is outweighed by the case against dispositions being causes.

I say instead that causal explanations by appeal to dispositional properties have a dual character. When you explain by citing a dispositional property you make two claims together. First, you say that what was caused was caused by the disposition's categorical basis. If the categorical basis of the O-ring's fragility at low temperatures did not cause the Challenger disaster, Feynman was wrong. Secondly, you say that the kind of inputs and outputs distinctive of the disposition in question lie in the causal path to what happened.[16] One part of what Feynman was saying was that the categorical basis of fragility caused the disaster, and the other, more interesting part, was that what happened resulted from the kind of output distinctive of fragility: under pressure the O-ring broke, and its breaking lead to the disaster. Or consider the difference between explaining a death in terms of a metal's opacity as opposed to its thermal conductivity or its electrical conductivity. Although the operative property of the metal will be the same in each case, in the first it will, say, have lead to the obstruction of someone's vision at a crucial moment; in the second the property will have caused, as it might be, a surge of heat that burnt someone to death; and in the third the death will be by electrocution. In each case the causally operative internal property of the metal will be the same; the differences lie in how this property interacts with different inputs to cause different outputs, and the right disposition to cite in a causal explanation will be settled by the nature of the inputs and outputs.

4. EXPLANATION BY INDIVIDUAL BELIEF CONTENTS (BACK TO THE MAIN PLOT)

We started by noting that we might explain one bit of behaviour by reference to Harry's believing that there is coffee to the right, and another by reference to his believing that there is beer to the left. Radical holism cannot account for the difference by citing different individual beliefs. There are no such animals as individual beliefs according to radical holism. But the lesson of explanation by appeal to dispositions is that we can have distinct causal explanations without having distinct causes. The explanation of Harry's right-moving behaviour by reference to his believing that there is coffee to the right can be importantly distinct from the explanation of his left-moving behaviour by reference to his believing that there is beer to the left, as it had better be, without supposing that the two explanations involve distinct individual beliefs. Explanation by distinct belief contents need not be explanation by distinct individual beliefs. Both bits of behaviour can be caused by the one system of belief and yet the explanations be distinct. For radical holism must, we saw, take an essentially behavioural approach to content.

The content of a system of belief is somehow or other the product of the total story about how the system moves a body through the world.[17] In particular, there will be the inputs and outputs that make it true that Harry's

[16]We can think of cases where the distinctive outputs are what is being explained as a special case: we want fragility to explain (albeit in a rather uninformative way) breaking as well as matters downstream from breaking.

[17]Part of the story about how the system moves the body through the world will be that it does not do it the way a blockhead does. If it did, there would be no content at all. But I set aside the important question of precisely how radical holism should rule out

system of belief represents coffee to the right, and the distinct inputs and outputs that make it true that Harry's system represents beer to the left. Radical holists can and should see the distinctness of explanations by appeal to different individual belief contents as deriving from the distinctness of the input-output pairings associated with the different contents. In particular what makes it true according to radical holism that Harry's right moving behaviour is explained by his believing that there is coffee to the right is the conjunction of: a) his right moving behaviour is caused by how his system of belief represents things as being b) how his system of belief represents things as being includes that there is coffee to the right and c) his right moving behaviour is, or is caused by, part of the pattern of inputs and outputs that make it true that his system of belief represents that there is coffee to the right. The difference between an explanation in terms of his belief that there is coffee to the right and one in terms of his belief that there is beer to the left is now transparent. It lies in the difference between the pattern of inputs and outputs responsible for his system of belief representing that there is coffee to the right as opposed to the pattern responsible for the very same system representing that there is beer to the left. It lies, that is, in the same place as the differences between explanations in terms of the opacity, thermal conductivity and electrical conductivity of metals. What causes Harry's right moving and what causes his left moving behaviour is the very same complex internal structure according to radical holism; the difference is in whether what gets caused is, or is caused by, something definitive of the structure representing inter alia that there is coffee to the right or its representing that there is beer to the left. There are many questions that might be asked about this story on behalf of radical holism. I will conclude by offering answers to two of the more obvious ones. The first concerns the type-type identity theory of mind; the second concerns certain overdetermination cases.

5. RADICAL HOLISM AND THE TYPE-TYPE IDENTITY THEORY

Although radical holism must, it seems to me, give some kind of dispositional cum functional treatment of content, the most plausible version does not treat the content of a subject's system of belief as a dispositional property. For then content would not be a cause, and I—and thousands—take it to be obvious that how a subject takes things to be literally moves that subject around the world. True, I must admit that the observation that dispositional properties are not causes opens up an interesting alternative position on content. It is obvious that dispositional properties can causally explain. This is so whether or not you accept the two-stranded account I gave of how they do so. We must

blockhead. Also, we could as noted earlier insist that the history of how structures of the kind in question moved bodies around in the past is important.

therefore acknowledge a notion of causal relevance that includes but goes beyond causal efficacy or production or, as we are calling it most simply, causing. Dispositional properties are causally relevant to various outcomes without actually causing them. In putting it this way I am taking a stand against the view that if dispositional and functional properties are not causes then they are, as Ned Block, for instance, sometimes puts it, epiphenomenal.[18] To be epiphenomenal is to be outside the causal story about how things come about. Dispositions are not epiphenomenal in this sense. The causal story about how a fragile glass comes to break includes its fragility: by the time you have explained how the glass's categorical nature gets together with the impact to cause breaking, you have said inter alia that the glass is fragile. Although dispositional properties do not cause, they are an ineliminable and integral part of many stories about what does cause. This is why they should be counted among the potentially causally relevant properties of one or another happening, including behaviour.

The interesting, alternative position on content then is that it is causally relevant to behaviour without actually causing behaviour. All the same, I think we should regard this position as a fall back one. The clear intuition is that the content of belief drives behaviour.[19] If this is right, belief content is a neurophysiological property that drives behaviour and gets to be belief content, and gets to be the belief content that it is, in virtue of how it does the driving; in virtue, that is, of how it moves a subject around the world. Thus radical holism is best thought of, it seems to me, as a style of type-type identity theory. But instead of making a claim like 'Belief that there is beer to the left in Harry = neuro-state B' it makes a claim like 'A rich taking of things to be thus and so in Harry = huge, complex neuro-state N.' There are no identity claims for individual beliefs for the good reason that there are no individual beliefs according to radical holism.

The identity claims made by radical holism can only be contingently true. Although radical holism is very different from more individualistic views of belief, the powerful case for multiple realisability applies equally to both. In consequence, it must be allowed that Harry's taking things to be thus and so might have been a different neuro-state or indeed not a neuro-state at all, just as the most dangerous poison might have been a different structure from the one it in fact is. The obvious question then is what do I say against common objections to contingently true type-type mind-brain identities. In outline—all that is possible here—I say the following.

True, the early identity theorists offered as illustrations of what they had in mind when they said that mental states are brain states, scientific identities

[18] Block (1990).

[19] In Jackson and Pettit (1988) we are neutral as between the view that content drives behaviour and the view that it is causally relevant to behaviour without driving it. My reasons for now preferring the driving position are spelt out in a little more detail in Jackson (1995).

like 'Lightning = electrical discharge', identities that are now widely held to be necessary a posteriori. But that was because it was then widely believed that these identity sentences are contingent. They could have made the points they wished to make in terms of indubitably contingent type-type identity statements like 'The most dangerous poison for humans = plutonium' (which is, I understand, true on some ways of measuring toxicity).

True, many early identity theorists left out qualifications like 'in Harry' that I included in the two examples just given, and this gave the impression that it was an essential part of their view that the same neuro-kind was the same mental state type in everything. It may even be that some early identity theorists thought that this was an essential feature of their view. But it was not and is not.

True, many on being presented with examples of what seem to me—and thousands—to be clear examples of contingent type-type identities respond by saying that we must be reading phrases like 'the most dangerous poison' as definite descriptions and Russell teaches us that definite descriptions are not singular terms. But, they then say, the examples cannot really be identity statements. This seems to me to reflect an unfortunate terminological decision in the philosophy of language. 'Singular term' is a good phrase for a bunch of words intended by the conventions of the language to apply to one thing, as W. V. Quine says;[20] and this is true independently of the success of Russellian accounts of definite descriptions. But in any case at most what is shown is that type-type identity theorists chose a bad name for their theory.

6. THE RSG QUESTION

Ramsey, Stich and Garron (1990) have pointed out that the following kind of situation is possible. Harry believes that P. Harry believes that Q. Harry acts in a way that might be rationalised by reference to his believing that P, but equally it might be rationalised by his believing that Q. But only, say, his believing that P, in fact causally explains what he does. Perhaps Harry wants access to a phone and also to a computer. He believes, truly let's suppose, that the best way to access a phone is to enter room 107, and that the best way to access a computer is to enter room 107. He intentionally enters room 107. Clearly, either belief rationalises his entering the room, but consistently with this it might be that his belief that entering room 107 is the best way to access a computer is what in fact causally explains his action.

Let's call a situation of this kind a RSG situation. They are clearly possible. The RSG question for any theory of belief is how it accounts for the possibility of RSG situations.

The individualists' answer to the RSG question for our example is that Harry's belief that entering room 107 is the best way to access a phone is

[20] Quine (1960), §20.

a different state from his belief that entering room 107 is the best way of accessing a computer; where what is meant is not just that the contents *entering room 107 is the best way of accessing a phone* and *entering room 107 is the best way of accessing a computer* are distinct, but that there are distinct states with those contents. As there are two states it is possible that one and not the other does the causing. Radical holism cannot give this answer.

Radical holism has to answer the RSG question by insisting that the difference between Harry's action being explained by his believing that entering room 107 is the best way of accessing a phone and its being explained by his believing that entering room 107 is the best way of accessing a computer will show up somehow or other in the relevant, actual or possible, inputs and outputs. If, as we supposed, it is his believing that entering room 107 is the best way of accessing a computer which explains his entering the room, his body will be oriented towards where he takes the computer to be in the room rather than where he takes the phone to be; or perhaps his movements will be that bit more sensitive to information concerning computers than to information concerning phones; or whatever. A fair and obvious question is whether we can describe an RSG case where there is no relevant difference whatever describable somehow or other in terms of inputs and outputs. Perhaps, it might be urged, there could be a determinate fact of the matter as to whether Harry's behaviour is causally explained by the computer belief content or by the phone belief content, and yet there be no difference in actual and possible behavioural responses. I think the jury is still out on this possibility and that the verdict it will return will depend on how representation in the head takes place. The possibility is posteriorto whether individualism or holism is the right story about mental representation. It is not a prior datum with which to assess theories of mental representation.[21]

Philosophy Program
Research School of Social Sciences
Australian National University
Canberra

REFERENCES

Block, Ned (1990) "Can the Mind Change the World" in G. Boolos, ed, *Meaning and Method: Essays in Honor of Hilary Putnam*, Cambridge: Cambridge University Press.

Davidson, Donald (1967) "Causal Relations", *Journal of Philosophy*, 64: 691–703.

Davidson, Donald (1993) "Thinking Causes" in John Heil and Al Mele, eds, *Mental Causation*, Oxford: Clarendon Press.

Field, Hartry (1981) "Mental Representation" in Ned Block, ed, *Readings in Philosophical Psychology*, vol. II, pp. 78–114.

[21]This is a revised and slightly expanded version of the paper read to 10th International Congress for Logic, Methodology and Philosophy of Science, in Florence on August 22, 1995.

Fodor, Jerry (1987) *Psychosemantics*, Cambridge, Mass.: M.I.T. Press.
Hume, David (Treatise) *A Treatise of Human Nature.*
Jackson, Frank (1995) "Essentialism, Mental Properties and Causation", *Proceedings of the Aristotelian Society*, XCV: 253–268.
Jackson Frank and Philip Pettit (1988) "Functionalism and Broad Content", *Mind*, 97: 381–400.
Lewis, David (1986) *On the Plurality of Worlds*, Oxford: Basil Blackwell.
Quine, W.V. (1960) *Word and Object*, Cambridge, Mass.: M.I.T. Press.
Ramsey, William, Stephen Stich and Joseph Garon (1990) "Connectionism, Eliminativism and the Future of Folk Psychology", in *Philosophical Perspectives*, vol. 4, pp. 499–533.
Stich, Steve (1990) *The Fragmentation of Reason*, Cambridge, Mass.: M.I.T. Press.

JAEGWON KIM

WHAT IS THE PROBLEM OF MENTAL CAUSATION?

1. SAVING MENTAL CAUSATION

Giving an account of mental causation — in particular, explaining how it is possible for the mental to causally affect the physical — has been one of the central problems in the philosophy of mind over the past decade or so. The problem of course is not new: Descartes famously was confronted by many of his contemporaries — for example, Gassendi and Princess Elizabeth of Bohemia — with the same question. But this does not mean that Descartes' problem is our problem. For his problem, as his contemporaries saw it, arose from his substantival dualism, a dualism of material and mental substance. But, at least for most of us, that is not the source of our worries about mental causation. Few of us now believe in the existence of substantival minds or some kind of mind-stuff that is ontologically independent of material bodies.

Philosophical problems do not arise in a vacuum. Typically, they emerge when we come to see a conflict among the assumptions and presumptions that we explicitly or tacitly accept, or commitments that command our presumptive respect. The gravity of a philosophical problem, therefore, depends on two related questions: first, how deep and irrevocable is our commitment to the assumptions that give rise to the problem?; second, how easy or difficult is it to bring the conflicting assumptions into an acceptable reconciliation? The process of reconciliation may require serious modifications to our original assumptions and commitments. Short of abandoning the entire framework of the existing commitments, which philosophical revolutionaries are apt to recommend but which often is far easier said than done, compromises must be negotiated; there are no free solutions to philosophical problems.

In this paper, I want to set out, in what to my mind is the simplest and starkest way, how our problem of mental causation arises. In saying this, I do not want to imply that there is a *single* problem of mental causation. I believe that several different sets of assumptions and principles that many people find plausible to varying degrees can make mental causation prima facie problematic. In referring to "our problem" of mental causation, I am suggesting, although I don't want this to become a contentious focus of this talk, that it is a problem that arises for anyone with the kind of broad physicalist outlook that many philosophers, including myself, find compelling and prima facie unavoidable. I will not try to give a description, even a general sketch, of just what such a physicalism is committed to. That will be a difficult job to do with any kind of precision, and it is not needed for the present

M.L. Dalla Chiara et al. (eds.), Structures and Norms in Science, 319–329.

purposes. In the course of my talk, however, bits and pieces of physicalism, some large and some small, will dribble out, and give you at least a partial outline of the kind of physicalism that is in play here.

Let me begin with a few words about why we want to save mental causation — why it is important to us that mental causation does exist (some will say that its existence is an ultimate, nonnegotiable commitment). First, the very possibility of human agency seems to demand that our mental states, in particular, beliefs and desires, have physical causal effects; in voluntary actions our beliefs and desires, or intentions and decisions, must somehow cause our limbs to move in appropriate ways, and thereby cause objects around us to be rearranged. That's how we manage to cope with our surroundings, write philosophy papers, build bridges and cities, and destroy the ozone layers. Second, the possibility of knowledge requires mental causation: perception, arguably our sole window on the world, requires the causation of perceptual experiences by physical objects and events around us; reasoning, by which we acquire new knowledge from the existing fund of what we already know, involves causation of new beliefs by old beliefs; and memory is a complex process involving interactions between experiences, their physical storage, and retrieval in the form of belief. If you take away perception, memory, and inference, you pretty much take away all human knowledge. To move on: the possibility of psychology as a causal-theoretical science that generates law-based explanations of human behavior evidently depends on the reality of mental causation: mental phenomena must be capable of functioning as indispensable links in causal chains leading to physical behavior. A science that invokes mental phenomena in its explanations is prima facie committed to their causal efficacy; for anything to have an explanatory role, its presence or absence in a given situation must make a difference — a *causal difference* when the explanation of an empirical phenomenon is at issue.

2. DILEMMA 1: SUPERVENIENCE AND MENTAL CAUSATION

I will try to generate the problem of mental causation through a pair of interlocking dilemmas. The first dilemma begins with the doctrine of mind-body supervenience:

(i) Either mind-body supervenience holds or fails.

This is an obvious tautology, setting forth the two horns of the dilemma to be constructed. But what does mind-body supervenience assert? For our purposes, it may be stated as follows:

> *Mind-body supervenience*: For every mental property M there is a physical property P such that necessarily whenever something instantiates P at t it instantiates M at t (such a P is called a "base"

or "subvenient" property of M). Moreover, nothing can instantiate M at t unless it instantiates some physical base property of M at t.

Note that a base property is "necessarily" sufficient for the supervenient property; the necessity involved here is standardly taken to be at least *nomological necessity* — so that if mind-body supervenience holds, it holds in all worlds that are similar to ours in fundamental laws of nature. Note, further, that according to our formulation, any given mental property can have *multiple physical bases*; thus, mind-body supervenience is consistent with a phenomenon often noted, namely that mental properties are "multiply realizable" in diverse physical substrates.

Returning to (i), we pursue the second horn first, quite briefly:

(ii) If mind-body supervenience fails, there is no visible way of understanding the possibility of mental causation.

Jerry Fodor writes: "If mind/body supervenience goes, the intelligibility of mental causation goes with it".[1] To my knowledge, Fodor has never explained why he thinks this (and not just once), but the simplest and most obvious reason for the physicalist to accept (ii) lies, I believe, in her commitment to *the causal closure of the physical domain*. The principle of the physical causal closure says this: If you pick any physical event and trace its causal ancestry or posterity, that will never take you outside the physical domain. That is, no causal chain will cross the boundary between the physical and the nonphysical. The interactionist dualism of Descartes is in clear violation of this principle. If you reject this principle, you are ipso facto rejecting the possibility of a complete and comprehensive physical theory of all physical phenomena (never mind a physical theory of *all* phenomena). And you would be saying that any complete explanatory theory of the physical domain must invoke nonphysical causal agents. It is safe to assume that no serious physicalist could accept such a prospect.

Now if mind-body supervenience fails — that is, if the mental domain floats freely, unanchored in the physical domain, the causation from the mental to the physical would obviously result in the violation of the physical causal closure. Mind-body supervenience grounds each mental phenomenon (or, more broadly, mental property) in the physical domain by providing for it a set of physical conditions that are (at least) nomologically sufficient for it. A corollary is the thesis that no mental phenomenon can occur, or no mental property can be instantiated, unless an appropriate physical base condition is present. Every mental event, be it a sensation like pain or itch, or an intentional state like belief and desire, must have a physical basis: it occurs because an appropriate physical basis is present, and it would not occur if such a basis was absent.[2] These claims can be put in equivalent forms — equivalent as far as

[1] *Psychosemantics* (Cambridge: The MIT Press, 1987), p. 42.

[2] On content externalism, wide-content states will not supervene on internal physical properties of the subject, but physicalists will not deny that they supervene on his extrin-

the arguments of this paper are concerned — in the idioms of "emergence" or "realization". Instead of saying that mental phenomena are supervenient on the physical phenomena, one may want to say that they are "emergent from", or "realized" or "implemented" by, physical-biological processes. But as these terms are standardly understood, both the emergence thesis and the physical-realization thesis entail mind-body supervenience: When the same physical conditions are present, the same mental phenomenon will emerge, or be realized. So if the argument works with mind-body supervenience as a premise, it should work for those who prefer the "emergence" or "realization" formulation of the mind-body relation.[3]

In any case, mind-body supervenience at least brings mental phenomena within the ambit of the physical: in a clearly understandable sense, the physical determines the mental, and in that sense the mental does not constitute an ontologically independent domain that projects causal influences into the physical domain. Now, it is another question whether or not mind-body supervenience brings the mental *close enough* to the physical to allow mental causation to circumvent the constraint of the physical causal closure.[4] But we can skirt this issue here: if it fails in this regard, our conclusion is established in any case: Mental causation is prima facie not possible. What I want to do here is to pursue another line of argument leading to the same conclusion.

(iii) Suppose, then, an instance of mental property M causes another mental property M^* to be instantiated.

So this is a case of mental-to-mental causation, one in which an instance of a mental property causes an instance of another mental property. We may take "instances" or "instantiations" of properties as events, states, or phenomena. For simplicity's sake, I will often speak of one property causing another property; this is to be understood to mean that an *instance* of the first causes an *instance* of the second.[5] Returning to our argument: it follows from (ii), the supervenience premise, that:

(iv) M^* has a physical supervenience base, P^*.

We now ask the following critical question: Where does this instance of M^* come from? How does M^* get instantiated on this occasion? There appear to be two possible answers:

---

sic/relational properties. For the present paper, we will ignore the issues that arise from content externalism.

[3] For details see my "'Downward Causation' in Emergentism and Nonreductive Physicalism", in *Emergence or Reduction?* ed. A. Beckermann, H. Flohr, and J. Kim (Berlin: De Gruyter, 1992).

[4] See on this issue "Postscripts on Mental Causation" in my *Supervenience and Mind* (Cambridge: Cambridge University Press, 1993).

[5] Strictly speaking, this doesn't go far enough: it must further be the case that one instance causes another instance *in virtue of the fact that the first is of an F-instance and the second is a G-instance.*

(v) M^* is instantiated: (a) because, ex hypothesi (iii), M caused M^* to be instantiated; (b) because P^*, the physical supervenience base of M^*, is instantiated on this occasion.

There clearly is a tension between these two answers: under the assumption of mind-supervenience, M^* occurs because its supervenience base, P^*, occurs, and as long as P^* occurs, M^* must occur no matter what other events had preceded this instance of M^* — in particular, regardless of whether or not an instance of M preceded it. This puts the claim of M to be a cause of M^* in jeopardy: P^* alone seems fully responsible for, and capable of explaining, the occurrence of M^*.[6] The only question that is left seems to be this: What caused P^* to be instantiated on this occasion? And it would seem that the following is the only way of reconciling the claim of M to be a cause of M^* with the fact that M^* has P^* as its supervenience base:

(vi) M caused M^* by causing P^*. That is how this instance of M caused M^* to be instantiated on this occasion.

There is a plausible general principle involved here, which can be stated this way: *To cause a supervenient property to be instantiated, you must cause its base property (or one of its base properties) to be instantiated.* To relieve a headache, you take aspirin: that is, you causally intervene in the brain process on which the headache supervenes. That's the only way you can do anything about your headache. To make your painting more beautiful, you must physically alter the painting — that is, change the physical supervenience bases of the aesthetic properties you want to alter. There is no direct way of making your painting more beautiful or less beautiful; you must work on it physically if you want to change it aesthetically, and there is no other way.

But note what (vi) asserts: it says that a mental property, M, causes a physical property, P^*, to be instantiated. This of course is a case of mental-to-physical causation. So what our argument has shown so far is this: *Under the mind-body supervenience assumption, mental-to-mental causation implies, or presupposes, mental-to-physical causation.* So the question that we now face is whether we can make sense of mental-to-physical causation — that is, under the premise of mind-body supervenience.[7]

Going back to (vi): we realize that on the assumption of mind-body supervenience, it follows:

(vii) M itself has a physical supervenience base, P.

We must now compare M and P in regard to their causal status with respect to P^*. When we reflect on this point, I believe we begin to see reasons for

---

[6] This argument is based on what I have called "the principle of causal/explanatory exclusion"; see, e.g., my "Mechanism, Purpose, and Explanatory Exclusion", reprinted in *Supervenience and Mind* (Cambridge: Cambridge University Press, 1993).

[7] We could have begun with (vi) as our initial premise of mental causation. The point of starting with (iii) is to show that the argument applies to mental-mental causation as well as to mental-physical causation. On the assumption of mind-body supervenience, the former is as problematic, if my argument succeeds, as the latter.

taking P as pre-empting the claim of M as a cause of P^*. If you take causation as grounded in nomological sufficiency, P qualifies; since P is sufficient for M and M is sufficient for P^*, P is sufficient for P^*. If you choose to understand causation in terms of counterfactuals, again there is good reason to think that P qualifies as a cause of P^*: If P hadn't occurred, M would not have occurred (we may assume, without prejudice, that no alternative physical base of M was present on this occasion), and given that if M had not occurred P^* would not have occurred, we may reasonably conclude that if P had not occurred, P^* would not have either.[8] Moreover, it seems not feasible to think of this situation as involving a causal chain from P to P^*, with M as an intermediate causal link. For the relation from P to M is not happily thought of as a causal relation; in general, the relation between base properties and supervenient properties is not happily construed as causal.[9] For one thing, the instantiations of the related properties are wholly synchronous, whereas causes are standardly thought to precede their effects; second, it is difficult, perhaps even incoherent, to imagine a causal chain, with intermediate links, between the subvenient and the supervenient properties. What intermediate stages could link the beauty of a painting and its physical properties? What intermediate events could causally connect a mental event with its subvenient physical base? Are such intermediaries themselves mental or physical?

It seems to me that the most natural way of viewing the situation is this:

(viii) P caused P^*, and M supervenes on P and M^* supervenes on P^*.

This explains the observed regularities between M-instances and M^*-instances, and M-instances and P^*-instances.[10] These regularities are not accidental; in a clear sense, they are law-based, and may even be able to support counterfactuals. However, if we understand the difference between genuine, productive causal processes on the one hand and the noncausal regularities that are observed because they are parasitic on them, we are in a position to understand the picture I am recommending here. In the case of supposed M-M^* causation, the situation is rather like a series of shadows cast by a moving car: there is no causal connection between the shadow of the car at one instant and its shadow an instant later, each being an effect of the moving car. The moving car represents a real causal process; but the series of shadows it casts, however regular and lawlike it may be, does not constitute a causal process.[11] Hence, we have:

---

[8] This of course is not to assume transitivity for counterfactuals in general.

[9] One philosopher who claims that the base properties "cause" the supervenient properties is John Searle, in his *The Rediscovery of the Mind* (Cambridge: The MIT Press, 1992).

[10] Note, however, that these regularities are likely to be restricted in generality. The reason is that M's alternative supervenience bases cannot be counted to cause P^* and hence M^*.

[11] On the distinction between "causal process" and "pseudo-process" see Wesley Salmon, *Scientific Explanation and the Causal Structure of the World* (Princeton: Princeton University Press, 1984).

(ix) The M-to-M^* and M-to-P^* causal relations are only apparent, arising out of a genuine causal process from P to P^*

Thus, our first dilemma:

(x) If mind-body supervenience fails, mental causation is unintelligible; if it holds, mental causation is again unintelligible. Hence, mental causation is unintelligible.

3. DILEMMA 2: REDUCTION AS A WAY OUT

One good question to raise about the foregoing argument is this: Wouldn't the same argument show that all properties that supervene on basic physical properties are epiphenomenal, that their causal efficacy is unintelligible? There seems to be excellent reason to think that biological properties, say, are supervenient on fundamental physical properties, and if mind-body supervenience could be shown to put mental causation in jeopardy, wouldn't the very same considerations do the same for biological properties? But no one seems to worry about biological causation, and there evidently seems no reason to start worrying. If so, shouldn't we conclude that there must be something wrong with the argument of the preceding section?[12]

I will not deal with these questions in detail here.[13] As I see it, however, the heart of the issue involved here is this: With properties like biological and chemical properties, we are much more willing to accept a reductionist solution to the problem. That is, if the "higher-order" properties can be reduced to basic physical properties, in some appropriate sense of reduction, there is no *independent* problem of the causal efficacy of the reduced properties. This means that the argument of the preceding section is incomplete rather than materially flawed: there is a potential escape route via reductionism.

This reductionist way out is exactly what the second argument is designed to block. Now, I expect that this comes as no surprise to most people, since mind-body reductionism has been out of favor for many years and serious reductionists are just about as rare as, if not rarer than, those who believe in substantival minds. What turned the tide against reductionism, toward the end of the 1960s, is the so-called "multiple realization" argument,[14] which is by now a familiar piece of conventional wisdom in the philosophy of mind.

[12] Several philosophers have raised exactly these questions (though not necessarily directed against our first argument); e.g., Lynne Rudder Baker, "Metaphysics and Mental Causation", in *Mental Causation*, ed. John Heil and Alfred Mele (Oxford: Clarendon Press, 1993); Robert Van Gulick, "Three Bad Arguments for Intentional Property Epiphenomenalism", *Erkenntnis* 36 (1992); Louise M. Antony, "The Inadequacy of Anomalous Monism as a Realist Theory of Mind", in *Language, Mind, and Epistemology*, ed. G. Preyer, F. Siebelt, and A. Ulfig (Dordrecht: Kluwer, 1994).

[13] For more details see my "Mental Causation: What? Me Worry?", forthcoming in *Philosophical Issues*, and "The Idea of 'Levels' of Properties", in preparation.

[14] First advanced by Hilary Putnam, in "The Nature of Mental States" (1967), reprinted in Putnam's *Philosophical Papers* II (Cambridge: Cambridge University Press, 1975).

Briefly, the argument goes like this: any mental state is realizable in a large and diverse variety of physical-biological mechanisms, and hence cannot be reductively identified with any one of them. Thus, for example, the neural state that realizes pain in humans must be vastly different from the neural state realizing pain in octopuses or reptiles. When we consider all nomologically possible realizers of pain, in biological and nonbiological systems, we can see, the argument goes, that pain, as a mental kind, cannot be identified with any single physical kind.

I do not believe that the multiple realization argument is sufficient to defeat reductionism. In my view, it is a tacit presupposition of this argument that a mental property is instantiated in virtue of the fact that an appropriate physical realizer of the property is instantiated on that occasion, and this leads, naturally and straightforwardly, to the view that mental properties, though not uniformly reducible to physical base, are nonetheless reducible — "locally reducible", as we might say — to their multiple physical realizers.[15]

The paradigm of reduction that has tacitly been assumed as a basis for the reductionist/antireductionist debate during the past three decades is the model of inter-theoretic reduction introduced by Ernest Nagel in the 1950s and '60s.[16] The heart of this model is the idea of "bridge laws", laws connecting the predicates of the theory being reduced to coextensive predicates in the base theory. In the material mode, this amounts to the requirement that to reduce a set of properties (say, mental properties) to another set (say, neural-biological properties), each property to be reduced must be provided with a nomologically coextensive property in the base set. This requirement is exactly what the multiple realization argument claims we cannot meet in the mind-body case: For each mental property there is a set of properties each of which is nomologically sufficient for it, but none that is both necessary and sufficient — that is, none that is nomologically coextensive with it.

I believe that finding a nomologically coextensive physical property for a mental property is neither necessary nor sufficient for reduction. It isn't necessary because, briefly again, local reductions of the sort I just mentioned, namely reductions that are restricted to a given species or structure, do not require it. It isn't sufficient because, although the nomic coextensivity between mental property M and physical property P may suffice for the *derivation* of laws involving M from laws involving P, it does not deliver the *reduction* of M to P. For the latter, we must have an identity, "$M = P$". Unless we are in a position to make the identification, two essential desiderata of reduction remain unfulfilled. First, as long as M and P remain distinct properties, there are facts involving M that are distinct from, or "over and above" as people

[15]For details see my "The Myth of Nonreductive Materialism" (1989) and "Multiple Realization and the Metaphysics of Reduction" (1992), both reprinted in *Supervenience and Mind* (Cambridge: Cambridge University Press, 1993).

[16]In particular, see Nagel, *The Structure of Science* (New York: Harcourt, Brace & World, 1961), chapter 11.

used to say, facts involving P, and a complete inventory of the world must include these M-facts in addition to the P-facts. In this sense, there has been no reduction in our ontology. Second, a correlation of the form "M iff P" cries out for an explanation. Even if the correlation is "local" and species-specific, like "In humans, pain occurs to a person iff his/her c-fibers are activated", the same explanatory question arises: Why is it that pains occur in humans just when their c-fibers are activated? Why don't sensations of itch occur instead? Why don't pains occur under other types of neural conditions? Reduction must be explanatory, and when the emergentists claimed the mental to be emergent but irreducible, it is because they thought these explanatory questions are in principle impossible to answer; their worry had nothing to do with the availability of bridge laws.

But what does genuine reduction require? The following, I believe, states a more realistic view of how reduction works:

(i) To reduce a property Q, Q must first be *functionalized* — i.e., must be given a *relational construal* in terms of its *causal/nomological role*. The reduction consists in finding a property Q^*, characterized in a lower-level theory, that fills the specified causal/nomological role.

Let's look briefly at a couple of examples. First, the stock example of reducing temperature: to reduce temperature we must first construe it in terms of its characteristic causal role, perhaps something like this: temperature is that magnitude of a physical system which increases when the system is in contact with another with a higher degree of that magnitude, which decreases when in contact with a system with a lower degree of that magnitude, which, when sufficiently high, causes objects to burn or melt, which, when sufficiently low, turns liquids into solids, and so on (well, you get the idea). And a reduction is accomplished when we have identified a physical quantity, characterized in a more fundamental physical vocabulary, that fills this causal role. It turns out that, in the cases of gases, the average translational kinetic energy of the constituent molecules has exactly this causal role, and, moreover, we have a theory that explains why this property has these causal powers. We can then reductively identify temperature in gases with the mean kinetic energy of their molecules. We can say something similar about the gene and the DNA molecules: the gene is that factor in a biological system that is causally responsible for the transmission of heritable characteristics, etc., and we find that the DNA molecule is just the mechanism that performs these causal tasks. Again, we have an explanation, at least in convincing outlines, of why and how the DNA molecules are able to perform these causal functions. The foregoing gives only a rough picture of how we should think of reduction, but this will suffice for our present purposes. [17]

---

[17]This view of reduction goes back to David Armstrong's *A Materialist Theory of the Mind* (London: Routledge and Kegan Paul, 1968). For more details on reduction see my "The Mind-Body Problem: Taking Stock after 40 years", forthcoming in *Philosophical Per-*

On this view of reduction, then, the reducibility of mental properties is crucially tied to the possibility of a functional conception of mentality — that is, mind-body reductionism and the functionalist view of mental properties share the same fate: they sink or swim together. This differs sharply from the widely shared view that functionalism about the mind is a form of antireductionism, or at least is consistent with it. In any case, let us proceed with our argument:

(ii) Hence, to reduce mental property M, M must first be functionalized.

If M is the property of experiencing pain, M must be reconstrued in some such fashion as "the property that typically is caused to be instantiated by tissue damage and that typically causes cryings, wincings, and escape behavior". Are mental properties functionalizable? I think that depends on the kind of mental property we have in mind; however, I am prepared to defend, although not here, the view that phenomenal mental properties, or "qualia", cannot be construed as extrinsic/relational properties. If there are any intrinsic/nonrelational properties, qualia must be among them. If a property is intrinsic, it isn't extrinsic and we can't make it extrinsic by some magical philosophical manipulation. Thus:

(iii) If M is a phenomenal property, M is an intrinsic property and cannot be functionalized. Hence, M cannot be reduced to its physical supervenience base.

From which it follows:

(iv) M, therefore, remains distinct from its physical supervenience base, and its relationship to its physical supervenience base must be considered basic and "brute".

This of course means that the escape route through reductionism is blocked, and:

(v) We are back with Dilemma 1.

This, then, is the problem of mental causation for the physicalist. Some will be willing to jettison (iii), either by insisting on the functionalization of qualia or by eliminating them outright. From a metaphysical point of view, however, it is difficult to see a real difference between functionalization of qualia and their elimination. For in fuctionalizing qualia, we are in effect denying their status as intrinsic properties, and if this is what we are doing, we might as well admit that there are no such things as qualia. At least, the considerations of this paper enable us to better understand why some physicalists have embraced eliminativism: if our arguments are in general correct, to keep qualia would, prima facie, amount to retaining causally inert, hence theoretically useless, entities in one's ontology.

---

spectives. See also Joseph Levine, "On Leaving Out What It's Like", in *Consciousness* ed. Martin Davies and Glyn W. Humphreys (Oxford: Blackwell, 1993).

Let me finally note that to run the second argument, it is also possible to consider substituting for (iii) a premise that denies the functionalizability of intentional mental properties, or any other distinguished set of mental properties that you believe resist functionalization. In this sense, the second argument can be thought of as an *argument schema* for antireductionism. It seems to me, however, that qualitative mental properties give us the strongest antireductionist argument. I believe that it is still an open question whether or not intentionality can be explained in causal/physical terms (including non-question-begging biological terms).

Brown University
Providence, RI

BRIAN P. MCLAUGHLIN

CLASSICAL CONSTITUENTS IN SMOLENSKY'S ICS ARCHITECTURE

In a recent paper,[1] Paul Smolensky replies to an article that Jerry Fodor and I published in 1990.[2] I shall use this opportunity to make a partial reply; partial, because I lack the space here to address all of the issues Smolensky raises in his paper—a paper that spans sixty-seven pages of print. I shall confine myself to making one key observation and to attempting to clarify a few issues, leaving a fuller discussion for another occasion.[3]

I. SOME PRELIMINARIES

To place my disagreements with Smolensky in theoretical perspective, he and I share a commitment to a representational theory of mind: the cognitive architecture of (at the very least) normal humans includes a system of mental representation. Moreover, we both share the view—long championed by Fodor and Zenon Pylyshyn—that the way to explain the systematicity and productivity of thought-capacities is by appeal to the compositional structure of that system of mental representation. Further, we agree that Fodor and Pylyshyn (1988) shows that connectionist systems of representation that contain only "local representations"—i.e., representations realized by activated individual units in a network—lack that compositional structure. We disagree, however, over whether an architecture that postulates representations realized by activation patterns distributed over units in groups of units will offer an *adequate alternative* to a classical or "language of thought" cognitive architecture. The question is empirical; for all I know, some such architecture might do that. Unfortunately, Smolensky has given us no reason to think that his architecture will.

Some terminological points are useful at the outset so as to avert possible misunderstanding. I take vectors to be geometrical entities. Smolensky (1995, p.281) insists that they should be thought of as (ordered?) lists of numbers. I shall, then, understand them here to be (ordered) lists of numbers. Smolensky sometimes uses the phrase 'activity vector' to mean a pattern of activity over

---

[1] Smolensky (1995).

[2] Fodor and McLaughlin (1990).

[3] It should, of course, go without saying that I shall speak here strictly for myself. I wish to note, nonetheless, that I wrote this paper while in Europe during a period in which Fodor was working on his John Locke lectures. (We have not had the opportunity to discuss Smolensky's reply to our paper.) Smolensky's spirited reply deserves, I feel, a prompt response.

M.L. Dalla Chiara et al. (eds.), Structures and Norms in Science, 331–343.

the units of some group of units, and sometimes to mean a list numbers. I shall assume that lists of numbers index patterns of activity over units, and that such patterns are states of the group of units in question. I shall use 'pattern of activity' and 'activation pattern' to mean such states, and 'activity vector' to mean an ordered list of numbers indexing such a state. Smolensky characterizes the relationship between symbols and activation patterns in several non- equivalent ways. He tells us that activation patterns "represent" symbol (1995, p.231);[4] that they "are" symbols (p.234, p.261); that symbols are "borne by" activation patterns (p.232); that activation patterns "realize" symbols (p.236); and that they are the "micro-structures" of symbols (p.264). I won't try to sort through all this here; thankfully, how, exactly, all this is to be sorted out will not matter for present purposes. Smolensky employs the notion of realization (see, especially, p.237), and taking the relationship between activation patterns and symbols to be that of realization makes best sense of what he says. Thus, I shall understand him to mean that certain activation patterns, ones with semantic values, realize symbols. (As Smolensky acknowledges, not all activation patterns in his architecture will have semantic values (p.241).) So, then, (ordered) lists of numbers index states which are patterns of activation over groups of units; and when such patterns have semantic values, they realize symbols. Smolensky uses 'representation' and 'symbol' interchangeably. I shall follow him in this. Moreover, I shall follow him in counting symbols or representations in a network as distributed when they are realized by patterns of activity over groups of units, rather than by individual activated units.

I shall now supply a little background. As Fodor and Pylyshyn (1988) points out, while localized connectionist representations can represent things that have syntactic structure, they themselves lack syntactic structure. Since they lack syntactic structure one cannot appeal to their syntactic structure in order to explain such phenomena as systematicity and productivity. An adequate cognitive architecture must yield explanations of such phenomena. And connectionists who postulate such representations have failed to indicate how such phenomena can be explained without appeal to the syntactic structure of representations. With all this, Smolensky agrees (see, e.g., p.285); but one of the main points of Smolensky (1988) is that while Fodor's and Pylyshyn's critique of connectionist architectures succeeds for ones that postulate only local representations, it fails to apply to connectionist architectures with distributed representations.

Fodor and McLaughlin (1990) raises the issue of whether any of the distributed representations Smolensky postulates are *themselves* syntactically structured. One philosopher insists that this is simply the wrong sort of question to ask. We are, he informs us, in midst of a revolution in cognitive science;

---

[4]Hereafter, unless I indicate otherwise, all pages references in the body of the text will be to Smolensky (1995).

Smolensky's architecture is at the cutting edge of that revolution, issuing in the new paradigm; and the notions of the new paradigm are incommensurate with those of the old. Revolutionaries should thus not be expected to answer questions framed in the vocabulary of the old guard. This author informs us that for these reasons

> the best answer to the received question about Connectionist representations [Are they syntactically structured?] is that they are *neither* syntactically structured *nor* not syntactically structured. Neither answer characterizes them acceptably. We need to set aside that particular vocabulary, which is embedded in and effectively limited to the Classical framework, and develop a new set of conceptual tools which will be sufficient to comprehend the emerging connectionist alternative. (Van Gelder (1991), p.372) (emphasis his)

I hate to keep being pedantic. But, surely, the connectionist revolution does not run so deep as to overturn classical logic.

In answer to our question, Fodor and I claimed that Smolensky's distributed representations lack the kind of syntactic structure that complex representations in a Classical architecture possess. According to Classicism (or the language of thought hypothesis), mental representations with syntactic structure contain distinct representations as constituents, where the constituency relationship is a part/whole relationship.[5] In the papers Fodor and I discussed,[6] Smolensky maintained that his distributed representations are (realized by) patterns of activity over groups of units in a network; that some have constituent representations; but that it is *not* the case that the patterns of activity that are constituent representations actually occur when the patterns of which they are constituents occur; the constituents representations are, rather, he claimed, "imaginary." To illustrate his idea, a representation of, for instance, Sandy's loving Kim might be (say) the activation pattern (2,4,6,3,1,0), and the representation of Sandy might be (say) the activation pattern (5,8).[7] The pattern (5,8) is not a part of the pattern (2,4,6,3,1,0), and indeed the latter pattern can occur in a network without the former pattern ever occurring. Smolensky employed a notion of constituency according to which the representation of Sandy can be a constituent of the representation of Sandy loves Kim, even though the activation pattern realizing the latter can occur without the activation pattern realizing the former ever occurring. He thus spoke of such constituents as being "imaginary."

So as not to quarrel over the word 'constituent', Fodor and I partially characterized a notion of classical constituent as follows: A type of representation R is a classical constituent of a type of representation $R^\star$ only if $R^\star$ cannot have

[5] For further discussion see Fodor and McLaughlin (1988); and McLaughlin (1993a) and (1993b).

[6] Smolensky (1987), (1988).

[7] Please attach no significance to these specific lists of numbers; I am trying to keep things as simple as possible here.

been tokened unless R has been.[8] We argued that since activity patterns that are distributed representations can be tokened without the activity patterns that Smolensky maintains are their constituents having (ever) been tokened, his would-be constituents are not classical constituents. We then proceeded to point out that since Classicism explains systematicity (in part) by appeal to mental representations with classical constituents, Smolensky cannot available himself of the Classical explanation of systematicity. And we noted that he offered no alternative explanation in its place. We concluded that despite his appeal to distributed representations, he failed to meet Fodor's and Pylyshyn's challenge of specifying a kind of connectionist architecture that could explain systematicity and productivity without implementing a Classical cognitive architecture.

Some of Smolensky's philosophical supporters respond by agreeing that his distributed representations lack classical constituents, and thus classical syntactic structure as well, and by insisting that is what is new and exciting about them.[9] They maintain that Smolensky's distributed representations have constituent representations, albeit non-classical ones. What is new and exciting about Smolensky's architecture, they tell us, is that it explains such cognitive phenomena as systematicity without appeal to representations with classical constituents, and thus in a truly novel way. Unfortunately, however, these supporters fail to tell us *how* systematicity is to be explained without appeal to classical constituents.[10]

These philosophical supporters eschew talk of imaginary constituents, and argue, by way of examples involving waves, that the constituents of something need not be classical, let alone be parts of it. Following them, Smolensky's introduces his own wave example in his response to Fodor and McLaughlin (1990), an example that is characteristic of others I have seen in the literature by his supporters. He asks us to consider the

> superposition of sound waves: the pressure wave encoding *The Sounds of Silence* is, after all, a superposition which mixes up altogether the Simon and Garfunkel constituents—yet we have no trouble believing that in a meaningful sense those constituents are real (p. 241).

The point of the example is that real constituents need not be classical constituents.

The example does not establish this, however, since it is not an example of constituency. Consider the case in which one is listening to Simon and Garfunkel sing *The Sounds of Silence* live in concert. Neither Simon's singing and Garfunkel's singing is a constituent of the sound waves that will strike one's ears; rather, these acts will be causes of those sound waves. What, then, about

[8] Fodor and McLaughlin (1988), p.186.

[9] See, e.g, Van Gelder (1991, p.367, p.378), which cites McLaughlin (1987) on the constituency issue.

[10] See McLaughlin (1993b) for further discussion.

the sound of Simon's singing and the sound of Garfunkel's singing? The sound of Simon's singing and the sound of Garfunkel's signing will likewise not be constituents of the sound waves striking one's ears. Rather, those sound waves will themselves count as the sound of Simon's singing and as the sound of Garfunkel's singing (among other things); and they will so count because Simon's singing and Garfunkel's singing will both be (pragmatically) relevant causes of them. To spell this out: when Simon and Garfunkel sing together, they produce sound waves which merge into a single sound wave.[11] That single sound wave counts as the sound of Simon's singing and the sound of Garfunkel's singing in virtue of the fact that Simon's singing and Garfunkel's singing are among its relevant causal sources. (Roughly, a sound wave counts as the sound of X if X contributes in some pragmatically relevant way to emitting or producing it.) The sound wave carries information about these causal sources, and our auditory mechanisms, marvelous mechanisms that they are, can extract that information by generating representations of them. Thus, when we hear *The Sounds of Silence* our auditory mechanisms generate auditory experiences of Simon's singing, and of Garfunkel's singing. But the fact that a wave carries information about X does not make X a constituent of the wave. Consider light waves.[12] The light waves that stimulate our retinas carry information about the spectral reflectance distribution of facing surfaces; and our visual mechanisms, marvelous devices that they are, can extract this information by generating a representation of the spectral reflectance distributions of such surfaces. But the spectral reflectance distributions of facing surfaces are dispositional properties of the surfaces that are in no sense of the term 'constituent', constituents of the radiant fluxes stimulating our retinas; rather, they (or their categorical bases) are among the causes of the radiant fluxes stimulating our retinas.

Fodor and I drew a flat-footed distinction between activation patterns that actually occur in a network and those that could but do not; and, as mentioned above, we noted that some of the activation patterns that Smolensky counts as constituents of certain other activation patterns actually fail to occur when the activation patterns he alleged them to be constituents of occur, and thus that they are not classical constituents of them. Consider, once again, the activation pattern (5,8). That pattern actually occurs only if some unit is at level of activation 5 and some other unit is at level of activation 8. Smolensky's notion of constituency counts the pattern (5,8) as a constituent of activation patterns that can occur without any unit being at level of activation 5 and without any unit being at level of activation 8. Thus, as Fodor and I correctly pointed out, his notion counts as constituents activation patterns that, in a perfectly straightforward sense, *actually fail to occur* when the activation patterns he alleged them to be constituents of occur. Indeed, since the alleged

[11]For the record, sound waves do not really superimpose since they are not continuous. It is simply a theoretically useful idealization to treat them as superimposing.

[12]Light waves really superimpose.

constituents do not actually occur, the activation patterns alleged to have (contain?) them as constituents do not even carry information about them. For something can carry the information that the pattern (5,8) occurred only if the pattern (5,8) occurred. Smolensky's wave example and the related other wave examples in the literature have no bearing whatsoever on these points.

That said, I want now to correct a mistake that Fodor and I made, one that unintentionally contributed to misleading Smolensky's philosophical supporters. We were quite right in claiming that Smolensky employs a non-classical notion of constituency. We were, however, mistaken in thinking that his distributed representations invariably lack classical constituents. Smolensky mischaracterized his distributed representations, and Fodor and I unfortunately accepted his characterization as correct.

While Smolensky had, prior to Fodor and McLaughlin (1990), produced some technical papers laying out some of the mathematics of his architecture, those papers did not provide the relevant details for evaluating his metaphysical claims about the status of constituent representations in his architecture. Fodor and I took him at his word about their status, and responded in the way sketched above. Metaphysics is, of course, not Smolensky's thing. We should have pressed him to make explicit the mathematics behind his metaphysical claims. But we did not. Mea culpa. In his response to Fodor and McLaughlin (1990), and in a series of published papers and technical reports since (1990),[13] Smolensky spells out the mathematics of his architecture in considerably more detail. He makes explicit features of it that directly bear on the issue of whether his distributed representations can have classical constituents. It is now quite apparent that it is simply false that his distributed representations invariably lack classical constituents. In fact, any architecture that meets his conditions for being an ICS architecture (an "integrated connectionist-symbolic architecture") will postulate distributed representations with classical constituents. New and exciting Smolensky's ICS architecture may be, but if it is, that is *not* because its complex representations lack classical constituents.

Unfortunately, in his response to Fodor and McLaughlin (1990), Smolensky continues mistakenly to assume that his distributed representations invariably lack classical constituents; and he has joined the chorus of his philosophical supporters in maintaining that this is part of what is new and exciting about his architecture. ICS architectures, he insists, contain representations with constituents, albeit of a non- classical sort (p.235). My main aim here is to correct this mistaken assumption. It is truly unfortunate that the notion of non-classical constituency has become a centerpiece of discussion. Since ICS architectures contain distributed representations with classical constituents, my hope is that future discussions will waste no more time on the notion of non- classical constituency.

[13] See, e.g., Smolensky et.al. (1992).

In what remains, I shall proceed as follows. In Section II, I shall present the features of Smolensky's ICS architecture that are relevant to the present discussion. Then, in III, I shall say why there are distributed representations with classical constituents in ICS architectures. Finally, I shall close in IV with some brief remarks about topics for further discussion.

II. ICS ARCHITECTURES

Smolensky tells us that the fundamental principles of ICS architectures that concern representations are these:

> Rep*pdp*: In all cognitive domains, representations are distributed patterns of activity.
>
> Nativism*ics*: Some crucial structure in connectionist networks does *not* arise from learning.
>
> Rep*ics*: In core parts of higher cognitive domains, representations are vectors which have the structure of recursive tensor product representations.
>
> Rep*sym*: In core parts of higher cognitive domains, representations are symbol structures.
>
> Sem*sym*: The semantic interpretation of these symbolic representations are compositionally derived from their syntactic structure. (p.261)

ICS architectures share principle Rep*pdp* with any connectionist architecture that includes distributed representations. Principles Rep*sym*, Sem*sym* are shared with Classical architectures. Rep*ics* is a principle that Smolensky correctly regards as "a novel integration" of symbolic and connectionist features (p.232). He employs Rep*ics* to reconcile Rep*pdp* with Rep*sym* and Sem*sym*.

Smolensky explicates Rep*ics* by indicating ways to map binary-branching Lisp trees into vectors. He characterizes Lisp trees as dividing into roles and things which can fill them. More specifically, on his most frequently used formulation (others are possible), he characterizes Lisp trees in terms of two roles: the role of being the left branch of a tree node and the role of being the right branch of a tree node. His example will help to illustrate the mapping procedure (see pp.236–239). Letting the symbol L represent 'loves', the symbol S represent 'Sandy', and the symbol K represent 'Kim', consider the binary-branching Lisp tree that expresses the proposition that *Sandy loves Kim.* Using r_0 to label a vector for the role of being the left branch of a tree node and r_1 to label a vector for the role of being the right branch, Smolensky tells us

> The two vectors r_0 and r_1 are the fundamental building blocks in terms of which the representations for all positions in binary trees are recursively defined... Obviously, it will not do for r_0 and r_1 to be

one and the same vectors; we would have no way of distinguishing left from right in that case. Indeed, since we are working in a vector space, what is required is that r_0 and r_1 be *independent* vectors: one cannot simply be a multiple of the other, as in $r_0 = 2r_1$. (p.237)

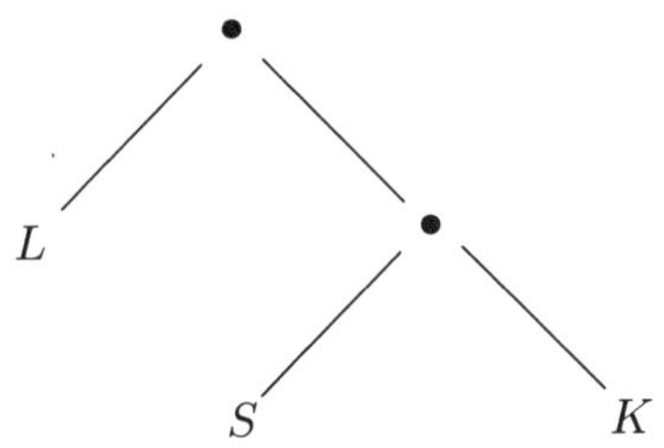

The Lisp tree-structure itself is, he tells us, "realized by" the pattern of activation indexed by the following activity vector:

$$r_0 \otimes L + r_1 \otimes [r_0 \otimes S + r_1 \otimes K]$$

The operations are, he says, vector addition (superposition) and the tensor product (p.238).

Here, then, we have an actual mathematical formula for doing the mapping, and thus something solid on which to assess Smolensky's claims about the metaphysical status of constituents of distributed representations.[14]

III. CLASSICAL CONSTITUENTS IN ICS ARCHITECTURE

Unfortunately, Smolensky's formula does not work. Anyone familiar with linear algebra will immediately see that it is incoherent. Smolensky's tensor products cannot be superposed since they are tensors of different rank.

Unsurprisingly, Smolensky is well-aware of this. In a footnote, he says:

> I have ignored a minor complication needed to make this account formally correct. As the example of p [Sandy loves Kim] shows, a symbol at depth one in the tree (L) will appear in a second-order tensor product, while a symbol at depth two will appear in a third-order product. These vectors have different numbers of elements, and cannot be added according to the definition of + given in the text. (p.282, n.16)

Smolensky goes on to tell us, however, that

> The difficulty is immediately resolved: a representation of a tree is a vector comprised of sub- vectors, each listed one after the other: the

[14]It should be noted that none of Smolensky's philosophical supporters to date explain how, exactly, on his view, binary Lisp trees are to be mapped into vectors. That is of course unsurprising since prior to the papers discussed here, he never told them in sufficient detail.

> first represents symbols at depth 0; the next, symbols at depth one; the next, symbols at depth two... Each subvector has more elements than the one before. Constituent superposition is done separately in each subvector. Formally, the superposition operation required is the 'direct sum' $\oplus$ rather than the simple sum $+$. A representational scheme can be devised which does not separate the representations of different levels in the tree, but it is less simply defined; see Smolensky et.al. (1992). (p.282, n.16)

I shall first discuss the direct sum proposal, and then the proposal alluded to in the last sentence of the above quotation.

Consider, once again, the example of the Sandy-loves-Kim tree. Suppose that we assign r_0 the vector (1,0), r_1 the linear independent vector (0,1), L the vector (2,4), S the vector (6,2), and K the vector (7,3). On the direct sum proposal, the Lisp tree $[L, [S, K]]$[15] is, then, paired with the vector (2,4,0,0,6, 2,0,0,0,0,7,3).[16]

Notice that the activation pattern that realizes the distributed representation of the tree contains as a part the activation pattern that realizes 'Sandy', the one that realizes 'loves', and the one that realizes 'Kim'. The distributed representation for the tree thus has *classical* constituents. The distributed representation for the tree cannot have been tokened without the activity pattern for 'Sandy' having been tokened, the activity pattern for 'Kim' having been tokened, and the activity pattern for 'loves' having been tokened; indeed, as I mentioned, these constituents are literally parts of the representation of the tree. The vectors (1,0) and (0,1) are the standard bases for a two-dimensional vector space. Whatever vector space we use, if we use standard bases of that space for the roles and employ the direct sum proposal, then the distributed representations of trees will have classical constituents.

Turn to the proposal Smolensky alludes to in the footnote quoted above. Smolensky et.al. spell out the gist proposal as follows:

> The crux of the idea is to add to the fundamental role vectors $\{r_0, r_1\}$ of the stratified representation a third vector v which serves basically as a place holder, like the digit 0... [use] as many v_s as necessary to pad the total tensor product out to produce a tensor of some rank $D + 1$. Now, atoms at all depths are represented by tensors of the same rank...
>
> Trees up to depth D can now be represented with complete accuracy (assuming the three vectors $\{r_0, r_1, v\}$ are linearly independent). The Stratified representation... can be straightforwardly embedded as a special case of this new fully distributed representation by map-

[15] This non-standard notation is Smolensky's. See p.236.

[16] If Smolensky intends us to use simple vector sum at each level and direct sum across levels, then the activation pattern for the tree will be (2,4,0,0,6,2,7,3) (Here and in the text above, I use F (filler) $\otimes$ R (role), rather than $R \otimes F$. That makes no difference. He uses $F \otimes R$ himself in Smolensky et.al. (1992).

ping $r_0 \mapsto (r_0, 0)$, $r_1 \mapsto (r_1, 0)$ and by setting v equivalent to (0,1) where 0 is the zero vector with the same dimensionality as r_0 and r_1. (Smolensky, et.al. (1992), p.19)

Smolensky's strategy is to use the dummy vector to add however many zeros are needed in order to always end up with tensors of the same rank, and thus tensors that can be superposed.

We need not linger over the mathematical details here. Suffice it to note that, on this second proposal, if assignments to r_0 and r_1 remain the same as before, we once again have classical constituents. Our role vectors go over to (1,0,0), and (0,1,0), respectively; and the dummy vector, v, is (0,0,1). Keeping L, K, and S as before, the distributed representation for $[L, [S, K]]$ will contain the patterns (2,4), (7,3) and (6,2) as parts.

Smolensky et.al. (1992) points out that Lisp trees need not be characterized in terms of the two roles we have been discussing. There are other ways of employing the filler-role idea. But the above point generalizes. However the filler-role idea is employed, if the role vectors are standard bases for the space they are in, then the distributed representations will have classical constituents.

Consider a different way of employing the filler-role idea suggested to me (in conversation) by Tim Maudlin, a way that is much cheaper than any of Smolensky's proposed ways in terms of resources measured by the number of units required to realized trees of some depth D.[17] We can take the roles to be positions in trees, having as many roles as we need to represent trees of some given depth D. We assign each role a standard base for as many positions as we choose to represent. Thus, consider the next tree. We can assign r_1 the vector (1,0,0,0,0,0,0,0,0,0,0,0,0,0); r_2 the vector (0,1,0,0,0,0,0,0,0,0,0,0,0,0);... and r_{14} the vector (0,0,0,0,0,0,0,0,0,0,0,0,0,1). On Maudlin's proposal, the number of positions for a tree of depth D is $2^{D+1} - 2$. In contrast, Smolensky et.al's proposal requires 3^{D*a} units. This proposal will require twenty times more units for a tree of depth ten than will Maudlin's. On Maudlin's proposal, of course, the distributed representation once again have classical constituents.

Of course, on any of these ways of employing the filler-role idea we could use nonstandard bases for the vector space. Consider, again, Smolensky et.al.'s proposal. We could, as they note, choose $\{r_0, r_1, v\}$ to be three linearly independent vectors in three-dimensional space, each with non-zero components along all three coordinate axes.

[17]He should not, of course, be held responsible for my presentation of (what I take to be) his idea.

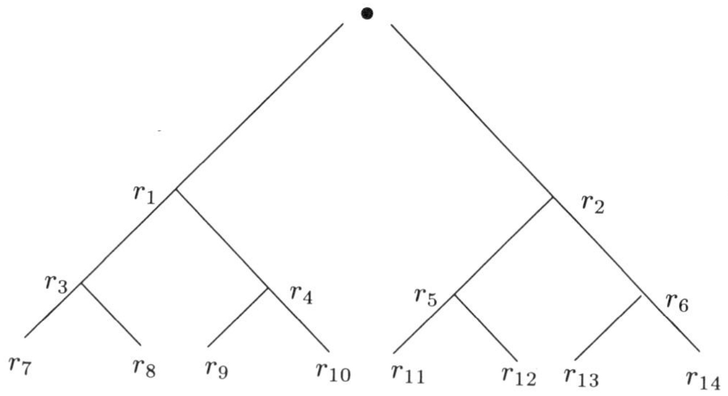

There can be no mathematical reason for doing this rather than choosing standard bases, or for choosing standard bases rather than such bases. Whatever independent bases we choose, the proposals will be mathematically equivalent to one that chooses standard bases. Moreover, using non-standard bases does not, of course, yield a different architecture; shifts in bases do not make for shifts in architecture. Any architecture that meets Smolensky's conditions for being an ICS architecture will have distributed representations with classical constituents.[18]

IV. SOME BRIEF REMARKS IN CONCLUSION

Smolensky claims that (1) Classicism assumes systematicity, rather than explaining it (p.269), and that (2) his ICS architecture actually explains it (p.263). I shall very briefly comment on (1) and (2) in turn, and then close by listing some points.

What Smolensky seems to have in mind by the surprising claim that Classicism assumes systematicity is something like this: Classicism postulates a system of mental representation which generates all and only mental representations that express the contents of (architecturally) possible thoughts and concepts of the subject possessing the system; Classicism does not, however, "derive" the principles of the system "from more fundamental principles" (p.269). That is right. Classicism *postulates* such a system of mental representation. However, it then appeals to this system to explain (among other things) systematic relationships among thought-capacities.[19] Classicism does not purport to explain what Smolensky says it merely assumes; and what Classicism purports to explain, namely systematic relationships among thought-

[18] My point can be put this way: Any ICS architecture will have an instantiation in which the distributed representations with constituents have classical constituents.

[19] This is all spelled out in detail in several places. See, e.g., McLaughlin (1987); Fodor and Pylyshyn (1988); McLaughlin (1993a) and (1993b).

capacities, is unmentioned by Smolensky. He says nothing at all about such relationships.

If, as Smolensky appears to be claiming, possession of an ICS architecture could explain why a subject possesses such a system of mental representation, the explanation would be welcomed by Classicists. An implementation architecture for a Classical architecture would have to do that. Unfortunately, Smolensky has not said how his architecture does that. What he has done is this: (a) indicated some ways that either binary Lisp trees or propositions expressed by them can be mapped into vectors, and, (b) shown how some functions that can be computed in Lisp only by a long series of serial steps can be computed by a single matrix multiplication in an ICS architecture. That is all very interesting (I really mean that), but Smolensky's doing (a) and (b) does not suffice for him to explain what he says he has explained. A proper discussion of this will, however, have to await another occasion.

I shall now make some "quick points" in closing. First, I find Smolensky's Gödel number analogy quite apt (see p.241). Second, while Smolensky sometimes appears to suggest otherwise, distributed representations in ICS architectures have causal effects: patterns of activation over hidden units will causally mediate between input and output patterns. Third, while the basic algorithms governing the behavior of networks with ICS architectures will concern only individual units, and thus not activity patterns, this is consistent with the activity patterns that count as symbols participating in algorithmic causal processes; it is just that the symbolic algorithms will not be basic algorithms of the network. Fourth, Smolensky is of course quite right that when functions whose computations in Lisp require a series of steps are computed in ICS architectures, the relevant Lisp algorithms are not being executed; but it does not follow from that that no higher-level, non-basic symbolic algorithms are being executed by means of the execution of the basic algorithms governing the behavior of the network. Fifth, it is *consistent* with Classicism that mental representations are realized by activation patterns and that symbolic processes are implemented by matrix multiplication operations. Sixth, I myself find ICS architectures of interest as potential implementation architectures for Classical language of thought architectures. (I get goose-bumps of excitement when he says things like, "In ICS, we brazenly dare to commit to a theory of how such realizations [realizations of symbols in the brain] take place" (p.236)) It is apparent that Smolensky finds revolting the idea that his architecture may be such an implementation architecture. But I shall continue to try to warm him up to it by pointing out that implementation is not the "slavish" (p.227) enterprise he thinks it is. (Or perhaps it would help if I dropped the term 'implementation' and employed his term 'reduction'.) Finally, Smolensky and I are not as far apart as he makes it sound. In addition to the points of agreement noted in the second paragraph of this paper, I too voted for Dukakis.[20]

Rutgers University

REFERENCES

Fodor, J. & McLaughlin, B.P. (1990). Connectionism and the problem of systematicity: why Smolensky's solution does not work. *Cognition* **35**, 183–204.

Fodor, J. & Pylyshyn, P. (1988). Connectionism and cognitive architecture: a critical analysis. *Cognition*, **28**, 3–71.

McLaughlin, B.P. (1987). Tye on connectionism. *Spindel Conference on Connectionism, Southern Journal of Philosophy*, **26**, 185–193.

McLaughlin, B.P. (1993a). The classicism/connectionism battle to win souls. *Philosophical Studies*, **70**, 45–72.

McLaughlin, B.P. (1993b). Systematicity, conceptual truth, and evolution. In C. Hookway and D. Peterson (eds.) *Royal Institute of Philosophy, Supplement*, No.34, reprinted in C. Hookway and D. Peterson (eds.) *Philosophy and Cognitive Science*, Cambridge, Cambridge University Press, pp.217–34.

McLaughlin, B.P. & Warfield, T.A. (1994). The allure of connectionism reexamined. *Synthese* **101**, 365–400.

Smolensky, P. (1987). The constituent structure of mental states: a reply to Fodor and Pylyshyn. *Spindel Conference on Connectionism, Southern Journal of Philosophy*, **26**, 137–160.

Smolensky, P. (1988). Connectionism, constituency and the language of thought. In B. Loewer and G. Rey (eds.) *Meaning in Mind: Fodor and His Critics*, Oxford, Blackwell.

Smolensky, P. (1995). Reply: constituent structure and explanation in an integrated connectionist/symbolic cognitive architecture. In C. Macdonald and G. Macdonald (eds.) *Connectionism: Debates on Psychological Explanation*, Oxford, Blackwell, pp. 223–290.

Smolensky, P., Legendre, G. and Miyata, Y. (1992). *Principles for an integrated connectionist/symbolic theory of higher cognition.* Technical Report CU-CS-600-92, Department of Computer Science, University of Colorado at Boulder.

Van Gelder, T. (1991). Classical questions, radical answers: connectionism and the structure of mental representation. In T. Horgan and J. Tienson (eds.) *Connectionism and the Philosophy of Mind*, Dordrecht: Kluwer, pp.355–381.

[20] I wish to thank Paul Smolensky and Georges Rey for helpful discussions of this paper at a joint symposium that Smolensky and I did before the University of Maryland Cognitive Science Group. I wish also to thank members of the audience at the International Philosophy and Methodology of Science Meetings in Florence, Italy. Thanks are also due to Stephen Ellis and Barry Ward; and very special thanks are due to Tim Maudlin for what amounted to hours of helpful discussion about the issues addressed in this paper.

14
Foundations of Linguistics

DAVID DOWTY

NON-CONSTITUENT COORDINATION, WRAPPING, AND MULTIMODAL CATEGORIAL GRAMMARS

Syntactic Form as Logical Form

1. COORDINATION VS. WRAPPING: THE DILEMMA

When contemporary categorial grammar is viewed as a theory of natural language syntax and semantics, the feature which most distinguishes it methodologically from other current linguistic theories is its assumption that ('surface') syntactic form *is* logical form, i.e. that no other significant levels of analysis ('LF', F-structure, etc.) need be appealed to to yield an adequate semantics from the syntax. Empirically, one of this theory's very distinctive achievements in syntax has been a successful analysis of so-called non-constituent coordination, a construction which has resisted an analysis that is adequate in both semantics and syntax in any other theory (Steedman, 1985; Dowty, 1985[1988]). However, these two important features have appeared to come into irresolvable conflict in a kind of analysis called 'Wrapping': this paper will show, among other things, that this conflict is only apparent.

Coordination of normal or 'true' constituents as in (1) is a familiar feature of natural language syntax:

(1) [Mary ate rice yesterday] and [John ate beans today]. *S and S*
John [ate rice] and [drank tea] yesterday. *VP and VP*

But many languages, including English, also rather freely allow coordinations of parallel sequences of words that are not traditional syntactic constituents and do not obviously seem to constitute units of the compositional semantic interpretation; the sequences *rice yesterday*, *beans today* in (2) exemplify such NON-CONSTITUENT COORDINATION (NCC):

(2) John ate [rice yesterday]and [beans today]. *NP Adv and NP Adv*

(3) Mary gave [a book to John] and [a record to Bill]. *NP PP and NP PP*

With the resources of the Associative Lambek Calculus, the kind of verb phrase occuring in (3) can be derived not only in the obvious way shown in (4), but also via the less obvious derivation (5), which builds the "non-constituent" *a book to John* as a unit.

M.L. Dalla Chiara et al. (eds.), Structures and Norms in Science, 347–368.

(4)

$$\dfrac{\dfrac{\textbf{Mary}}{np} \qquad \dfrac{\dfrac{\dfrac{\textbf{gave}}{(vp/pp)/np} \quad \dfrac{\textbf{a book}}{np}}{vp/pp}\,E/ \qquad \dfrac{\textbf{to John}}{pp}}{vp}\,/E}{s}\,\backslash E$$

(5)

$$\dfrac{\dfrac{\textbf{Mary}}{np} \qquad \dfrac{\dfrac{\textbf{gave}}{(vp/pp)/np} \qquad \dfrac{\dfrac{\dfrac{\textbf{a book}}{np}}{((vp/pp)/np)\backslash(vp/pp)}\,TR \qquad \dfrac{\dfrac{\textbf{to John}}{pp}}{(vp/pp)\backslash vp}\,TR}{((vp/pp)/np)\backslash vp}\,FC}{vp}\,FC}{s}\,\backslash E$$

Here, *TR* indicates TYPE RAISING and *FC* FUNCTIONAL COMPOSITION, two rules which are derivable as theorems in the Lambek Calculus **L**. By means of the Curry-Howard correspondence with lambda terms, (5) is guaranteed to yield the same semantic interpretation as (4). The grouping in (5) is the key to deriving the NCC examples in (1): if there is a means of coordinating two consecutive instances of any category with *and*, we will have derivations such as (6) and (7) (from (Dowty, 1985)[1988]), where the ellipses in (6) indicate subderivations like those of *a book to John* in (5) above:

(6)

$$\dfrac{\dfrac{\textbf{Mary}}{np} \qquad \dfrac{\dfrac{\textbf{gave}}{(vp/np)/pp} \qquad \dfrac{\dfrac{\textbf{a book to John} \\ \vdots}{((vp/np)/pp)\backslash vp} \quad \dfrac{\textbf{and}}{Conj} \quad \dfrac{\textbf{a record to Bill} \\ \vdots}{((vp/np)/pp)\backslash vp}}{((vp/np)/pp)\backslash vp}\,C}{vp}\,\backslash E}{s}\,\backslash E$$

(7)

$$\dfrac{\dfrac{\textbf{John}}{np} \quad \dfrac{\dfrac{\textbf{ate}}{vp/np} \quad \dfrac{\dfrac{\dfrac{\dfrac{\textbf{rice}}{np}}{(vp/np)\backslash vp}TR \quad \dfrac{\textbf{yesterday}}{vp\backslash vp}}{(vp/np)\backslash vp}FC \quad \dfrac{\textbf{and}}{Conj} \quad \dfrac{\dfrac{\dfrac{\textbf{beans}}{np}}{(vp/np)\backslash vp}TR \quad \dfrac{\textbf{today}}{vp\backslash vp}}{(vp/np)\backslash vp}FC}{(vp/np)\backslash vp}C}{vp}\backslash E}{s}\backslash E$$

In (7), C is the COORDINATION SCHEMA. Two ways of introducing cross-categorial Boolean coordination into a categorial grammar are currently in common use: (i) a SYNCATEGOREMATIC schema as used above, and (ii) a CATEGOREMATIC method of assigning the conjunctions **and, or** to the category schema $(X\backslash X)/X$, for all categories X. In either case, a generalized cross-categorematic semantics for the Boolean connectives is assumed, as proposed in (Partee and Rooth, 1983) and elsewhere.

1.1. *'Wrapping' as a syntactic operation in Categorial Grammar*

In the 1980's, Emmon Bach (1980, 1984) and other linguists proposed that there is a mode of combination in natural language which Bach called RIGHT WRAP, in which an expression of a functor category is combined with its argument by "wrapping" the functor expression around the argument, i.e. placing the argument 'to the right of the first word' of the functor expression, as shown in the wrapping of the phrasal transitive verb (category $(s\backslash np)/np$) *persuade to leave* in (8) and *give to Mary* (same category) in (9). Under this view, wrapping is always used with phrasal transitive verbs in English.[1]

(8)

$$\dfrac{\dfrac{(vp/np)/vp_{inf} : \textbf{persuade} \quad vp_{inf} : \textbf{to leave}}{vp/np : \textbf{persuade to leave}}/E \quad np : \textbf{John}}{vp : \textbf{persuade John to leave}}/E : RW$$

[1]The derivations here have to be exhibited in the form of LABELED DEDUCTIONS, with each categorial type labeled by the expression that is its model-theoretic interpretation, since the discontinuity in a derived string *persuade John to leave* cannot be reconciled with the derivation of a linear sequence of types identifiable with the desired syntactic structure.

(9)

$(vp/np)/pp$: **gave** pp : **to John**
————————— $E/$
vp/np : **gave to John** np : **a book**
————————— $/E : RW$
np : **Mary** vp : **gave a book to John**
————————— $E\backslash$
s : **Mary gave a book to John**

The conflict between Wrapping and the categorial NCC analyses can now be seen from comparison of the handling of the direct and indirect objects in (9), vs. in (5) above. Specifically:

- The wrapping analysis requires an indirect object to be combined with a verb before a direct object, resulting in the (Curried) function-argument configuration:
  [[**give**(**to John**)](**a book**)](**Mary**)
- The NCC analysis requires a direct object to be combined with a verb before an indirect object, with function-argument configuration:
  [[**give**(**a book**)](**to John**)](**Mary**)

The various other wrapping configurations (e.g. *persuade Mary to leave*) give rise to the same problem.

One researcher who has directly confronted this dilemma is Mark Steedman (1993, *i.a.*). Steedman's solution to it consists in:

- Postulating a distinction (like that drawn in Government-Binding theory) between *surface syntactic form* (generated by a categorial grammar) and *logical form* ("LF").
- Taking the order of combination of arguments in surface structure to be:
  [[**give**(**a book**)](**to John**)](**Mary**)
- But making the order of combination of arguments in Logical Form:
  [[**give**(**to John**)](**a book**)](**Mary**)
- Adding translation rules (relating s-structure to LF) that "interchange" the order of two arguments of multi-place predicates

Steedman thus eschews a Wrapping analysis in the English syntax and, indeed, makes it a methodological assumption of his theory, the so-called ADJACENCY PRINCIPLE, that only normal concatenation combines expressions.

As in Montague Grammar, Steedman's translation rules take the form of lambda-calculus expressions assigned compositionally to each word and constituent. But in contrast to Montague's practice, the syntactic FORM of the translations, *after* lambda-conversion reductions are performed on them (and not just their model-theoretic interpretations), plays a crucial role in Steedman's theory (cf. below), in the same way that the syntactic form of LFs play a non-redundant role in GB theories, because much depends on the newly "unpacked" argument order at this level.

What is at stake in this dilemma is the following: Because the natural language NCC phenomenon is one where categorial grammar has appeared to have a major descriptive advantage over all other linguistic theories, the theory would lose important empirical support if its NCC analysis could no longer be maintained because of the need for Wrapping. On the other hand, if a bifurcation between s-structure and LF (as in Steedman) is adopted, then categorial grammar abandons its major distinguishing feature on the theoretical side, a very strong version of the position that that a natural language's (surface) syntactic form *is* its logical form.

2. THE LINGUISTIC MOTIVATION FOR WRAPPING

In view of this dilemma, it is important to ask just what the linguist's motivation for wrapping is. In fact, the sum of arguments for wrapping is quite compelling as a whole, though we can survey them only very briefly here.

2.1. *General Categorial Definitions of Grammatical Functions*

The Wrapping analysis is intimately tied to the way that grammatical functions (*subject, object,* etc.) are naturally defined in a categorial grammar:

- Assume that the verbal subcategories of all languages take the form: s, s/np, $(s/np)/np$, $((s/np)/np)/np$, (or $s\backslash np$, $np\backslash(np\backslash s)$ etc., (as well as s/x, $(s/np)/x$, etc., for $x \neq np$, e.g., s/vp, $(s/np)/vp$, etc.).
- A SUBJECT (S) is any np that combines with s/np (or $s\backslash np$) to form s.
- A DIRECT OBJECT (DO) is any np that combines with $(s/np)/np$ to form s/np. (Similarly for types with variant slash directions.)
- An INDIRECT (*or* OBLIQUE) OBJECT (IO) is any np that combines with $((s/np)/np)/np$ (etc.) to form $(s/np)/np$.

These definitions can (and should) be taken as independent of the NP case morphology, verbal agreement morphology, or the word orders that manifest them overtly, in different ways, in individual languages: see (Dowty, 1982) for discussion.

A similar 'hierarchy' of grammatical functions, is also posited in Relational Grammar, Lexical Functional Grammar, and Head-Driven Phrase-Structure Grammar, where it is called the *obliqueness hierarchy* (cf. "o-command"). In these theories, however, the hierarchy is stipulated simply as a linear ordering on grammatical functions (S, O, etc.) which are taken as primitives. But two things follow from the categorial definitions which do not follow from such a stipulation. First, a HIERARCHY OF DISTRIBUTION is also a consequence of the categorial characterization but not the other one: the categorial account predicts that a verb may have a S alone, or a S and a DO, or a S, DO and IO, etc., but not a DO without a S or an IO without a DO and/or S.

Secondly, the fact that local anaphors take their antecedents "higher" (but not lower) in the grammatical function hierarchy is predicted by the categorial definitions, assuming the simplest syntax/semantics for local anaphors in this theory (as functions mapping n-place into $(n-1)$-place predicates) but not predicted in this way by other theories.

The point relevant to Wrapping is that such a characterization of grammatical functions is only possible for English (and most other rigid-word-order languages) if Wrapping is a means of combining arguments with functors. Otherwise, the 'wrong' NP would be defined as DO in sentences with three-place verbs (similarly with transitive-complement verbs such as *persuade*), and thus properties of DOs common to sentences with three-place verbs and those with transitive (two-place) verbs would no longer be predicted, nor would the binding superiority of DOs over obliques (cf. below).

2.2. *Distributional Motivations for Wrapping*

There are also motivations for wrapping independent of grammatical functions; one class of these involves distribution and constituency:
Phrasal Verbs that must be analyzed as single lexemes: Among the English VERB-PARTICLE CONSTRUCTIONS (*Mary threw the suitcase down*) are a great many examples that do not have compositional meanings (10) but must be recorded as a semantic unit, a DISCONTINUOUS LEXEME:

(10) a. look up, live down, throw up, mull over, *etc.*
b. look *vs.* look up; crack *vs.* crack up *vs.* crack down

Moreover, these combinations are also unpredictable in distribution, both in that certain verbs take some but not other particles (cf. *finish off, finish up* vs. **finish on, *finish down*) and some items are "verbs" only in combination with a particle, not by themselves (*egg on, ante up, bog down* but **egg*, **ante*, **bog* are not verbs).

At least two other classes of phrasal verb constructions in English Verb-adjective have these same problematic properties: (i) FACTITIVE constructions, also called RESULTATIVES, such as *hammer (the metal) flat, let (the dog) loose, knock (him) unconscious* (contrast *hammer flat* vs. **hammer round; wipe clean* vs. **wipe damp*, also *shake loose* vs. **wrench loose, knock unconscious* vs **slug unconscious*), and (ii) combination of verb-plus-prepositional-phrase which are non-compositional and restricted in distribution, such as *take NP to task, put NP to shame, put NP to the test,* etc.

It is reasonable to treat all these phrasal combinations syntactically as single basic expressions (lexemes) of category *vp/np*, each being assigned a semantic interpretation *qua* unit, but with each analyzed phonologically as consisting of two (or more) words. If this is done, then if the wrapping operation is uniformly associated with the functor type *vp/np* in English (and if if Wrapping is defined so as to be sensitive to word boundaries), the correct word order for these cases will result, in exactly the same way as is done for *persuade to leave, give to Mary* etc.

Instances of Wrapping in English other than phrasal transitive verbs: Bach (1980, 1984) noted that there is a variety of other constructions where a functor phrase of some kind, one which is clearly a constituent in distributional and semantic terms, 'wraps' around its object. Three of these are: (i) *Tough*-adjective phrases (*He is an easy man to please,* —compare with *This man is easy to please*), (ii) Comparative complements (*She is more intelligent than he is*), and (iii) Degree-modifier complements (*It is too hot to eat*).

The reasoning here (and also in the cases from languages other than English below) is that in this data we see evidence from a variety of other constructions, and from multiple natural languages, that wrapping is a mode of syntactic combination that languages make use of in various ways in any case. Hence postulating wrapping for English phrasal verbs does not enlarge the repertoire of syntactic processes that universal grammar provides for natural languages.

Verb Phrases in VSO languages: In several languages with the word order V(erb)S(ubject)O(bject), for example Irish and Indonesian, the combination of verb and DO has been observed to behave syntactically like a VP 'constituent' in a number of ways. This is explained if VPs in these languages "wrap" around subjects, just as phrasal verbs wrap around objects in English.

"Wackernagel phenomena" As first observed by the 19th century linguist Wackernagel, many languages require a word or particle to appear as second word (or second constituent) in a clause, often cliticized to the first word (phonologically attached). This phenomenon is readily described with Wrapping operations; cf. (Hoeksema and Janda, 1988).

2.3. *'Superiority' of Direct Object over Indirect Object:*

Once the theoretical explanation of the above linguistic phenomena in terms of a categorial Curried-function account is accepted, then a variety of more abstract semantically-related properties of grammatical functions in a clause are also immediately predicted, because of the strong identification of syntactic structure with logical form in a categorial grammar.

Facts about English DO vs. IO provide a particularly striking illustration of these abstract properties, striking because the kinds of patterns in question, which go under rubrics such as "binding theory", "anaphoric domains", "scope", etc. in Government-Binding theories, are assumed to be explained in such theories in terms of asymmetries of tree-structure and relations such as *c-command*, yet there exists no real motivation for any tree-structural asymmetry between the two objects in 'internal dative' sentences (*Mary gave John a book*).

It is important to note that I assume throughout this paper that the immediate post-verbal NP is the grammatical DO in both types of English sentences with three-place verbs (11a,b); this assumption is motivated in the categorial approach by a variety of facts (involving passive, etc.) and coincides with most Relational Grammar and Lexical Functional Grammar analyses of this class:

(11) a. [Mary $]_S$ gave [a book $]_{DO}$ [to John $]_{IO}$

b. [Mary $]_S$ gave [John $]_{DO}$ [a book $]_{Obl-O}$

Barss and Lasnik (1986) cite six different ways in which the immediate post-verbal argument of a three-place verb behaves as grammatically "superior" to the other object NP, in a way that is elsewhere attested for structurally "superior" vs. "inferior" argument (e.g. S vs. DO, S vs. IO, S or DO vs. arguments in embedded clause). To illustrate, here are three of these:

Anaphoric binding by the first vs. second NP: only the former can be the anaphoric controller of the other NP (NB in both sentence types):

(12) I showed John$_i$ to himself$_i$ in the mirror.
*I showed himself$_i$ to John$_i$ in the mirror.
I showed John$_i$ himself$_i$ in the mirror.
*I showed himself$_i$ John$_i$ in the mirror.

Quantificational binding of an anaphor internal to the other NP:

(13) I denied each worker$_i$ [his$_i$ paycheck].
*I denied [its$_i$ owner] each paycheck$_i$.

WH-superiority (accessibility to extraction in multiple WH-questions):

(14) Who(m) did you give ____ which book?
*Which book did you give who(m) ____?

By reversing the hierarchical relationship of arguments between syntactic form and Logical Form, Steedman does indeed correctly correlate this last collection of facts with generalizations about superiority relations among other arguments, since in his account (as in GB) binding generalizations are assumed to be described on the syntactic configuration at LF. From this point of view alone, the difference between the two analyses might seem to be a trade-off: Steedman's analysis has an added complexity in mapping syntax to logical form but a simpler syntax, while the Wrapping analysis involves a more complex syntactic operation but a simpler relation between syntax and logical form.

The important difference is that Steedman's analysis fails to connect these "binding" generalizations with the distributional generalizations cited earlier about phrasal transitive verbs—that discontinuous *verb+particle*, resultative *verb+adjective*, etc. have idiosyncratic meanings and distributions as if they were single lexical items,—and fails to connect these with any purely syntactic manifestations of the hierarchy of grammatical functions. On the categorial wrapping analysis, however, in which the hierarchy of arguments in syntax and logical form is always taken to be identical, the distributional and binding generalizations are necessarily connected, as both follow from the hierarchical relation of DO to IO (Oblique) arguments. Since Steedman's adjacency principle has other counterexamples in many natural languages anyway (cf. §2.2), no objection to Wrapping of DOs can be raised on general linguistic grounds.

3. NON-CONSTITUENT COORDINATION IN ENGLISH

Another important linguistic question to address, in order for the significance of the NCC vs. Wrap dilemma to be fully assessed, is how general the pattern of non-constituent coordination in English is. If NCC turned out to be a marginal, highly limited phenomenon, the problem might be more easily dismissed. But in fact, a very wide variety of NCC examples are found.

Some examples of RIGHT-SIDED NCC (coordinated sequences appear at the right periphery of the complete constituent they belong to) are in (15); in each case the coordinated non-constituent groups are underlined:

(15) a. John ate rice slowly and beans quickly *NP Adv and NP Adv*
b. John ate rice on Monday and beans on Tuesday *NP PP and NP PP*
c. John ate rice slowly on Monday and quickly on Tuesday *Adv PP and Adv PP*
d. Mary gave Ray a book and Kip a record *NP NP and NP NP*
e. Susan persuaded Fay to drive and Mary to walk *NP Inf and NP Inf*
f. The trips though the mountains by bus and across the lake by boat both sound exciting. *PP PP and PP PP*

An added complexity is that NCC can 'split' constituents across two levels of structure as in (16), where brackets indicate the 'natural' constituents, underlining indicates the sequences actually coordinated:

(16) a. Mary [drove [to Chicago] yesterday] and Detroit today.
b. John [talked [to Helen] about Mary] and Bill about Meg.
c. John [wants [to visit Chicago] very much] but Lisbon hardly at all.

LEFT-SIDED NCC, traditionally called RIGHT-NODE RAISING (RNR) in the linguistic literature, is exemplified by (17):

(17) a. John caught and Mary ate the fish. *NP V and NP V*
b. I sold Peg but gave Ray a copy of my book. *[V NP and V NP] NP*
c. Mary persuaded John and convinced Meg to help her wash the car. *[V NP and V NP] Inf*
d. John gave to Mary but Ray ordered at the store a copy of the latest Ann Rice novel. *[NP V PP and NP V PP] NP*

Examples can be constructed with Left and Right NCC in the same sentence:

(18) a. John gave and Mary sold a book to Roy and a record to Sally. *NP V and NP V; NP PP and NP PP*
b. John gave Mary and sold Peg a book yesterday and a record today. *V NP and V NP; NP Adv and NP Adv*

From such examples, we might wonder whether any sequence of words at all, anywhere in a clause, can serve as a non-constituent conjunct. In fact, the adoption into a syntax for English of both the associative Lambek calculus **L** and a fully cross-categorial coordination schema predicts a very strong version of this conjecture, what might be called the hypothesis of *Lambek Coordinate closure*:

(19) **Lambek Coordinate Closure Hypothesis**: English is fully closed under right-sided, left-sided, and two-sided NCC: If

$$A_0, \ldots, A_i, \ldots, A_k, \ldots, A_n \Rightarrow C$$

is a well-formed type sequent in English, where $A_0 \ldots A_n$ are any types, then the type sequent

$$A_0, \ldots, A_i, \ldots, A_k, \mathbf{Conj}, A'_i \ldots, A'_k, A_{k+1}, \ldots, A_n \Rightarrow C$$

is also well-formed (syntactically and semantically), where $i \geq 0$, $k \leq n$, and each A'_j is the same type as A_j, for $i \leq j \leq k$.

This prediction[2] is made because (i) **L** is "closed under rebracketing" (if $A_0, \ldots, A_n \Rightarrow C$ is a possible derivation under any grouping of $A_0, \ldots, A_n$ at all, then all possible (contiguous) groupings of $A_0, \ldots, A_n$ lead to derivations of C), and (ii) the generalized coordination schema permits two constituents of any possible type to be coordinated (and provides an appropriate semantics).

However, left-sided NCC is often supposed to be more restricted than Right-Sided NCC; (Moortgat and Morrill, 1994) infer, from a few bizarrely unacceptable left-sided NCC examples, that English freely allows right-sided NCC, but that left-sided NCC ("right-node raising") is very limited: see that paper for an ingenious analysis based on this assumption about the data.

In fact, the linguistic data is much more complicated than this, though space does not permit me to discuss it in full here (see instead (Dowty, 1996) for this). Briefly the situation is, I believe, that right-sided and left-sided NCC are both subject to several kinds of restrictions: (i) SYNTACTIC ISLAND CONSTRAINTS, which are familiar from the study of WH-extraction (e.g. the Complex NP Constraint, Sentential Subject Constraint), significantly limit otherwise possible NCC of both left-sided and right-sided varieties (as was noted already for some cases in (Dowty, 1985)); (ii) There is a general constraint on NCC (which I do not attempt to treat in this paper) requiring parallelism between the respective members of the two sequences of types forming the conjuncts, a requirement not induced by the Lambek calculus analysis (cf. note 2). For example, *?John met and spoke for a moment with Susan* is awkward at best, because *met* is not parallel in the appropriate way to *talked for a moment with*, even though both strings could be analyzed in type vp/np and are thus predicted to coordinate by the unadorned associative-Lambek analysis. (iii) A Right-Node-Raising Construction may exist independently of left-sided NCC—i.e. a rightward extraction construction, which would also be responsible for the special prosody of certain 'left-sided NCC' examples.

[2] Actually, a slightly stronger prediction than (19) is made by **L**: (19) implies that the total number of types k in the subsequence on the left side of the conjunction must equal that of the matching subsequence on the right, but in fact **L** would also predict cases where a single type A_j within the sequence on one side is matched by an arbitrarily long sequence on the other side that can be proved to reduce to the same type A_j. But as pointed out below, there is in fact a fairly strong requirement of parallelism between the two sides of an acceptable NCC.

When these factors are all taken into account, there is no clear motivation for treating right-sided and left-sided NCC in different ways.

4. MULTIMODAL CATEGORIAL GRAMMARS

Though numerous formal extensions of traditional categorial grammars have been proposed over the years which permit non-concatenative operations like Wrap, I have elected to employ here a non-concatenative framework which has recently received much attention, MULTIMODAL CATEGORIAL LOGIC (Moortgat and Oehrle 1994, Moortgat 1995, i.a.). Although the version in (Moortgat and Oehrle, 1994) is not the most recent or perhaps ultimately the most adequate formulation of a multimodal system, I have chosen it because I think is likely to be familiar to a wider range of readers.

The standard technical formulation of the Associative Lambek calculus **L** (cf. Lambek 1958) is as a calculus of syntactic *types*, and not as a system for generating linguistic expressions (or trees) directly. Instead, linguistic expressions (strings of words) are viewed as the formal model-theoretic *interpretation* of the calculus of types, where an interpretation function v maps each each syntactic type into the set of linguistic expressions that belong to it, so as to meet these conditions:

(20) $v(A \circ B) = \{z | \exists x \exists y[z = (x^\frown y) \wedge x \in v(A) \wedge y \in v(B)]\}$
$v(C/B) = \{x | \forall y \forall z[(z = (x^\frown y) \wedge y \in v(B)) \rightarrow z \in v(C)]\}$
$v(A \backslash C) = \{y | \forall x \forall z[(z = (x^\frown y) \wedge x \in v(A)) \rightarrow z \in v(C)]\}$

As the only means of combining expressions is concatenation operation (denoted "$\frown$"), any discontinuous combinations are obviously excluded.

Of course, other categorial systems have different deductive properties and thus different classes of objects as their proper model-theoretic interpretations. For example, in models for the systems **AB** (classical Adjukiewicz/Bar-Hillel categorial grammars) and **NL** (the non-associative Lambek calculus), linguistic expressions are binary-branching trees; in **L** they are (as mentioned) strings, and in **LP** (Lambek calculus with unrestricted permutation) they are multi-sets. Still, any one of the systems as it stands only permits a single view of expressions. The innovation in multimodal systems is to permit a more complex, algebraic model of linguistic expression that 'mixes' tree-like, string-line and multiset-like properties.

A simple linguistic example may make motivation for this clearer: Although the order of most constituents in an English VP is relatively rigid, many adverbs can appear almost anywhere in a VP (albeit with certain limitations I will ignore for the moment). Suppose we could deal directly with this situation by treating the "rigid" constituents of the VP as forming a string, but somehow simultaneously allow an adverb to combine with a VP as if it formed a multiset with its constituents, so that all the different adverb positions merely counted as alternative "linearizations" of one and the same expression. This effect is exactly what a multimodal system allows us to achieve.

4.1. *Ingredients of a multimodal system: types and interpretation*

Here I sketch a MULTIMODAL categorial logic and its interpretation, following (Moortgat and Oehrle, 1994), only in a brief way; the reader must be referred to that paper and subsequent papers by these authors for full details. I also note that I adopt in this paper a NATURAL DEDUCTION presentation of their system, whereas their own presentation is in the Gentzen-sequent format. Natural deduction derivations resemble more closely the usual formats of derivations in linguistic theories, so this format should be easier to follow for readers that are not fluent in reading Gentzen proofs. However it should be emphasized that the theory itself is not affected by this change in format. First, the set of type constructors and their interpretations are relativized to modes:

(21) Rather than a set of only three type constructors, the series of constructors $A \circ_i B, A/_iB, A\backslash_iB$, for each mode $i \in I$, is employed.

(22) The linguistic interpretation for each mode i is given in terms of an ACCESSIBILITY RELATION R_i modeling the relevant kind of linguistic combination for that mode: R_ixyz is read "putting together expressions x and y in mode i yields expression z." (This relation is so-named because of its formal parallel to KRIPKE-FRAMES for modal logic, although here it is a 3-place rather than a 2-place relation.)

(23) An INTERPRETATION v (for each $i \in I$) then will meet these conditions:

$$v(A \circ_i B) = \{z|\exists x\exists y[R_i(x,y,z) \;\wedge x \in v(A) \;\wedge y \in v(B)]\}$$
$$v(C/_iB) \;= \{x|\forall y\forall z[(R_i(x,z,y) \;\wedge\; y \in v(B)) \rightarrow z \in v(C)]\}$$
$$v(A\backslash_iC) \;= \{y|\forall x\forall z[(R_i(x,y,z) \;\wedge\; x \in v(A)) \rightarrow z \in v(C)]\}$$

Note that associativity is assumed in traditional derivations in **L**, so that a sequent is normally written as a string, e.g.,

$$np,\ (np\backslash s)/np,\ np\ \Rightarrow s$$

In a multimodal system, however, the question of what (Gentzen sequent) rules are applicable could depend both on which pair of the three antecedent terms was combined first and which of the several modes of combination was used to combine them. Therefore the traditional notation for schema as in (24) is replaced by the schema format illustrated in (25), which designates "an occurrence of type $A/_iB$ and proof Δ combined via mode i, somewhere within proof Γ."

(24) $\Gamma_1, A/B, \Delta, \Gamma_2$

(25) $\Gamma[(A/_iB, \Delta)^i]$

The set of LOGICAL INFERENCE RULES is also relativized to mode, as follows:

(26)

$$\frac{}{A \Rightarrow A}\ Ax$$

$$\frac{(\Gamma, B)^i \Rightarrow A}{\Gamma \Rightarrow A/_i B}\ R/_i \qquad \frac{\Gamma \Rightarrow B \quad \Delta[A] \Rightarrow C}{\Delta[(A/_i B, \Gamma)^i] \Rightarrow C}\ L/_i$$

$$\frac{(\Gamma, B)^i \Rightarrow A}{\Gamma \Rightarrow A\backslash_i B}\ R\backslash_i \qquad \frac{\Gamma \Rightarrow B \quad \Delta[A] \Rightarrow C}{\Delta[(\Gamma, A\backslash_i B)^i] \Rightarrow C}\ L\backslash_i$$

$$\frac{\Gamma \Rightarrow A \quad \Delta \Rightarrow B}{(\Gamma, \Delta) \Rightarrow A \circ_i B}\ R\circ_i \qquad \frac{\Gamma[(A, B)^i] \Rightarrow C}{\Gamma[A \circ_i B] \Rightarrow C}\ L\circ_i$$

$$\frac{\Gamma \Rightarrow A \quad \Delta[A] \Rightarrow C}{\Delta[\Gamma] \Rightarrow C}\ Cut$$

The STRUCTURAL RULES (rules not involving the introduction or elimination of a slash but some other manipulation of a sequent) play an especially important role in multimodal systems; again, each one is specific to a mode of combination. For example, a structural rule specifying that a certain mode c is commutative would take the form (27) in its 'official' format:

(27)

$$\frac{\Gamma[(\Delta_2, \Delta_1)^c]}{\Gamma[(\Delta_1, \Delta_2)^c]}\ Comm$$

Structural rules can however be stated more perspicuously (and henceforth will be) in a format in which the mode of combination is designated by an infixed binary operator: in this format, the Commutativity rule (27) would instead take the form (28), and rules asserting that a mode of combination a is left and right associative would appear as (29); following (Moortgat and Oehrle, 1994), these are called axioms for the operations:

(28) Commutativity:

$$\frac{B \bullet_c A}{A \bullet_c B}\ Comm$$

(29) Associativity:

$$\frac{A \bullet_a (B \bullet_a C)}{(A \bullet_a B) \bullet_a C}\ Assoc1 \qquad \frac{(A \bullet_a B) \bullet_a C}{A \bullet_a (B \bullet_a C)}\ Assoc2$$

Some interactions between different modes would follow necessarily as consequences of the interpretation assigned to them (i.e., given that we want the inferential rules to be sound and complete with respect to this interpretation). Suppose $\bullet_{nc}$ is to be a non-commutative mode of combination and $\bullet_c$ is a commutative combination (that is otherwise similar to $\bullet_{nc}$), then the axiom (30) would be adopted. In other words, non-commutative combination is "less informative" than commutative combination, and one should be able to validly infer a less informative statement from a more informative one.

(30) $$\frac{(A \bullet_{nc} B)}{(A \bullet_{c} B)}$$

But the real linguistic descriptive potential of a multi-model system lies in the possibility of specifying, via axioms, other interactions between two modes besides obvious ones like (30).

For example, Moortgat and Oehrle's analysis of Dutch (and also mine for English below) involves both an ADJACENT MODE of combination (i.e. ordinary concatenation, denoted below by $\bullet$), and a DISCONTINUOUS MODE of combination (denoted by $\bullet_w$); this second mode is used to achieve both wrapping and infixing (as in the Dutch case below). Four possible axioms that specify commutativity and associativity between two modes (termed here MIXING AXIOMS) are listed in (31, 32), and from these the specific axioms that define the properties of Wrapping or Infixing in a given language will be chosen:

(31) **Mixed Associativity:**

$$\frac{A \bullet_w (B \bullet C)}{(A \bullet_w B) \bullet C} \text{ M-Assoc 1} \qquad \frac{(A \bullet B) \bullet_w C}{A \bullet (B \bullet_w C)} \text{ M-Assoc 2}$$

(32) **Mixed Commutativity**:

$$\frac{(A \bullet_w (B \bullet C))}{(B \bullet (A \bullet_w C))} \text{ M-Comm 1} \qquad \frac{(A \bullet B) \bullet_w C}{(A \bullet_w C) \bullet B} \text{ M-Comm 2}$$

To complete the technical apparatus of (Moortgat and Oehrle, 1994), we need, finally, to introduce a SORTING of types:

(33) **Type Sorting:** Each type $A_i, A/_iB, A\backslash_iB, A \circ_i B$ may be divided into SORTS $(A_i)_s, (A/_iB)_s, (A\backslash_iB)_s, (A \circ_i B)_s$, for each $s \in S$.

Sorts will be used to distinguish the following PROSODIC CLASSES of expressions (or what I would term MORPHO-PHONOLOGICAL CLASSES):

- WORDS: w
- CLUSTERS: c (such as clusters of clitic(s) plus host). Sort c subsumes sort w, indicated: $c \sqsubseteq w$ (i.e. a word is trivially also a cluster).
- PHRASES: ph. Also, $ph \sqsubseteq c \sqsubseteq w$. (A single word or cluster is also trivially a syntactic phrase, an idea familiar elsewhere in categorial grammar.)

In addition to those inferences of one sort from another licensed by the subsumption of sorts specified above, structural rules called SORT INCLUSION RULES may be stipulated to allow inference of one sort from another in particular situations (see for instance (37) below). Besides the expected applications of these sorts for morpho-phonological phenomena, they will play a key role in infixing and wrapping.

Before proceeding to English Wrapping and NCC, I will illustrate the potential of this kind of system by briefly reviewing Moortgat and Oehrle's treatment of Dutch verb clusters, a much-discussed phenomenon in the linguistic literature because of the non-nested word order occurring in it.

The Dutch verbal categories, mode interaction axiom, and sort inclusion axiom needed for Moortgat and Oehrle's analysis are in (34)–(36) respectively:

(34) Categories for Dutch verbs (i is the sort of "infixing triggers"):

drinken	$np\backslash vp$	*(non-infixing)*
laten, zag(en)	$((vp\backslash np)/_w vp)_i$	*(infixing mode, infix-trigger sort)*

(35) Mode Interaction Axiom postulated for Dutch:[3]

$$\frac{(A\bullet_w(B\bullet C))}{(B\bullet(A\bullet_w C))}\ MComm$$

(36) Specific Sort Inclusion rule postulated for Dutch:

$$\frac{((A)_i\bullet_w(B)_c)_{ph}}{((A)_i\bullet(B)_c)_c}\ Inc$$

A familiar example of a Dutch subordinate clause with non-nested word order, (37), can now be derived as in (38) and (39) (the derivation is split into two parts here for perspicuity):

(37) dat Jan Marie de kinderen zag laten bier drinken
"that Jan saw Marie let the children drink beer"

(38) *Initial, logical steps of derivation of (37):*

$$\frac{np\text{: }\mathbf{bier}\qquad vp\backslash vp\text{: }\mathbf{drinken}}{vp\text{: }(\mathbf{bier}\bullet\mathbf{drinken})}\ \backslash E$$

$$\frac{(vp\backslash np)/_w vp\text{: }\mathbf{laten}_i\qquad vp\text{: }(\mathbf{bier}\bullet\mathbf{drinken})}{np\backslash vp\text{: }(\mathbf{laten}_i\bullet_w(\mathbf{bier}\bullet\mathbf{drinken}))}\ /_iE$$

$$\frac{np\text{: }\mathbf{de\ k.}\qquad np\backslash vp\text{: }(\mathbf{laten}_i\bullet_w(\mathbf{bier}\bullet\mathbf{drinken}))}{vp\text{: }(\mathbf{de\text{-}kinderen}\bullet(\mathbf{laten}_i\bullet_w(\mathbf{bier}\bullet\mathbf{drinken})))}\ \backslash E$$

$$\frac{(vp\backslash np)/_w vp\text{: }\mathbf{zag}_i\qquad vp\text{: }(\mathbf{de\text{-}kinderen}\bullet(\mathbf{laten}_i\bullet_w(\mathbf{bier}\bullet\mathbf{drinken})))}{vp\backslash np\text{: }(\mathbf{zag}_i\bullet_w(\mathbf{de\text{-}kinderen}\bullet(\mathbf{laten}_i\bullet_w(\mathbf{bier}\bullet\mathbf{drinken}))))}\ /_iE$$

$$\frac{np\text{: }\mathbf{M.}\qquad vp\backslash np\text{: }(\mathbf{zag}_i\bullet_w(\mathbf{de\text{-}kinderen}\bullet(\mathbf{laten}_i\bullet_w(\mathbf{bier}\bullet\mathbf{drinken}))))}{s\backslash vp\text{: }(\mathbf{Marie}\bullet(\mathbf{zag}_i\bullet_w(\mathbf{de\text{-}kinderen}\bullet(\mathbf{laten}_i\bullet_w(\mathbf{bier}\bullet\mathbf{drinken})))))}\ \backslash E$$

$$\frac{np\text{: }\mathbf{J.}\qquad s\backslash vp\text{: }(\mathbf{Marie}\bullet(\mathbf{zag}_i\bullet_w(\mathbf{de\text{-}kinderen}\bullet(\mathbf{laten}_i\bullet_w(\mathbf{bier}\bullet\mathbf{drinken})))))}{s\text{: }(\mathbf{Jan}\bullet(\mathbf{Marie}\bullet(\mathbf{zag}_i\bullet_w(\mathbf{de\text{-}kinderen}\bullet(\mathbf{laten}_i\bullet_w(\mathbf{bier}\bullet\mathbf{drinken}))))))}\ \backslash E$$

$$\frac{s/s'\text{: }\mathbf{dat}\qquad s\text{: }(\mathbf{Jan}\bullet(\mathbf{Marie}\bullet(\mathbf{zag}_i\bullet_w(\mathbf{de\text{-}kinderen}\bullet(\mathbf{laten}_i\bullet_w(\mathbf{bier}\bullet\mathbf{drinken}))))))}{s'\text{: }(\mathbf{dat}\bullet(\mathbf{Jan}\bullet(\mathbf{Marie}\bullet(\mathbf{zag}_i\bullet_w(\mathbf{de\text{-}kinderen}\bullet(\mathbf{laten}_i\bullet_w(\mathbf{bier}\bullet\mathbf{drinken})))))))}\ /E$$

[3]Moorgat and Oehrle also require a mixed associativity axiom for Dutch, though it is not needed for the example discussed below

(39) *(Subsequent steps using structural rules:)*

$$\vdots$$

$$\frac{}{s' : (\mathbf{dat}\bullet(\mathbf{Jan}\bullet(\mathbf{Marie}\bullet(\mathbf{zag}_i\bullet_w(\mathbf{de\text{-}kinderen}\bullet(\mathbf{bier}\bullet(\mathbf{laten}_i\bullet_w\mathbf{drinken})))))))}\ MComm$$

$$\frac{}{s' : (\mathbf{dat}\bullet(\mathbf{Jan}\bullet(\mathbf{Marie}\bullet(\mathbf{de\text{-}kinderen}\bullet(\mathbf{zag}_i\bullet_w(\mathbf{bier}\bullet(\mathbf{laten}_i\bullet_w\mathbf{drinken})))))))}\ MComm$$

$$\frac{}{s' : (\mathbf{dat}\bullet(\mathbf{Jan}\bullet(\mathbf{Marie}\bullet(\mathbf{de\text{-}kinderen}\bullet(\mathbf{bier}\bullet(\mathbf{zag}_i\bullet_w(\mathbf{laten}_i\bullet_w\mathbf{drinken})))))))}\ MComm$$

$$\frac{}{s' : (\mathbf{dat}\bullet(\mathbf{Jan}\bullet(\mathbf{Marie}\bullet(\mathbf{de\text{-}kinderen}\bullet(\mathbf{bier}\bullet(\mathbf{zag}_i\bullet_w(\mathbf{laten}_i\bullet\mathbf{drinken})_c))))))}\ Inc$$

$$\frac{}{s' : (\mathbf{dat}\bullet(\mathbf{Jan}\bullet(\mathbf{Marie}\bullet(\mathbf{de\text{-}kinderen}\bullet(\mathbf{bier}\bullet(\mathbf{zag}_i\bullet(\mathbf{laten}_i\bullet\mathbf{drinken})_c)_c)))))}\ Inc$$

In (39), **laten** can first commute to become adjacent to **drinken** (thanks to the infixing operation $\bullet_w$), then **zag** undergoes a similar rearrangement. Since **drinken**, a word, is of the sort w and thus automatically of sort c, the sort inclusion axiom is applicable here to convert **laten drinken** to a cluster, and similarly for **zag laten drinken**. This results in a fully 'linearized' string, i.e. only the regular concatenation operation $\bullet$ appears in it.

5. ENGLISH WRAPPING ANALYSIS: DIRECT OBJECTS

The same apparatus suffices to describe "wrapping" of English direct objects as "infixing" in the Dutch example, with minor adjustments. As I assume that English discontinuity triggers eventually form 'clusters' with their neighbors, just as Dutch verbs do (on this see below), I postulate the sort inclusion rule (42) for English in addition to distinctive categories for wrapping verbs (40) and mode interaction rules (41). (As with the Dutch example, all sorts in the derivation below are assumed to be *ph(rase)* unless labeled otherwise, and individual words are automatically also clusters as well as phrases.)

(40) English Wrapping verb categories:

build	$vp/_w np$
give	$(vp/_w np)/pp$
persuade	$(vp/_w np)/Inf$

(41) English Mode Interaction Rules:

$$\frac{A\bullet_w(B\bullet C)}{(A\bullet_w B)\bullet C}\ \text{M-Assoc 1} \qquad \frac{(A\bullet B)\bullet_w C}{(A\bullet_w C)\bullet B}\ \text{M-Comm 2}$$

(42) English Sort Inclusion Rule: $$\frac{((A)_i\bullet_w B)_{ph}}{((A)_i\bullet B)_c}\ Incl$$

A simple transitive verb phrase wrapping example is produced as follows:

(43)

$$\dfrac{np : \textbf{Mary} \qquad \dfrac{\dfrac{(vp/_w np)/pp : \textbf{gave}_i \qquad pp : \textbf{to-John}}{vp/_w np : \textbf{gave}_i \bullet \textbf{to-John}}/E \qquad np : \textbf{a-book}}{vp : (\textbf{gave}_i \bullet \textbf{to-John}) \bullet_w \textbf{a-book}}/_w E}{\dfrac{\dfrac{s : \textbf{Mary} \bullet ((\textbf{gave}_i \bullet \textbf{to-John}) \bullet_w \textbf{a-book})}{s : \textbf{Mary} \bullet ((\textbf{gave}_i \bullet_w \textbf{a-book}) \bullet \textbf{to-John})} MComm2}{s : \textbf{Mary} \bullet ((\textbf{gave}_i \bullet \textbf{a-book})_c \bullet \textbf{to-John})} Incl}\backslash E$$

Thus as in Dutch, a discontinuity trigger introduces an argument via the operation $\bullet_w$; the commutativity axiom for this operation allows the argument to "permute" (perhaps recursively) to a target position where the discontinuity operation is eliminated and a cluster is formed with the wrapping trigger. The difference is that in English the argument will migrate "toward" the discontinuity trigger, rather than the discontinuity trigger migrating toward its target as in Dutch, and this mandates choosing the MComm2 version of the mixing axiom instead of MComm1.

The claim that an English verb always forms a morpho-phonological cluster with its object in some sense is relatively novel and might seem ad hoc. However, there are various independent motivations for this, two of which are as follows: It has long been argued in the literature on cliticization that an English pronominal direct object it behaves phonologically like a traditional clitic or "leaner" (except when stressed); cf. (Zwicky, 1986). The view taken here is thus that a verb-object 'cluster', as a phrasal sort, always exists, but that the phonological manifestations of it may only be apparent when the object is a pronoun. Secondly, it is a well-known peculiarity of English syntax that no other constituent can intervene between a verb and its DO in English (whether the DO is a pronoun or not): the one exceptional position in which an adverb *cannot* appear is between verb and object (cf. **Someone has been eating supposedly the cake*), and the existence of a cliticization cluster would naturally predict this fact.

6. NCC WITH WRAPPED DIRECT OBJECTS

The correct interaction of non-constituent coordination and wrapping now follows immediately. In (44) is shown the derivation of a phrase like **to John yesterday**, which will be a subpart of several subsequent derivations. Here, instead of Functional Composition I have used a Lambek derivation, hypothesizing a type $(vp/_w np)/pp$ which is later removed by $\backslash I$ (hence its empty linguistic label $[\ ]_i^1$).

(44)

$$\dfrac{\dfrac{(vp/_{w}np)/pp : [\ ]^1_i \quad pp : \textbf{to-John}}{vp/_{w}np : ([\ ]^1_i \bullet \textbf{to-John})}\,/E \qquad \dfrac{vp\backslash vp : \textbf{yesterday}}{\vdots \atop (vp/_{w}np)\backslash(vp/_{w}np) : \textbf{yesterday}}}{vp/_{w}np : (([\ ]^1_i \bullet \textbf{to-John}) \bullet \textbf{yesterday})}\,\backslash E$$

$$\dfrac{\dfrac{vp/_{w}np : (([\ ]^1_i \bullet \textbf{to-John}) \bullet \textbf{yesterday})}{vp/_{w}np : ([\ ]^1_i \bullet (\textbf{to-John} \bullet \textbf{yesterday}))}\,Assoc\bullet}{(vp/_{w}np)\backslash(vp/_{w}np) : (\textbf{to-John} \bullet \textbf{yesterday})}\,\backslash I^1$$

Such a phrase could, on the one hand, be used to derive a non-coordinated sentence (the derivation would look like the last four steps of like (43)), but alternatively, two such phrases can be coordinated, then combined with verb and then object as in (45):

(45)

$$\dfrac{\vdots \atop (vp/_{w}np)\backslash(vp/_{w}np) : (\textbf{to-John} \bullet \textbf{yesterday}) \quad \textbf{and} \quad \vdots \atop (vp/_{w}np)\backslash(vp/_{w}np) : (\textbf{to-Bill} \bullet \textbf{today})}{(vp/_{w}np)\backslash(vp/_{w}np) : ((\textbf{to-J} \bullet \textbf{ystrdy}) \bullet \& \bullet (\textbf{to-B} \bullet \textbf{tdy}))}\,C$$

$$\dfrac{vp/_{w}np : \textbf{gave}_i \quad (vp/_{w}np)\backslash(vp/_{w}np) : ((\textbf{to-J} \bullet \textbf{ystrdy}) \bullet \& \bullet (\textbf{to-B} \bullet \textbf{tdy}))}{vp/_{w}np : (\textbf{gave} \bullet ((\textbf{to-J} \bullet \textbf{ystrdy}) \bullet \& \bullet (\textbf{to-B} \bullet \textbf{tdy})))}\,\backslash E$$

$$\dfrac{vp/_{w}np : (\textbf{gave} \bullet ((\textbf{to-J} \bullet \textbf{ystrdy}) \bullet \& \bullet (\textbf{to-B} \bullet \textbf{tdy}))) \qquad np : \textbf{a-bk}}{vp : ((\textbf{gave}_i \bullet ((\textbf{to-John} \bullet \textbf{yesterday}) \bullet \textbf{and} \bullet (\textbf{to-Bill} \bullet \textbf{today})) \bullet_w \textbf{a-book}))}\,/_{w}E$$

$$\dfrac{vp : ((\textbf{gave}_i \bullet ((\textbf{to-John} \bullet \textbf{yesterday}) \bullet \textbf{and} \bullet (\textbf{to-Bill} \bullet \textbf{today})) \bullet_w \textbf{a-book}))}{vp : ((\textbf{gave}_i \bullet_w \textbf{a-book}) \bullet (\textbf{to-John} \bullet \textbf{yesterday}) \bullet \textbf{and} \bullet (\textbf{to-Bill} \bullet \textbf{today}))}\,MCm$$

$$\dfrac{vp : ((\textbf{gave}_i \bullet_w \textbf{a-book}) \bullet (\textbf{to-John} \bullet \textbf{yesterday}) \bullet \textbf{and} \bullet (\textbf{to-Bill} \bullet \textbf{today}))}{vp : ((\textbf{gave}_i \bullet \textbf{a-book})_c \bullet (\textbf{to-John} \bullet \textbf{yesterday}) \bullet \textbf{and} \bullet (\textbf{to-Bill} \bullet \textbf{today}))}\,Incl$$

Notice that the DO **a book** combines logically with the full phrasal verb **gave to John yesterday and to Bill today**, but this DO can form a cluster and abandon the discontinuity mode only when it becomes adjacent to the trigger-sort word $\textbf{gave}_i$.

A different way that object Wrapping can interact with *NCC* is in *VPs* like **give a book to John and a record to Susan**, where there is not a single direct object (as above) but one in each conjunct; this is shown in (46) and (47). A key detail here is that because the mode by which the hypothesized type $(vp/_{w}np)/pp$ is related to its argument in the 5th line of (46) is the discontinuous mode (note operation $\bullet_w$ in label), this type can only be removed (in the last line) by a slash-elimination rule of the same mode, according the multimodal logical rule stated in (38)—i.e. by $\backslash_w I$. Thus the resulting type must be $((vp/_{w}np)/pp)\backslash_w vp$ and not $((vp/_{w}np)/pp)\backslash vp$. As a consequence, when a coordination of two instances of this type is combined with $\textbf{give}_i$ in the fourth line of (47), the operation used must be $\bullet_w$ (and not

$\bullet$). After an application of *MAssoc* in the fifth line, the Inclusion axiom will then yield the cluster $(\textbf{give}_i \bullet \textbf{a-book})_c$, parallel to the verb-object cluster in other examples.

(46)

$$\dfrac{\dfrac{\dfrac{\dfrac{\dfrac{(vp/_w np)/pp : [\;]^1_i \quad pp : \textbf{to-John}}{vp/_w np : [\;]^1_i \bullet \textbf{to-John}}\,/E \qquad np : \textbf{a-book}}{vp : ([\;]^1_i \bullet \textbf{to-John}) \bullet_w \textbf{a-book}}\,/_w E}{vp : ([\;]^1_i \bullet_w \textbf{a-book}) \bullet \textbf{to-John}}\,MComm}{vp : [\;]^1_i \bullet_w (\textbf{a-book} \bullet \textbf{to-John})}\,MAssoc}{((vp/_w np)/pp)\backslash_w vp : \textbf{a-book} \bullet \textbf{to-John}}\,\backslash_w I$$

$$\vdots$$

(47)

$$\dfrac{\dfrac{\dfrac{(vp/_w np)/pp : \textbf{give}_i \qquad \dfrac{\dfrac{\dfrac{\vdots}{((vp/_w np)/pp)\backslash_w vp : \textbf{a-bk}\bullet\textbf{to-J}} \quad \dfrac{\vdots}{((vp/_w np)/pp)\backslash_w vp : \textbf{a-rcd}\bullet\textbf{to-S}}}{((vp/_w np)/pp)\backslash_w vp : (\textbf{a bk to J.})\bullet\textbf{and}\bullet(\textbf{a rcd to S.})}\,C}{((vp/_w np)/pp)\backslash_w vp : (\textbf{a bk})\bullet(\textbf{to J. and a rcd to S.})}\,\bullet Assoc}{vp : (\textbf{give}_i) \bullet_w (\textbf{a-book}\bullet(\textbf{to John and a record to Susan}))}\,\backslash_w E}{vp : (\textbf{give}_i \bullet_w \textbf{a-book})\bullet(\textbf{to John and a record to Susan})}\,MAssoc}{vp : (\textbf{give}_i \bullet \textbf{a-book})_c \bullet (\textbf{to John and a record to Susan})}\,Inc$$

6.1. *"Head Wrap"*

Whereas Bach (1984) had assumed that Wrapping should be formalized so as to always place an argument "after the first word" of a phrase, Pollard (1984) observed that this would not give correct results when a phrase that wraps around its head, such as *easy to please*, itself has a pre-head modifier, as in *extremely easy to please*: cf. (48b). Rather the wrapped-in argument must be positioned after the modifier's head, as in (48c), hence Pollard redubbed the corrected wrapping operation "Head Wrap".

(48) a. This person is [extremely easy to please].
b. *He is an [extremely person easy to please] .
c. He is an [extremely easy person to please].

What is of interest here is that the correct word order (48b) results automatically in the present multimodal analysis and does not require explicit appeal to the "head" of a phrase (as it had to be in Pollard's (1984) formulation.) To treat these examples, we need only the additional category assignments (49) (the obvious ones in this framework).[4] The correct word order (47c) is produced as in (50).

(49) **easy** $((n/_w n)/(inf/np))_i$
extremely, very $(n/_w n)/(n/_w n)$

(50)

$$\dfrac{np/n : \mathbf{a} \quad \dfrac{\dfrac{(n_w/n_w)/(n_w/n_w) : \mathbf{very} \quad \dfrac{(n/_w n)/(inf/np) : \mathbf{easy}_i \quad inf/np : \mathbf{to\text{-}please}}{(n_w/n_w) : (\mathbf{easy}_i \bullet \mathbf{to\text{-}please})}/E}{n_w/n_w : (\mathbf{very} \bullet (\mathbf{easy}_i \bullet \mathbf{to\text{-}please}))}/_wE \quad n : \mathbf{person}}{n : ((\mathbf{very} \bullet (\mathbf{easy}_i \bullet \mathbf{to\text{-}please})) \bullet_w \mathbf{person})}/E}{np : (\mathbf{a} \bullet ((\mathbf{very} \bullet (\mathbf{easy}_i \bullet \mathbf{to\text{-}please})) \bullet_w \mathbf{person}))}/E$$

$$\dfrac{np : (\mathbf{a} \bullet ((\mathbf{very} \bullet (\mathbf{easy}_i \bullet \mathbf{to\text{-}please})) \bullet_w \mathbf{person}))}{np : (\mathbf{a} \bullet (((\mathbf{very} \bullet \mathbf{easy}_i) \bullet \mathbf{to\text{-}please}) \bullet_w \mathbf{person}))}MAssoc$$

$$\dfrac{}{np : (\mathbf{a} \bullet (((\mathbf{very} \bullet \mathbf{easy}_i) \bullet_w \mathbf{person}) \bullet \mathbf{to\text{-}please}))}MComm$$

$$\dfrac{}{np : (\mathbf{a} \bullet ((\mathbf{very} \bullet (\mathbf{easy}_i \bullet_w \mathbf{person})) \bullet \mathbf{to\text{-}please}))}MAssoc$$

$$\dfrac{}{np : (\mathbf{a} \bullet ((\mathbf{very} \bullet (\mathbf{easy}_i \bullet \mathbf{person})_c) \bullet \mathbf{to\text{-}please}))}Incl$$

Although a derivation stage (51) can be reached here by applications of $\bullet Assoc$ and *Mcomm*, observe that $\bullet_w$ cannot be subsequently converted to $\bullet$ from this structure, because there is no adjacent expression of sort i to the left (as the sort inclusion rule requires), so the incorrect order (48b) cannot be produced.

(51) $np : (\mathbf{a} \bullet ((\mathbf{very} \bullet_w \mathbf{person}) \bullet (\mathbf{easy}_i \bullet \mathbf{to\text{-}please})))$

7. REMARKS ON THE MOORTGAT-OEHRLE 1994 SYSTEM AS A THEORY OF WRAPPING

Although the grammar format of (Moortgat and Oehrle, 1994) as it stands treats a number of parameters as formally independent, the exploitation of these devices in their analysis of Dutch verb clusters and in English head wrapping in this paper always turns out to conform to a single pattern of interaction among these parameters:

[4] Moortgat and Oehrle do include a treatment of attributive *tough*-phrases, but not one involving infixing/wrapping but rather extraposition, which has very different properties in their system. However, that treatment is less desirable on empirical grounds, as it predicts that sentences like **An easy person arrived to please* should be grammatical.

(52) An expression is assigned to a cluster-trigger sort (i.e., a sort to which an inclusion rule applies that converts it to a cluster) if and only if it combines with one of its arguments by a wrapping/infixing (syntactically discontinuous) mode. Only basic expressions, not complex phrases, belong to the trigger sort. Similarly, an inclusion axiom relates a phrasal sort to a cluster only if it is formed by the discontinuous mode and is "triggered" by one of its two operands being of this special sort.

Adherence to this pattern (52) results in what I might term the BOOMERANG EFFECT: whenever two elements are combined via a discontinuous mode, the 'lexical functor' which heads the discontinuity-licensing type $A/_wB$ and the argument type B must eventually come to be contiguous and be joined into a prosodic cluster. ("Boomerang", because the argument which is 'set loose' from rigid word order by the discontinuous mode must sooner or later meet up with the lexical head which triggered the discontinuity.) The boomerang effect implies that wrapping will necessarily be 'head wrap' rather than 'right wrap' or 'left wrap', and that phonological clustering will accompany this kind of discontinuity.

To be sure, not all instances of discontinuity in natural language are of this kind. Discontinuities of extraposition (e.g., relative clauses and other modifiers that appear some distance to the right of their governing heads) are a quite different class of discontinuous phenomena and even certain kinds of "second-position" infixing have distinct properties. Nevertheless, discontinuities of this 'boomerang' type may well turn out to constitute a well-defined natural linguistic class of syntactic phenomena that will bear further study as such when the full typology of kinds of syntactic discontinuity is better understood. English syntax has, besides the two cases already discussed, at least two more where head wrapping occurs and where there is also evidence of phonological clustering (cliticization) and unexpected syntactic uninterruptability: (i) Preposition-plus-complement wraps around object (*We voted [with the chairman absent]*), and (ii) Subject-Auxiliary inversion, i.e. VP wraps around subject (*Has she been sampling the punch already?*).

If such a natural empirical class of phenomena do exist, then then the task for the theoretician is to determine exactly what formal theory of 'infixing' and 'cliticization' sorts and 'discontinuity modes' provides the best way to describe this class. Undoubtedly, a more elegant theory of the phenomena can be found that does not require stipulating the coocurrence of two sorts and discontinuity modes as it is done in (Moortgat and Oehrle, 1994) and in this paper. Indeed, more recent formulations of multimodal logics such as (Moortgat, 1995) seems to represent progress in this direction (and I hope to investigate this issue in a subsequent paper).

Department of Linguistics
Ohio State University
Columbus

REFERENCES

Bach, Emmon: 1980, 'In Defense of Passive', *Linguistics and Philosophy* 3:297–342.

Bach, Emmon: 1984, 'Some Generalizations of Categorial Grammars', In F. Landman and F. Veltman, eds. *Varieties of Formal Semantics,* Foris, Dordrecht, pp. 55–80.

Barss, Andrew, and Howard Lasnik: 1986, 'A Note on Anaphora and Double Objects', *Linguistic Inquiry* 17:347–354.

Dowty, David R.: 1982, 'Grammatical Relations and Montague Grammar', In *The Nature of Syntactic Representation*, ed. by Pauline Jacobson and Geoffrey Pullum, Reidel, Dordrecht, pp. 79–130.

Dowty, David R.: 1985, 'Type Raising, Functional Composition, and Non-Constituent Coordination'. Presented at the University of Arizona Conference on Categorial Grammar and Linguistic Theory, Tucson, May 1985. [published in Richard T. Oehrle *et al.*: 1988, pp. 153–198.]

Dowty, David R.: In Press, 'Toward a Minimalist Theory of Syntactic Structure,' In *Syntactic Discontinuity,* ed. Harry Bunt, De Gruyter.

Dowty, David R.: 1996, 'Wrapping and Coordination in Natural Language and in Multimodal Categorial Grammars', Unpublished paper, Ohio State University.

Hoeksema, Jack, and Richard Janda: 1988, 'Implications of Process Morphology for Categorial Grammar', In Oehrle *et al.*, eds.: 1988, pp. 199–248.

Lambek, Joachim: 1958, 'The Mathematics of Sentence Structure', *American Mathematical Monthly* 65:154–170.

Moortgat, Michael, and Richard T. Oehrle: 1994, 'Adjacency, Dependence and Order', In *Proceedings of the Ninth Amsterdam Colloquium,* pp. 447–466.

Moortgat, Michael and Glyn Morrill: 1994, 'Heads and Phrases: Type Calculus for Dependency and Constituent Structure', Unpublished ms., University of Utrecht.

Moortgat, Michael: 1995, 'Multimodal Linguistic Inference', Unpublished ms., University of Utrecht. [To appear in *Journal of Language, Logic and Information.*]

Oehrle, Richard T., Emmon Bach, and Deidre Wheeler, *eds.*: 1988, *Categorial Grammar and Natural Language Structures,* Kluwer, Dordrecht.

Partee, Barbara and Mats Rooth: 1983, 'Generalized Conjunction and Type Ambiguity', *Meaning, Use and Interpretation of Language,* ed. R. Baeuerle, Christoph Schwarze, and Arnim von Stechow, Walter de Gruyter, Berlin, 361–383.

Pollard, Carl J.: 1984, *Head Grammars, Generalized Phrase Structure Grammars, and Natural Language,* Ph.D. dissertation, Stanford University, Stanford, California.

Steedman, Mark J.: 1985, 'Dependency and Coordination in the Grammar of Dutch and English', *Language* 61.3:523–568.

Steedman, Mark J.: 'Surface Structure', unpublished paper, University of Pennsylvania.

Zwicky, Arnold M.: 1986, 'The Unstressed Pronoun Constraint in English', *Ohio State University Working Papers in Linguistics,*, 32: 114–124.

TERENCE PARSONS

MEANING SENSITIVITY AND GRAMMATICAL STRUCTURE

The topic of this talk is the semantics of natural language. This is an area of investigation that today merges work in linguistics with work in philosophy. In linguistics there is a going enterprise in the study of the syntax of natural language, coupled with a methodological goal of studying semantics as well. In philosophy there is a going enterprise in the study of semantical issues, without a great deal of regard for the syntax of natural language. The two enterprises — the study of syntax by linguists and the study of semantics by philosophers, or by linguists working within philosophical frameworks — proceed mostly on parallel tracks, informed by one another, but not frequently very tightly linked. Part of the success of Montague Grammar was that Montague's essays linked them completely. But Montague was a layman in syntax, and so Montague Grammar remains mostly a semantical enterprise. There are a number of people who would try to link these two enterprises more closely. That is the topic of my talk today — a problem about whether and how that linkage might be established. The particular issue I will focus on I call 'meaning sensitivity'.

Frege held that in certain contexts, such as 'that'-clauses, words come to refer to the meanings that they normally express. He called these 'indirect contexts' because in them one speaks of things indirectly, only by speaking directly of the meanings of words; I call such contexts "meaning sensitive" without endorsing Frege's analysis of them in terms of a kind of indirect reference. Philosophers who do semantics usually take meaning sensitivity to be an essential element of semantical structure. The question I want to focus on is how this is related to syntactical structure.

In a nutshell, I want to defend the view that logical form is the same as grammatical form. The contrary view is well-entrenched in the philosophical heritage of the last century. Frege set the tone in his *Begriffsschrift*, when he wrote:

> In my first draft of a formula language I was mislead by the example of [ordinary] language into forming judgements by combining subject and predicate. I soon became convinced, however, that this was an obstacle to my special goal and led only to useless prolixity. (end of section 3)

Russell nailed the point down in *Introduction to Mathematical Philosophy*, stating:

M.L. Dalla Chiara et al. (eds.), Structures and Norms in Science, 369–383.

> Mislead by grammar, the great majority of those logicians who have dealt with [the question of"unreality"] have dealt with it on mistaken lines. They have regarded grammatical form as a surer guide in analysis than, in fact, it is. (page 168)

Frege and Russell held the views they did because they did not possess sophisticated theories of grammatical form. We have sophisticated theories, and it is an exciting and natural idea to want to base a semantics for natural language on such a theory of syntax.

The question can be posed initially in terms of the contrast between meaning sensitivity and opacity. A sentential context is opaque if codesignative terms cannot necessarily be substituted for one another within it without affecting the truth value of the whole, or if existential quantification of a term within the context is not valid. Philosophers, influenced by Quine, tend to think in terms of opacity when formulating semantical theories. I see opacity as an important semantic phenomenon, which is typically a *symptom* of meaning sensitivity, but not definitive of it. For example, I see the contents of 'that'-clauses as automatically meaning sensitive, as a matter of grammar. In a that-clause following the word 'thinks' one then has meaning sensitivity accompanied by the symptom of opacity:

Mary thinks that **snow is white**

But since that-clauses are automatically meaning sensitive, as a matter of grammar, the embedded clause of:

It is true that **snow is white**

is in a meaning sensitive context, even though this is *not* an opaque context. It is not opaque because you *can* intersubstitute coextensive terms in it, and because you can existentially generalize on singular terms. This is because of the special meaning of the word 'true' which in effect cancels out the potential opacity created by meaning sensitivity. I want to explore the idea that meaning sensitivity is systematically correlated with the syntactic structure of natural language. So, for me, meaning sensitivity is the potential for opacity caused by syntactic structure. The occasional lack of opacity of meaning sensitive contexts then needs explanation in terms of the special meanings of other items with which they interact, such as the special meaning of 'true' in 'it is true that snow is white'.

What guarantee do we have that the semantics of natural language can be based on a theory of syntax of the sort that linguists currently practice? What if Frege and Russell are right, and current linguistic syntax just plain ignores important semantic phenomena? I do not think that these questions can be answered *a priori*. So in other work I have tried to describe a semantics for natural language based on its linguistic syntax. That is the job I have been working on for many years now, and my talk today focusses on one of the impediments to it: is it the case that every time you have a meaning sensitive context you can point to a linguistic construction that is causing the meaning sensitivity?

There are two forms that this hypothesis might take. The weak version holds that in each language there is a correlation between syntactic structures that create meaning sensitivity and those that do not. The strong version holds that it is possible to identify types of syntactic structures that create meaning sensitivity in *all* languages. This stronger hypothesis does not follow from the weaker. A reason to believe the weaker is that meaning sensitivity is an important aspect of the meanings of constructions, that these need to be learned in order to master the language, and a correlation of meaning sensitivity with syntactic structure would explain how this is possible. But the structures could be different from language to language and they could still be learnable. I won't try to address the stronger hypothesis here. Indeed, I'll only explore the weaker version, and that only for English.

This version works relatively flawlessly for certain generators of meaning sensitivity in English. That is, it is plausible to see "automatic" meaning sensitivity in complement phrases (CP's) formed with 'that', 'for' ('for the meat to cool'), and 'whether', and in certain '-ing' constructions such as 'Brutus's violently stabbing Caesar'.

For example, here are some complement clauses whose structures are well-known, even if not well understood:

NonInterrogative, Finite:

> Mary believes [that [Sam will be early]]
> Mary believes [– [Sam will be early]]

NonInterrogative, Nonfinite:

> She is anxious [for [them to arrive on time]]
> She asked [– [them to arrive on time]]
> She intended [– [– to arrive on time]]

Interrogative, Finite:

> I don't know [whether [she will be on time]]
> I don't know [if [she will be on time]]

Interrogative, NonFinite:

> She wondered [whether [– to leave early]]

Then there are some that are called "exceptional clauses":

> I believe [the president to be right]
> I consider [him to be conscientious]
> He was reported [– to be in pain]

along with more problematic clauses that apparently have fairly complicated unpronounced structures:

> John persuaded Mary [– [– to resign]]
> John promised Mary [– [– to resign]]

Without going into detail, I assume that all of these structures create meaning sensitivity for their contents.

Next we have clauses that use the progressive or gerund forms of verbs. These come in two sorts. The meaning sensitive ones preserve verbal and adverbial structure:

> MEANING SENSITIVE: **Her violently stabbing Caesar** amazed the onlookers.

They are not to be confused with gerundive clauses that refer to events or states. These lose the verbal and adverbial structures:

> EVENT SENSITIVE: **Her violent stabbing of Caesar** was her last evil deed.

The difference is that in the progressive form adverbs remain adverbs (we have 'violently', not 'violent') and the direct object remains in direct object position instead of being the object of the preposition 'of'. The other phrases refer to events or states, and they are not meaning sensitive.

If the example is sufficiently short the two constructions will be spelled and pronounced alike:

> **Her running** ...

This is ambiguous between a meaning sensitive progressive construction and a non-meaning sensitive gerund construction.

State predicates seem to yield the same ambiguity with 'ness' instead of 'ing':

> MEANING SENSITIVE: **His nakedness** offended her
>
> (She was offended that he was naked.)
>
> EVENT SENSITIVE: **His nakedness** offended her
>
> (She didn't care whether he was naked, but seeing it grossed her out.)

So, summing up, there are a host of constructions in English that are known in virtue of their syntactical structure to automatically produce meaning sensitivity or event/state sensitivity. If the whole language were like this, things would be nice.

Problem Case I:

There are a number of apparent counterexamples to the hypothesis. There are cases in which the same grammatical structure does or does not yield meaning sensitivity depending on the choice of verb — at least, this is how it seems. One example of this is with certain embedded clauses that linguists, for want of a better term, call "small clauses." These may not contain tense, or 'to', or modals, or auxiliaries. They have the bare forms:

> NP VP — Mary saw [Brutus][stab Caesar] — *event sensitive*
> NP AP — I consider [him][intelligent] — *? ? ?*

NP NP — I consider [her][a fool] — — *meaning sensitive*
NP PP — I want [him][off this plane] — *meaning sensitive*

Small clauses of the form NP VP are event sensitive; they refer to events, and are not meaning sensitive. Example:

Mary saw [Brutus][stab Caesar].

is equivalent to

Mary saw the stabbing of Caesar by Brutus.

This is not meaning sensitive; if Caesar is the emperor and if Mary saw Brutus stab Caesar then she saw Brutus stab the emperor.

Small clauses of the form NP NP or NP PP are meaning sensitive. Suppose there are no unicorns and no chimeras. It could still be true that I consider her a unicorn though I don't consider her a chimera. Likewise, I could want him in the living room without wanting him in the dining room even if unbeknownst to me the living room in this house is the dining room.

Small clauses of the form NP AP are problematic, since whether they are meaning sensitive or not seems to depend on what verb they follow, thus conflicting with the idea that meaning sensitivity is purely a matter of grammatical structure. For example,

She imagined [him][naked]

contains a meaning sensitive context; it is equivalent to

She imagined him to be naked

but

She saw [him][naked]

does not contain a meaning sensitive context. This is exactly what is not supposed to happen on the hypothesis under discussion. If the small clauses following 'imagine' and 'saw' have the same syntactic structure, then you can't get meaning sensitivity in one case and not in the other. How can this be explained?

Problem Case II:

The second problem case is more well-known; it involves the meaning sensitivity of the direct objects of *intensional* verbs. Normally the direct object is not a meaning sensitive position:

Not Meaning Sensitive: I gave you **a horse**.

But in other cases it is:

Meaning Sensitive: I promised you **a horse**.

Verbs such as 'promise' are often called intensional verbs for this reason.

The Three Known Solutions:

There are three known ways to deal with these problems.

First, there is Montague's approach to NP's that are direct objects of verbs, which is to declare all direct object positions meaning sensitive, and rely on the particular meanings of verbs to cancel this out. This conforms with the letter of the hypothesis under investigation, though not its spirit, since the cancelling out is the default case. Almost nobody likes this approach, though it is not at all easy to say in detail specifically what is wrong with it.

The other popular approach in the philosophical literature is to propose particular analyses of particular verbs, such as:

Mary sought a unicorn = Mary tried to find a unicorn.

These sorts of analyses are all *ad hoc*, and this does not fit well with current views of language learning since there is no evidence that learners are ever taught such paraphrases. The paraphrases are not constant across the small but apparently open class of verbs that require such analysis. This is no more than to say that this approach treats the problem cases as idioms. But they are more systematic than idioms. The idea of analyzing 'seek' into 'try to find' might be adequate for certain philosophical purposes, such as when an explication is needed. But that does not address the semantical problem of the actual meaning of 'seek' in English.

Finally, there is a recent idea explored by Ede Zimmerman (1992) and partially endorsed and criticized by Friederike Moltmann (forthcoming). Notice that this sentence can be read either transparently or opaquely:

Mary sought a unicorn.

The transparent reading is given by 'Some unicorn is such that Mary sought it', and this is true only if there is a specific unicorn that Mary sought. On the opaque reading Mary is just out unicorn hunting, with no particular unicorn in mind, and this is consistent with the fact that there are no unicorns.

Zimmerman notes the oddity of trying to read sentences like:

Mary sought every unicorn.

opaquely. The transparent reading is 'Every unicorn is such that Mary sought it'. He believes that there is no opaque reading, and tries to explain this by proposing that the problematic direct object positions really contain predicates, not NP's; it is only an accident that the indefinite article is used before 'unicorn'; the sentence 'Mary was seeking a unicorn' might as well read 'Mary was unicorn seeking'.

This is a very compelling idea, but it seems to run afoul of the fact that most of the prohibited readings actually exist. The sentence 'Mary sought every unicorn' has a reading given by the opaque reading of 'Mary tried to find every unicorn'. This is unnatural, but it exists, as do others, such as 'Mary sought three unicorns'. Although we need to explain what is odd about these constructions, we can't use an analysis that prohibits them entirely. (Zimmerman and Moltmann indeed have ways of dealing with them, which I will not pursue here.)

The Structural Ellipsis Solution

I want to explore a different approach. It is a version of the second approach, the one that uses *ad hoc* analyses of these contexts. It is what you get by requiring that the analyses not be *ad hoc*, but instead be systematic. The requirement is simply that the paraphrases introduce no lexical items from open classes. The meaning sensitive uses of these constructions are all cases of *structural ellipsis*; they are short for slightly longer phrases that add *structure without content*. Or something close to that. Specifically, I want to propose that the missing content be confined to the grammatical helping verbs 'be' and 'have', and words whose sole purpose is to mark grammatical structure (like the infinitival use of 'to'), and regular lexical formatives, in particular inchoatives and causatives.

Hamlet Ellipsis

The idea is to see what minimal grammatical structure is needed to achieve the required meaning sensitivity while getting the meaning right and while introducing as little additional structure as possible. So let us start with the simplest cases, which are psychological verbs allow paraphrases using 'be', on the pattern:

She imagined him naked = She imagined him to be naked.

The idea is that this structure:

She imagined [him] [naked]

is elliptical for this one:

She imagined [[him] to be [naked]]

which is a clausal structure, and has the extra grammatical formatives 'to' and 'be'. This analysis sees the initial problematic structure as being elliptical for one with a closely related, slightly more complex, grammatical structure that is already known to be meaning sensitive. I imagine that speakers learn that it is acceptable to use the shorter structure for the longer one with psychological verbs.

In the early days of transformational grammar there would have been a transformation proposed to generate 'She imagined him naked' from 'She imagined him to be naked' by deleting 'to be'; this would probably have been called HAMLET deletion. So let me call my proposal Hamlet ellipsis: the sentence 'she imagined him naked' is elliptical for 'she imagined him *to be* naked'. Ellipsis is not a rule of syntactic generation; the uttered sentences are not derived syntactically from their elliptical expansions. Instead, as a matter of language learning we learn to produce the shortened versions as abbreviations of the expanded versions.

Hamlet ellipsis is nice in that it seems to unify the treatment of all meaning sensitive small clauses. For if you apply it to the NP NP and NP PP forms you get examples like this:

I consider [her][a fool] = I consider [her] to be [a fool]

I want [him][off this plane] = I want [him] to be [off this plane]

These are pretty clearly synonymous. One can speculate then that all of these constructions can shorten to small clauses, as stylistic variants, and the only thing left over to discuss is the independent fact that most small clauses of the form NP VP and NP AP and NP PP already have event/state readings which are different.

How far can Hamlet ellipsis take us? Well, it also applies, though not so neatly, to the direct objects of intensional verbs. In fact, 'imagine' is one of these as well. It occurs with a simple direct object, as in:

Mary imagined a unicorn.

There are two ways to apply Hamlet deletion to such examples. One is to see this terse sentence as itself elliptical for something discoverable only from context, like:

Mary imagined a unicorn prancing around

or

Mary imagined a unicorn just standing there.

Then *these* examples could be seen as Hamlet ellipsis, e.g.

Mary imagined a unicorn to be just standing there.

This approach can be summed up by the formula:

Mary imagined a unicorn =
For some F, Mary imagined a unicorn to be F-ing

But there is a simpler (and, I think, better) way to handle these. This is to see

Mary imagined a unicorn

as short for

Mary imagined a unicorn to be.

This is stylistically bad in English, though grammatical. It is more natural to read it with an added expletive 'there':

Mary imagined there to be a unicorn.

I prefer the shorter form for the theory because it has less structure, and because it is not a theoretical disadvantage for our problem sentences to be elliptical for unnatural constructions; they only need to be elliptical for grammatical constructions.

I suggest then that we consider applying Hamlet ellipsis to all small clauses with intensional readings, and to all direct object constructions with what Moltmann calls creation verbs: 'paint', 'draw', 'imagine', 'plan', 'conceive', ... and what she calls epistemic resultative verbs, such as 'see' (in the *imagine* sense), 'recognize', 'discriminate', and epistemic 'find' (meaning *discover*).[1]

[1]There is a complication for what Moltmann calls resultative verbs with direct objects alone. These all take small clauses as well, and the challenge is to say how they are related.

Hemingway Ellipsis

How far can Hamlet ellipsis take us? Farther than we would like to go. It yields perfectly wrong results for some verbs with direct objects, which are the other sort of case at which I am aiming. For example, by Hamlet ellipsis

I want a giraffe

is short for

I want to be a giraffe.

This is not quite right. But it's close, and a slight variation will do the job. What is missing is not the helping verb 'be' but the other helping verb: 'have'. The required analysis is:

I want a giraffe = I want to have a giraffe.

Instead of Hamlet ellipsis of 'to be (or not to be)' we have Hemingway ellipsis of 'to have (and have not)'. Hemingway ellipsis would apply to the verbs that Moltmann calls verbs of absence, including 'need', 'lack', 'owe', 'promise', 'seek', 'want', and 'look for'.

This proposal is similar to one from Generative Semantics; that idea is to propose that 'I want a giraffe' is generated syntactically from 'I want to have a giraffe' by a transformation called 'have'-deletion. (See Dowty 1979, 245.) This does not fit with current work in syntax, and it is not what I propose. Instead I am talking about ellipsis, a process where, given certain words, we reconstruct something longer that is meant. An example is when we respond to the question:

"How many horses does Fred have?"

with:

"Two."

Our 'Two' is elliptical for 'Two horses', and even more elliptical for 'Fred has two horses'. I am imagining that it is a constraint on ellipsis that the longer reconstructed version be constructed in a simple fashion from materials explicitly supplied in context (as in the case of the question) or from the short version itself, and I am supposing that we can take liberties with this only to the point of supplying the helping verbs 'be' and 'have'. In the case of words like 'want' and 'need' I am supposing that this is so regular a process that we learn it when we learn the words. But I don't go as far as Dowty (1979, 260) in suggesting that the extra structure be a part of the meaning of the word itself, so that the word 'need' has two meanings, a simple one when it occurs with an infintival phrase, and a more complex one when it occurs with

The verbs are 'appoint', 'hire', 'elect', 'find' (resultative). We know how to handle 'They elected Mary president' by Hamlet ellipsis; it is elliptical for 'They elected Mary to be president'. But what about 'They elected three cabinet members'? Clearly this is short for 'They elected three people to be cabinet members'. The trick is to formulate this generally, since it looks *ad hoc*.

a simple direct object. Rather, it has only one meaning, and we reconstruct the rest of the phrase.

Problems with the Analyses

There are hosts of problems with this proposal. First, a simple one. On this analysis

I promised you a horse

is short for

I promised you a horse = I promised you to have a horse.

But this looks as if it should mean that I promised you that *I* would have a horse, when we need it to mean that I promised you that *you* would have a horse. It doesn't mean that. We must keep in mind that Hemingway ellipsis ellipses a *minimal* structure, one in which 'you' is the subject of 'have':

I promised [– [[you] to have [a horse]]],

not the *additional* complicated structure needed to supply a missing empty subject to be controlled by the subject 'I':

I promised you [– [– to have [a horse]]].

The desired reading of 'I promised you to have a horse' is not a favored one in English, but we are talking about constructions that are avoided by the use of ellipsis, not explicitly used, and there is no reason to expect them to be unmarked.

Further questions now arise as to whether more grammatical structure is needed than this. (These questions arise on any analysis.) The structures I have in mind are inchoative and causative formations. These are very systematic in English; from

the door is open

we have the inchoative:

the door opened (= the door *became* open)

and the causative inchoative:

Mary opened the door (= Mary *caused* the door to *become* open)

In the promise case it appears that the inchoative is needed. For consider:

I promised you to have a horse.

And suppose that you already have a horse. Then how can we distinguish the case in which you keep the horse and the promise is thereby fulfilled from the case in which you keep it and the promise is not thereby fulfilled? The latter case seems to require the *inchoative* of 'have':

I promised you to *come to have* a horse

which is not fulfilled if you just keep the one you have, because you never get one.

This reasoning is not conclusive, because it brings in the extraneous notion of fulfilling a promise. But it hints at a difficulty. Let's see if we can link this with the meaning of the statement about promising. Suppose you already have a horse but don't know it. I promise you a horse for your birthday tomorrow. And tomorrow I reveal that you now have a horse. You then discover that you already had this horse when I made the promise, and you look back on my promise from a new point of view. To be specific, suppose I actually said:

I promise you a horse for your birthday.

Question: was this promise kept? I think that there is no clear answer, and if we think about why that is it is because the promise could be taken in two different ways — which means, apparently, that it is ambiguous. I either promised that you would have a horse, or I promised that you would acquire a horse. This is the contrast between the noninchoative and inchoative readings. For notice that in the construction:

I promise you to come to have a horse

the scope of 'a horse' can come all the way outside, in which case there is a particular horse that I promise you that you will come to have, or all the way inside, in which case I promise you that you will become in the state of having a horse; this is a state you cannot become in if you already have a horse. But it can also come inside the embedded clause but not inside the inchoative, with the meaning:

I promise that there is a horse you will come to have.

which is the appropriate meaning when I promise that you'll get a horse you don't already have in the face of the fact that you already have one.

How far does this sort of complication go? Suppose I promise you a horse, knowing that you don't have one but that your mother intends to give you one. And suppose that she does give you one. You do acquire a horse, but I had nothing to do with it. *Then* is my promise fulfilled? Again, there are two ways to take it, and this time we seem to need to introduce a causative inchoative versus a noncausative inchoative to capture this:[2]

I promise you to come to have a horse
I promise to cause you to come to have a horse

So good so far, but it is not clear that we can keep things so neatly in hand. Suppose you get a horse because I blackmail a local rancher into giving you one. Then I cause you to get a horse. But you didn't get it from me. And you may very well have expected me not just to *get* you a horse but to *give* you a

[2]I am not supposing that 'I promise you a horse' is elliptical for the sentence 'I promise to cause you to come to have a horse'. I am supposing that it is elliptical for something like 'I promise to get you a horse', with the understanding that 'x gets y z' has a semantical analysis along the lines of 'x does something that causes a getting by y of z'. And the simpler 'x gets y' has the analysis 'there is a state of having y & there is a becoming in that state of x'. For analyses of these constructions, see Parsons 1990.

horse — from me. Is this yet another meaning of 'promise you a horse'? If so, we seem to need to add some additional structure that is arguably not just a matter of grammatical formatives:

I promise to cause you to come to have a horse *from me.*

What if in fact I said 'I promise to *give you a horse*'. Then I promised you a horse, but the promise wasn't kept. There are clearly two options here. One it to expand the stock of ellipses to account for these examples, by adding the grammatical formative "source", to yield 'from me'. The other option is to not do this, and to point out that the sentence 'I promised you a horse' does not say anything, even elliptically, about the source of the horse. (This is independent of the fact that if I *say* I will give you a horse then that promise is not kept unless the horse you get is from me.)

There is another sticky problem to deal with, raised by Moltmann. Consider:

I need exactly five pipes.

There is a transparent reading which we are ignoring: that there are five specific pipes and I need each of them. The catch is, there seem to be two opaque readings:

I need to have five pipes and I need to have no more than five.

and

I need to have five pipes and five is the maximum number of pipes that I need, though I have no need to have only five.

Moltmann offers this as a problem for such analyses. But I think it has a simple solution, perhaps even a glib one. 'Exactly' is a word whose meaning is a recipe: exactly means 'at least, and not more than'. The 'at least' and the 'more than' go in the sentence exactly where the 'exactly' goes, but the scope of the 'and not' may vary. And that is the key to the solution. The sentence:

I need exactly five pipes

is elliptical for

I need to have exactly five pipes,

and this has two construals due to the fact that the meaning of 'exactly' allows the negation to occur in two different places:

I need to have at least five pipes AND to have NO more than five pipes.

I need to have at least five pipes AND I DO NOT need to have more than five pipes.

We are not through with problems yet, for our proposal suggests other ellipses that are more problematic. First, notice that it sanctions:

Mary sought a unicorn = Mary sought to get a unicorn

This is a little odd, but, on reflection, it seems to me to be exactly right. In fact, the traditional paraphrase:

Mary sought a unicorn = Mary tried to find a unicorn

seems to leave something out: it leaves out the *seeking*. And the more systematic paraphrase restores that. But now what about:

I owe you a horse = I owe to get you a horse,

which is not clearly even grammatical?

It is too bad we cannot appeal to etymology here, for the word 'owe' derives from 'ought', and there is nothing particularly awkward about:

I owe you a horse = I ought to get you a horse

But that isn't right either, because it leaves out the *owing*. I don't owe you a horse just because I ought to give you a horse; I only owe you a horse because of some previous commitment. So the ungrammatical paraphrase contains something vital. The best I can do here is to speculate that the ellipsis here has become ossified and retained even when the full form is archaic. But it is interpretable. In fact, if someone struggling with the language were to utter this to you, you would get the meaning exactly right. So maybe I can use (perhaps, misuse) a suggestion from Higginbotham 1985, that we need to take account of not just what is grammatical in English, but of what *could* be grammatical, and 'I owe to get you a horse' certainly is that. Otherwise we are stuck with seeing 'owe' as a real idiom that needs learning. I don't know if that is right or not. Details of this kind of paraphrase need to be developed and assessed.

Meaning Intonation

Let me turn now to a quite different phenomenon, one that is either a glaring counter-example to the hypothesis under consideration or a confirming instance. I call it *meaning intonation*, and it is illustrated by the italicized words in the example:

'Schnee' means *snow*.

The word 'snow' in this sentence is clearly in a meaning sensitive context. But what is its syntactic structure here? An idiosyncrasy of these contexts (shared with quotation) is that the context may be occupied by a phrase of any grammatical structure whatsoever, and the result is automatically construed as a Noun Phrase. Examples are:

'etwas unanständig' means *something obscene*.
'zu essen' means *to eat*.
'essen' means *eat*.

In the first example, you know something special is going on because the construction is ambiguous. If the direct object is read without meaning sensitive intonation, it means one thing — it means that 'etwas unanständig' itself is an obscene phrase, which is false; if the direct object is read with meaning sensitive intonation then the direct object tells you correctly what the phrase 'etwas unanständig' means. On both of these construals the direct object is a Noun Phrase. In the second example there is only one construal; you can only read it grammatically as if 'to eat' is a Noun Phrase. On any normal interpretation, 'to eat' is not a Noun Phrase at all, so something special is going on here. The italics indicates the conversion of a phrase of any category into a definite singular Noun Phrase, a Noun Phrase that denotes the meaning of the phrase as usually used. At least that is what seems to be going on. And if that is right, then there is a syntactic process that produces Noun Phrases of these very special sorts from grammatical phrases of any category whatever. And this is a syntactic construction that creates meaning sensitivity, in keeping with the hypothesis under consideration. (The same thing also happens with quotation, at least as used by philosophers; a phrase of any syntactic category is converted into a definite Noun Phrase by quotation marks, a Noun Phrase that denotes the phrase contained between the marks.)

This is how I see things, but I might be wrong. I am not in a position to tell syntacticians how syntax works. And linguists might want to deny that there is any such syntactic construction as the one I have been discussing. It is in fact quite unlike any syntactic rules normally studied, since it takes any grammatical unit at all as input and produces uniform output. So one might want to deny that there is any syntactic process here at all; there is just an unusual use of words. If so, the hypothesis under consideration is falsified by this phenomenon. I leave that unsettled; instead I want to explore the consequences of the phenomenon, however it is construed. In particular, I want to try to use it to address a problem posed by Moltmann.

Counting Ships

We need to worry about overkill. To illustrate this, let me consider an example of Moltmann's, the construction:

I counted 28 ships.

This can mean transparently that there were 28 ships that I counted, or it has a meaning sensitive reading on which I counted and came up with the answer *28 ships*; the latter reading is consistent with there being no ships at all. What worries me about this example is not that I have no solution for how to construe the example; I have two solutions without a clear way to choose between them. One solution employs what I am calling meaning intonation. If asked what answer I got on the exam, I might say 'eat beans', or 'I got *eat beans*', or 'I got the answer *eat beans*'. This is a clear case of meaning sensitive intonation, because 'eat beans', normally construed, is not an NP at all, and so something special must be done to make it the direct object of 'got'. But there seems little difference between 'I counted, and I got *28 ships*' and 'I counted 28 ships'. So perhaps the meaning sensitivity here is best accounted for by meaning intonation.

On the other hand, the previous proposal seems to work here as well. 'I counted 28 ships' is just elliptical for 'I counted there to be 28 ships'. So perhaps the phenomenon is ellipsis.

The worry, of course, is that meaning intonation works for *all* of the contexts with which we have been dealing, and perhaps renders the ellipsis solution irrelevant. Not completely irrelevant; we still need to explain the synonymy of 'I want him off this ship' and 'I want him to be off this ship', which the ellipsis solution does automatically.

Hosts of additional constructions provide at least apparent counter- examples to the hypothesis, and require discussion.

Department of Philosophy
University of California at Irvine

REFERENCES

Dowty, David. *Word Meaning and Montague Grammar.* Reidel, Dordrecht, 1979.
Higginbotham, James. "On Semantics," *Linguistic Inquiry* 16 (1985).
Moltmann, Friederike. "Intensional Verbs and Quantifiers," *Natural Language Semantics* (forthcoming).
Montague, Richard. *Formal Philosophy.* Yale University Press, New Haven, 1974.
Parsons, Terence. *Events in the Semantics of English.* M.I.T. Press, Cambridge MA, 1990.
Zimmerman, Ede."On the Proper Treatment of Opacity in Certain Verbs," *Natural Language Semantics* 2 (1992).

15
Foundations of Social Sciences

ROBYN M. DAWES

QUALITATIVE CONSISTENCY MASQUERADING AS QUANTITATIVE FIT*

ABSTRACT.
Much of physical science is based on the ability to control and negate the effects of variables not being studied in order to examine the quantitative effects of variables of interest. In contrast, "soft" psychologists and other social scientists who study people and institutions and societies can never negate the effects of variables they are not studying. The results are generalizations of a *ceteris paribus* nature, occasionally without even a grudging understanding that the effect of variables not studied — often incapable of being controlled — is critical in determining the "significance" and effect size of the variables that are studied in any particular context. The result is that the generalizations in most psychological and social science contexts are of a very simple, usually qualitative, nature, although ironically phrased in quantitative terms. (The ubiquitous linear model is an example; even though it is specified in quantitative terms involving slope and intercept, it will be "fit" whenever variables have a merely monotonic relationship, and we feel validated whenever we have predicted the right direction.)

Psychologists and social scientists feel most comfortable when their results are consistent across *qualitatively diverse* situations; the results concerning the "clinical versus statistical prediction" controversy are a prime example. Our current formal theories of science and statistical generalization, however, hardly justify such comfort. For example, standard statistical models of hypothesis testing, effect estimation, or Bayesian inference all involve random sampling from a population, which is an absurdity in social science contexts other than those such as polling likely voters. Moreover, there is no precise theory about what constitutes a "conceptual replication," which is in fact preferred to an exact one.

In psychology and allied social sciences we often act as if we have solved the problem of induction. Our reports follow a script appropriate to having solved that problem. We test statistical hypotheses, estimate parameters, perhaps compute confidence intervals, and perform meta-analyses. These scripts are appropriate for such activities as assessing voter opinion based on a random sample from a well-defined population of voters, or evaluating a new experimental treatment for a carefully defined problem afflicting a particular type of person by randomly assigning subjects who satisfy a "protocol" to experimental or control group.

It is easy to point out that the script is often phony in a number of respects. Aside from pretty clear deviations from it (how often is our sampling "really" random?), various components of the script itself can be challenged. For example, a whole forest of paper has been generated by criticism of null hypothesis testing. For starters, null hypotheses are never true, and then at the

*I would like to thank my colleagues Cristina Bicchieri and Baruch Fischhoff for comments on an earlier version of this paper. They are not to be held responsible —in any way— for its many remaining deficiencies.

M.L. Dalla Chiara et al. (eds.), Structures and Norms in Science, 387–394.

end of the trail people (sure to be criticized) equate failure to reject them with their acceptance. In the middle, we generate a sampling distribution based on the hypothesis (which we don't believe) that all variation is random, locate a statistic we have computed from our sample in this sampling distribution (we could equally well locate other computed statistics in other distributions), and note the probability that we sample a value this or more extreme; in other words, we begin by assuming nothing is there and end by concluding that if indeed there is nothing there we have sampled a pretty unusual value of it. While some of us (Dawes, 1992b) believe that null hypothesis testing is simply used as a *tool* — the best available one while far from an optimal or very satisfactory one — for establishing directionality (zero's being the extreme of the other direction), most derogate it, but ironically continue to use it, if only to please journal editors.

The rest of the script appears to make sense, but only if we are interested in ending up with an estimate of *a* value characterizing a population. Are we?

It is trivial to point out that the problem of justifying induction has not been solved. Even where our script can be best supported, there is the problem of temporal induction. What I want to point out here is that in many areas of psychology and allied social science we are most confident of our findings where the deviation from the script is *greatest*, not least. That is, we are often most confident in areas where we couldn't possibly specify a population — without vague reference to ill defined entities like "problem solving," "learning," or "expertise," or to potentially well-defined entities like "people" or "societies" from whom we cannot even begin to pretend we sample randomly. Moreover, we are often not interested in an effect but in a family of effects whose quantitative values are not nearly so interesting as are some of their qualitative aspects — such as direction — that might be consistent across contexts. It is this qualitative consistency that we find most convincing, and also the best basis for making a prediction of the future (for example that when clinical versus statistical predictions of whom will be forcibly retired from a police force are compared, we can predict that the latter will be superior, as occurs; see Inwald, 1988). But we evaluate this qualitative consistency across contexts *as if* we were evaluating quantitative consistency both within and across contexts, and then — ironically perhaps — we have more faith in our results the *worse* the fit between this quantitative consistency script and what we actually do and find.

One such inductive inference — a "success story" in the view of many — involves the comparison of clinical with statistical prediction Studies from a wide variety of contexts have examined the synthesis of psychologically incomparable factors to predict important human outcomes — such as staying out of jail on parole, successfully completing graduate school, or behaving violently; these studies uniformly indicate that a statistical ("actuarial") synthesis is superior to one based on the "clinical judgment" of experienced experts (Dawes, Faust, and Meehl, 1989). This contrast between clinical "versus" statistical prediction was first introduced to a wide audience by Meehl's influential book (1954), in which he reported and analyzed about 20 comparative studies. Roughly 30 years later, Meehl was about to conclude that (1986, pg. 373): "There is no controversy in social science that shows such a large body of qualitatively diverse studies coming out uniformly in the same direction as this one [the relative validity of actuarial versus clinical prediction]. When you're pushing 90 investigations [now closer to 150, see Grove and Meehl, in press], predicting everything from the outcome of football games to the diagnosis of liver disease, and when you can hardly come up with a half-dozen studies showing even a weak tendency in favor of the clinician[1], it is time to draw a practical conclusion."

A crucial phrase in the quote from Meehl is "*qualitatively diverse studies.*" I propose that in fact we, in the areas generally termed "social science," are more confident of a finding the more *diverse* the contexts in which it appears, the more *diverse* the subjects studied in these contexts, and the more *diverse* the ways in which it is assessed. "Conceptual replication" is preferred to precise replication, by those of us who cannot give a precise definition of what constitutes "conceptual" replication. If we had sampled only a few types of predictive judgments about a few types of human population we would not be confident in reaching a "clinical versus statistical" conclusion. It is precisely the diversity of the areas and of the populations that engenders confidence — and also makes it hardest to justify the standard quantitative script.

A related area is behavioral decision making, which provides hypotheses about judgmental bias and studies that may explain the poor performance of clinical intuition unencumbered by scientific findings or systematic thought. This area is as well based on conceptually replicated findings from qualitatively diverse contexts involving different types of judgment in diverse contexts — some applied some experimental (Dawes, 1988). Psychotherapy outcome studies (Dawes, 1984, Chapter 2) provide another example. Insofar as possible, each good one is based on an "randomized trials" model — with true random assignment to experimental and control (usually wait list control) group. So far so good for certain therapists in certain clinics or universities. Generalizing across studies to answer such questions as "does psychotherapy work" — or more specifically, "what types of psychotherapy work for what types of problems" — truly involves generalizing across "apples and oranges."

[1]The studies with the "weak tendency" turned out to have two characteristics in common. All were in medical or business contexts, and all involved comparisons where the predictions of the actuarial models were based on a subset of the information available to the clinician (such as results of a "clinical examination"). (This "extra information" is apparently of no help to "expert" psychologists or people working in allied areas such as personnel selection — perhaps in part because feedback about "errors" in these areas is either nonexistent, or at the best difficult to obtain and even then highly probabilistic — and perhaps in part because errors of which judges are in fact aware cannot be "corrected" on the basis of a model of correct judgment; see Dawes, 1994, Chapter 4).

The point is that in order to generalize about fruit it is necessary to generalize across apples and oranges.

Our big problem is that the "philosophy" providing a rationale for our beliefs, in fact even for our analyses of our results, is quite at variance with both our beliefs and our analyses. Again consider analyses in the area of comparing clinical with statistical prediction. True, we tend in each comparison of actuarial versus clinical prediction to test the null hypothesis that the two methods of prediction yield equally accurate results. We do use this much-maligned technique, and we engage in other statistical analyses that might be easier to justify — but the analyses occur in *only one prediction context.* Even if we have somehow heroically sampled our clinicians randomly from that context, we have only one type of prediction to study. Do we really, as in attempting to predict the votes in a particular area — or nationwide — then generalize only to the population from which we have sampled? No. (Or if that is all the generality in which we're interested, most of us would quickly lose interest.) What we're really interested in is the domain of predictions of important human outcomes when these predictions are based on factors that do not form meaningful patterns or gestalts (e.g., scores on tests carefully validated somewhere else combined with grade point averages providing summaries of past performance yet somewhere else again, number of arrests since adolescent and number of violations of prison rules since incarceration). What have we found out about generalization across contexts by employing the usual tools of random sampling from a well-defined population to test an inference *separately* for each k = 1 context?

These same questions arise in evaluating effect sizes — expressed, perhaps, as confidence intervals about these estimates of these sizes in the population. But effect sizes, again, of what? Nowhere in the clinical versus statistical area, for example, do we have any expectation that the degree of predictability in one context of one method will be the same as that in another context, even using "the same" method. (Consider, again, the question of police termination; prior to reading Inwald's article I had no idea whatsoever how well it could be predicted, just confidence that the statistical synthesis of the predictive variables — whatever they would turn out to be — would be superior to the clinical synthesis, as it was.) And even the definition of "the same" method is questionable (e.g., a regression equation with 3 variables?, with 4?, with or without product terms?) The experts are different, have different expertise, and so on. And then there is the whole problem of what we mean by the domain itself — of problems involving the prediction based on psychologically incomparable inputs of important human outcome. The same observations may be made about "decision making tasks," and "psychotherapies." My point here is both that we do have expectations based on these diversity of studies and that these expectations are born out empirically. But where is their normative justification?

One possibility is to define each task or group of experts as a "subpopulation" and then ask whether there are interaction effects that require that main effects must be modified in light of the subpopulation under consideration. Unfortunately, the answer — following this approach — is *a priori* yes (at least in the population, an inability to establish such an interaction here or there being simply a matter of deficient power); moreover, the fact the answer is yes is irrelevant to our conclusion. Put differently, it is true that in applying the standard script we may hypothesize interactions, but their hypothesized or discovered existence neither justifies nor calls into question our generalization across diverse contexts. If we are willing to limit our concern sufficiently, our statistical analysis — including interactions — may determine how much we expect the actuarial prediction to be superior *in this here situation or a narrowly defined class of situations*, but the whole point is that "this here" situation is not the focus of our concern. It is not our focus even as applied people, because what we are then interested in is applying what we know to a new situation, we must rely on generality. To claim such generality because we can "parametrize" such problems (e.g., clinical versus actuarial prediction) to the point that application to a new area can result from quantitative analyses involving interaction-by-contexts terms is, I claim, incompatible with generalization as it is actually practiced. At the least, those who endorse this possibility don't provide methods for achieving it, or examples of success.

Another possibility is to consider each context as constituting a sample of size n = 1 and seek generalization on that basis (which incidentally does *not* preclude meta-analyses as opposed to "vote counts" in order to seek generality, although once again it is difficult to justify *either* as involving estimation of a parameter common to all contexts). The problem with this contexts as sample points approach is that we have no real definition of "predictive contexts" analogous to one of "voters." (Or of decision making tasks, or of therapies for qualitatively different sorts of human "distress.") Certainly, I can *describe* the contexts in which any finding occurs, but what is required for sampling contexts is a definition that can yield a list from which to sample, or at least the way of devising such a list even if it is not concretely provided. And even if I could succeed at that, sampling can never be random. We are all, for example, dependent of finding "convenience" samples of diverse contexts in which people ("clinical experts," "decision makers," "psychotherapists," "patients") are willing to be studied. (And some of us have increasing difficulty in finding such people as the — unpopular — results of our earlier studies become public. "The subjects in your study could not have been competent neuropsychologists because any that were competent would have refused to be in your study" — actual quote.)

Our dependent measures are often no more easily justified than are our sampling of independent variables, tasks, and subjects. For example, in the clinical versus statistical context, we're not just interested in "predictive success" measured in *one* way that we all agree constitutes it. In fact, one striking (and persuasive) characteristic of the results in this area is that as critics propose different outcome measures (e.g., success rate as opposed to a correlation coefficient) and these measures are investigated to answer these critics, the overall conclusion remains unchanged. That reinforces our belief in the generalization. (In contrast, would we be equally happy with a qualitative Newtonian physics involving "force" and "acceleration" defined "every which way but up?"). The point again is that our faith in the results in this area is enhanced by the *violations* from the procedures we generally believe to enhance scientifically valid generalizations: for example, clear (operational?) definitions of domain and measures, random sampling from a well-defined population, clear comparability and differences between contexts, quantitative as opposed to qualitative generalization. Interestingly, epidemiology has similar characteristics, with epidemiologists asserting that they are more confident of findings the more diverse and noncomparable the populations in which they are found (Taubes, 1995).

The problem is that most of us who study human judgment and decision agree with Meehl. Does that mean that we are misguided? A mother of a famous physicist widely regarded as a genius has been quoted as saying that if her son was one of the greatest geniuses in the country, then the country was seriously in need of God's help. Are we to conclude, in a similar fashion, that if the clinical versus statistical prediction findings are among the clearest in social science, then social sciences — psychology in particular — is in need of divine intervention. (Admittedly, some practitioners do happily assume the role of demigod, basing their work on their own clinical intuition while derogating what "science" there may be that could form a basis for "application.")

An alternative possibility is that our conclusions have some validity, that they are epistemologically justified, but that their justification is just different from that found elsewhere (Dawes, 1992a). Much of physical science, for example, is based on the ability to control and negate the effects of variables not being studied in order to examine the quantitative effects of variables of interest. In contrast, "soft" psychologists and other social scientists who study people and institutions and societies can never negate the effects of variables they are not studying. For example, in experimental work, the basic elements studied — people — come to the laboratory with presuppositions, goals, and even ideas about what the experiment is all about, perhaps coupled with explicit or implicit tendencies or desires to please of confound the experimenter. Field studies are subject to even more extraneous influences, although of a different nature. The results are generalizations of a *ceteris paribus* nature, occasionally without even a grudging understanding that the effect of variables not studied — often incapable of being controlled — is critical in determining the "significance" and effect size of the variables that are studied in any particular context. The result is that the generalizations in most psychological and social science contexts are of a very simple, usually directional, nature, although ironically phrased in quantitative terms. (The ubiquitous linear model is an example; even though it is specified in quantitative terms involving slope and intercept, it will be "fit" whenever variables have a merely monotone relationship, and we often feel validated when we have simply predicted the right direction.)

We can, following this line of reasoning, regard the usual logic of statistical inference as a *pragmatic* device used to establish a simple relationship in a particular context, and to regard qualitative generalizations based on contexts as forming the basis of generalization (as in "vote counts" or meta-analyses) rather than whatever happened to be the sample unit within the context. (In null hypothesis testing then, zero simply becomes the limit point of the opposite direction.) But since, again, there is no theory about a population of contexts to be sampled, we must regard even the meta-analysis (or vote count) as an approximation for what we would really do but (as yet anyway) have no means of doing: *applying an epistemologically rationalizable model for generalizing from "conceptual" replications across qualitatively diverse contexts.*

How good are the standard models whose assumptions we routinely violate as tools for achieving this generalization for which we have no better tools? I am not sure. Stating that applying them, evaluating them, and believing in their results all involve "fuzzy" conclusions yields no fresh insights.

Note, however, that all these standard models and techniques are based on the scientific imperative that the person claiming to have a reasonable "scientific" answer to a question must be willing to rise to the challenge of "show me"; moreover, the response to this question must involve "potential disconfirmability" in the sense that it involves confirmation of a prediction the negation of which is not *a priori* absurd (Dawes, 1992b). At least in the social sciences, we can accept Aristotle's "law of the excluded middle" to the point of believing we have provided information supporting A only when not-A is a reasonable possibility, or at the least a possibility. The question is whether the standard tools we use — which involve assumptions of constancy of context, population sampled, manipulation, measure, and so on — in fact rise to the "show me" challenge even when all these assumptions are violated. It is sometimes difficult to assess the effect of violation of assumptions even when the assumptions are of a numerical nature, and these are structural. Perhaps by strict adherence to the "show me" and "compare" principles it would be possible to build an epistemological justification that does not yet exist for what we do.

I hope that at the least this paper may spur some examination of the discrepancy between the justification for what "we" who study the psychology of judgment and prediction actually believe and the standard justifications for scientific, or even coherent, belief. Perhaps this paper will be of some help in motivating people to develop new justifications or modify old ones appropriately. Conversely, it could persuade the reader that the activities of many (even highly regarded?) psychologists and social scientists, and their (our) confidence in the generalizations obtained, have no normative justification whatsoever.

Carnegie Mellon University
Department of Social & Decision Sciences
Pittsburgh PA

REFERENCES

Dawes, R. M. (1988). *Rational Choice in an Uncertain World.* San Diego, CA: Harcourt, Brace, Jovanovich.

Dawes, R. M. (1992a). Comment: Quandary: Correlations coefficients and contexts. In Montada, L., S. H. Filip, and M. J. Lerner (Eds.), *Life Crises and Experiences of Loss in Adulthood.* Hillsdale, New Jersey: Lawrence Erlbaum and Associates, Inc. 521-529.

Dawes, R. M. (1992b). The importance of alternative hypotheses — and hypothetical counterfactuals in general — in social science. William James Book Award Address, division #1 of the American Psychological Association, August 16, 1991. *The General Psychologist*, 28, No. 21, spring, 2-7.

Dawes, R. M. (1994). *House of Cards: Psychology and Psychotherapy Built on Myth.* New York: The Free Press.

Dawes, R. M., Faust, D., and Meehl, P. E. (1989). Clinical versus actuarial judgment. *Science*, 243, 1668-1674.

Grove, W. M., and Meehl, P. E. (in press). Comparative efficiency of formal (mechanical algorithmic) and informal (subjective, impressionistic) prediction procedures: The clinical/statistical controversy. *Psychology, Public Policy, and Law.*

Inwald. R. E. (1988). Five-year follow-up study of departmental terminations as predicted by 16 pre-employment psychological indicators. *Journal of Applied Psychology, 73*, 703-710.

Meehl, P. E. (1954). *Clinical versus statistical prediction: A theoretical analysis and a review of the evidence.* Minneapolis, MN: University of Minnesota Press.

Meehl, P. E. (1986). Causes and effects of my disturbing little book. *Journal of Personality Assessment*, 50, 370-375.

Taubes, G. (1995). Epidemiology faces its limits. *Science*, 269, 164-169.

DANIEL M. HAUSMAN

WHY DOES EVIDENCE MATTER SO LITTLE TO ECONOMIC THEORY?

Let me begin with two apologies. First, this essay[1] is not fully self-contained. It addresses shortcomings in the position I defend in my 1992 book, *The Inexact and Separate Science of Economics*, and in some parts it relies on arguments made there. Second, my title is misleading. It suggests that it is useful and sensible to generalize about economics as a whole. I doubt this. Economists do many different things. In parts of economics evidence matters as much as it possibly could anywhere. It would have been more accurate to give this paper the title, "Why Does Evidence Matter So Little to General Equilibrium Theory and Theoretical Orthodox Microeconomic Theory?" But that would have been a very long title. Let me call this branch of economics — that is, general equilibrium theory and theoretical orthodox microeconomics — "*equilibrium economics*". The name may be confusing, because other economists are concerned with equilibria, too, but I cannot think of a better one. My question is: Why does evidence matter so little to equilibrium theory?

Since most people think that evidence ought to matter quite a lot in empirical science, most people would take the question, "Why does evidence matter so little to equilibrium theory?" as implying that one thinks that equilibrium economics is not an empirical science or that there is something wrong with it. These implications are provocative, because equilibrium economics is central to contemporary economics, and the most renowned economists have contributed to it. It seems arrogant for an outsider to maintain that this discipline is not an empirical science or that there is something wrong with it. Many who have read *The Inexact and Separate Science of Economics* have interpreted it as an apology for equilibrium economics, and they will be surprised at the implied criticism in my title today. But I did not mean my book to be an apology for orthodox economics. I meant it instead to be a "nuanced" and "balanced" appraisal — the kind of appraisal that may be irritating to those who like unqualified assertions and sharply defined positions. Sometimes I criticized, and sometimes I defended orthodox theory, and when the book was finished, many readers thought I was mainly defending it. I did not state my conclusions clearly, and there were mistakes in them. In this essay I hope to

[1]I am grateful to audiences at the London School of Economics, the University of Paris I, the Catholic University of Louvain-la-Neuve, the University of Helsinki and the meetings in Florence for many criticisms and suggestions. I am especially indebted to Michel deVroey, Philippe Mongin, Uskali Mäki, and Don Ross.

M.L. Dalla Chiara et al. (eds.), Structures and Norms in Science, 395–407.

remedy the failures and to argue for an explicit critique of the confirmational practice of equilibrium economics.

My concern in this essay is with *the methods of theory appraisal* — the ways of judging or assessing theories — that equilibrium economists employ. In the book I maintained that *there is nothing wrong with them.* But this mistaken assertion exaggerated my position. In this essay I want to explain how I can defend equilibrium theorists from criticisms that they accept an unjustifiable theory of confirmation at the same time that I maintain that equilibrium economics is poorly supported by the evidence and unlikely to provide the best way forward.

Section 1 explains what equilibrium economics is. Section 2 describes how economists respond to successes and failures of predictions concerning market phenomena and defends their responses from unreasonable criticisms. Section 3 explains what I find mistaken in the way economists assess theories, and why I maintain that equilibrium economics is poorly confirmed and that theorists should seek alternatives.

1. WHAT IS EQUILIBRIUM ECONOMICS?

A correct understanding of the structure of equilibrium economics is crucial to any accurate appreciation or appraisal of the methods of theory assessment economists use. I do not find it helpful to employ well-known categories such as "paradigm," "research programme," or "research tradition" introduced by philosophers such as Thomas Kuhn (1970), Imre Lakatos (1970), or Larry Laudan (1970). In part this is because a description of equilibrium economics in terms of these categories is too *thin.* Features that are peculiar to economics — such as the importance of the theory of rationality, the fact that explanations of individual choices are both causal and reason giving, or the links between equilibrium economics and particular theories of welfare and ethics — are arguably of more importance than is identifying the elements that, according to Kuhn for example, constitute its "disciplinary matrix." In addition the general categories are too restrictive. Borrowing an argument of Kevin Hoover's (1991), one might say that works in equilibrium economics all bear a "family resemblance" to each other. Most of the central lawlike principles are the same. Particular simplifications and mathematical techniques are common. But if one insists on identifying some set of features that is common to *every* piece that is clearly an instance of equilibrium economics, one will wind up with something too weak to distinguish equilibrium economics from distinct enterprises.

In my view a set of lawlike generalizations or principles, which I call "equilibrium theory," is the theoretical heart of equilibrium economics. Although work in equilibrium economics may deny *some* of the generalizations in this set, anything that denies *most* of them is not equilibrium economics. These generalizations consist of consumer choice theory, the theory of the firm, and the claim that economies come to equilibrium. This last thesis is seldom stated as an axiom. On the contrary, in most models it is a theorem. But it is not something that just happens to be provable. For most equilibrium economists, it is a condition of adequacy on models that they imply the existence of equilibrium. Various simplifications and mathematical techniques are also common in equilibrium economics, but these are less central to its content, and a change in these would not be regarded as a change in what equilibrium economics *maintains*.

Consumer choice theory is built around a thin theory of rationality—which says that people ought to rank alternatives consistently and that they ought to choose from among the available alternatives whatever they most prefer. These *normative* claims form the core of a *positive* theory as soon as one adds the proposition that people are (at least to a considerable extent) rational. Equilibrium economics incorporates a theory of rationality, because its explanations of individual choice behavior are both causal and reason-giving. A richer theory of rationality—expected utility theory—is gradually becoming a standard part of equilibrium theory.

The theory of consumer choice adds two generalizations concerning the character of preferences to the claim that individuals choose rationally. These two generalizations are that individuals prefer more commodities to fewer and that their preferences show diminishing marginal rates of substitution. Self-interest is an implication of the first of these. The theory of the firm adds generalizations about relations between inputs and outputs in production (the law of diminishing returns) and about the motivation of firms. One among many disturbing features of equilibrium theory is that the motivational claims of consumer choice theory and of the theory of the firm—that individuals are utility maximizers and firms are profit maximizers—sometimes make conflicting predictions. Choices that maximize utility for managers of firms do not necessarily maximize profits for the firms they manage.

Equilibrium theory plus simplifications concerning information, the divisibility of commodities, and so forth is as central to welfare economics as it is to positive economics. If individuals are self-interested and well-informed and have consistent preference rankings, then they will prefer whatever makes them better off. So it is easy to take welfare to be the satisfaction of preferences. If welfare is just preference satisfaction, then it is difficult to see how to make interpersonal welfare comparisons. How can one compare where Marie is in her preference ranking with where Pierre is in his? The central features of standard welfare economics thus all follow from features of positive economics—the notion of welfare, the rejection of interpersonal comparisons, the consequent concern with notions of Pareto superiority and Pareto optimality, and the infatuation with perfect competition. The tight link equilibrium theory forges between positive and normative economics makes equilibrium theory more resistent to refutation and revision.

Equilibrium theory also influences economic methodology. For many equilibrium economists, phenomena belong to the subject matter of economics if and only if the most important causal factors governing the phenomena are those captured in equilibrium theory. Economics simply *is* the application of equilibrium theory. Of course economists—even equilibrium economists—also define economic phenomena to concern production, distribution, exchange and consumption of commodities and services. So when they regard equilibrium theory as defining economics they are implicitly maintaining that the causal factors captured in equilibrium theory are the "major causes" responsible for the phenomena that have traditionally been taken to be the subject matter of economics. Equilibrium economists suppose that it is possible to have a single theory that spans the entire domain of economics. This is an extravagant assumption to which we will return later. As I argue below and in the book, this fact explains many of the questionable features of the confirmational practice of equilibrium economists.

What is most important for the purposes of this essay is that equilibrium theory is incorporated into positive equilibrium economics. Partial equilibrium theories consist of equilibrium theory plus simplifications and a focus on only one or two markets. General equilibrium theories (whether intended for practical application or abstract inquiry) consist of equilibrium theory plus similar simplifications and a focus on the interdependencies in whole economies. Notice that on this account general equilibrium theory is *not* the fundamental theory of orthodox economics. Equilibrium theory is the fundamental theory, and general equilibrium theories are particular applications of it. In both partial and general equilibrium theories, further generalizations are made concerning, for example, the character of preferences or technology or the structure of knowledge. Not all of the generalizations that make up equilibrium theory play a part in every partial or general equilibrium model, but most of equilibrium theory is always there, and economists are extremely reluctant to add any behavioral generalizations that alter or add to, rather than merely further specify, the causal factors cited in equilibrium theory. It is remarkable but fair to say that *equilibrium theorists believe that equilibrium theory captures all the major causes of economic phenomena.* They believe that, at a certain level of abstraction and approximation, equilibrium economics is complete.

2. HOW EQUILIBRIUM ECONOMISTS APPRAISE THEORIES

Many of the generalizations of equilibrium theory are plausible. The actions of firms do, to a considerable extent, aim to increase net returns. People generally want more commodities. But these generalizations are not universal laws. Firms sometimes have other aims. People do not always want more. To comprehend just what plausible content such generalizations have requires some fancy philosophical analysis. This analysis occupies a large part of Chapter 8 of my book, and I shall not repeat it here. Let us for now just *assume* that there is some way to interpret the claim that "there is a good deal of truth" in many of the basic generalizations of equilibrium theory and that one can be justified in having some confidence in these generalizations.

In saying this, I am taking for granted some claims about theory appraisal that other philosophers have criticized. Karl Popper (1967) and his followers insist that scientists should *never* regard hypotheses as more than conjectures, and they deny that decisions about theory development, assessment, and modification should be guided by the degree of confidence scientists have in various propositions. The Popperian position is mistaken. It is crucial to science and human practices as we know them that people act on their appraisals of how securely established various propositions are. One part of my view of the appraisal of equilibrium theory is that equilibrium economists are justified in having a good deal of confidence in its basic generalizations.

Most of the claims that economic theories make, unlike the basic generalizations, are not, however, obviously plausible or justified by everyday experience. For example, suppose the legislature is considering increasing a tax on industrial property that is owned mainly by monopolies. One might suppose that monopolies will immediately pass along the expense to consumers. But according to the standard theory of the firm, pricing and output decisions are determined by marginal costs, which are not influenced appreciably, at least in the short run, by property taxes. So monopolies will not pass the tax expense on to consumers. This is a surprising prediction, and it seems to be testable. I do not know whether this prediction of the theory of monopolistic firms has been tested or what the results of such the tests may have been.

Testing such predictions is difficult. Consider some of the things that must be done:

1. One needs to decide which firms are monopolies and which are not. (In this case, however, similar predictions apply, although under more restrictive assumptions, to competitive firms. So this may not be as serious a problem.)
2. One needs either a "natural" experiment in which business property tax rates were changed or one must try to draw inferences by comparing prices in different locations where tax rates differ.
3. One must identify "the" prices the firms charge.
4. One must distinguish the change in the prices monopolistic firms charge from overall changes in price levels.
5. One must deal with the effects of all other causal factors.

Suppose that such tests are actually carried out, and suppose that the data establish unambiguously whether the prediction is true. What significance should such findings have for economists?

According to standard philosophical models of theory assessment, the theory of the firm is confirmed by data indicating that property taxes had no effect on prices and disconfirmed by data indicating that property tax increases led to price increases. Exactly how the reasoning is supposed to go and how significant the confirmation or disconfirmation might be are obviously matters of philosophical controversy. But there is broad agreement that theories are disconfirmed by predictive failures and confirmed by predictive successes, particularly when what is predicted is surprising.

Yet "in fact" economists' commitment to the theory of the firm is scarcely affected by such test results. This is a sociological observation, and you may reasonably ask for my evidence. I confess with some embarrassment that I do not have any very good evidence. The claim is not, however, very controversial. It is hard to find instances of contemporary economists who have changed their mind about any fundamental feature of economic theory as a result of empirical evidence.[2] What happens, I maintain, is that if the theory passes such tests, equilibrium economists are pleased, but they are not much more confident that their theory of the firm is correct than they were before. If the theory fails such tests — as it often does — economists are unperturbed. It looks as if they are not behaving the way that responsible scientists ought to behave. It looks as if they need a course in introductory philosophy of science. Indeed it looks a great deal as if they are following an objectionably dogmatic scientific method that many have attributed to John Stuart Mill.

In his essay, "On the Definition of Political Economy and the Method of Investigation Proper to It" (1836) and later in *A System of Logic* (1843, esp. Bk. 3, ch. 10–11, Bk. 6, ch. 3, 6, and 9), Mill defended a "deductive" method of appraisal. The deductive method is called for when the phenomena are both "mechanical" and depend on many different causes. "Mechanical" phenomena are, in Mill's terminology, in some sense the "sum" of the effects of various causes, as total acceleration in mechanics depends on the vector sum of component forces. If phenomena are mechanical, then one can explain them in terms of a sum of individual causes. This is fortunate, because, owing to the causal complexity, the inductive methods of "direct experience" are not applicable. For example, Mill argues that one cannot tell what the effects of tariffs are by comparing economic outcomes with and without tariffs, because there are too many other significant causal factors (1843, bk. 6, ch. 7, sec. 3).

The sort of "inexact" deductive method that economists must employ is, Mill maintains, exemplified by the practice of those who study tides (1843, bk. 6, ch. 3, sec. 13). First one establishes the laws of important individual causal factors or, if one is lucky, one can borrow these laws from some other discipline. In the case of tides, one starts with Newton's laws of motion and his laws of gravitation that tell one how the sun and the moon act on ob-

[2]Philippe Mongin has argued that this has not always been true. In the so-called "full-cost" and "marginalist" controversies of the 1940's and 1950's, the commitment of many economists to the standard theory of the firm was shaken by the results of surveys. See Mongin 1992. He does not question my claim that the theoretical commitments of contemporary equilibrium economists are very little influenced by evidence.

jects on the earth. Such laws could never be well-confirmed by observations of complex phenomena such as tides, but it is possible to test them with respect to simpler circumstances and to establish them firmly. One then deduces the consequences of these established laws for the tides. One can be reasonably confident that these consequences concerning tides correctly state how tides would behave if the only significant causal factors affecting them were the gravitational attraction of the sun and moon, for these consequences follow deductively from well-established laws and specifications of relevant initial conditions. The results of one's deductions may not, however, fit the observations very well, because one has left out many causal factors. If there is a reasonable fit, then one can claim to understand the tides and to have a good theory of them. If there is a persistent bad fit, then one needs to "add in" further causal factors. Thus the importance of the assumption that one is dealing with mechanical phenomena. Nothing in the fit between observations and what one deduces with the help of Newton's laws from facts about the locations of the sun and moon casts doubt on Newton's laws or increases our confidence in their accuracy—they are already established.

This inexact deductive method resembles the practice of equilibrium economists, and it appears to be unjustifiable. Even if tests of Newton's theory by astronomical data or by experiment are of much higher quality than are tests against the data involving tides, one should not maintain that tide data could *never* influence our assessment of Newton's laws at all. And the generalizations of equilibrium economics are, of course, hardly Newton's Laws. They are more precarious and less well confirmed. So it seems as if equilibrium theorists are guilty of a serious mistake concerning the character of science. In Mark Blaug's view (1992), for example, economists need to reread their Popper and start practicing the falsificationist methodology so many of them preach. In Blaug's view, there is a serious moral failing in the comparative indifference economists show to the outcomes of market tests.

I disagree with Blaug. I think that the reactions of economists to market data do not reflect commitment to a unjustifiably dogmatic view of theory assessment. On the contrary, I think that once one takes into account the weaknesses of tests such as the one discussed above, one can defend the reactions of equilibrium economists to failures of market predictions. Here is the argument: To derive testable predictions about changes in monopolist's prices, one conjoins to the theory of the monopolistic firm a number of additional statements. Many of these are unlikely to be true. One's subjective degree of belief in most of these additional assumptions individually (let alone in their conjunction) is much less than one's subjective degree of belief in diminishing returns or in the claim that firms are (roughly) profit maximizers. Because the links between theory and data are so weak, and the theory itself is already plausible, predictive successes and failures have little effect on one's appraisal of the theory. Little effect is not, however, no effect. On some occasions the links are stronger and the data more striking than on other occasions, and there is no excuse here for refusing to pay attention to market data. Indeed I think that Mill himself was only pointing out that data concerning complex phenomena such as markets or tides have comparatively little confirmational worth, rather than issuing a methodological injunction against drawing any inferences from them.

There are thus two competing explanations for why market data do not lead economists to lose their confidence in equilibrium theory. One is Blaug's: Equilibrium economists hold a mistaken and unreasonably dogmatic view of theory appraisal. Another is mine: their reaction to market disconfirmations results from: (1) the view that the disconfirmation of a prediction should lead one to question the most dubious of the claims needed to derive the prediction, (2) a recognition that tests against market data require many implausible assumptions, and (3) confidence that there is a good deal of truth to the generalizations of equilibrium theory. All three of these are justified. There is no mistake in refusing to modify equilibrium theory in the face of market disconfirmations. No doubt some economists would flunk introductory philosophy of science (just as some physicists or chemists would), but the reactions to market data are justifiable.

Which of these explanations — Blaug's or mine — is correct? For evidence concerning which hypothesis better explains why the theoretical commitments of equilibrium economists are so little influenced by market data, I examined in my book some reactions to higher-quality and more pertinent data generated from experimentation. Although hardly proof of my explanation, I pointed out that one can find essays by respected economists in prestigious journals that conclude that the data apparently disconfirm central propositions of equilibrium theory. I cannot prove that equilibrium economists do not fail to take data seriously, as Blaug alleges, but my interpretation of their behavior is at least as consistent with what I know of their practice and, I suggest, far more plausible.

3. WHAT'S WRONG WITH EQUILIBRIUM ECONOMICS?

In this way I defend equilibrium theorists from the accusation that their reaction to market data reveals a commitment to an unjustifiable theory of confirmation. Moreover, I maintain that the evidence does not refute equilibrium theory or make it unreasonable to accept that theory as a basis for both policy and further research. I find no general fault with economists' reaction to market data, I agree that market evidence does not bear forcefully on theories, and I agree that some confidence in equilibrium theory is justified by everyday experience. One of the reasons why evidence matters so little in economics is thus a good reason: market data are of little evidential worth. Thus readers have seen me as an apologist for equilibrium economics. But I am not.

Economists who understand how weakly market data are connected to economic theory, yet who want to bring evidence to bear on their theories, should do three things. First, they should refine their techniques of gathering and analyzing market data to make them more relevant to the assessment of theory. Economists have certainly done this, but it is hard to make much progress in this task. Second, economists should devote the lion's share of their assessment efforts to the gathering of other sorts of data that bear more forcefully on economic theories. Such work might include surveys, observational studies, and controlled experiments. This work is not easy to do, and the difficulties of performing it constitute a crucial reason why so little is done. But even when it is possible to experiment or to perform evidentially valuable surveys and observation studies, there is enormous resistance in the discipline to undertaking them. Experimentation has lately become respectable, but observational studies and surveys are still viewed with deep suspicion if not contempt. The central message in Milton Friedman's extremely influential essay, "The Methodology of Positive Economics" (1953) is that the *only* relevant data are market data. Surveys are, Friedman maintains, worthless and irrelevant.

The determined resistance of the discipline to seeking evidence from case studies and surveys is, in my view, badly mistaken. If one is going to employ a deductive method and to hold that conclusions of economics are credible because they follow from credible premises, then one ought to be testing those premises as fully and harshly as possible. Milton Friedman advises exactly the opposite course. He mistakenly claims that the realism of assumptions is irrelevant — that is, that it does not matter whether one's basic premises are true and that only the accuracy of one's predictions concerning market data matters. His position narrows the frontier in which theory and data may come into contact, when what is needed is exactly the reverse. Market data, like all data concerning the interplay of many different causal factors, bear weakly on theories. The only way to get good evidence is to test separately the components of theories — that is, their "assumptions". Fortunately, it seems to me that in recent years equilibrium economists have become willing to consider a wider range of evidence. Survey reports now occasionally appear in major economics journals. Experimental economics has become more important. These changes are in the right direction and should be encouraged. The second reason why evidence matters so little in economics is that economists have refused to seek the sort of data that could confirm or disconfirm their theories.

The motivation for this unjustifiable refusal to do observational studies and surveys does economists little credit. As economists in the 1930's and 1940's noticed, observational studies and surveys appeared to refute the central proposition of the theory of the firm, that firms attempt to maximize profits. Although these surveys and studies were flawed, they showed clearly enough that firms do not always equate marginal cost and marginal revenue. Indeed, in the course of condemning all such studies and defending the orthodox theory of the firm, Fritz Machlup concedes explicitly that "the assumption of consistently profit-maximizing conduct is contrary to fact" (1956, p. 488). Such a concession poses a problem for equilibrium theorists, since their fundamental theory says that firms are profit maximizers. Machlup and Friedman deal with the problem by insisting that all that matters is how well the predictions of the theory agrees with market data. Friedman adds that it is a mistake even to ask whether the assumptions are realistic. In a nutshell, economists endorsed a ban on surveys and observational studies, because they didn't like what they found.

Popperians would, of course, join me in criticizing this. But I differ from Popperians about what should have been done when such empirical weaknesses were recognized. A Popperian view seems to leave economists with no alternative to rejecting their theory and looking for a replacement. In contrast I have argued (1992, ch. 8) that one can legitimately make use of "inexact" claims, which are not true, but which nevertheless have some truth to them and are capable of telling one testable and significant things about the world.

I said above that there are *three* things that someone who recognizes the inadequacy of market tests should do. The first is to try to improve on techniques of gathering and analyzing market data. The second is to seek other data. Economists do well on the first and miserably on the second. The third is to use the results of high quality tests to improve and to modify theories. Economists do not score well here either. The third reason why data matters so little in economics is that economists make little use of the results of the few good tests they have.

One reason for this is that economists are overconfident concerning how well supported equilibrium theory is by the data. Although everyday experience does provide reason to believe, for example, that people generally prefer more commodities to fewer, I deny that such generalizations are as well established as are claims about the chemical properties of elements or the laws of electromagnetism. This denial requires defense, for many distinguished economists have disagreed. For example, Roy Harrod claims (with some plausibility),

> ...the laws in question are deducible from a single simple principle, itself based on experience, but on an experience far wider than that vouchsafed by the study of markets and prices and extending back to the earliest phases of man's self-conscious existence—namely, the Law of Diminishing Utility or the Law of Demand... The experience is so broad that the principle may be taken as an axiom of the highest possible degree of empirical probability. (1938, pp. 386–7)

The plausibility of Harrod's claims depends, I think, on an equivocation. The principle that has "the highest possible degree of empirical probability" is, I think, not a principle that justifies any policy or predictive conclusions. It is something like "*most* people's preferences for additional units of some commodity *eventually* diminish with the quantity of the commodity the individuals have." What economists need, in contrast, are reliable generalizations about whether marginal utilities or rates of substitution are, given current consumption bundles, actually diminishing, and I doubt that any such principle has "the highest possible degree of empirical probability." Furthermore, confidence in equilibrium economics requires not only confidence in the basic principles of equilibrium theory but also the belief that these basic principles capture the most *important* causal factors responsible for economic outcomes. It is hard to see how claims about the relative *importance* of the factors singled out by equilibrium theory could be strongly supported by everyday experience. Even if physicists are justifiably confident in the basic law of gravitation, they cannot conclude that gravity will be a major causal factor in explaining the behavior of a neutron within a particular apparatus.

There is another and, I think, more important explanation for why economists have not been able to learn more from data that does bear on their theories. Consider the following findings. Experimental subjects are given a choice between two bets. The first (the "P-bet") has a very high probability of winning a small prize and a low probability of losing a small amount. The second (the $bet) has a low probability of winning a high prize and a large probability of losing a moderate amount. Two-thirds of the subjects who state a preference for the P-bet are willing to pay *more* for the $bet — that is, they are willing to pay *less* for the bet they claim to prefer. Preference reversals in the opposite direction (preferences for the $bet with higher prices for the P-bet) are rarely observed. This behavior systematically violates even the very weakest consistency requirements. This phenomenon was first reported in a 1971 paper by Lichtenstein and Slovic, who had predicted it before they observed it. In a 1979 paper in the *American Economic Review*, David Grether and Charles Plott replicated these results in the course of an unsuccessful attempt to show that the phenomenon was spurious or would not apply in settings with stronger incentives. They concluded that the phenomenon was indeed real and that it did apparently refute central propositions of the standard theory of choice.

All this is excellent scientific work. Grether and Plott cannot be accused of overconfidence, of refusing to pay attention to potentially disconfirming evidence or of dismissing the evidence as due to some interference or other. Yet listen to what they conclude:

> The fact that preference theory and related theories of optimization are subject to exception does not mean that they should be discarded. No alternative theory currently available appears to be capable of covering the same extremely broad range of phenomena. (1979, p. 634)

What takes the bite out of the disconfirming evidence is the implicit methodological *requirement* that any alternative worth serious consideration and investigation must "be capable of covering the same extremely broad range of phenomena." Grether and Plott and most of the other economists who have studied the preference-reversal phenomena reject out of hand any consideration of hypotheses like those proposed by the psychologists who predicted the preference-reversal phenomena. For these psychological hypotheses have only a very narrow scope. Rather than invoking scope to help choose among theories that are equally in accord with the data, Grether and Plott invoke it as a *constraint* which leaves equilibrium theory as the only contender in the field. Such a methodological constraint is unjustified, as is the view, upon which it depends, that equilibrium theory captures all the major causal factors.

Evidence winds up mattering so little because of overconfidence and because of an unjustified commitment to a particular style of theorizing. As I already suggested, many economists believe that equilibrium theory provides, at a certain level of resolution, a complete theory of the whole economic domain. Equilibrium theorists rarely state this thesis explicitly, for on the face of it, such a claim is highly implausible. But equilibrium economists show their commitment to this implausible view when they reject as *ad hoc* any behavioral generalizations about individual agents that are not part of equilibrium theory and are not further specifications of the causal factors with which equilibrium theory is concerned. For example, as part of his explanation for the existence of depressions, Keynes offered the generalization that the marginal propensity to consume out of additional income is less than one. This psychological generalization is consistent with equilibrium theory, but independent: it does not follow from equilibrium theory. Instead, it purports to identify an *additional* important causal factor. For precisely this reason, regardless of its truth or falsity, many equilibrium theorists have rejected it as *ad hoc* (Hands 1988). If those theorists believed that economists need to uncover additional causal factors, then they would not reject attempts to do so.

Critics are right to fault orthodox economists for paying too little attention to data, but wrong to focus on the response to market data. Economists are right to take the claims of equilibrium theory to have considerable plausibility. They are right not to be overly impressed with successes or failures of tests against market data. But the fact that market data bear so little on equilibrium theory is a curse, not a boon. Even though it permits economists rationally to hold on to their basic theory more or less come what may, the price of this security is an inability to learn from these data. So long as economists focus mainly on market data, they are unlikely ever to find any stronger confirmation for equilibrium economics or for a superior successor.

No one can predict whether economists would do any better if they did not confine themselves to market data. It may be that the problems involving experimentation, surveys and observation studies would prevent economists from gathering informative data, no matter how they tried. And even if they were to gather powerful and relevant data, there might be no way to combine what one learns about specific causal factors into theories that would be useful in answering the questions that economists want to answer. I am only pleading that this work be tried.

Though right to be unimpressed with the fate of market predictions, economists are wrong not to seek other data more assiduously. They are wrong too to think that their theory is well-established, and they are wrong to insist that any competitor worth considering must span the entire domain of economics. Evidence matters so little to economic theory, mainly because of these mistakes.

University of Wisconsin
Department of Philosophy
Madison WI

REFERENCES

Blaug, Mark. *The Methodology of Economics: Or How Economists Explain.* 2nd. ed. Cambridge: Cambridge University Press, 1992.

Friedman, Milton. "The Methodology of Positive Economics." In *Essays in Positive Economics.* Chicago: University of Chicago Press, 1953, pp. 3–43.

Grether, David and Charles Plott. "Economic Theory of Choice and the Preference Reversal Phenomenon." *American Economic Review* 69 (1979): 623–38.

Hands, D. Wade. "Ad Hocness in Economics and the Popperian Tradition." in Neil de Marchi, ed. *The Popperian Legacy in Economics.* Cambridge: Cambridge University Press, 1988, pp. 121–39.

Harrod, Roy. "Scope and Method of Economics." *Economic Journal* 48 (1938): 383–412.

Hausman, Daniel. *The Inexact and Separate Science of Economics.* Cambridge: Cambridge University Press, 1992.

Hoover, Kevin. "Scientific Research Program or Tribe? A Joint Appraisal of Lakatos and the New Classical Macroeconomics," in Neil de Marchi and Mark Blaug, eds. *Appraising Economic Theories: Studies in the Methodology of Research Programs.* Aldershot: Edward Elgar, 1991, pp. 354–94.

Kuhn, Thomas. *The Structure of Scientific Revolutions.* 2nd. ed. Chicago: University of Chicago Press, 1970.

Lakatos, Imre. "Falsification and the Methodology of Scientific Research Programmes," in Imre Lakatos and Alan Musgrave, eds. *Criticism and the Growth of Knowledge.* Cambridge: Cambridge University Press, 1970, pp. 91–196.

Laudan, Larry. *Progress and its Problems.* Berkeley: University of California Press, 1977.

Lichtenstein, Sarah and Paul Slovik. "Reversals of Preference Between Bids and Choices in Gambling Decisions." *Journal of Experimental Psychology* 89 (1971): 46–55.

Machlup, Fritz. "Rejoinder to a Reluctant Ultra-Empiricist." *Southern Economic Journal* 22 (1956): 483–93.

Mill, John Stuart. "On the Definition of Political Economy and the Method of Investigation Proper to It." (1836) Repr. in *Collected Works of John Stuart Mill.* Vol. 4. Toronto: University of Toronto Press, 1967.

Mill, John Stuart. *A System of Logic* (1843). London: Longman, Green & Co., 1949.

Mongin, Philippe. "The 'Full Cost' Controversy of the 1940's and 1950's: A Methodological Assessment." *History of Political Economy* 24 (1992): 311–56.

Popper, Karl. *The Logic of Scientific Discovery* (rev. edn) London: Hutchinson & Co. London: Routledge & Kegan-Paul.

DAN SPERBER

SELECTION AND ATTRACTION IN CULTURAL EVOLUTION*

Suppose we give ourselves the goal of developing mechanistic and naturalistic causal explanations of cultural phenomena. (I don't believe, by the way, that causal explanations are the only ones worth having; interpretive explanations, which are standard in anthropology, are better at answering some of our interrogations.) A causal explanation is mechanistic when it analyses a complex causal relationship as an articulation of more elementary causal relationships. It is naturalistic to the extent that there is good ground to assume that these more elementary relationships could themselves be further analysed mechanistically down to some level of description where their natural character would be wholly unproblematic.

The kind of naturalism I have in mind aims at bridging gaps between the sciences, not at universal reduction. Some important generalisations are likely to be missed when causal relationships are not accounted for in terms of lower level mechanisms. Other valuable generalisations would be lost if we paid attention only to lower level mechanisms. If we want bridges, it is so as to be able to move both ways.

Social sciences explanations are sometimes mechanistic, but they are hardly ever naturalistic (with a few exceptions in demography or in historical linguistics). They fail to be naturalistic if only because they freely attribute causal powers to entities such as institutions or ideologies the material mode of existence of which is left wholly mysterious. If we want to develop a naturalistic program in the social sciences, we must exert some ontological restraint and invoke only entities the causal powers of which can be understood in naturalistic terms.

Here is a proposal: let us recognise only human organisms in their material environment (whether natural or artificial), and focus on these organisms' individual mental states and processes, and on the physical-environmental causes and effects of these mental things.[1] And here is how, having so restricted our ontology, we might approach the social. A human population is

*I am grateful to Ned Block, Gloria Origgi, John Maynard Smith, and Eliott Sober for useful comments on earlier versions of this paper.

[1] Of course, it is contentious that mental things can be naturalised. If they cannot, if no bridges can be thrown between the psychological and the neurological levels, then this reduces to a proposal to bridge the social with the psychological and the ecological sciences. If, as I believe, a naturalistic program in psychology is well under way, then this is a proposal to bridge the social and the natural sciences through psychology and ecology.

M.L. Dalla Chiara et al. (eds.), Structures and Norms in Science, 409–426.

inhabited by a much wider population of mental representations, i.e., objects in the mind/brain of individuals such as beliefs, fantasies, desires, fears, intentions, etc. The common physical environment of a population is furnished with the public productions of its members. By public production I mean any perceptible modification of the environment brought about by human behaviour. Productions include bodily movements and outcomes of such movements. Some productions are long lasting, like clothes or buildings, other are ephemeral, like a grin or the sounds of speech.

Typically, public productions have mental representations among their causes and among their effects. Mental representations caused by public productions can in turn cause further public productions, that can cause further mental representations, and so forth. There are thus complex causal chains where mental representations and public productions alternate. Given that public productions are likely to have many mental representations among their causes and conversely — a point to which I will revert —, the image of a causal lattice or a causal network might be more appropriate. I will keep using the standard phrase "causal chain," it being understood that every link in this chain may be attached to many, both up and down the causal path.

Of particular interest are causal chains from mental representations to public productions to mental representations etc., where the causal descendants of a representation resemble it in content. The smallest ordinary such causal chain is an act of successful communication. Typically, the public productions that are involved in communication are *public representations* such as linguistic utterances. Public representations are artefacts the function of which is to ensure a similarity of content between one of their mental causes in the communicator and one of their mental effects in the audience.

Communication is one of the two main mechanisms of transmission, imitation being the other. Transmission is a process that may be intentional or unintentional, co-operative, or non-co-operative, and which brings about a similarity of content between a mental representation in one individual and its causal descendant in another individual. Most mental representations are never transmitted. Most transmissions are a one-time local affair. However, it may happen that the recipient of an act of transmission becomes a transmitter in turn, and the next recipient also, and so on, thus producing a long chain of transmission, and a strain of mental representations (together with public representations in cases of communication) linked both causally, and by similarity of content. Fast moving rumours, and slow moving traditions are paradigmatic examples of such cultural causal chains.

When you have a strain of representations similar enough in content to be seen as versions of one another, it is possible and often useful to produce yet another public version to represent in a prototypical manner their partly common content. Thus we talk of *the* belief in metempsychosis, *the* recipe of the Yorkshire pudding, or *the* story of King Arthur, each identified by a content. These are, of course, abstractions, at least as much so as *the* zebra, *the* Doric order, or *the* Russian peasant. It is tempting to see all the concrete representations that can be identified by means of a prototypical version as having the same content, with only negligible variations, as imperfect replicas of one another, but replicas nevertheless. Once this is done, it is but one step to seeing all tokens of the "same" representation as forming a distinct class of objects in the world, just as all zebras are commonly seen as forming a natural kind. Granting such unity to strains of representations makes it possible to use, in order to develop a causal explanation of culture, one of intellectual history's most powerful tool: the Darwinian idea of selection.

On this approach, cultural representations are self-replicating representations. They replicate by causing those who hold them to produce public behaviours that cause others to hold them too. Occasionally representations "mutate", possibly starting a new strain. The task of explaining the contents and evolution of a given culture can be seen, then, as one of finding out which representations are more successful at replicating, under what conditions, and why. Versions of this idea have been defended by, among others, Donald Campbell (1974), Richard Dawkins (1976, 1982), Cavalli-Sforza & Feldman (1981), Lumsden & Wilson (1981), Boyd & Richerson (1985), and William Durham (1991).[2] Dawkins has coined the name "memes" for cultural replicators and the success of this word has been such that it could be seen as confirming, or at least illustrating, the very idea of a meme. I will focus this discussion on Dawkins' memes, espoused in philosophy by Daniel Dennett (1991, 1995), and developed in anthropology by William Durham (1991). My argument extends unproblematically (mutatis mutandis, of course) to all the other proposals mentioned.[3]

Since Sperber (1985b -see also Sperber 1996), I have been arguing that there is a severe flaw in attempting to develop a naturalistic explanation of cultural evolution on the basis of the Darwinian model of selection. I am not moved by any reservation towards Darwinism. Quite the opposite, I believe that Darwinian considerations have a central role to play in the explanation of human culture by helping us answer the fundamental question: what biological, and in particular what brain mechanisms make humans cultural animals, with the kinds of culture they have. In other words, to characterise "human" in the phrase "human culture" we must draw on biology, hence on evolutionary theory, hence on the Darwinian model of selection. It is "culture" in the phrase "human culture" that calls for some different and, I believe, novel thinking, remaining however, as we will see, within the broad Darwinian approach.

[2]See Sober 1991.

[3]On the other hand, the view of cultural evolution put forward by Pascal Boyer (1993, chapter 9) and my own are very near. Boyer's arguments and mine are in part similar, and in part complementary. Boyer offers a detailed discussion of the models of Lumsden and Wilson (1981), Boyd and Richerson (1985) and Durham (1991). There is also a good deal of convergence with Tooby and Cosmides (1992). Two other original and important approaches, that of the cognitive anthropologist Ed Hutchins (1995), et that of the philosopher Ruth Millikan (1984, 1993) would deserve a separate discussion.

My two basic points have been 1) that representations don't in general replicate in the process of transmission, they transform, and 2) that they transform as a result of a constructive cognitive process. Replication when it truly occurs is best seen as a limiting case of zero transformation. These remarks of mine have been taken as an emphatic way of making a correct but unimportant point to the effect that replication is not perfect.[4] But after all, hasn't Dawkins himself pointed out that "no copying process is infallible," and that "it is no part of the definition of a replicator that its copies must all be perfect" (Dawkins 1982: 85)?

Dawkins however is aware of the looming problem:

> The copying process is probably much less precise than in the case of genes: there may be a certain 'mutational' element in every copying event [...]. Memes may partially blend with each other in a way that genes do not. New 'mutations' may be 'directed' rather than random with respect to evolutionary trends. [...]. These differences may prove sufficient to render the analogy with genetic natural selection worthless or even positively misleading (Dawkins 1982: 112).

Dawkins main interest and most relevant contribution are to point out that the mechanism of Darwinian selection is by no means limited to biological material but may apply to replicators of any substance and any kind.[5] Computer viruses, for instance, are successful replicators (alas!) of a non-biological kind.

Provided that certain conditions obtain, replicators will undergo Darwinian selection. The two main conditions are that there should be variations among replicators, and that different types of replicators should differ in their chances of being replicated. In the case of the selection of genes, the source of variation is random mutation, that are, actually, failures to replicate properly. For selection to operate on replicators capable of mutating, a further condition has to be fulfilled. It has to do with the rate of mutation. Obviously, if genes mutated, not just occasionally, but all the time, they wouldn't be replicators anymore, and selection would be ineffective. How much mutation is compatible with effective selection? Here is George Williams's answer:

> The essence of the genetical theory of natural selection is a statistical bias in the relative rates of survival of alternatives (genes, individuals, etc.). The effectiveness of such bias in producing adaptation is contingent on the maintenance of certain quantitative relationships among the operative factors. One necessary condition is that the selected entity must have a high degree of permanence and a low rate of endogenous change, relative to the degree of bias (Williams 1966: 22–23)

[4] This is how I seem to be interpreted by Dennett (1995: 357–359).

[5] A theme developed in detail by Ruth Millikan 1984.

In fact, interesting replicators — genes in the biological case — can be characterised as entities replicating well enough to undergo effective selection.[6]

In the case of genes, a typical rate of mutation might be one mutation for one million replications. With such low rates of mutation, even a very small selection bias is enough to have, with time, major cumulative effects.[7] If, on the other hand, in the case of culture there may be, as Dawkins acknowledges, "a certain 'mutational' element in every copying event," then the very possibility of cumulative effects of selection is open to question.

There are, of course, bits of culture that do replicate. Some people copy chain-letters. Medieval monks copied manuscripts. Many traditional artefacts are replicas. Thus a pot may be copied by a potter, and some of her pots may be copied by other potters, and so on for many generations of pots and potters. This slow manual reproduction process has been superseded in Modern times by more and more sophisticated technologies such as printing, broadcasting, or email forwarding, which allow massive replication. However the number of artefactual replicas of a would-be cultural item is only a poor and indirect indicator of its genuine cultural success. Waste paper baskets and their electronic counterparts are filled with massively replicated but unread junk, while some scientific articles read by only a few specialists have changed our cultural world. The cultural importance of a public production is to be measured not by the number of its copies in the environment but by their impact on people's minds.

The most blatant cases of replication are provided by public productions rather than by mental representations. However, when the a public replica is produced by an individual rather than by a machine, this production is caused by an intention or a plan of the individual, i.e. a mental representation. Mental representations causing the production of public replicas can themselves be seen as mental replicas of mental representations. Jane's mental representation of a pot caused her to make a pot in conformity to this representation. This pot was seen by John and caused in him the construction of a mental representation identical to Jane's. John's representation caused him to produce a pot identical to Jane's pot, etc.

The question then arises as to whether the true memes are public productions — pots, texts, songs, etc. — that are both effects and causes of mental representations, or, as Dawkins (1982) argues, mental representations that are both causes and effects of public productions. With either option, however, there are similar problems. To begin with, most cultural items, be they mental

[6]Williams goes as far as to suggest that, "In evolutionary theory, a gene could be defined as any hereditary information for which there is a favourable or unfavourable selection bias equal to several or many times its rate of endogenous change. The prevalence of such stable entities in the heredity of populations is a measure of the importance of natural selection" (Williams 1966: 25).

[7]See Wilson & Bossert, 1971: 61–62.

or public, have a large and variable number of mental or public immediate ascendants.

Leaving aside mechanical or electronic reproduction, cases of new items produced by actually copying one given old item are rare. When you sing Yankee Doodle, you are not trying to reproduce any one past performance of the song, and the chances are that your mental version of the song was the child of the mental versions of several people. Most potters producing quantities of near-identical pots are not actually copying any one pot in particular, and their skill is typically derived from more than a single teacher (although there may be one teacher more important than the others, which complicates matters still).

Generally, if you are serious in describing bits of culture — individual texts, pots, songs or individual abilities to produce them — as replications of earlier bits, then you should be willing to ask about any given token cultural item: Of which previous token is it a direct replica? In most cases, however, you will be forced to conclude that each token is a replica not of one parent token, nor (as in sexual reproduction) of two parent tokens, nor of any fixed number of parent tokens, but of an indefinite number of tokens some of which have played a much greater "parental" role than others.

You might want then to envisage that this process of synthetic replication of a variable number of models is done by a natural equivalent of a morphing program (i.e. a program that takes, say, the image of a cat and that of a man as input, and produces the image of a creature somewhere between the cat and the man as output). Just as in a morphing program, different inputs can be given different weights: you can have your cat-man more like a cat or more like a man, and Jill's skill and her pots may be more like Joan's than like Jane's, though still owing both to Joan's and Jane's skills and pots.

The model that comes to mind now is less immediately reminiscent of Darwin notion of selection than of the notion of "influence" much used in History of Ideas and in Social Psychology. In the case of selection, genes succeed or fail to replicate, and sexual organisms succeed or fail to contribute half the genes of a new organism. Thus descent relationships strictly determine genic similarity (ignoring mutations). Influence, on the other hand, is matter of degree. Two pottery teachers may have shared the same pupils and have therefore the same number of cultural descendants but their common cultural descendants may be much more influenced by one teacher than by the other. The resulting pots too may be descendants of both teachers' pots, but more like the pots of the one than like the pots of the other.

There are nevertheless commonalties between the meme model and the influence model. Both involve an idea of competition. Both define a measure of success, in terms of the number of descendants in one case, in terms of the degree and spread of influence in the other case. Both predict that the most successful items will dominate culture, and that culture will evolve as a result of differences in success among competing items. The meme model might be seen as a limiting case of the influence model: the case where influence is either 100% or 0%, i.e. where descendants are replica. Actually, formal models of influence in Social Psychology tend to concentrate on this limiting case (e.g. Nowak, Szamrej, & Latane 1990).

Both the meme model and the influence model see human organisms as agents of replication or synthesis, with little or no individual contribution to the process of which they are the locus. At most, the replicating agent may, to some extent, choose what to replicate, and the synthesising agent may choose not only what inputs to synthesise, but also the weights to give to the different inputs. Among the factors of either reproductive or influential success, then, there is the attraction that various possible inputs hold for the agents. However once inputs (and weights in the case of synthesis) have been chosen, the outcome of a successful process of replication or synthesis is wholly determined. Moreover, in these two views, mental representations involved in cultural transmission never contain more information that the inputs they are supposed to represent or synthesise.

The influence model is right, against the meme model, in treating replication in cultural transmission not as the norm but as a limiting case (of 100% influence). Both are wrong however in assuming that, in general, the output of a process of transmission is wholly determined by the inputs (and weights, in the case of influence) accepted or chosen by the receiving organism. The relative stimulus-drivenness involved in both models is not the norm of cultural transmission; it also is a limiting case. Not much of culture is transmitted by means of simple imitation or averaging. Medieval monks copying manuscripts — apparently perfect examples of cultural replication — understood what they copied , and, on occasion, corrected what they took to be a mistake in earlier copying on they basis of what they understood. In general, human brains use all the information they are presented with not to copy or synthesise it, but as pieces of more or less relevant evidence to construct representations of their own.

Let me give three brief illustrations.

Consider your views on President Clinton. They are likely to be very similar to the views of many, and to have been influenced by the views of some. However, it is unlikely that you formed your own views simply by copying, or by averaging over other people's. Rather, you used your own background knowledge and preferences to put into perspective information you were given about Clinton, and to arrive by a mixture of affective reactions and inferences at your present views. The fact that your views are similar to many other people's may be explained, not at all by a copying process, and only partly by an influence process; it may crucially involve the convergence of your affective and cognitive processes with those of many people towards some psychologically attractive type of views in the vast space of possible views on Clinton.

Take languages as a second illustration (see also Boyer 1993: 281). Languages are, at first blush, superb examples of memes: complex skills transmitted from generation to generation and similar enough across individuals to allow communication. However, as Noam Chomsky argued long ago (Chomsky 1972, 1975, 1986), a language such as "English" is an abstraction to which correspond, in speaker's minds, mental grammars, and, in the environment, linguistic utterances. Individuals never encounter other people's grammar, or representations of other people's grammars. Individual learners develop their own grammar on the basis of a large but limited set of linguistic utterances. Different individuals encounter very different sets of utterances. Acquiring a language does not consist in imitating these utterances. In fact, most utterances are never repeated. New utterances are not derived either by averaging over, or recombining old ones.

Clearly, what happens in language acquisition is that utterances are used as evidence for the construction of a mental grammar. How constraining is this evidence? Chomsky argued — quite convincingly — that the linguistic evidence available to the child vastly underdetermines the grammar. Moreover, many utterances, being grammatically defective, are bad evidence for the grammar to be constructed. Given this underdetermination, and given the differences in the inputs available to different children, the fact that children do each develop a grammar, and moreover that, within the same community, these grammars converge, raises a deep problem. Again we owe to Chomsky at least the general form of the solution: There is a domain-specific genetically specified language acquisition device in every child's mind. In the vast space of possible uses of the stimulation provided by linguistic utterances, children are attracted towards their use as evidence for grammar construction. In the vast space of logically possible grammars, they choose among just a few psychological possibilities, and they end up converging towards the one grammar psychologically available in the vicinity of the evidence they have been given. Just as it does not matter on which side of the trough you drop the ball, it will roll to the bottom, it does not matter which French utterances a French child hears, she will construct a French grammar.

As a third example, take Little Red Riding Hood. As good an example of a meme as you will ever get. Here, there is no question that many individuals who hear the tale do aim at retelling it, if not verbatim, at least in a manner faithful to its content. Of course they don't always succeed, and many of the public versions produced by one teller for the sake of one or a few hearers differ from the several standard versions.

For instance, suppose an undergifted teller has the hunters extract Little Red Riding Hood from the Big Bad Wolf's belly, but forgets the grand mother. Meme theorist might want to argue — and I would agree with them — that such a version is less likely to be replicated than the standard one. The meme theorists' explanation would be that this version is less likely to have descendants. This is indeed plausible. There is another explanation, however, which is also plausible: hearers whose knowledge of the story derives from this defective version, are likely to consciously or unconsciously correct the story when they tell it, and, in their narrative, to bring the grand mother back to life too. In the logical space of possible versions of a tale, some versions have a better form, that is, a form seen as being without missing or superfluous parts, easier to remember, and more attractive. The factors that make for a good form may be rooted in part in universal human psychology, and in part in a local cultural context. In remembering and verbalising the story, tellers are attracted towards the better forms. Both explanations, in terms of selection and in terms of attraction, may be simultaneously true, then; the reason why defective versions have fewer replicas may be both that they have fewer descendants, and that the descendants they have are particularly unlikely to be replicas.

I hope the general idea is now clear: There is much greater slack between descent and similarity in the case of cultural transmission than there is in the biological case. Most cultural descendants are transforms, not replicas. Transformation implies resemblance: the smaller the degree of transformation, the greater the degree of resemblance. However resemblance among cultural items is greater than one would be led to expect by observing actual degrees of transformation in cultural transmission. Resemblance among cultural items is to be explained to some important extent by the fact that transformations tend to be biased in the direction of attractor positions in the space of possibilities.

How, then should cultural transmission be modelled? Isn't a Darwinian selection model still the best approximation, to be corrected maybe, but not discarded? To try and answer, let me review the case by means of simple and sketchy formal considerations.

Imagine a population of items that are individually capable of begetting descendants, and that have a limited life-span. Let us imagine that these items come in 100 types with relationships of similarity among the types such that we may represent the space of possibilities by means of a 10 by 10 matrix. Imagine some initial stage (which might, for instance, have been experimentally contrived) where we have a random distribution of say 10,000 items among the 100 types. Suppose we examine our population after a number of generations, and observe a different distribution. While he overall size of the population is roughly the same and there is still a scatter of items across the space of possibilities, now some types are much better represented than others. More specifically, we observe that items tend to be concentrated at, and around two types (see figure 1). Imagine that repeated observations show this distribution to be roughly stable.

A well-know kind of explanation of such a state of affairs would be that some of the types were initially better at replicating, and increased in numbers until a kind of ecological equilibrium was reached, where the more successful types can keep up a higher representation than others. Suppose however that we investigated the manner in which items in this population actually beget descendants, and discovered that an offspring is *never* of the same type as its parent! Rather, the offspring is always of one of the eight types adjacent

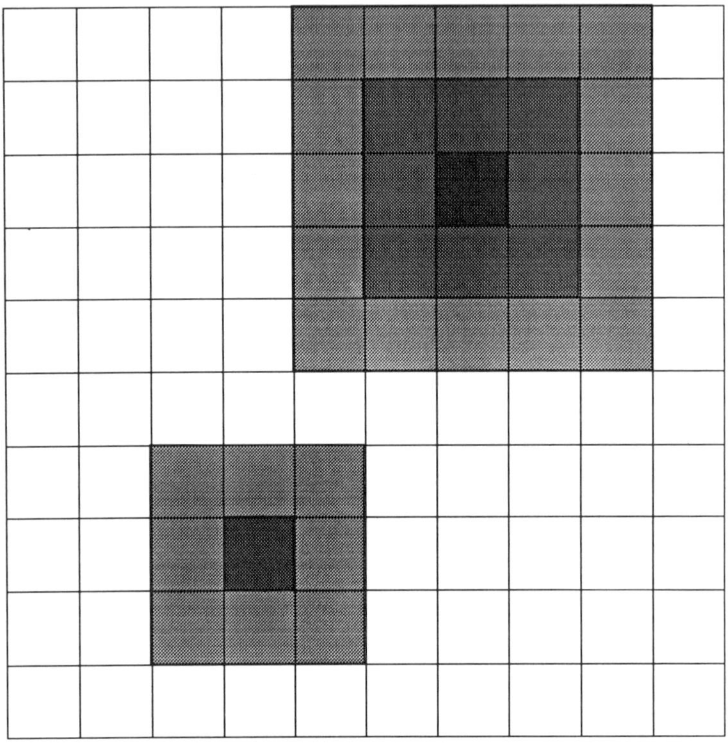

Fig. 1. The space of possibilities: After a few generations, density is greater in the two shaded areas.

in the matrix to that of its parent (fig 2). Ecological equilibrium among differently endowed replicators cannot be the explanation then, since this is a transformative, rather than a reproductive descent system.

An alternative explanation would start from the assumption that the eight possibilities for the offspring of parent of a given of type are not equiprobable. A parent is more likely to beget a transform that differs from it in a given direction. Suppose that the differences in transformation probabilities are such that the matrix has two attractor points.[8] If we trace the descent line of a given item, then, it will not look like a true random walk on the space of possibilities, but will seem, rather, to be attracted towards one or the other of these attractors (figure 3). If the departure point of a descent line is far from the attractors, then it is likely that the arrival point will be near one of them.

[8]Sophisticated notions of attractors ("strange attractors" in particular) have been developed in complex systems dynamics and may well turn out to be of future use in modelling cultural evolution, but a very elementary notion of an attractor will do for the present purpose.

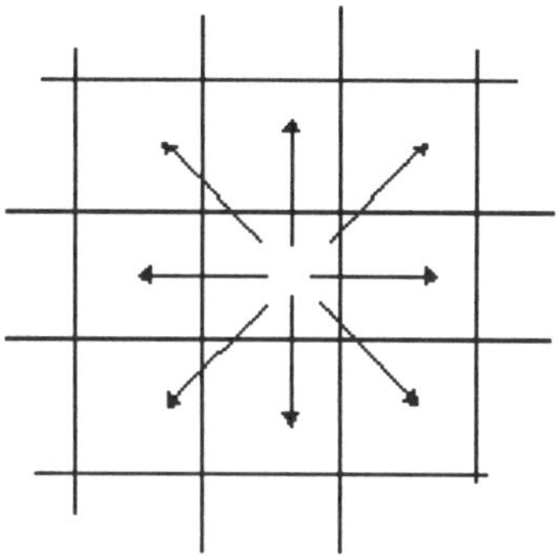

Fig. 2. An item begets descendants of neighbouring types

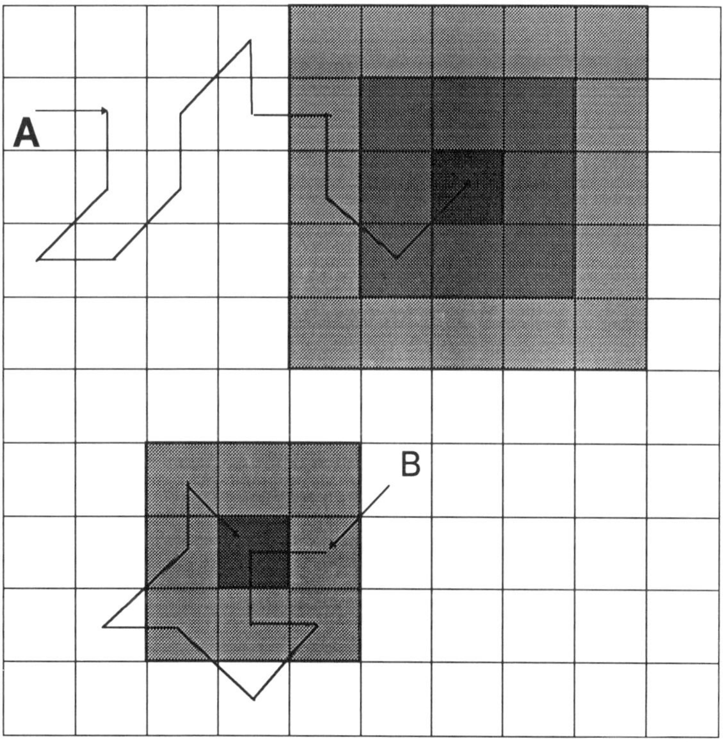

Fig. 3. Descent lines tend to move towards an attractor (line A), or to stay in its vicinity (line B).

It the departure point is near an attractor, then it is likely that the whole line will stay in its vicinity.

If items begot replicas, differences in initial reproductive success and ecological equilibrium would explain the observed distribution. Since items beget transforms, differences in transformation probabilities provide a better explanation.

Transformation and replication can combine. For instance, all the types might have at all times an equal probability of, say, 1/9 of replicating instead of transforming. In this case, although some replication would occur, the difference in distribution between the types would entirely be explained by differences in the probabilities of given transformations. Or the probability of an item replicating rather than transforming might differ according to its type. We could, then, in principle have a dual explanation invoking both reproductive success and attraction. However, in such a case, for the sake of generality and simplicity, replication is better considered as zero transformation, and reproductive success in a given region as defining, or contributing to define, that region as an attractor.

The multiplicity and varying number of "parents," or sources, for the same item, which is, as we noted, a typical aspect of cultural evolution, is also more naturally handled in terms of attraction. The generation of new items in a space of possibilities with attractor regions is to be expected somewhere between existing items and nearby attractors. A simple distance metric is not to be expected, however. The actual mechanisms of generation determine both which items will be transformed, and in which manner.

The attraction model easily incorporates the influence model as a special case: the case where the space of possibility does not exhibit an attractor in the vicinity, and where a simple metric will predict where new items will emerge.

Note that an attractor, as I have characterised it, is an abstract statistical construct, like a mutation rate or a transformation probability. To say that there is an attractor is just to say that, in a given space of possibilities, transformation probabilities form a certain pattern: they tend to be biased so as to favour transformations in the direction of some specific point, and therefore clustering at and around that point. An attractor is not a material thing, it does not physically "attract" anything. To say that there is an attractor is not to give a causal explanation; it is to put in a certain light what is to be causally explained, namely a distribution of items and its evolution, and to suggest the kind of causal explanation to be sought, namely the identification of genuine causal factors that bias micro-transformations.

The existence of attractors is to be explained by two kinds of factors: psychological and ecological. The environment determines the survival and composition of the culture-bearing population; it contains all the inputs to the cognitive systems of the members of the population; it determines when, where and by what medium transmission may occur; it imposes constraints on the formation and stability of different types of public productions. The mental organisation of individuals determines which available inputs get to be processed, how they are processed, and which information gets to guide behaviours that, in turn, modifies the environment.

Psychological factors interact with ecological factors at several levels corresponding to different time scales: that of biological evolution, that of social and cultural history, that of the cognitive and affective development of individuals, and that of micro-processes of transmission.

It is within the time scale of biological evolution that emerges a species endowed with mental capacities that make cultural transmission possible. The role of biology is not just to make cultures possible, without affecting their character and content. The picture of the human mind/brain as a blank slate on which different cultures freely inscribe each its own worldview, the picture of worldviews as integrated systems wholly determined by socio-cultural history, these pictures, which many social scientists still hold, are incompatible with our current understanding of biology and psychology.

The brain is a complex organ. Its evolution has been determined by the environmental conditions that could enhance or hamper the chances our ancestors had to have offspring throughout phylogeny. There are good reasons to believe that the brain contains many sub-mechanisms or "modules" which evolved as adaptations to these environmental opportunities and challenges (Cosmides & Tooby 1987, 1994, Tooby & Cosmides 1989, 1992). Mental modules, i.e., adaptations to an ancestral environment, are crucial factors in cultural attraction. They tend to fix a lot of cultural content in and around the cognitive domain for the processing of which they are specialised (Sperber 1994).

Evolutionary pressures are likely to have favoured not only the emergence of specialised mental mechanisms, but also some degree of cognitive efficiency within each of these mechanisms and in their mutual articulation. At any given time, humans perceive more phenomena than they are able to pay attention to, and they have more information stored in memory than they can exploit. Cognitive efficiency involves making the right choices in selecting which available new information to attend to, and which available past information to process it with. The right choices in this respect consist in bringing together input and memory information the joint processing of which will provide as much cognitive effect as possible for as little mental effort as possible.

Deirdre Wilson and I have argued that the effect/effort balance in the processing of any given piece of information determines its degree of *relevance* (Sperber & Wilson 1986/1995). We claim that human cognitive processes are geared towards the maximisation of relevance. Most factors of relevance are highly idiosyncratic and have to do with the individual's unique location in time and space. Some factors of relevance however are rooted in genetically determined aspects of human psychology. Thus, the processing of stimuli for which there exist a specialised module is comparatively less effort demanding and therefore potentially more relevant. For instance, from birth onwards, humans expect relevance from the sounds of speech (an expectation often disappointed, but hardly ever given up).

It is plausible that individuals are equipped so as to tend to optimise the effect/effort balance not just on the input side, but also on the output side. Public productions, from bodily movements, to speech, to buildings, even when they are modelled after some previous productions, are likely to be attracted towards forms where the intended effect can be achieve at a minimal cost.

Human culture has been around long enough for biological evolution to have been affected by it in return. Gene-culture "co-evolution" (Boyd & Richerson 1985, Durham 1991, Lumsden & Wilson 1991) helps explain in particular the existence in humans of abilities that are specifically geared towards cultural interaction such as the language faculty (Pinker & Bloom 1990, Pinker 1994). Gene-culture co-evolution is, however, too slow a process to explain cultural changes in historical time.

Generation after generation humans are born with essentially the same mental potential. They realise this potential in very diverse ways. This, then, is due to the different environments, and in particular to the different cultural environments, in which they are born. However, from day one, an individual's psychology becomes is enriched and made more specific by cultural inputs. Each individual quickly becomes one of the many loci among which is distributed the pool of cultural representations inhabiting the population. The cultural history of a population is both that of its pool of cultural representations, and that of its cultural environment. These macro-ensembles — the pool of representations and the environment — evolve as an effect of micro-processes where the causes belong to the environment and the effect to the pool, or conversely.

In general, the phrase "cultural environment" is used very loosely and refers to an ensemble of meanings, values, techniques, etc.. The cultural environment so understood has little to do with the physical environment. Its ontological status is at best very vague; its causal powers are mysterious. By "cultural environment," I mean an ensemble of material items: all the public productions in the environment that are causes and effects of mental representations. The cultural environment so understood blends seamlessly in the physical environment of which it is a part. The causal powers it exerts on human minds are unproblematic: public productions affect sense organs in the usual, material way. They trigger the construction of mental representations the contents of which are partly determined by the properties of the triggering stimuli, and partly by preexisting mental resources.

Cultural attractors emerge, wane, or move, some rapidly, others slowly, some suddenly, over historical time. Some of these changes have standard ecological causes: overexploited ecological niches lose their economic attraction; rarely walked paths become overgrown; some practices tend to increase, and others to decrease, the size of the populations that might be attracted to them, etc.

Most historical changes of attractors, however, are to be explained in terms of interactions between ecological and psychological factors of attraction of a kind specific to cultural evolution. The cultural environment causes at every instant the formation of mental representations some of which themselves cause public productions, and so on. This process modifies the relative density of mental representations as well as that of public productions in different areas of the space of possibilities. In particular, density tend to increase in the vicinity of attractors. An increase in the density of public productions in the vicinity of an attractor tends to reinforce the attractor, if only because of the increase in probability that attention will go to these more numerous productions. On the other hand, an increase in the density of mental representations in the vicinity of an attractor may weaken the attractor. The repetition of representations having the same tenor may decrease their relevance and bring individuals either to lose interest in them, or to reinterpret them differently.

Established practices (in matters of dress, food, etiquette, etc.) act as strong attractors. At the same time, because of their expectability, established practices often are low on relevance, while manifest departures from established practices often are an easy way to attract attention and achieve high relevance. Once public productions massively converge towards some cultural attractor, they may foster the emergence of nearby competing attractors. This is illustrated in a dramatic way by the rapid turnover of fashions that quickly lose their power because of their very success.

When, on the contrary, one encounters practices that remain stable for generations, one may suppose that they somehow maintain a sufficient level of relevance in spite of repetition, and try to see whether such is indeed the case, and why. A repetitive practice may remain relevant because its effects are. This is, for instance, the case with technological practices, the economic effects of which are important to peoples welfare or even survival. A repetitive practice may remain relevant because it is in competition with other practices and the choice of one rather than the others by a given individual at a given time may be quite consequential. This is, for instance, the case with practices used to assert one's belonging to some minority. A repetitive practice may remain relevant because different individuals are in competition for the right of engaging in it, and because success in this competition is consequential. This is, for instance, the case with ritual practices marking promotion to some desired status. A repetitive practice may remain relevant because, without hardly modifying its public form, it lends itself to different interpretation according to the agent, the circumstances, the stage in the life cycle. This possibility of renewed re-interpretation is typical of religious practices (see Sperber 1975).

Within the time scale of individual life cycles, ecological and psychological factors also interact in a specific manner. At different stages of their psychological development, individuals are attracted in different directions. Initially, the main psychological factors of attraction are genetically determined, but experience, that is, the cognitive effects of past interactions with the environment, becomes an increasingly important factor of attraction.

For much of childhood, information that allows the child to develop competencies for which she has an innate disposition is attended to and used for this purpose. The child becomes a competent speaker, a competent climber, thrower, catcher, eater, drinker, a competent manipulator of objects, a competent recogniser of animals, a competent predictor of other people's behaviour, and so on. In all these domains, new information easily achieves relevance because it meets the not yet satisfied needs of specialised modules. As basic competencies are acquired, however, attraction shifts to new information relevant in the context of the already acquired basic knowledge. It shifts in particular to cultural information, for instance religious representations, that seems to challenge basic competencies. It shifts also to information relevant to the various goals that the individual has acquired the ability to conceive and pursue.

The contribution of individuals to cultural transmission varies throughout their life. Not only do they transmit different amounts and different contents, they also transform what they transmit in different directions and transmit to different audiences according to their stage in life. The amplitude of transformations also varies with the age and social role of the communicator and with those of the audience. In some configurations, a relatively conservative communication appears more relevant. In other configurations, the search for relevance demand innovation. From the point of view of individuals, cultural attractors seem to move following a path that, actually, combines historical changes and the individuals' own movement in their life cycle and in their social relationships.

It is the micro-processes of cultural transmission that make possible gene-culture coevolution, and that bring about the historical evolution of culture, and the cultural development of individuals. My argument has been that these micro-processes are not, in general, processes of replication. I am not denying that replications occur and play a role in cultural evolution. I am arguing that replications are better seen as limiting cases of transformations. Constructive cognitive processes are involved both in representing cultural inputs and in producing public outputs. All outputs of individual mental processes are influenced by past inputs. Few outputs are mere copies of past inputs. The neo-Darwinian model of culture is based on an idealisation, which is good scientific practice. However this idealisation is itself based on a serious distortion of the relevant facts, and this is where the problem lies.

The neo-Darwinian model and the ideas of replication and selection seemed to offer an explanation of the existence and evolution of relatively stable cultural contents. How come, if replication is not the norm, that among all the mental representations and public productions that inhabit a human population and its common environment, it is so easy to discern stable cultural types such as common views on Bill Clinton, tellings of Little Red Riding Hood, English utterances, and also shake-hands, funerals, and pick-up trucks? For two reasons. Firstly, because, through interpretative mechanisms the mastery of which is part of our social competence, we tend to exaggerate the similarity of cultural tokens and the distinctiveness of types (see Sperber 1985a, chapter 1, 1993). Secondly because, in forming mental representations and public productions, to some extent all humans, and to a greater extent all members of the same population at a time are attracted in the same specific directions.

Even if it contrast with the neo-Darwinian models of culture put forward by Dawkins and others, the model of cultural attraction that I have outlined is, quite obviously, of Darwinian inspiration in the way it explains large-scale regularities as the cumulative effect of micro-processes. The culture of a given population is described as a distribution of mental representations and public productions. Cultural evolution is explained as the cumulative effect of differences in frequency between different possible transformations of representations and of productions in the process of transmission. In the study of cultural evolution, borrowing the Darwin's selection model, is not the only way, and may not be the best, to take advantage of Darwin's most fundamental insight.

CNRS and École Polytechnique
Paris

REFERENCES

Boyer, Pascal (1993). *The naturalness of religious ideas.* Berkeley: University of California Press.

Boyd, Robert & Peter J. Richerson (1985). *Culture and the evolutionary process.* Chicago: The University of Chicago Press.

Cavalli-Sforza, L.L. & M.W. Feldman (1981). *Cultural transmission and evolution: A quantitative approach.* Princeton University Press.

Chomsky, Noam (1972). *Language and mind.* New York: Harcourt Brace Jovanovich.

Chomsky, Noam (1975). *Reflections on language.* New York: Pantheon.

Chomsky, Noam (1986). *Knowledge of language: Its nature, origin, and use.* New York: Praeger.

Cosmides, L. & J. Tooby (1987). From evolution to behavior: Evolutionary psychology as the missing link. In J. Dupré (Ed.) *The latest on the best: Essays on evolution and optimality.* Cambridge Mass.: MIT Press.

Cosmides, L. & J. Tooby (1994). Origins of domain specificity: the evolution of functional organization. In L.A. Hirschfeld & S.A. Gelman (eds), *Mapping the mind: domain specificity in cognition and culture.* New York: Cambridge University Press, 85–116.

Dawkins, Richard (1976). *The selfish gene.* Oxford: Oxford University Press.

Dawkins, Richard (1982). *The extended phenotype.* Oxford: Oxford University Press.

Dennett, Daniel (1991). *Consciousness explained.* Boston: Little, Brown.

Dennett, Daniel (1995) *Darwin's dangerous idea: Evolution and the meaning of life.* New York: Simon & Schuster.

Durham, William H. (1991) *Coevolution: Genes, culture and human diversity*. Stanford: Stanford University Press.

Hutchins, Edwin (1995). *Cognition in the wild*. Cambridge, Mass: MIT Press.

Lumsden Charles J. & E.O. Wilson (1981). *Genes, mind and culture*. Cambridge, Mass: Harvard University Press.

Millikan, Ruth Garrett (1984). *Language , thought, and other biological categories*. Cambridge, Mass: MIT Press.

Millikan, Ruth Garrett (1993). *White Queen psychology and other essays for Alice*. Cambridge, Mass: MIT Press.

Nowak, Andrej, Jacek Szamrej & Bib Latane (1990). From private attitude to public opinion: A dynamic theory of social impact. *Psychological Review. 97*. 362–376.

Pinker, S. (1994). *The language instinct*. New York: Morrow.

Pinker, S. & P. Bloom (1990). Natural language and natural selection. *Behavioral and brain sciences*, 13 (4), 756–758.

Sober, E. (1991). Models of cultural evolution. In P. Griffiths (ed.), *Trees of life: Essays in the philosophy of biology*. Kluwer. 17–38.

Sperber, Dan (1975a). *Rethinking symbolism*. Cambridge: Cambridge University Press.

Sperber, Dan (1985a). *On anthropological knowledge*. Cambridge: Cambridge University Press.

Sperber, Dan (1985b). Anthropology and psychology: towards an epidemiology of representations (The Malinowski Memorial Lecture 1984). *Man* (N.S.) 20, 73–89.

Sperber, Dan (1993). Interpreting and explaining cultural representations. In Gisli Palsson (ed.), *Beyond boundaries: Understanding, translation and anthropological discourse*, Oxford: Berg. 162–183.

Sperber, Dan (1994). The modularity of thought and the epidemiology of representations. In L. A. Hirschfeld & S. A. Gelman (eds), *Mapping the Mind: Domain specificity in cognition and culture*, New York: Cambridge University Press, 39–67.

Sperber, Dan (1996). *Explaining culture: A naturalistic approach*. Oxford: Blackwell

Sperber, Dan & Deirdre Wilson (1986/1995). *Relevance: Communication and cognition*. Oxford: Blackwell. *Second edition* 1995, Oxford: Blackwell.

Tooby, J. & Cosmides L. (1989). Evolutionary psychology and the generation of culture, Part I: theoretical considerations. *Ethology & sociobiology*, 10, 29–49.

Tooby, J. & Cosmides L. (1992). The psychological foundations of culture. In J. Barkow, L. Cosmides & J. Tooby (Eds.) *The adapted mind: Evolutionary psychology and the generation of culture*. New-York: Oxford University Press.

Williams, George C. (1966). *Adaptation and natural selection*. Princeton: Princeton University Press.

Wilson, E. & Bossert W. (1971). A primer of Population biology. Sunderland, Mass: Sinauer.

Symposium 2
Semantics and Semiotics

SEMANTICS AND SEMIOTICS

Every word in the title of this symposium is polysemic. I will consider some of their meanings.

The word "semantics" occurs in a broader and narrower meaning. In the broader sense, its meaning coincides with that of "semiotics" (Husserl, Kotarbiński), in which case the semantics-semiotics opposition disappears and the word "semantics" is as polysemous as "semiotics." In the narrower sense, semantics is a branch of semiotics —whose other branches include syntactics and pragmatics (Morris 1938; Carnap 1942)— and its task is the study of relations between sign and its meaning and referent, in abstraction from the sign user.

The term "semiotics" is understood variously: as the name of a doctrine of signs, synonymous with "logic" (John Locke); as a name of the quasi necessary, or formal, science of signs (C.S. Peirce); as a name (in the form of "semiology") of a science studying the life of signs in social life, a part of social psychology and hence also general psychology (F. de Saussure); as a name for logic of language (Husserl; Ajdukiewicz used the name "logical semiotics" here). In all the above cases we have to do with theoretical semiotics.

The word "semiotics" also occurs in other meanings however.

Firstly, semiotics(1) are the syntactic, semantic and pragmatic properties of the sign, three "dimensions" (Morris 1938) of semiosis, i.e., of the process in which something functions as a sign, semiosic relations of the sign to other signs (syntactics), to the objects to which the sign is applied (semantics), to its interpreters (pragmatics).

Secondly, semiotics(2) are considerations about semiotics(1), i.e., about semiosic properties and semiosis; their description and analysis.

Thirdly, semiotics(3) are considerations about semiotics(2). Usually semiotics(2) and semiotics(3) combine and intertwine to jointly form theoretical semiotics as a science whose branches are syntactics, semantics and pragmatics.

Fourthly, semiotics(4) is the totality of semiotic methods, e.g. the methods of sign interpretation, sign formalization, etc.

Lastly, semiotics(5) is tantamount to applied semiotics, i.e., the applications of semiotics(4) methods, especially of the method of interpretation, in the study of some branch of culture or part of nature.

Theoretical semiotics is a science, discipline or domain of considerations. Its structure is at least two-level. On the level of semiotics(A) we have considerations on semiosis, thus cognition and communication, and on semiosic

M.L. Dalla Chiara et al. (eds.), Structures and Norms in Science, 429–430.

properties of the sign, code or language. On the higher level of semiotics(B), the description, analysis and criticism concern semiotics(A), and are thus metatheoretical, or, putting it generally, of a metalogical or metascientific nature. Semiotics(A) studies the semiosic properties of the sign to arrive at generalizations concerning semiosis through the analysis of maximally variant kinds of concrete signs — linguistic, iconic, musical, mimic, etc. Semiotics(B) analyses and criticizes the concepts and theses occurring on the lower level of semiotics(A), and also constructs concepts and poses problems for the latter. Semiotics(B) tries to answer questions such as: What is meaning in general? What is indirect cognition, i.e., cognition by means of signs? What is communication?

Since the object of theoretical semiotics is semiosis —which in turn consists in cognition through the mediation of signs of that to which they refer, and in communicating that cognition— theoretical semiotics may be regarded as a science of cognition and communication, activities inseparable from one another. In theoretical semiotics, semiotics(A) is often superimposed not only by metatheoretical considerations but also by the meta-metatheoretical, etc. In view of this, on successive levels we have metalanguages of higher and higher order.

Logical semiotics is theoretical semiotics treated from the viewpoint of general logic, methodology and philosophy of science. The branches of logical semiotics are logical syntactics, logical semantics, and logical pragmatics. Logical semiotics is not always a formalized axiomatic system; it also occurs e.g. in the form of a science consisting of an inductive and deductive part. It deals with all kinds of signs, not just with language.

Jerzy Pelc
Department of Logical Semiotics
Warsaw University

UMBERTO ECO

ON MEANING, LOGIC AND VERBAL LANGUAGE

1. LOGIC, SEMIOTICS, PHILOSOPHY OF LANGUAGE

I do not know why the members of the organizing committee of this conference have so cordially invited me — and not as an auditor but as a speaker. I am not a logician and I am not a philosopher of science. I suppose I have been brought here as a representative of the curious genus of semioticians, and probably the organizers thought that it would have been wise and witty to provide logicians with an ostensive example of what should not be done — in the sense in which, according to an example by Charles Sanders Peirce, a drunkard can be exhibited by the Salvation Army in order to advertise by contrast the advantages of temperance.

In this century we have witnessed a sort of cold war between two sorts of semantics. Roughly speaking, on one side there are linguists, lexicographers and semioticians who privilege an intensional semantics, primarily concerned with the meanings culturally assigned to given expressions (independently whether they are verbal or not); for them the referential function of any semiotic system is a pragmatical consequence of its (intensional) set of semantic rules. To follow Strawson (1950), "mentioning or referring is not something an expression does: it is something that someone can use an expression to do." On the other side there is the analytic tradition (or the "analytic paradigm") according to which philosophers of language are mainly interested in a truth conditional semantics.

The reason for such a difference is that linguists primarily study grammars, that is, set of rules governing a given natural language. General linguists, on their turn, investigate the general rules (if any) governing every natural language. Linguistics being a branch of semiotics, and semiotics studying every sign system (be it verbal or not), the same division of labor holds for semiotics too. There are *specific semiotics* that study a specific sign-system, such as iconic signs, road signals, ostensional and gestural communication and so on. Even though many of these systems are more elementary than a verbal language (the rules of a traffic lights language can be outlined very shortly and with traffic lights is practically impossible to generate ambiguous or poetic expressions), we can equally speak of a grammar of traffic lights, of road signals or of the American Sign Language. These grammars are descriptive, frequently prescriptive and to some extent also predictive, since they can successfully tell how the user of a given sign system, under normal circumstances, will generate or interpret messages.

M.L. Dalla Chiara et al. (eds.), Structures and Norms in Science, 431–448.

Then there is a *general semiotics*, which investigates not only the cognitive or cultural rules that govern every possible sign system but also all the cases in which communication takes place without, beyond, before and against any system. In other words a general semiotics studies systems and rules as well as processes — including those processes which don't follow any rule or can generate new rules. Such an endeavor has been charged with a sort of speculative bulimia but, if general semiotics has a right to exist, it has it inasmuch as it is able to unify, to subsume under a single set of categories all the particular cases in which human (and even non human) beings use phonations, sounds, gestures, natural or artificial objects and every other form of behavior in order to refer to objects, classes of objects, external states of affairs or internal mental states that are not perceivable during a communicative interaction. If it is the same physio-psychological apparatus that makes us capable of understanding the sentence *It will rain tomorrow*, as well as the cloud which announces a rainshower not yet experienced (or both the sentence *it was a dark and stormy night* and a painting representing a dark and stormy night), then there must be theoretical instruments by which it will be possible to investigate from a unified point of view these different processes of perception, elaboration, prediction and dominion over something that semiosis does not place before our eyes.

In this sense a general semiotics could be defined as a philosophy *of languages*. If however one means by philosophy of language the so-called analytical paradigm, a first fundamental difference should be stressed. Usually the analytical philosophy of language takes grammars for granted. Given the canonical example

1 the present king of France is bald,

an analytical philosopher is mainly interested in ascertaining if there is a king of France and if he is bald. The fact that (1) is more understandable than (2)

2 the bald of king present France is

is taken as intuitive. By the way, the use of the word *intuitive* by analytical philosophers is a revealing symptom of their scarce interest in grammars. No linguist would say that the meaning of a given sentence is intuitive, since a linguist's job is just to explain why (on the grounds of which rules) a given expression looks as intuitively meaningful.

For many historical reasons, most of contemporary semiotics developed under the influence of linguistic research and was thus largely involved in a reflection about grammars, to the point that its philosophical reflection becomes heavily enmeshed in grammatical description.

Now, a grammar has to take into account also a lexical component. Let me use such a term in its widest sense: there are not only linguistic lexicons; there is for instance a sort of lexical competence that allows one to distinguish between the Italian and the British flag, or between two different sounds emitted by my computer, one informing me that I am on the verge of getting a successful connection with my remote server on Internet, and the other saying that something wrong occurred. A lexical competence has little to do with questions of truth. If — in the course of a battle in the Balkans area — a Serbian commando waves a Croatian flag, it is certainly important to know that we are witnessing a case of lie and that it is not the case that those soldiers are Croatian but only pretend to be so; however, semiotically speaking, they are certainly asserting (mendaciously) that they are Croatians and it is grammatically (and lexically) important to understand that and on the grounds of which semiosic conventions they are certainly asserting that. If we are unable to explain on the grounds of which lexical rules they are asserting that, we are also unable to ascertain that they are lying, that is, that by some meaningful semiosic strategy they are referring to what is not the case.

So it happens that in the linguistic and *semiotic* framework by semantics people mean a intensional semantics, a theory of meaning, and frequently a structural description of the ways in which a given semiotic system *cuts* and organizes experience into a system or many competing systems of meaningful units — and consider such a theory of meaning prior to any effort to use languages in order to mention (to refer to, to denote, to designate truly or falsely) states of the world. In the framework of the analytic paradigm, on the contrary, people mainly mean by semantics an extensional one, and are mainly interested in ascertaining how we refer to, designate or denote truly or falsely given states of affairs — to such an extent that frequently truth conditional theories of reference claimed to eliminate meaning as a disturbing, ghostly and unnecessary entity.

These two different approaches could honestly compete in a sort of open battle, which would require a common playground. Such a common playground exists: for instance, the holistic approach in the analytical philosophy of language has many points in common with structural semantics. However I was speaking of a cold war for many reasons: in a hot war the two competing forces know each other very well, or do their best to know the nature of their enemies. In a cold war it is enough to assert that the Others are representing the Empire of Evil and it is not so urgent to know what they really are. Obviously, every cold war needs secret agents and spies who know pretty well what happens in the opposite field, but they are supposed to conceal rather than revealing what they know. Happily there also are double agents, working for one side and keeping the other one informed. I am proud to belong to such genus, but, believe me, it is not an easy job.

What I would say today is that there has been, since the origins, an unfortunate marriage between aletic logic and the analytic paradigm. Fortunately such a marriage has been frequently undermined by some adulterous characters (let me mention at least people like Quine, Putnam, Davidson, not to mention the second Wittgenstein). But adultery does not mean divorce. Logicians can say that they have not to complain if many philosophers of language love them so much, but I am convinced that such a marriage succeeded in being embarrassing for logicians too.

2. SIX DOGMAS OF THE TRUTH CONDITIONAL PARADIGM

Let me, following a venerable tradition, to outline six dogmas of what I would call the Truth Conditional Paradigm (TCP), at least in its crudest format. No analytical philosopher is today taking TCP in such a crude format, but I am convinced that it still survives as a set of silent presupposition even when philosophers are challenging the early dogmas of analytical philosophy.

2.1. The first dogma says that *we mainly use natural languages for referential purposes*, that is, in order to assert what is the case. Certainly, it is well known — also from within the TCP — that we use natural language for many other purposes, to ask, to command, to convince and so on. Logicians know it very well, and were compelled to deal with many forms of non standard logic. However, philosophers were several times more indolent than logicians, as if they wanted to make logician's life simpler, while every logician, from the times of Aristotle, was encouraged to go on only when philosophers reflecting upon natural languages showed them that the life of language is more complicated than the picture offered by any previous formal logic.

I must say that the lexical semantics of linguistic origins has frequently failed in paying sufficient attention to the problem of reference. On the contrary, especially under the influence of cognitive studies, from within the analytical milieu philosophers are today outlining a clear-cut division between the inferential aspect of lexical competence (which should enable speaker to draw inferences from sentences or to translate words into paraphrases or definitions) and referential competence, which should provide the ability to map lexical items onto the world, in order to pick up their referent (see for instance Marconi 1995). This notwithstanding, many followers of the TCP can be charged with indolence because from the very beginning they have mixed up a theory of reference with a theory of logical truth.

2.2. This decision depends on the second dogma, which reads: *we use language in order to tell the truth*. Nobody can deny that we mainly use language in order whether to tell the truth and to make people believe we are telling the truth, but only in the same sense in which we mainly use music in order to express feelings or to arouse aesthetic enjoyment (without any referential purpose); in spite of this we also use music to tell what is the case, or to precisely order something, as it happens with trumpet signals in the army.

From Plato, Aristotle and on, we have been told that we can assert something true or false only by complete propositions. The word *dog* in itself does not assert anything; only by uttering *dogs are animals*, or *my dog is barking* one can assert something true or false. But to focus one's attention on the truth functional aspects of language leads one to the presupposition that a good philosophy of natural language is only provided, to stick only to Aristotle, by *De Interpretatione* or by the *Analytics* and not also by *Poetics* or *Rhetoric*.

2.3. Such a strict link posited between meaning acquisition and declarative sentences has encouraged the growth of the third dogma, that is, that *semantics cannot be but truth conditional*. In this line of thought the TCP has interpreted Tarskian semantics, which is a logical and formal one, as a model for the way we use language for our everyday affairs.

It would be wrong to mistake Tarski with a Medieval Aristotelian tradition according to which truth was not only a correspondence between statements and reality but also as reflection-like phenomenon (a true thought is a picture of a true state of affairs). When Tarski said that a statement is true when it asserts (or refers to) what is the case, he was not thinking of a sort of psychological reflection theory; he was not saying that, in order to decide whether a statement is true or not, we must know the corresponding state of affairs. Otherwise, given the well known example

(i) "snow is white"
is true if and only if
(ii) snow is white

we would know pretty well what a kind of logical and linguistic entity is (i) (it is a verbal sentence in an object language L, that can be uttered and which conveys a proposition) but we would have no idea of what (ii) is or could be. If it only were another sentence in a metalanguage L^2, then the Tarskian assertion should be translated as

(i) the statement (in L) "snow is white"
is true if and only if
(ii) "snow is white" is a true statement in L^2

which would produce a sorites of unending recursive definitions. But if (ii) is not a sentence, it cannot even be a state of affairs because a state of affairs is not a sentence, nor a proposition, but something that can at most be denoted by a proposition. If we take the Tarskian example naively we are as naive as Saussure's editors who, in order to define the *signified* (the meaning) of the *signifier* (or word) *tree* drew a diagram of the kind shown in fig. 1.

It is evident that the upper image of a tree is not the meaning of the word *tree*, not even if it claimed to represent the corresponding mental image. It is nothing more than an unbound variable.

Likewise in Tarski (ii) stands conventionally for the assignment of a truth value to a proposition. The Tarskian state of affairs is not what makes a proposition to be true, but rather something to which a true proposition corresponds, or everything is expressed by a true proposition (cf. McCawley 1981: 161). The fact that a scientist can decide to implement protocols in order to check by physical experiences whether or not what we call *snow* is really

Fig. 1.

white, has nothing to do with logic. It is a matter of experimental knowledge. Logically speaking we can correctly assert that

(i) "x is the highest prime number"
is true if and only if
(ii) x is the highest prime number

without disturbing the Tarskian theory even though (ii) would embarrass a mathematician.

Within the Tarskian framework, to say that *snow is white* is true if and only if snow is white is an acceptable formal criterion (since it means that in the formal calculus a state of affairs is just an assignment of truth values to a proposition); on the contrary, to say that the same statement refers to (designates, denotes) the empirical fact or to the universal law that snow is white is a radically different matter. A semantics of natural languages based upon a Tarskian notion of truth is a semantics according to which statements can or cannot refer to what is *really* the case in a given possible world; but such a semantics does not have any criterion in order to ascertain how the world goes and not even how can we use words in order to pick up objects or world states. At most it can say that *snow is white* is a true statement because it confirms that, according to a series of universally accepted meaning postulates, snow has the property of being white. But if so, an allegedly extensional and truth conditional semantics becomes an intensional one (where *true* does not mean "what is the case" but rather "what has been conventionally and culturally assumed to be the case for the sake of the argument").

I have nothing to object à propos such a position, since we frequently use language this way. I only remark that an alleged truth conditional semantics of natural languages is incapable to explain what we are referring to when uttering a declarative sentence, except the fact that we assign a given truth value to the corresponding proposition. But this is not the purpose of a theory of reference, whose aim is to grant the correspondence between statements and states of affairs. In a theory of reference a complete (intensional) semantic competence should in some way also provide instructions about how to identify in a given world the possible referent of an expression (at most one can argue whether referential competence is a part of the lexical one, or they are mutually independent abilities, governed by different parts of our brain, and the questions seems to me still open). When a theory of reference takes a Tarskian definition of truth as its model, its concern for realism brings to an excess of idealism.

2.4. In formal logic every calculus can be done without recurring to examples borrowed from natural languages. Logicians use verbal examples for the sake of clarity, but frequently these examples can be misleading. One of the most common trompe-l'oeil effects is that, since logical formulas can be interpreted by sentences uttered in natural languages, then it seems that a verbal sentence is the only way to interpret or translate a logical formula. On the contrary, computers prove that a formalized language can be interpreted by variations of iconic arrangements, shades of color and so on, and vice versa.

This trompe-l'oeil effect has lead the TCP to its fourth dogma: *meaning is always a linguistic business*. In other words, semiosis is so strictly identified with verbal language that, even when admitting that also other semiotic systems can convey any meaning or can state something about the world, this can be ascertained only by interpreting non verbal expressions in verbal terms.

Let me take as an example the beautiful *Languages of Art* by Nelson Goodman (1968). Here a philosopher of language tries to legitimate the existence of visual languages and to elaborate new adequate semiotic categories where logical and linguistic categories are unable to distinguish certain fundamental differences. I am thinking of the sections on "Examples and illustrations", the discussion of autographical and allographical arts, and so forth. There Goodman is starting to outline a *visual semiotics*.

Yet such a remarkable endeavor is undermined by the silent but influential presence of the fourth dogma. When Goodman deals with the representational character of images, he fails to free himself from the idea that meaning is mainly a verbal affair.

When Goodman asks if a gray-tone painting representing a landscape, and which certainly denotes a landscape, also denotes the property of grayness or is denoted by the predicate "gray", he tells us nothing about the signification that the color gray in this painting may have for one who views it. He seeks only to make a phenomenon of visual communication interpretable in linguistic terms, and he thereby betrays his purpose of elaborating a semiotics of the visible. When he asks whether a red object *exemplifies* the property of *redness* (and in this case one is also obliged to wonder if the object also exemplifies coextensive properties, such as trilaterality or triangularity) or whether it exemplifies the predicate *red* (in which case the problem arises of whether it also exemplifies the predicate *rouge* for a French speaker), or whether it exemplifies the denotation of this same predicate, Goodman tells us nothing about the signifying function that, for example, a red object has for a viewer of a film who has just watched a bloody scene a few seconds earlier. To ask what the red color exemplifies in a priestly garment has nothing to do with the liturgical signification assumed by red color in that context.

One of the most interesting chapters of Goodman's book is the one on the denotation of images, Goodman makes subtle distinctions between a "man-picture" and a "picture of a man," and asks multiple questions about the denotative modality of a painting which represents both the Duke and Duchess of Wellington. This painting could be said to denote at one and the same time both the couple and, in part, the Duke; and could thus be seen as a "two-person picture" and also, in part, as a "man-picture," yet would not be a representation of the Duke as two people... and so on. This is a challenging and even amusing set of questions that arise only if the painting is understood as the equivalent of a series of verbal sentences — and therefore, as the visual parasite of a natural language. But (except when a painting is used for identification purposes, as a passport photo, or as a historical-document) the viewers of the portrait never worry about these problems; nevertheless for them the portrait *means* something.

In the framework of visual semiotics (see for instance Calabrese 1981) one is rather concerned with oppositions concerning the framing of the image, the position of the hands, the relationship between the figure and the background, the direction of the gaze, and consequently the relationship between a portrayed person who looks outside the painting, and seems to know that he/she is being looked at by a viewer, and another one who is looking at something inside the painting; and so on. In certain artistic periods the fact that a single person, portrayed among many, looks directly into the eyes of the viewers had a precise coded meaning, since it meant that that person was the painter himself. In other cases the direction of the gaze cannot be interpreted in verbal terms (it can be verbally described but not fully interpreted); this notwithstanding, this that gaze means something, at least in the sense in which *to be or not to be* means more than the raw fact that Hamlet is certainly saying *either p or q*.

Notice that I am not demarcating a difference between the semantic and the aesthetic functions of a painting; I am saying that either we grasp the semantic modes of a visual statement, or we are forced to deny to the visible the

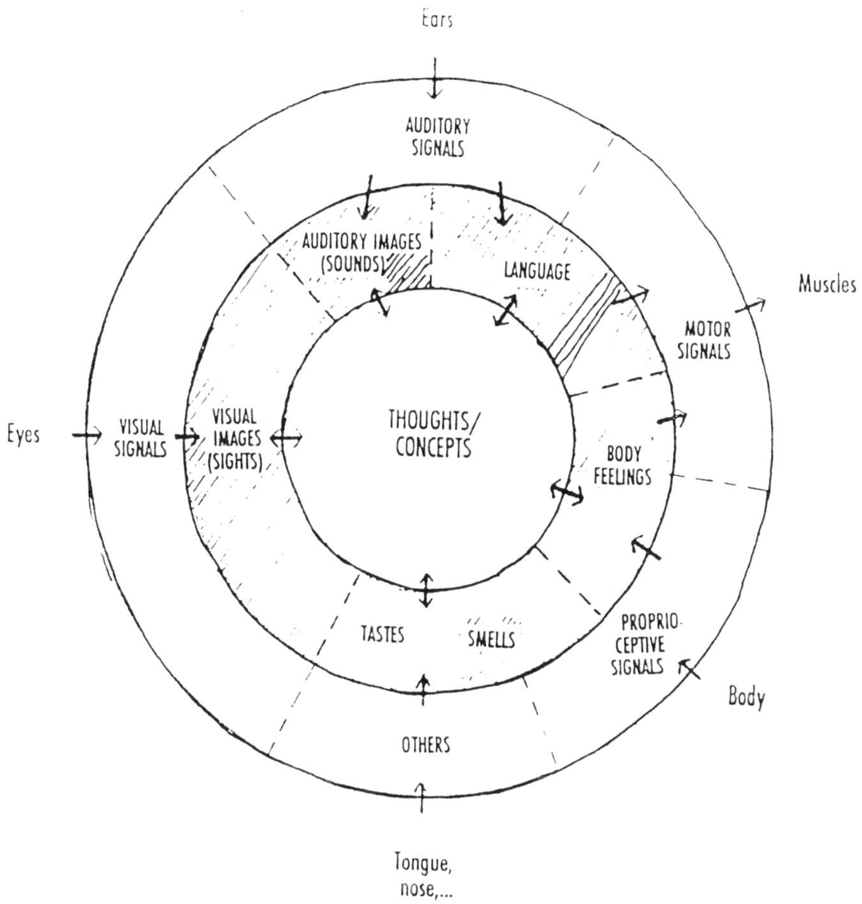

Fig. 2.

capability of communicating meanings; denying, too, that there are meanings not immediately interpretable in linguistic terms.

Recent cognitive approaches (see for instance Jackendoff) tend to undere-vualate the role played by language in cognition, as if though were "a mental function completely separate from language" which "can go on in the absence of language" (Jackendoff 1995). Thus the formation of concepts would result from a series of inputs among which verbal language plays a very partial role.

Leaving unprejudiced the role Jackendoff gives verbal language, let me remark that the other areas represented by Jackendoff are all concerned with semiosic, even though not verbal, phenomena. It is evident that the way these other inputs work is potentially different from the one followed by verbal language. From a mentalistic point of view it is certain that the verbal expression *it smells of incense* can evoke a less precise mental image than the direct experience of scent of incense. But also the word *dog* produces a mental image more general and imprecise than the one produced by the proper name of my own dog.

These differences are so blatant that it seems preposterous to waste time and paper to uncover them. I do not think however that it is trivial to see if and to what an extent these different phenomena have something in common. To do that we do not need to assume that we have mental images. The whole business can be solved on a merely cultural an intersubjective ground.

The sign that something acts as a sign is given by the possibility of interpreting it. According to Charles Sanders Peirce the word *dog* is a sign in so far as it can be interpreted by other signs, belonging to different semiotic systems, acting as the *interpretants* of the previous one. So the word *dog* can be interpreted by a verbal definition, by the ostension of a real dog, by the image of a dog, by a story referring to dogs ("a dog is that kind of animal that shows up in Lassie movies"), by the corresponding Egyptian hieroglyph or Chinese ideogram, by isolating a given position within the taxonomic tree, and so on *ad infinitum*.

Likewise the smell of a dog, or the acoustical experience of a barking dog can be interpreted in the same ways — in the very sense in which a painting representing a dark and stormy night can be interpreted either by the sentence *this is a dark and stormy night* or by quoting the famous opening lines of Snoopy's stories.

One could say that formal logic sticks to verbal language (and philosophies of verbal language to formal logic) because verbal language can be interpreted in formal terms while a sequence of sounds, images or body feelings cannot. But in Virtual Reality algorithms produce (therefore are interpretable by) tactile sensations, not to speak of images, and a scanner digitally translates images into a series of mathematical expressions.

Moreover Jackendoff, in his attempt to limit the role of verbal language in thought (Jackendoff 1987) states (quite correctly, I think) that the meaning of a word cannot only be analyzed in terms of other verbal features. No verbal definition of a human body can satisfactorily tell how it looks like: it is a matter of everyday experience that when one hears the word *man* (or, to be politically correct, *human being*) one does not immediately think of a mortal rational animal but rather of a visual stereotypical structure that in David Marr's terms (see Marr and Nishishara 1978) is called a 3D model (fig. 3).

(3) *A sample 3D model structure*

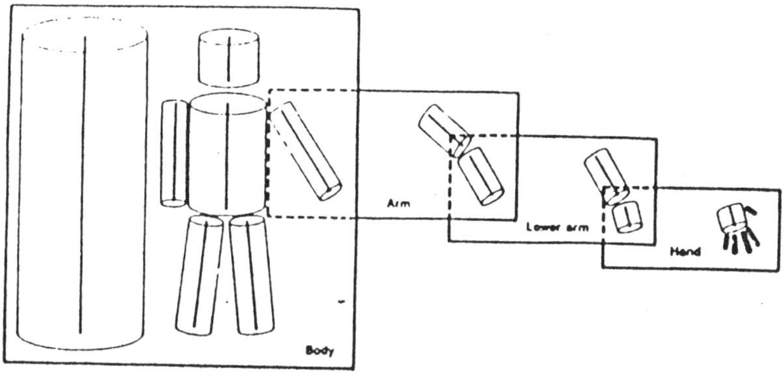

Fig. 3. A sample 3D model structure

Once again I am not so much concerned with what happens in our mind (in this case I think that we entertain a more refined visual interpretation than a simple 3D model) but with the way in which the expression *human being* can be interpreted by other semiosic means. It is certain that a 3D model seems to be a satisfactory public (not only privately mental) interpretant of *human being* — and if Champollion had interpretants of this sort associated to the hieroglyphs of the Rosetta Stone his job would had been easier.

If all this is true, then the many circles or bubbles that Jackendoff designed around the central core of thoughts (or of culturally recognized meaning) are perhaps alien to verbal language but not to semiosis as a more general phenomenon. In so far as the inputs belonging to the non verbal bubbles can be interpreted, they could also be the object of a formal calculus, in the same way as a 3D model can be geometrically analyzed and produced by an adequate algorithm. The area of phenomena that can become the object of a formal calculus, then, is larger than the area of phenomena that can be expressed by verbal statements.

2.5. The fifth dogma of the TCP is that meaning is not only a mere verbal affair but that *it can be expressed only by and in the context of verbal sentences*. Since sentences convey propositions and proposition can be interpreted in logical terms, this has more and more reinforced the strict link between analytic philosophy and logic, convincing philosophers that a logical representation is the only way to analyze meaning.

Now it is true that frequently the proper meaning of a word is explained only by a sentence. But it also true that a child can capture the meaning of a word by ostension. There are meanings that can be captured only by observing a system of opposition within the framework of a given language,

	1	2	3	4	5	6	7	8	9	et cetera
(a) Generation										
$G+2$	+	+								
$G+1$			+	+					+	
G_0					+	+				
$G-1$							+	+		
$G-2$										
(b) Sex										
m	+		+		+		+		+	
f		+		+		+		+		
(c) Lineage										
L_1	+	+	+	+			+	+		
L_2					+	+			+	
L_3										

Fig. 4.

and others that require the visual display of a conceptual space. If a child asks his mother what *sister* means, the mother can show him the meaning of the word by inserting it into a sentence (*suppose that Daddy and I have another child and that she is a girl; then she would be your sister*). But in order to know the meaning of *cousin, father-in-law* and so on, especially if we are dealing with the Bororo kinship system, an articulated visual structure of oppositions and dependencies would work better. (Fig. 4.)

Let us consider the case of syncategorematic or pseudo-syncategorematic terms like adverbs and prepositions. Try to explain by verbal instances the meaning of such words as *beneath, under, upon* and so on. The best way I know (in any case the best way to teach a non native English speaker about their use) is the one proposed by John Wilkins in his *Essay towards a Real Character* (1668) (fig. 5).

This representation displays three interesting semiotic features: (i) it is an example of visual procedure for the construction of concepts; (ii) it defines the meaning of an expression by contrast within a common field of oppositions (in this case the oppositions are soft, but in case of other lexical items the system of oppositions can be rigidly structured, as it happens with kinship relations); (iii) it can offer challenging opportunities for an analysis in terms of fuzzy logic (how much a point ought to be close to the center of the central sphere in order to instantiate a case of *within-ness*?)

Let me take another example, concerning names of countries and nationalities. Given a good taxonomy it seems easy to define natural kinds like *camel* or *horse* by dictionary-like atomic properties. For the sake of the argument I

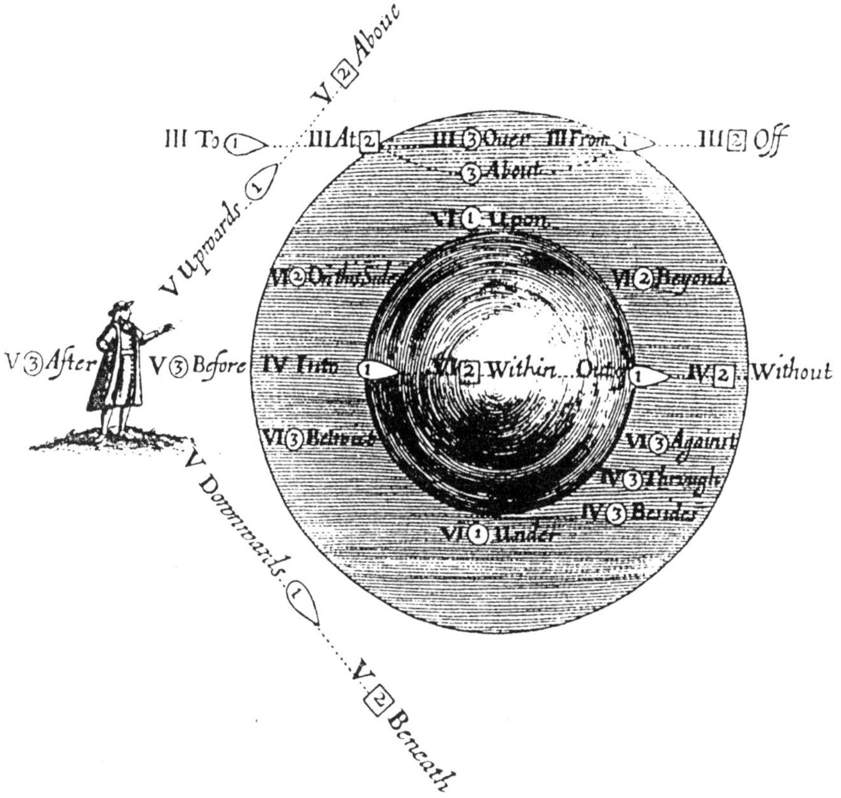

Fig. 5.

even accept the idea that proper names of persons do not have meaning but only reference (I do not believe it, I think that proper names are overcharged with meaning, since to speak properly of Napoleon in infinite contexts one should know the whole list of his properties, including parental relations, date of birth, color of hairs at the age of twenty and perhaps even his genetic code). In any case we cannot say the same for names of countries.

Try to tell somebody what was the meaning of the term *Poland* in different linguistic contexts related to different centuries. Poland is a country which underwent many political ordeals and was very broad in certain historical periods and practically disappeared in others. The changing design of its borderlines had for centuries a chewing-gum like nature. Thus, if it is possible for a tense logic to deal with expressions like

PkA (=A was true k time units ago)

if A stands for *Byalistok belongs to Poland*, in order to ascertain if A was true at a time k, we need to fix the reference of the term Poland in several different times. If the meaning of a term can be expressed only by the verbal sentences in which the term truly or felicitously occurs, then we would need the complete list of all the true statements about Poland uttered in the course of the last millennium. Usually the best way to learn the different meanings of the name *Poland* is a series of geographical representations — in each of them Poland being exactly (at a given time) what is not a contiguous country. It goes without saying that a cinematic representation, or a CD-rom would be more effective than a historical atlas.

In the last months I had many problems in understanding what a newspaper meant when mentioning Serbians — even without considering the difference between being religiously, linguistically, ethnically, politically or culturally Serbian. I understood something when I saw a map designing the geographical position of the Bosnian Serbian, Croatian Serbian and Serbian-Serbian settlements. Only at that point I could successfully disambiguate many verbal statements about Serbians. Unfortunately this kind of visual information is not enough to fix the proper meaning of various occurrences of the word *Serbian*: for instance I do not know how to call the president of the Bosnian-Serbian Republic, Karadzic, born in Montenegro and for a long time belonging to the cultural milieu of Sarajevo; and I wonder what one means when labeling as Serbian a citizen of the Serbian republic, born in Belgrade from Croatian catholic parents and unable to read Cyrillic. Meaning is indeed a matter of multimedial information.

2.6. The desperate attempt to ignore such a multimedial nature of meaning acquisition has provoked the long lasting idea that *meaning should be represented in the form of a dictionary rather in the form of an encyclopedia.* It is since 1951, with Quine's *Two dogmas of empiricism* that the canonical and straightforward division between analytic and synthetic properties has been cast in doubt; it is at least after the first experiences in Case Grammar and later in Artificial Intelligence that we know that the meaning of a word can be provided also by frames, stereotypes, and even short stories and sequences of actions. A strict hierarchy of necessary atomic properties linked down-top by relations of entailment can hold at most for scientific taxonomies. Certainly, if there was a perfect Adamic language, Adam did not give names to the furnishing of the Garden of Eden in a dictionary form (*a tiger is a feline mammal*) but rather under the form of frames (*a tiger is that kind of black striped big cat that does so and so and that I can kill or tame in such and such way*).

In spite of this, every time one tries to translate meaning into declarative sentences, the temptation of fixing at least some atomic and quasi-necessary properties seems unavoidable.

Recently Diego Marconi has proposed to analyze different sentences by scoring by D (dictionary-like) the properties that must be taken as universal, necessary and constitutive of the linguistic competence, and by E (encyclopedia-like) those that are particular, contingent and non constitutive of linguistic competence. (It goes without saying that here such terms as "necessary" and "contingent" are taken more in a pragmatic than in a logical sense). Thus *bachelors are unmarried* has 3 D properties (universal, necessary and constitutive of competence) while *George arrived yesterday at 3 p.m.* displays only 3 E properties. In an intermediate situation we can find statements like *the atomic number of gold is 79* (universal, necessary, non constitutive of competence).

I find this proposal excellent in order to make interpretative decisions in certain contexts, but the same criterion is not equally cogent in many other circumstances.

Let us suppose that the dictionary format circumscribes the purely linguistic competence, that is, the kind of competence one needs in order to understand the elementary meaning of a term (and of the sentence in which it is used) even though one does not share any other kind of knowledge about the world. Let us suppose that scientific taxonomies are a good example of dictionary-like competence: thus a cat is certainly a felid carnivorous mammal, or so, and one cannot accept a sentence like *I saw yesterday an ovine ungulate cat* as semantically well formed.

I do not think that normal human beings have the chance of speaking frequently this way, but certainly such statements can undergo a logical translation, since the asserted properties are linked by relations of entailment (if felid then carnivorous, but not if carnivorous then ungulate). This is the reason why this kind of statements is *la tarte à la crème* of the followers of TCP.

Now, a short exploration on the history of dictionaries and encyclopedias proves that in the Hellenistic and medieval bestiaries a cat was described by scattered elements of world knowledge (for instance, he was mainly the enemy of rats and had phosphorescent eyes); in the dictionaries of the XVI and XVII (like the *Dizionario della Crusca*) cats or dogs were mainly defined as "well known animal (*animal noto*)" due to the difficulty of providing a satisfactory definition, description, or instruction about how to identify the corresponding referent.

In the XVII century Wilkins (in his above mentioned *Essay*) tries to outline by allegedly taxonomic criteria an inventory of the whole universe. His criterion recalls the Porphyrian Tree of Aristotelian tradition. Wilkins designed a table of 40 Major Genera subdivided into 251 characteristic Differences. From these he derived 2,030 Species which appear in pairs. For instance starting from the Major Genus of Beasts, after having divided them into viviparous and oviparous, and after having subdivided the viviparous ones into *whole footed, cloven footed* and *clawed*, Wilkins arrives at the species Dog-kind and Cat-kind.

Figure 6 is a summary and schematic taxonomy I have extrapolated from the tables. There are recorded other distinguishing features, that is, that

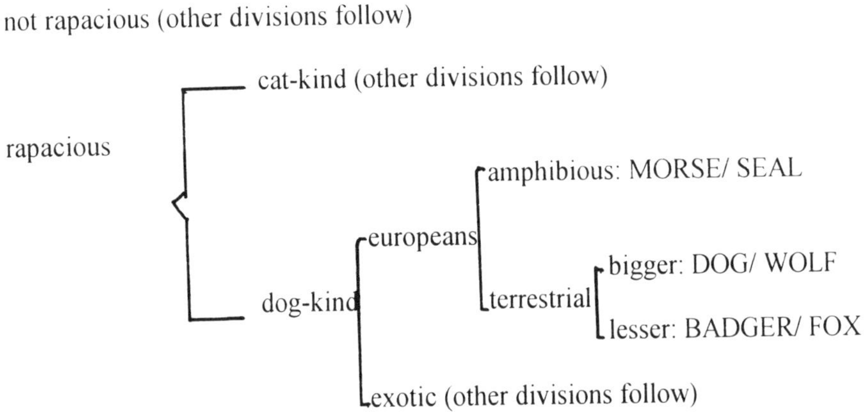

Fig. 6.

clawed viviparous animals have toes at the end of their feet; the rapacious ones have generally "six short pointed incisors, or cutting teeth, and two long fangs to hold their prey"; the head of the dog-kind beasts is oblong, while the head of the cat-kind animals is roundish; the larger of the dog-kind fall into two further groups, dogs which are "noted for tameness and docility" and wolves, noted for "wildness and enmity to sheep". It is evident that Wilkins' tables are more structured as an encyclopedia than as a dictionary, since they are stuffed with features that depend on world knowledge; in fact they are not logically structured into a system of mutual entailments (as I have tried to prove in my last research on *The Search for the Perfect Language*, Eco 1995). In any case this system does not provide any chance to tell the difference between cat and dog, and the detail concerning the form of their head is insufficient.

Acceptable taxonomies of natural kinds were established only from the XVIII century on, and only for these taxonomies a cat is definitely a felid carnivorous, the shape of its head being irrelevant.

Thus, from the birth of the Homo Sapiens to the XVIII century there was nothing similar to a dictionary-like competence. The curious predicament that follows from a dictionary-like theory of meaning is that — if such a dictionary format is constitutive of a purely linguistic competence — from Homo Sapiens to Linnaeus people were not endowed with any sort of linguistic competence. As a consequence Julius Caesar, Vergil, Dante, Machiavel, Shakespeare, Descartes were probably speaking but nobody, not even themselves, were able to understand what they were saying. Or, they were able to utter *this is my cat*, and to identify a cat among the furniture of the external world, but were unable to make paraphrases, to find synonyms or to draw inferences about cats. Since such a conclusion is absurd, this means that it is absurd to maintain that in order to understand language we must postulate that (at any cost) certain meaning features are more necessary than others.

It is not absurd to assert that, in order to support a formalized representation, a natural language must be postulated as a very simple grammar with a lexicon based upon a system of necessary features; but this is a fiction that logicians need in order to analyze basic forms of reasoning. Natural languages worked pretty well even when nobody was in the position to assign or to isolate such necessary features, and any logical analysis of a true natural language must give up with ideal languages and face the fuzzy nature of our ways of speaking.

I would like to finish my presentation with example provided by C.S. Peirce apropos of the possible meanings (or *interpretants*) of the word *lithium*.

> If you look into a textbook of chemistry for a definition of *lithium* you may be told that it is that element whose atomic weight is 7 very nearly. But if the author has a more logical mind he will tell you that if you search among minerals that are vitreous, translucent, gray or white, very hard, brittle, and insoluble, for one which imparts a crimson tinge to an unluminous flame, this mineral being triturated with lime or whiterite rats-bane, and then fused, can be partly dissolved in muriatic acid; and if this solution be evaporated, and the residue be extracted with sulphuric acid, and duly purified, it can be converted by ordinary methods into a chloride, which being obtained in the solid state, fused, and electrolyzed with half a dozen powerful cells will yeld a globule of a pinkish silvery metal that will float on gasolene; and the material of *that* is a specimen of lithium. (*Collected Papers*, 2.330)

It is difficult to distinguish, in this pseudo-definition, dictionary-like from encyclopedia-like properties. It is sure that, due to the semantic division of labor, people can understand statements about lithium without knowing its dictionary-like properties (let us say, "atomic number") as well as many encyclopedia-like properties (I trust Peirce, but I had never learned that lithium looks pinkish-silverly). I would not even be able to tell the difference between a specimen of lithium and a specimen of Uranium. But I think that, in order to describe the full meaning of the term *lithium*, we need all the above instructions, and perhaps more. In spite of this, we cannot say which of them is strictly necessary, in whatever context, in order to assume that we have correctly outlined the meaning of that word.

The whole of my remarks was not aimed at demonstrating that — as Strawson (1950) says — "neither Aristotelian nor Russellian rules give the exact logic for any expression of ordinary language, for ordinary language has not exact logic." This assertion could be accepted by every reasonable logician; moreover, if one invents a formal language in order to overcome the unavoidable ambiguities of natural languages one has decided that no ideography can account for the whole of linguistic usages and that there is not, for instance, a formal logic of poetry. In spite of this evident difference between formalized and natural languages, I am convinced that the study of natural languages has been enormously enriched by the efforts of logicians, from Aristotle's *Prior Analytics* to contemporary modal logic. In an old paper of mine I demonstrated that in order to criticize many irrationalistic approaches to philosophy or politics (even in order to make evident the weakness of the arguments used by the *Protocols of the Elders of Sion*) the *modus ponens* is still a strong and indispensable intellectual weapon.

What I tried to suggest is that, probably, the unbreakable marriage between formal logic and the TCP in philosophy of language can have refrained some logicians to face certain questions that are and should still be within their range of interest and competence, but that cannot by made evident by an analysis of verbal language performed by mere verbal means and in crude truth conditional terms.

Maybe tomorrow semiotic analysis of an interactive, hypertextual and multimedial CD-rom could raise more logical problems than so many reflections on the baldness of the present king of France.

Milano

REFERENCES

Calabrese, Omar 1981 "La sintassi della vertigine: Sguardi, Specchi e Ritratti". *VS* 29.

Eco, Umberto 1995 *The Search for the Perfect Language*. Oxford: Blackwell.

Goodman, Nelson 1968 *Languages of Art*. New York: Bobbs-Merrill.

Jackendoff, Ray 1987 "On Beyond Zebra: The relation of linguistic and visual information". *Cognition* 26.

Jackendoff, Ray 1955 "How Language Helps Us Think". *Pragmatics and Cognition* (forthcoming).

McCawley, James D. 1981 *Everything that Linguists have Always Wanted to Know about Logic*. Chicago: University of Chicago Press.

Marconi, Diego 1995 "On the Referential Competence of Some Machines". *Artificial Intelligence Review* (forthcoming)

Marconi, Diego 1995 "On the Structure of Lexical Competence". *Proceedings of the Aristotelian Society* (forthcoming)

Marr, D. and Nishishara, K. 1978 "Visual information processing: Artificial intelligence and the sensorium of sight". *Technology Review* 81.

Quine, W.V.O. 1951 "Two Dogmas of Empiricism". *Philosophical Review* 50.

Strawson. P.F. 1950 "On Referring". *Mind* NS 59.

DAGFINN FØLLESDAL

SEMANTICS AND SEMIOTICS

The relation between semantics and semiotics might seem straightforward: semantics is the study of the meaning and reference of linguistic expressions, while semiotics is the general study of signs of all kinds and in all their aspects. Semiotics comprises semantics as a part. Charles Morris (whom Jens Erik Fenstad mentioned in his opening speech) in *Foundations of a Theory of Signs*, one of the volumes of the Encyclopedia of Unified Science, in 1938, divided semiotics into three branches: syntax, semantics and pragmatics.

There has been astonishingly little contact between semanticists working within philosophy and linguistics and semioticists. As my co-symposiast Umberto Eco just mentioned in his paper, he almost feels out of place here in a gathering of philosophers. Philosopher semanticists are studying in great depth questions concerning meaning and reference, in particular those that relate to indexicals, demonstratives and other singular terms. This vast literature tends to be only cursorily referred to by semioticists. Semioticists, on the other hand, have brought to light a large variety of communicative phenomena that seem to be unknown to semanticists.

It is my opinion that this broad field of communication processes is highly relevant to semantics. And conversely, semantics has much to contribute to semiotics. The main aim of my paper is to show this through some examples that have also been a central concern of my co-symposiast: (1) my main focus will be on encyclopedia-oriented semantics, in particular applied to singular terms. Towards the end I will briefly discuss also (2) the nature and role of the background, and (3) perception and meaning.

1. ENCYCLOPEDIA-ORIENTED SEMANTICS

Let us consider encyclopedia-oriented semantics first. Among semioticists as well as among philosopher semanticists there are many different views concerning the appropriate approach to semantics. On both sides, the traditional view is that the meaning of a term can be given by a limited set of dictionary entries. The total information connected with a term makes up a much larger set of sentences, potentially infinite, a sample of which we find in encyclopedias. The sentences in the dictionary are presumably analytic while those in the encyclopedia are synthetic.

The analytic/synthetic distinction has, as we know, been challenged by Quine, and with it the distinction between dictionary and encyclopedia. My co-symposiast and I both agree with Quine in rejecting these distinctions. The

M.L. Dalla Chiara et al. (eds.), *Structures and Norms in Science*, 449–457.

basic question is: What form should semantics take when the analytic/synthetic distinction is rejected?

There is agreement that singular terms have reference, except for odd cases of vacuous singular terms, and that they refer to ordinary objects, concrete and abstract. Where disagreement sets in, pertains to meaning and to how reference is determined. What is meaning? Does meaning play any role in the determination of reference? What connection, if any, is there between the meanings of singular terms and meanings of other kinds of terms?

Charles Saunders Peirce's definition of lithium, one of my co-symposiast's favorite examples, hints in the direction of an encyclopedic notion of meaning: anything that gives us information about the object(s) that an expression applies to, is part of that expression's meaning. The information may be theoretical, or it may be practical, it may help us develop a certain practical ability that, for example, enables us to produce a specimen, or put a specimen to use.

My co-symposiast has argued that the meaning of a term is the collection of all this information:

> I can speak in terms of significance or content without resorting to any Platonic or mentalistic notion of meaning. In order to say what a classical Greek author meant by *kairos* I have only to list all the definitions of that term, all the myths that speak about such human possibility, all the tragedies that represent human beings facing unique opportunities, and the like. The content of the term *kairos* is the ideal ensemble of all the interpretations given of that concept in the most divergent ways and by different semiotic means. To satisfactorily understand the word kairos one must resort to the whole encyclopedia of Western culture.
> (Umberto Eco, "A View From Elm Street." In *On Quine*, edited by Paolo Leonardi and Marco Santambrogio. Cambridge: Cambridge University Press, 1995, pp. 28–29.)

I agree with Eco. However, there are complications. Let me mention four:

(i) Given that most terms require for their understanding the whole encyclopedia of our culture, some **structuring** and **weighting** within this encyclopedia is needed in order to assign different meanings to different terms. Eco has discussed some of the elements that must be taken into account in such an encyclopedic representation, such as frames, scripts, case grammar, instructional semantics, presuppositions, experts' knowledge, stereotyped knowledge and the Porphyrian tree. (Umberto Eco, *The Limits of Interpretation*, Bloomington: Indiana University Press, 1990, p. 238; see also Umberto Eco, *A Theory of Semiotics*, Bloomington: Indiana University Press, 1976.) However, there are also **normative** factors that enter into the structuring of encyclopedic representations. I shall come back to this in my discussion of the semantics of singular terms.

(ii) Not only expressions, what Peirce calls 'interpretants', are needed to understand a word. Interpretation also involves nuances of stress, tone, facial and bodily expression, etc., as well as activities, habits and skills that are never expressed in words, and that we often cannot put into words. All these, partly non-linguistic, **activities** that play a role in communication are important for understanding linguistic expressions.

The great diversity of such evidence is a main reason why semantics cannot be separated from the other aspects of language: pragmatics, syntax and even phonetics. Semantics must be developed jointly with these. This has also been stressed repeatedly by Eco, and is hence another point of agreement between us.

(iii) One kind of activity that is highly important for meaning and communication is **perception**. Peirce claimed that the meaning of one expression can be expressed only by another expression whose meaning must be expressed by another expression and so on ad infinitum. "Lo, another infinite series!" Peirce exclaimed. However, there are not expressions all the way down, there is also perception, which plays an important role in the establishment of meaning. I shall return to this in discussing perception and meaning.

(iv) All these encyclopedic resources, together with the activities that play a role in communication, are important not only for understanding linguistic expressions, but also for **constituting** their meaning, or bringing about that they have the meaning they have. According to what I have called the MMM thesis, or the thesis of **man-made meaning**:

> Meaning (and the reference relation) is a product of public interaction.
>
> That is, meaning, unlike nature, was not there before public interaction began, waiting to be discovered, but it has been produced, and is continually being produced, through this interaction. [This production is a co-operative enterprise, where previous generations and specialists on various subjects (theoreticians and practitioners) have done their part, and created the semantic construction that the learner strives to master. The production process still goes on, through our introduction of new expressions and through our using old expressions in new ways, as for example in metaphors.]

This thesis is the main reason that translation is indeterminate and not just under-determined. Where all the evidence that can guide interpretation leaves off, there is leeway for alternative interpretations (under-determination). Given that the evidence **constitutes** meaning, there is no right or wrong in the choice between these interpretations: all interpretations that are compatible with all possible evidence are right (indeterminacy).

1.1. *Semantics for singular terms*

Many semioticists and semanticists ignore the distinction between singular terms and general terms. Others, like Frege and Carnap, aim for a unified semantics where singular terms, general terms and sentences are treated on a par: they are all assimilated to singular terms (Frege) or to general terms (Carnap). However, in order to get a satisfactory semantics for non-extensional contexts singular terms have to be treated differently from general terms. Terms that warrant inferences that are specific for singular terms (substitution, existential generalization, universal instantiation) have to be rigid designators, they must refer to the same object in all "possible worlds" where they refer at all.

1.2. *Causal theories of reference and their problems*

It is tempting to insure this by means of a causal theory of reference, since the causal relation is not affected by mistaken beliefs and other vicissitudes that perturb other potential reference relations. In fact, there are many who seem to think that notion of a rigid reference and the causal theory of reference are basically the same. However, it is important to distinguish the two. While I have argued, in my Ph.D. dissertation from 1961, that rigid designators are needed in order to make sense of quantification into non-extensional contexts, I think that there are decisive problems with causal theories of reference. I shall mention four:

(i) Some of them (not Kripke's, but, for example, Evans') require the causal relation to go all the way back to the object referred to, which obstructs reference to **abstract** entities and **future** events and objects.

(ii) Proponents of causal theories of reference, perception, etc., talk about causal **chains**. However, causal processes usually depend on multiple factors, that make it more appropriate to talk about causal trees. It then becomes highly unclear which of the many nodes in such trees our terms refer to.

(iii) Change of reference. It sometimes happens that singular terms **change** their reference. 'Madagascar' is a well-known example. Causal theories are notoriously unable to cope with such changes. Their proponents try to explain such changes by saying that a new baptism has taken place: the term is intentionally given a new reference. However, this was not the case with 'Madagascar', where one intended to refer to the old reference. And it does not fit numerous other cases of reference change due to confusion and mistaken identity.

(iv) Semantics, I maintain, rests upon **epistemology**. If the MMM thesis is right, the meaning and reference of linguistic expressions is a product of evidence that is publicly available in a linguistic community. Neither meaning nor reference is taken care of by some ontological relation, such as a causal relation, that guarantees meaning and reference independently of human interaction, human fallibility and evolving scientific theories.

1.3. *A normative theory of reference*

A natural way out of these quandaries is to revert to Frege's old view that singular terms have a **meaning**.

However, I will propose two changes from Frege:

(i) First, and here I agree with Eco, the meaning of a term, singular or general, is not expressed by some analytic definition, but consists of all the information we connect with the term, whether we find it in an encyclopedia or a dictionary, together with all the practical activities that help us find out what the term is true of or refers to.

(ii) Secondly, while Frege held that meaning determines reference, it is the **reference**, not the meaning, that concerns us in the normal use of singular terms. In using a singular term we take upon ourselves a commitment to keep on referring to the same object throughout the vicissitudes of changing insight and a changing world. The meaning of a singular term is designed to keep track of the reference through all these changes. While for general terms, such as 'red', their meaning is what matters, and their extension varies with circumstances, for singular terms it is their reference that matters and which dominates their meaning in a complex interplay with our changing theories and circumstances.

This interplay reflects the holistic inseparability of theory and meaning that underlies encyclopedia-oriented semantics. It also manifests the **rigidity** of singular terms. Rigidity, as I conceive of it, is a regulative idea, in the Kantian sense, a **normative** notion. There is no guarantee that singular terms preserve their reference, as the causal theories would insure. My argument for rigidity that I mentioned earlier, shows only that **if** quantification and other types of inference involving singular terms is to make sense, **then** the singular terms have to keep their reference from world to world and from time to time. If they cease referring to the same object, as they sometimes do, then quantification involving these terms ceases to make sense.

1.4. *Theories of identity*

Given that the meaning of a singular term is designed to help us keep on referring to the same object, some features of the object are more important to the meaning than others, mainly those features that enable us to recognize and keep track of just **this** object as distinguished from others that are easily mistaken for it, for example because they are very similar. For each kind of object, for example persons, the meanings of the singular terms referring to them have to contain as a main ingredient an explicit or tacit theory of identity for such objects.

My conclusions concerning singular terms is hence that they have meaning, that this meaning is encyclopedic, that it involves lots of tacit features and practical activities, and that its structure is dominated by factors that help us preserve the reference. Preserving the reference is the key commitment which we take on when we use singular terms. However, there is no guarantee of success. Rigidity is a regulative idea, and my view on reference can therefore be called a **normative** view as opposed to a causal view.

I shall now discuss briefly the remaining two items on my agenda: the background and perception. First the background.

2. THE BACKGROUND

In talking about the theory of identity that enters into the meaning of singular terms, I said that it might be explicit or tacit. The notion of tacit, or background knowledge is crucial in semantics, as well as in all other branches of philosophy. It is best known to analytic philosophers through the work of Polanyi and his followers, but Edmund Husserl had more to say on this topic than anybody else I know of. Umberto Eco also goes into this notion in several of his works, but he is drawing on other traditions (although he refers to Husserl in connection with perception, which we are coming to).

Here are some points from Husserl's theory of the background which I regard as particularly relevant for semantics and semiotics:

When we are experiencing an object, our consciousness is focused on this object, the rest of the world and its various objects are there in the background as something we "believe in" but are not presently paying attention to. The same holds for most of the inexhaustibly many features of the object itself. All these further features that we explicitly or tacitly attribute to the object, together with the world in which we take it to be, make up what Husserl calls the **background**, or the **horizon** of that experience. The various features of the object, which are co-intended, or also-meant, but not at the focus of our attention, Husserl calls the **inner horizon**, while the realm of other objects and the world to which they all belong, he calls the **outer horizon**:

To take a simple example of an item belonging to this outer horizon. If I had asked you as you were entering this room, what your expectations were, you might mention something about friends you expected to meet, a lecture you expected to hear and so on. It is highly unlikely that you would mention that you expected there to be a floor in the room. Yet, as I saw you confidently stepping in, I have every reason to believe that you expected that there would be a floor here. You did not think about it, your attention was directed to other things, but you had a disposition that you acted on, and also, if I had asked you whether you expected there to be a floor in the room, you might have wondered why I asked such a trivial question, but you would probably have answered yes.

Expectations and beliefs are dispositional notions. We count as beliefs not only thoughts that we are actively entertaining, but also those that we rarely think about, for example, that $2 + 2 = 4$. We do have a problem when we try to delimit exactly what beliefs we have. The method of questioning is not reliable. On the one side, it gives too much, remember how in the *Menon* a skilled questioner uncovers that the slave boy has the most unexpected geometrical beliefs. Plato took this as evidence for his theory of *anamnesis*. On the other, it yields too little. As Freud and others have taught us, we often sincerely deny that we have beliefs that seem all too apparently to underlie our actions.

The most reliable criterion, that we often fall back on, is to assume that people have those beliefs that best explain their actions, including their verbal activities. However, then a further problem is that the states we appeal to in order to explain people's actions are not exclusively cognitive states. Also various physical states are needed, and skills of various kinds that it is often hard to classify as mental. Thus while our arithmetical skills are presumably mental, our skills in swimming or walking can hardly be classified as mental. Then we have tricky intermediate cases, such as one's keeping a standard distance to partners in a conversation, where the standard may vary from culture to culture. Is keeping this distance a matter of a tacit belief that it is the proper distance? Or is it a matter of a bodily skill which is gradually acquired as one grows up in this culture? And what about the way we sign our name? Obviously, cognitive activities are involved in the process that brought us to sign it the way we do, we had to learn the alphabet, we had to learn our name, and so on. But also, in our semi-automatic way of signing it, bodily skills are involved to a great extent. Various personality traits play a role, as do certain general traits of our culture.

Opinions concerning examples such as these may vary, one of our problems is that we have no clearcut way to settle such issues, lacking a precise definition of what is to count as mental, what as physical. However, there is obviously an interplay here, both in the process that leads to the skill and in the skill itself, and any satisfactory theory of intentionality must heed such an interplay. We have to take into account in the background not merely the anticipations involved in seeing, hearing and the like, but also the anticipations involved in kinestesis and bodily movement, where when something "goes wrong" we are aware of it. We are familiar with this experience of "going wrong" from cases of misperception: we cannot always tell exactly what went wrong, but we are aware that something went wrong.

The horizon is of crucial importance for Husserl's concept of justification. What is particularly significant, is the hidden nature of the horizon. As we noted, the horizon is that which is not attended to. Usually, as in the case of the floor in this room example above, we have not even thought about it. Typically, we cannot even recall when we first acquired the corresponding "belief" or "anticipation." According to Husserl, there may never have been any occasion when we actually judged there to be a floor in some particular room. Still we have come to "anticipate" a floor, not in the sense of consciously expecting one, but in the sense that if we entered the room and there were none, we would be astonished. In this example we would easily be able to tell what was missing, in other cases our "anticipations" are so imperceptible that we just may feel that something has gone awry, but we may not be able to tell what it is.

Words like 'belief' and 'anticipate' are clearly not the proper ones here, since they have overtones of something conscious and thematized. Both English and German seem to lack words for what we want to get at here: Husserl uses the words '*antizipieren*' and '*hinausmeinen*' and also '*vorzeichnen*'.

The notion of background is important for semantics and semiotic in many ways. The encyclopedias of encyclopedia-oriented semantics contain not only thematized information. they consist largely of never-thematized and never-formulated items. They incorporate skills and activities that we do not even know how to formulate. To call them encyclopedias might therefore be misleading, we usually think of encyclopedias as books.

The notion of background is also very useful in analyzing presuppositions. Much has been written on the notion of presupposition, by semioticists, philosophers and linguists. Eco, in particular, has proposed a highly interesting treatment of presuppositions, in his 1987 article "Presuppositions", reprinted in his volume *The Limits of Interpretation* (1990), to which I referred earlier. Here Eco makes use of a distinction between background and foreground knowledge possessed by various participants in a conversation and the background and foreground information that is conveyed by an expression. The latter notion stresses not what is previously known, but what is assumed as unchallengable by the participants (p. 228). Eco also discusses the power presuppositions have to create new contexts (p. 232).

3. PERCEPTION

Finally a few words on **perception**. According to Husserl, every act has a **noema**, a complex structure of anticipations through which we experience objects. An act of perception is not a passive reception of physical impulses, but an active structuring of our surroundings into objects, with properties, with relations to other objects and with a spatio-temporal setting. This structure prescribes how the object will look from other angles, how it will feel, taste, smell, etc. The object is experienced as having a plurality of profiles, of *Abschattungen*, through which the object can be experienced in ever new ways, if the object is turning around or moved, or if we move around or use other of our senses. This inexhaustibly rich structure is akin to the meaning of linguistic expressions. As Husserl puts it in a manuscript to a projected third volume of his main work, *Ideen*,

> "the noema is nothing but a generalization of the notion of meaning (*Bedeutung*) to the field of all acts." *Ideen* III, 89, 2–4.

In my opinion, this close tie between perception and linguistic meaning throws light on many of the issues discussed in semantics and semiotics. Thus, for example, the meaning of a term is not a meager definite description that contains just enough information to fit a unique object. Instead it is like the noema of an act. The meaning, like the noema, is a rich structure which delineates an inexhaustible number of features of the object, most of which we have never thought about and will never think about. In predicative judgments we can affirm or deny that the object has various features. It is through this affinity between linguistic meaning and noema that there is continual interanimation between perception and meaning. It is also through this affinity that language ties on to the world. And it is this affinity that brings us out of the infinite series of expressions that Peirce thought was all we could fall back on in explaining meaning. It is this affinity between linguistic meaning and noema that brings these series to an end.

Departments of Philosophy
University of Oslo
and
Stanford University

Symposium 3
Logic and Philosophy of Science in the Far East

LOGIC AND PHILOSOPHY OF SCIENCE IN THE FAR EAST

Dear colleagues and friends,

I am glad to see that the 10th Congress of the DLMPS organizes a Special Symposium on Logic and Philosophy of Science in the Far East for the first time in the history of the Division. This is a very important event.

Logic and philosophy of science originated from the West. About one century ago they were introduced into Asia. In China this happened in the late 19th century when John Stuart Mill's work was translated and published in China. Now in almost every Asian country there are logicians and philosophers of science who are teaching and studying in universities or institutes.

Several international conferences on logic or the philosophy of science have been held in this region in recent years, and logicians and philosophers of science in this region have been more and more involved in international dialogues with colleagues of other regions. After many years of experiencing ups and downs, bitterness and sweetness, I think it is time to examine how logic and philosophy of science are developing in a culture that is different from the West: what topics are in focus, how are they approached, is there any distinctive feature, what are the problems or difficulties facing them etc.?

Logic and philosophy of science as an international endeavor should be spread to all cultures, and their development in the Far East should be promoted and supported. I hope that in the near future, there will be more meetings to be held in the Far East, and more exchanges between logicians and philosophers of science in this region and their colleagues from all over the world. It is very important not only for the development of logic and philosophy of science, but also for the understanding and friendship between them.

Renzong Qiu
Chinese Academy of Social Sciences
Department of Philosophy of Science and Technology
Institute of Philosophy
Beijing

M.L. Dalla Chiara et al. (eds.), Structures and Norms in Science, 461.

WANG LU

LOGIC IN CHINA

As is well-known, China is one of the three sources from which logic has originated. In ancient China logic was developing at about the same time when Aristotle was developing his logic. But it is less known what is the actual situation of logic research and development in China nowadays. In this talk I will try to give a general description of what is going on among our Chinese logicians.

I

One can divide the people doing logic research in China mainly into two parts, one includes teachers in the universities and the other involves logicians in special research institutes.

There are logic courses in the universities, especially for students in departments of philosophy, and also for students in the departments of literature, law, Chinese, and so on. But in general what the students study is not modern logic, but traditional logic. In the past fifteen years, there has been a reform of teaching materials. As a result, in teaching traditional logic some knowledge is introduced of modern logic, such as truth tables, basic ideas of the formal language of first order logic, etc. However, as the main body of textbooks of logic it is still traditional, students cannot get a systematic training in modern logic. In some universities, first order logic serves as a selected course for students, and since 1987 a logic section has been set up in the department of philosophy of Beijing University in which all the students follow a series of courses in modern logic, such as first order logic, modal logic, set theory, proof theory, philosophy of logic, and so on. It is clear that logic teaching is being improved in the universities.

As for the institutes where one can do special research without any teaching duties, I would like to introduce to you my own institute as an example, that is, the Institute of Philosophy, Chinese Academy of Social Sciences. There is a logic department that has now eight people. We do logic research according to individual interests. In general we concentrate on the following subjects: philosophical logic, philosophy of logic, history of Chinese logic, history of western logic, logic of natural language, etc. Sometimes we are also invited to go to universities to give some lectures, but generally only on some special problems and issues, rather than basic courses.

There is an association in China for all those who are engaged in logic teaching and research, with the title "The Chinese Association of Logic". It

M.L. Dalla Chiara et al. (eds.), Structures and Norms in Science, 463–472.

was founded in 1979 and now has more than 1000 members working in different provinces and cities. Every year one or two conferences are organized, so that the logicians can come together from various parts of China to have academic exchanges. In order to improve the level of logical training and research, since 1993 the Chinese Association of Logic has had an informal journal with the name “Logic Research”. It is informal because we can only have one issue come out each year owing to financial problems. But it is the only single special journal for logic. And we are making efforts to overcome the difficulties and turn it into a formal one.

Since 1987 the Chinese Committee of Logic, Methodology and Philosophy of Science has been founded by the Chinese Association of Logic and the Chinese Association of Dialectical Research. It participates in the Division of Logic, Methodology and Philosophy of Science, and the International Union of History and Philosophy of Science. Thus Chinese logicians have started international academic exchanges with logicians all over the world.

II

As for logic research itself, it should be admitted that we are, generally speaking, not advanced by international standards. But we have done a lot of work and made some achievements. In the following I will try, with an emphasis on logical research in connection with philosophy, to give a general description of current logic research in China.

One issue that attracted many people is the logic of natural language, or natural logic as it is called in the West. Professor Zhou Liquan holds that one's thinking takes place in natural language, so that formal logic should be connected with natural language, that is, we should study the logical forms in natural language in connection with grammatical rhetoric. He put forward this idea in the seventies and states it in great detail in his book *Logic: correct thinking and successful communication* (1994). He holds that a declarative sentence has two elements: one is the speech act of assertion, the other is the content or the object of the assertion, i.e. proposition. Thus a declarative sentence is not equivalent to a proposition. The logic of natural language is different from both traditional logic and mathematical logic in the sense that the latter concern themselves only with relations of truth and falsity between propositions, while the former deals with not only truth and falsity between propositions, but also the relations of truth and falsity between various kinds of sentences involving speech acts and propositions, such as declarative sentences, interrogative sentences, imperative sentences, and so on.

Logical research into natural language is useful in many respects. Here I shall only state two points. One is about the ambiguity of language, the other about the verb “be”.

II.1

A sentence of natural language can be ambiguous both in its structure and in its terms. For example:

1. Li loves his father, and he loves science more than his father.

It might mean

1'. Li loves his father, and he loves science more than he loves his father.

or

1". Li loves his father, and he loves science more than his father does.

So the sentence 1 is equivocal because 1' is different in meaning from 1". Here is another example:

2. A Baitouweng spoke at the meeting.

It is equivocal as *Baitouweng* might mean a kind of bird with white hair on its head (the Chinese bulbue) or a white haired old man. To solve these sorts of problem one must take the context into consideration. Thus a proposition can be seen as a function of context. In this way one might carry on a series of reasoning. For instance, for 2 one might proceed as follows:

a "Baitouweng" means a bird or a white haired old man.
b A bird cannot speak at a meeting.
c "Baitouweng" here refers to a white haired old man.
d Sentence 2 expresses the proposition "A white haired old man spoke at the meeting".

Since Zhou put forward his ideas many articles have been published in this respect, and very recently, even a book entitled *An Introduction to the Logic of Chinese Language* (by Chen Zongming, 1994) has appeared. In my view, it is just a beginning in the logic of the Chinese language, and there are a lot of problems too. We cannot say that we have achieved important results. Nevertheless we do arrive at a better understanding of the logical features of some constant linguistic elements in natural language, especially in Chinese. Let us take quantifiers as an example.

The theory of quantification is one of the important results of modern logic and with it we can analyze very complex quantifiers. But in natural language quantifiers are far more complex than what the theory of quantification provides of. First of all, it is very common for a Chinese sentence to contain no quantifiers. For instance

3. Friends are very friendly.

In this sentence the quantifier is omitted, but it is habitually taken to mean either "Every friend is friendly" or "For anything, if it is a friend, then it is friendly". So, in general, such a sentence can be viewed as a universal one.

4. Workers went out together.

In this sentence the quantifier is omitted too, but with the word "together" it means "All the workers went out". Thus from 3 and 4 it follows that whether a sentence is universal or not is not only determined by the quantifiers.

As for the quantifiers themselves in natural language, we can also see a great deal of differences. For instance:

5. Anyone (Renhe ren) who heard it was frightened.
6. All those (suoyou ren) who heard it were frightened.

"*Renhe*" (any) and "*suoyou*" (all) are words for universal quantifiers in Chinese. 5 and 6 can be, from the point of view of modern logic, analysed in the same form as

$$\forall x(\text{heard}(x) \longrightarrow \text{frightened}(x))$$

But there is a difference between the two sentences. In general, 6 means

> "There were some people who heard it and all those people were frightened"

while 5 literally might mean

> "If there were one who heard it, then he would have been frightened"

But it could be that there is no one at all who heard it. Thus the two words "Renhe" and "suoyou" differ in expressing universal meanings. And their difference can be seen more clearly when they appear in predicate position. For example:

7. This book will not harm any (renhe) student.
8. This book will not harm all (suoyou) the students.

According to a simplistic theory of quantification both of them might be analysed as follows:

$$\forall x(\text{student}(x) \longrightarrow \text{harm}(\text{book}, x))$$

In fact, however, 7 means that there is no student who will be harmed by this book, while 8 means that it is not the case that all the students will be harmed by this book, but it does not exclude the case that some students will be harmed by this book.

From the examples we can see that quantifiers in natural language are often different and more complex than logical ones.

II.2

In studying the logic of natural language we can obtain a better understanding of the relations between logic and language. Let us take the copula "be" as an example.

In studying the copula "be", Aristotle established his theory of predicables; logicians in Middle Ages set up the theory of suppositio; and modern logicians have shown that in a sentence like "S is P", "be" expresses at least the following relations:

a membership
b identity of individuals
c inclusion
d existence

To sum up, the usage of "be" in Indo-Europese languages is generally normative and follows definite grammatical rules, though the meaning of the term itself exhibits ambiguity.

On the contrary, the usage of the term "be" in Chinese is quite different from that in Indo-Europese languages. In ancient China, there was no copula "be", so logic in ancient China is quite different from that of Aristotle. And in modern Chinese, that is different from ancient Chinese, the verb "be" has by itself no ontological meaning at all, and its usage, except from being used as a copula, is always in an anomalous way. Let us examine the following sentences:

9. "*Shi ni de dong xi, wo jiu bu yao.*"
("Is your thing, I never ask for")
("I never ask for the thing that is yours")
10. "*Shi ni chuang de huo, hai shi ta chuang de huo?*"
("Is it your fault or his?")
11. "*Men qian shi shan.*"
("There is a mountain outside the door.")
12. "*Shi you na mo yi hui shi.*"
("It is true as it is")
13. "*Jiu shi shang dao shan, xia hou hai, wo ye zai suo bu ci.*"
("No matter how difficult and dangerous it is, I will try my best.")

It is quite obvious that "be" (*shi*) in these sentences has different meanings. In sentence 9 it functions as an expression of a quantifier, i.e. "For anything, if it is yours, I never ask for it". In 10, "shi ... hai shi" is equivalent to a disjunction, i.e. "Either it is your fault, or it is his fault". In 11 it means existence, i.e. "A mountain exists outside the door". Then in 12 and 13 it expresses the meaning of the sentence with an emphatic mood, but the "be" in sentence 12 also has a meaning of assertion, it equals "P, this is true".

Moreover, although the verb "be" in Chinese has gradually become a linking verb and modern Chinese grammatical books also definitely describe and define its function as a copula, in our daily language it is very often absent. For example, "His cheeks (are) like rose", "Crowds of people (are) on the street". In fact it is not easy to study the verb "be" in Chinese.

From these examples we can see that "be" in Chinese has many abnormal features. In short, it follows from the study here that the research of natural language is different from that of logic. What logic studies is the logical sense of the verb "be", which serves as a copula in the sentences "S is P", and therefore has a definite grammatical function. As modern Chinese also has this linking verb in its language now, it is possible for us to deal with it as a copula. But because of the intrinsic characters of Chinese, we often face two problems. One is that not all uses of "be" belong to the field of logic, that is, many "shi" are not copulas. In this case it is necessary to analyze different meanings of "be", so that we can tell what belongs to logic and what not. The other problem is that very often the verb "be" is not needed in a sentence, but for such a sentence the logical structure still remains. In this case, to reveal its logical structure we must try to bring out the corresponding elements in the sentence. This means that mere logical analysis is not enough for our study. What we need here is not only logical analysis, but also linguistic analysis.

It follows also from the above that logic has close links with language. First, logic deals with reasoning, but not with language. It handles the problems in reasoning through analyzing linguistic factors in sentences, but it does not solve language problems for the purpose of language. The necessity of reasoning in the sense of logic is determined by the form of the sentences and has nothing to do with their content. So one has to look for the most stable linguistic factors that have definite grammatic functions and can be generally applied, when one begins logical research with the analysis of language.

Second, a logical theory which is formed through research into a certain language as an object must reflect the character of that language and contain the limitations of that language in time. For example, Aristotle's logic has the character of Greek and it is limited by Greek grammar, so that it is not the same as logic in ancient China, and above all, it does not reflect the character of Chinese. One of the features of modern logic is the use of formal languages, which is a fundamental difference with natural language. So modern logic is different from Aristotle's logic. But it is a common instrument for all of us. Although natural language is very complex and formal languages have limitations as well, no one will give up their use and thereby go back to Aristotle.

III

Since China is historically one of the three sources of logic, the history of Chinese logic is a very important subject. It is generally recognized that the "Mo-Jing" (Mohist Canon) logic is the high point in the development of logic in ancient China. So research into the history of Chinese logic is mainly that of "Mo-Jing". There have been a great number of books and papers published in this field over the past twenty years. In my view it can be divided in two periods: before and after the end of the eighties. In the first period great effort was devoted to sorting out the material, digging out the logical problems, and making scientific assessments. In the second period, a reflection is made on the conclusions of the first period, including sharp criticism. In order to make the introduction easier, I will concentrate on the argumentation in this field.

"Mo-Jing" logic is said to be the first scientific and complete system of logic in ancient China, to be as good as Aristotle's logic, and to have achievements comparable to that in Greek Antiquity, etc. There are also some people querying these points of view. There are two problems here. One is about the interpretation of Mo-Jing logic; the other is about the comparison between Mo-Jing logic and Aristotle's logic. We first turn to the interpretation of Mo-Jing logic and especially to three main problems, that is, "Bian", "Ci", and "Shuo".

The interpretation of "*Bian*" (distinction or disputation), which is said to mean logic, depends on the interpretation of "Bi". It is most generally thought that "*Bi*" (that) means a pair of contradictory propositions, or the things that the subjects of the propositions represent, so that "zheng bi" (arguing for that) means the argument for the truth and the falsity of a pair of contradictory propositions with the same subject. But there is an opposite view that holds that these interpretations have a common point, that is, they try to interprete "bi" as something meaning contradictory propositions or in relation to contradictory propositions, i.e., the subjects of contradictory propositions. It seems that only in this way one could bring "Bian" into relation with forms of thinking. The defect of these interpretations lies in the fact that the interpreters have as a presupposition the consideration of the relations between propositions, of subjects and predicates, etc. But we cannot find convincing evidence in Mo-Jing that shows the ancient logicians deal with these problems as their interpreters do, one always has the feeling that these interpretations are carried on according to a definite pattern but not to the text.

"*Ci*" is generally interpreted to mean judgment, sentence or proposition. Here I will only deal with the problems in interpreting "universal proposition" and "particular proposition".

Some people hold that in ancient China the universal propositions are expressed by means of the word "*jin*" and the particular propositions by the word "*huo*". We can understand this interpretation in two senses. In one sense a sentence expressed by the word "jin" or the word "huo" corresponds to a universal proposition or a particular proposition. In the other sense, in ancient Chinese one uses nothing but the word "jin" and "huo" in expressing universal and particular propositions. Some scholars hold that as a quantifier, the word "jin" means "all", "whole" and "no exception"; but the word "jin" is not only a universal quantifier, but also a universal judgment. Some hold that Mo-Jing researches judgments in an all-round way and therefore puts forward the forms of universal and particular judgments.

There is an opposite point of view, that there is a common point in all these and especially in the last two interpretations above. Its adherents try to argue that Mo-Jing has a study of the forms of proposition. In fact we can see in Mo-Jing only some interpretations of the meaning of some singular words.

Very often we see several interpretations of one word. For instance according to one text, the word "huo" could be interpreted as meaning a particular judgment, but according to another it could be interpreted as meaning a modal proposition. This shows we cannot see the results the interpreters try to get.

The study of grammatical functions of words and of sentences, including the study of forms of sentences, is very important for logic, but it primarily belongs to grammatical research. Thus we can ask: Does Mo-Jing contain research into grammar, or does it have results preceding grammatical research?

"*Shuo*" is interpreted to mean proving the thesis by putting forward reasons, supports and arguments. In this interpretation what is the most important is to interprete "gu", "li", and "lei". Some hold that "gu" is the direct reason to set up a premise, corresponding to the minor premise in a syllogism, that "li" is the universal law, corresponding to the major premise in a syllogism, and that "lei" is an analogy by which the universal law or universal proposition is achieved. Some hold the same view for "li" and "gu", but think that "lei" aims at supplying "li" with instance and evidence of inductive meaning. Some hold that "gu" is the reason, corresponding to the premise of inference, "li" is the logical rule or law that an inference from premises to a conclusion must follow, and "lei" is the relation between universal and particular, corresponding to the dictum de omni et nullo of the classical syllogistic.

Again, an opposing view is that there is a common point in these interpretations: the inference forms of "shuo" correspond to Aristotle's syllogisms. In this way there are at least the following problems.

- Why should "li" correspond to the major premise in a syllogism? Why should "gu" correspond to the minor premise in a syllogism?
- Why should the inferences of "shuo" be formed with two premises as in Aristotle's syllogisms, instead of one premise or three premises? In the text of "Mo-Jing" we only see examples of inference with one premise.

The fact is that "Mo-Jing" has indeed a treatment of inference in a conscious way, but it does not do it from a formal point of view. The inference in Mo-Jing from premise to conclusion is determined by the content of the premise, but not by the form of the sentence. So we must be careful when we say that it has a scientific system of logic. We must understand what one means by "scientific" and by "logic".

Why does Mo-Jing not study the form of inference or why does its logic not develop in the direction of formal research? I think the fundamental reason lies in the fact that in ancient China there is no research in grammar and therefore no theory of grammar. Without grammatic analysis our knowledge about linguistic forms is nill. Under such circumstances it is impossible to find the logical forms hiding behind grammatic forms, especially in the primitive period of the development of logic.

As research is going deeper, an argument has recently sprung up about the object of the history of Chinese logic. There are several main views. Either, Mo-Jing's logic is interpreted as natural logic, or intentional logic, or informal logic, or formal logic, or a kind of semantics. Or, on the contrary there is a critical opinion that points out that such interpretations need careful and valid justifications.

The difficulty in the research of the history of Chinese logic lies in the fact that the ancient Chinese language is very ambiguous and therefore very difficult to understand. But this research is of significance since it can help us, especially in comparison with Western logic, to understand the conditions in which logic comes into being, so that we can have a better understanding of the nature of logic. And it also helps us to reveal relations between logic and language, since many problems in the history of Chinese logic involve natural language. In this direction, there is much to be done.

IV

In this section I would like to say something about the valuable work done by my colleagues and me.

In tense logic and paraconsistent logic we have achieved some original and creative new results. The creation of tense logic has begun by Xu Ming, who is now studying in the USA, after he wrote his MA. thesis with the title "On some U, S-tense logic" under the guidance of professor Zhou Liquan. Later Xu published his thesis in the *Journal of Philosophical Logic* (Vol. 17, 1988, pp. 181-202) in which he has set up the minimal system of the tense operators U (until) and S (since). Then Wang Xuegang, who is studying in Great Britain, published a paper entitled "The minimal System L_0'" in *The Notre Dame Journal of Formal Logic* (Vol. 8, 1991, pp. 39-57), in which he has established a minimal system for the tense operator G', H'. And finally Zhang Qingyu has constructed the minimal paraconsistent G', H'-tense logic through combining the result Wang Xuegang has achieved with Da Costa's system Cn. Zhang has also published a series of papers in which he has set up paraconsistent conditional systems and a modal system $CnG', n \in w$ (see "A Weak Paraconsistent Conditional Logic" in *The Journal of Non-classical Logic*, vol. 8, 1991, pp.39-57; "Paraconsistent Modal Logics $CnG', n \in w$ in *Boletim da Sociedade Paranaense de Matematica* (2.a serie), v. 12/13, 1991/2, pp.33-40; "The Weak Paraconsistent Conditional Logic $CnW, n \in w$", with Da Costa, in *Boletim da Sociedade Paranaense de Matematica*, 1994).

Over the past 15 years, I have worked on the history of Western logic, especially, the origin of logic, and the genesis of modern logic. The former brings with it Aristotle's logic, and the latter Frege's thought. I have written a book entitled *Aristotle's Logical Theory* (1991) and a book with the title *Frege* (1994). Besides I have translated Frege's works into Chinese (*A collection of Frege's Works*, 1994, Commercial Press, Beijing) and my translation of his famous book *The Foundations of Arithmetic* is forthcoming. I conclude by saying something about academic exchanges in research.

In China we have a great weakness in our research, namely, we lack academical exchanges with abroad. Through my experiences in Germany and Great Britain, I feel this problem very deeply. I hope in the future we willd have more and more exchanges of opinions about our research. In fact, academic exchanges among scholars from various countries are very helpful to our research. Here I can show you an example. Frege's theory of "Bedeutung" is very important. The English translation of it is generally "reference" and it is widely accepted. But there are a lot of problems with it. In general one has no difficulty in understanding the reference of a proper name, but one has great difficulty in understanding the reference of a sentence, especially with truth as the reference of a sentence. The Chinese translation of Frege's "reference" is generally "suo zhi". I say "reference" because the Chinese translation originates from the English translation, not from the German "Bedeutung". "Zhi" is a noun that literally signifies "finger" and as a verb means "pointing to". "Suo" is an auxiliary word and has no meaning by itself. But it signifies "what is pointed to" when linked with "zhi". So with "suo zhi" as the translation one seems to have no difficulty in understanding a concrete object as "what is pointed to", but have great difficulty to understand the true and the false as "what is pointed to", for with a finger we can indeed point to a person, a place and the like, but we cannot point to the true and the false, because they are not concrete objects and physical things. I found this problem in my research and asked myself if it is a problem from Frege or from the Chinese translation and therefore from the English translation. And my final answer is that it is a problem not from Frege himself but from the translation, and therefore many interpretations of Frege's "reference" are based on a misunderstanding of Frege. I cannot show my argument here in detail, but I can show through this example that scholars with different backgrounds and different languages may have different understandings of the same problem. With exchanges, we can therefore enhance and deepen our research.

Institute of Philosophy
Chinese Academy of Social Sciences
Beijing

HYAKUDAI SAKAMOTO

RECENT DEVELOPMENTS OF LOGIC, COMPUTER SCIENCE, AND PHILOSOPHY OF SCIENCE IN JAPAN

Generally speaking, Japanese studies of modern logic and philosophy of science started at the end of World War II with American stimulation. Even now, we are mostly following the trends of research in the United States. There are, I regret, only few original or creative works in the fields of logic and philosophy of science in Japan. Recently, however, the number of researchers is increasing in these fields. And we already have two well-organized academic societies for these fields: the Philosophy of Science Society, Japan (of which I am now President) and the Japanese Association for Philosophy of Science. Younger researchers are very active and productive in these two societies. I believe the future of this field is promising.

1. LOGIC

In the field of logic, Symbolic Logic was first introduced into Japan only after the war, by American scholars including W.V. Quine and also by some young Japanese scholars who studied in the United States. Initially, "symbolic logic" of Russell, and Logical Positivism of R. Carnap and H. Reichenbach, was studied by Shigeo Nagai, Takehiro Sueki, Arata Ishimoto. Later, the Quine-style method of logic was popularized by Shozo Ohmori, Hidekichi Nakamura, and others. The introduction of these kinds of new logic gave a revolutionary stimulus to the Japanese logical world, and was the starting point of Japanese modern logic.

I myself studied Algebraic Logic with A. Tarski at UC at Berkeley and I first introduced into Japan Lindenbaum Algebra and the Tarski-McKinsey-type topological approach to logic, and later the Rasiowa-Sikorski-type algebraic logic. But I failed to popularize it among Japanese logicians. Ten years later Yoshio Shimizu tried to revive the algebraic approach to logic, but this trend is still very weak among philosophers.

Akira Oide made a study of quantum logic, Natsuhiko Yoshida analyzed the logical theories of Hilbert and Gentzen, Mitsuhiro Okada is now studying the recent constructivist linear logic of Y. Girard.

The historical study of logic, especially Frege's logic, is fairly active. It is being pursued by Tatsuo Fujimura, Kazuyuki Nomoto, Takashi Iida, and Syn Tsuchiya. Arata Ishimoto and Toshiharu Waragai are attempting a renewal of Lesniewski's logic. In Intensional Logic and Theory of Semantics, K. Nomoto and Soshichi Uchii are pioneers. Especially, K. Nomoto started the study of

M.L. Dalla Chiara et al. (eds.), Structures and Norms in Science, 473–480.

Possible World Semantics. He and his group also closely follow the works of recent analytic semanticists such as Kripke, Davidson, Dummett, and Kaplan. S. Tutiya together with Tomoyuki Yamada is involved in the movement called Situation Semantics of the Stanford group.

Recently, the logical study of the grammar of the Japanese language, adopting Montague Grammar, was initiated by Hidehisa Sakai, and his method was followed by younger logicians. They are investigating new computer-assisted teaching systems of logic for the native speakers of the Japanese language.

So far, I have reported logical studies by philosophers. However, the main lines of mathematical logic have been studied rather by mathematicians working in the Foundations of mathematics. (In Japan, mutual communication and cooperation between philosophers and mathematicians on the topic of logic have been, to my regret, very rare and weak.) This field of study was first cultivated by the late mathematics professors Shigekatsu Kuroda, Katsuji Ono, Motokichi Kondo, and Joichi Suetsuna. However, after the War, Gaisi Takeuti played the most decisive role in developing "Foundations of Mathematics" studies in post-war Japan. Stimulated by Gentzen's consistency proof for arithmetic, Takeuti tried to prove the consistency of the real number system based on Hilbert's program, and he established "Takeuti's Fundamental Conjectures". His work had a great influence on the study of Proof Theory, not only in Japan but in the whole world. His work has fixed the tradition of Gentzen style proof theory in Japan. Takeuti himself proved the consistency of second order natural number theory, which has Π_1^1 comprehension, using the idea of "ordinal diagrams", sometimes called "Takeuti diagrams". The concept of "ordinal diagrams" is now studied by Mariko Yasugi. Toshiyasu Arai later showed that Takeuti's result was optimal. Also, Moto Takahashi proved, using model theoretic methods, that Takeuti's Fundamental Conjecture is valid.

Progress of Proof Theory in Japan promoted the study of theoretical computer science. Susumu Hayashi developed the PX system, which made a great contribution to the Japanese study of computer science.

Takeuti, taking ideas from Gödel's consistency proof for GCH, also started research of recursion theory on ordinals, and Toshiyuki Tsuge formalized this by a Kleene-style equational system. These works are viewed as predecessors of the theory of admissible sets by Platek and Barwise. Takeuti's work on recursion theory on the ordinals was directly connected with the theory of "large cardinals", and a new study of set theory was originated. Kanji Namba proposed a new "forcing" notion known as "Namba forcing". The importance of this forcing notion was clarified later by Shelah and others. In this connection, study of ideals on cardinals was also activated. Yuzuru Kakuta formulated the set theory ZF(aa) on stationary logic, a logic formalizing the normality of ideals, independently of the work of M. Kaufmann. Studies of large cardinals are still actively pursued by Shizuo Kamo, Hiroshi Matsubara and others.

Nobuyoshi Motohashi newly introduced "model theory" on a full scale, and he originated a new style of mathematical logic, which might be called "proof theoretic model theory", and his follower Akito Tsuboi started the study of "stability theory" along this line.

Research on non-Boolean logic has recently been active. Study of "Intermediate logic", which was originated by the late Katsuji Ono, has been deepened by the "MLG group". And modal logic and other substructural logics are also actively studied by Hiroakira Ono and his group.

2. COMPUTER SCIENCES

The Japanese studies of computer science, especially studies of AI, are now on a high world level. Kazuhiro Fuchi of ICOT led the national project of developing the Fifth Generation Computer System, starting in 1982. (The word "fifth generation computer" was originated by MITI (Ministry of International Trade and Industry) to accelerate studies of computer technology and AI, in both their hardware and software phases. Dr. Fuchi thought that the present Von Neumann type computer is too primitive to be intelligent. Instead, he attempted to make a high-level logical inference machine as a knowledge processing intelligence for AI. For this purpose, he studied the thinking process of the syllogism, i.e. one-place predicate logic, using the Prolog Language, and also he developed a new sort of PDP (Parallel Distributed Processing) system called MPP (Massive Parallel Processor), and thus, he succeeded in constructing a new type of computer system, and also establishing a new concept of Artificial Intelligence. His work is widely viewed as a successful application of logic to computer science. His system is now applied to develop the technology of "natural language processing" and high level "pattern recognition", and also a variety of "expert systems".

Hiroshi Kawano attempted to apply new advanced computer technology to the field of Esthetics. Based on the new computer metaphor of PDP and "Society of Mind" model, which has been proposed as a human interpretation of PDP by Marvin Minsky, Kawano developed a new post modern art paradigm of "Folk Esthetics" to take over traditional European Modernism in such a way as to apply Minsky's "agent-agency" mechanism to the "object-collage" method of Pablo Picasso and Marcel Duchamp. Kawano is also one of the first artists of computer graphics.

Cognitive Science is prospering. In 1983, the Japanese Association of Cognitive Science was founded, and there, computer scientists and psychologists are working in cooperation. However, only a few philosophers of science have joined in this work.

3. PHILOSOPHY OF SCIENCE

Before the War there existed some sorts of philosophy of science in Japan. However, they were either a mere metaphysical interpretation (of a Hegelian type) of physical concepts, or a queer metaphorical mixture of scientific ideas or mathematical ideas with some Asian traditional ideas such as the "nothingness" of Zen Buddhism. Philosophy of science in a modern sense started only after the war through the stimulation of the extraordinary progress and development of science and technology, together with the strong influence of Logical Positivism. Prof. Hideki Yukawa, the founder of the meson theory, showed some interest in the philosophy of science. Other scientists and mathematicians also attempted to open ways to the philosophy of science on the basis of their own scientific background. Joichi Suetsuna founded the Association for Philosophy of Science with an emphasis on mathematics and mathematical sciences; and Takeshi Shinohara, originally a biologist, organized the Japan Congress of Philosophy of Science and held an interdisciplinary general meeting annually. But after their retirement, philosophy of science has been studied mostly by philosophers. Seizo Ohe (Nihon University), Nobushige Sawada (Keio University), Shozo Ohmori (Tokyo University) were pioneers in establishing this trend. Since then, philosophy of the physical sciences has naturally been a central topic. Problems in the Foundations of Physics, Analysis of Scientific concepts, Nature of Scientific Explanation, and Paradigm of Scientific Revolution have been under constant scrutiny. Recently, debates on the Foundation of Quantum Mechanics have become very active. Last year we revived the EPR Paradox in the light of new developments in quantum physics. This was pursued in the workshop of the annual meeting of the Philosophy of Science Society. Toshio Ishigaki (Hokkaido University), Yutaka Tanaka (Mejiro-Gakuen University) Shingo Fujita (Tsukuba University) and Hiroshi Watanabe (Chuo University) participated in the discussion.

The philosophy of biology is also a popular topic in the philosophy of science. Hiroshi Nagai (Tsukuba University) wrote a valuable book on the philosophy of the biological sciences, called *Philosophical Foundation of the Theory of Life*. Following the development of the biological sciences, and having strong sympathy to the philosophy of biology of Prof. E. Sober of the University of Wisconsin, Yosaku Nishiwaki (Keio University) together with Soshichi Uchii (Kyoto University) are now working on the philosophical foundations of the theory of evolution.

4. PHILOSOPHY OF SOCIAL SCIENCES

Within the philosophy of the social sciences, Philosophy of Law, or Legal Philosophy, has a strong affinity with analytic trends in philosophy. Junichi Aomi of Tokyo University cultivated this new trend of legal philosophy. His student, Ryuichi Nagao (Tokyo University), has shown some interest in the philosophy of science. Setsuko Sato (Aoyamagakuin University), also Aomi's student, studied the Scandinavian Legal Realism of Axel Hägerström under Prof. Olivecrona of Upsala University, and is pursuing the philosophical analysis of fundamental legal concepts in the same orientation as recent analytic trends in philosophy. She recommends a new type of "natural law", considering the possible global crises in the next century. Hajime Yoshino (Meijigakuin University) is a pioneer in the application of computer technology to the legal field. He is developing an Expert System for Legal Process, applying the Fifth Generation AI technology to legal reasoning. Of the younger generation, Tatsuo Inoue (Tokyo University) Itaru Shimazu (Chiba University), Yasutomo Morigiwa (Nagoya University), Susumu Morimura (Hitotsubashi University) are now working on a new establishment of the concept of "Justice" and also working together to find new ways of relating "Social Consent Formation" to the concept of "justice". Seigo Obata (Himeji Dokkyo University) a young legal philosopher, attracted wide notice with his book *Judicial Judgment as a Speech Act*, introducing Austin's theory of Speech Acts to the field of legal philosophy.

Philosophy of economics, philosophy of sociology, etc., are desirable, but as yet no influential development has occurred.

5. PHILOSOPHY OF LANGUAGE

In Japan, after the War, the technique of "formal language analysis" of logical positivism was introduced first. The issue of the "verification theory of meaning" has been under constant discussion. Jiichiro Takeo (Kansai gakuin University) wrote many exquisite papers on this. He is an authority on the study of logical positivism. At the same time, Philosophy of Ordinary Language was introduced. Almost all books of G.E. Moore, G. Ryle, J.L. Austin, and P.F. Strawson have been translated into Japanese. Especially G. Ryle's analysis of "disposition" and "category mistake" are popular among young philosophers of language. Austin's theory of "speech acts" was studied by H. Sakamoto and S. Tsuchiya. Sakamoto attempted to extend the applicability of speech act theory not only to the philosophy of language but also to linguistics, jurisprudence, economics and other social sciences. This method is now welcomed by these fields. Wittgenstein's philosophy of language is admired by some philosophers of language. Takashi Fujimoto (Tokyo University), Hiroshi Kurosaki (Seijo University) and Masahiro Oku (Osaka University) have offered deep interpretations of his work.

Quine's influence, however, was so remarkable that the philosophy of language has always been thought as one of the central fields of philosophy of science. Nowadays some young philosophers even identify philosophy of science with philosophy of language, and they tend to treat the ontological problem from a linguistic point of view. They are primarily concerned with the "Truth Concept" or the "Incommensurability" issues which Quine developed or initiated in his lecture "Word and Object" as he says, at the American Seminar in Tokyo, in 1959. Takashi Iida's book *Summa Philosophia* is a thorough summary of the philosophy of language of Quinian type.

Nobuharu Tanji's (Tokyo Metropolitan University) recent paper "Quine on theory and language" in *British Journal of Philosophy of Science* is highly evaluated.

6. PHILOSOPHY OF MIND

The mind-body problem is still one of the central topics in the philosophy of science, under the persistent influence of Logical Positivism as well as that of developments in sciences concerning the brain process. Epiphenomenalism has been the most popular theory. Recently, however, some mind-body theorists have proposed several types of monistic identity theory to overcome Cartesian dualism towards the 21th Century. Monistic theory with radical materialism was advocated by Wataru Hiromatsu of Tokyo University, Phenomenal Monism of a Berkeley type is maintained by S. Ohmori.

Against these two types of Monism, Hyakudai Sakamoto advocates a Monism of a third type, which he calls "Ur-monism". This is a sort of neutral monism of a Russellian-type, or a sort of "anomalous monism" of the Davidson-type. Naturally, it does not intend to introduce any kind of metaphysical entities, but according to the theory, the monistic something — we call it "Ur-event" — might be just virtual, or fictional, therefore; it is non-existent ontologically. At the same time, we claim that this "ur-event" must be accompanied and supported by conceptual dualism, or descriptive dualism of the "mental" and "physical", in the sense that there exists a virtual image of the monistic entity (ur-event), and at the same time we are *a priori* given exactly two ways of describing it, i.e. mental and physical. According to the theory, the causal explanation of perception will be saved in the following sense: an ordinary explanation, such as, "A flower (of the external world) is the cause of perception (as an inner event) of the flower" should be interpreted as an abbreviation of the following full explanation. "Some event (ur-event 1) which can be described as the existence of the flower (in the physical world) is the cause of some other successive event (ur-event 2) which is described as "the flower image" (in the inner world - mind) And, therefore, causality here can be claimed to exist only as ur-event 1 — ur-event 2 causality. But this causality is only virtual, or something like virtual reality. We can describe it only as physical — physical causality or as mental — mental causality.

This kind of monism might be considered a type of classical "double aspect theory" (of Spinoza), or the revival of the traditional East-Asian idea of mind-body monism (as seen in the philosophy of Zen Buddhism) combining a fictional monistic entity (such as "emptiness" or "nothingness") and a linguistic dualism of description. We should now affirm and expect an important East-Asian influence in the 21th century on the post-modern philosophical paradigm on a global scale.

Keiichiro Kamino (Osaka City University), a leader of the research group of philosophy of science in Kansai District, is also at work on the mind-body problem from another point of view.

7. BIOETHICS

Philosophical foundations of bioethics are being pursued by a few philosophers of science. They take Bioethics as a part of the movement of TA (Technology Assessment) or the critical evaluation of modern life sciences and technologies. From this point of view, I organized and founded a new academic association for bioethics "The Japanese Association for Bioethics" on a large, interdisciplinary scale, including almost all kinds of sciences and technologies — natural sciences, social sciences, humanities, and even religion, and I presided over the Association for three years. I thought of bioethics as a sort of Unified Science, like the "Einheitswissenschaft" of the Wiener Kreis, or a new sort of "Ethical Social Engineering". I believe this idea of mine was understood and welcomed at least by some scientific scholars. We attempted to establish a new ethics to catch up with the development of modern sciences and technologies, especially of biosciences and biotechnologies. Here, amongst other things, I put special stress on the philosophical and ethical foundations of genetics, gene manipulation or the Human Genome Analysis Project as its most important issues. Most Japanese bioethicists, however, have followed Euro-American Bioethics, which is mainly concerned with medical ethics or legal adjustment of ethical social problems raised by the application of new biomedical science and technologies. But recently some of them have begun to doubt the "universality" of Euro-American bioethics. Instead, they hold that bioethics is essentially ethos dependent, and therefore, relative to culture. It attacks the western paradigm of the concept of "person" which gave the basis for the European idea of "fundamental human rights". They examine the possibility of adopting Oriental or Asian ideas such as Confucian ethical philosophy, to establish a future global bioethics different from the on-going Euro-American bioethics. A new movement of Asian bioethics is emerging. We are now scheduling a First International East-Asian Conference of Bioethics to be held this November in cooperation with Chinese and Korean bioethicists to investigate the possibility of an East-Asian proper bioethics.

Nihon University
Tokyo

ACKNOWLEDGEMENTS

In writing this report I was given much useful information as well as suggestions and advice by many academic colleagues, especially by Professor Kazuyuki Nomoto (philosopher), Professor Yuzuru Kakuta (mathematician),

Professor Kazuhiro Fuchi (computer scientist) and Professor Hiroshi Kawano (aesthetician).

SONG SANG-YONG

PHILOSOPHY OF SCIENCE IN KOREA, 1950 – 1995

Just a week ago Korea celebrated the fiftieth anniversary of her liberation from the Japanese colonial rule of thirty-five years. The period I am going to deal with roughly coincides with the post-liberation era. However, let me start with the prehistory of the philosophy of science in Korea.

Korea's first contact with Western philosophy was made in 1631 when a Korean envoy in Beijing brought home a philosophy book written by a Jesuit missionary. In 1872 a Korean named I In-jae wrote *Ch'olhak Kobyon* (*A Study of Philosophy*) utilising Japanese and Chinese sources. The first Korean who attended lectures in philosophy was Yu Kil-chun. He studied at Keio University in Tokyo, and later at Dunbar Academy in the United States in the 1880s.

Early in the twentieth century, Koreans began to study abroad, which culminated in the 1920s. The first graduate of the philosophy department came from Waseda University in Japan in 1917. Lee Kwanyong, the first Ph.D. in philosophy appeared at the University of Zürich in 1921. Korean students studied philosophy in various universities in Europe, including Paris, Vienna, Jena, Würzburg, Marburg, Berlin and Leipzig. Five out of them got Ph.Ds. Later three Ph.Ds were added in the United States. They attended the universities of Southern California, Chicago and Michigan. Less than ten philosophers came out of Japanese universities, viz. Tokyo, Kyoto, Tohoku, Waseda and Rikyo.

After the nationwide independence movement in 1919, there was a campaign by the Koreans to create a university of their own. The Japanese colonial government did not like the idea and in 1926 established Keijo Imperial University, the precursor of Seoul National University. But it was primarily for the Japanese. Given this handicap, only fifteen Korean philosophers had been produced from it by 1945.

In 1933 the Society of Philosophical Studies was formed by the graduates of universities. It had its journal *Ch'olhak* (*Philosophy*), but was forced to stop by the Japanese after three issues. The Philosophical Discussion Group at Keijo Imperial University which started in the same year was also active.

After the liberation in 1945, about fifteen philosophers began to teach at several universities. Due to the influence of Japan, Korean philosophy was dominated by German idealism and there were only two who could be classified as philosophers of science. The division of country, the fierce ideological struggle between the right and the left, and the war in 1950 were great blows

M.L. Dalla Chiara et al. (eds.), Structures and Norms in Science, 481–485.

to the weak philosophy community in Korea. Some excellent philosophers with Marxist leanings went to the north.

It was only in 1953 that the Korean Philosophical Association was organised. However, it could publish barely two issues of its journal till 1957. While it was in a stagnant state, another group called Ch'olhak Yonguhwe (The Society of Philosophical Studies) emerged with young philosophers in 1963. It consisted mainly of graduates of Seoul National University and published immediately its organ *Ch'olhak Yongu* (*Philosophical Studies*).

It was not until the middle of 1950s when philosophy of science began to be taught at Korean universities. In 1953 Kim Choon Sup came back home with a Ph.D. from Columbia University. Though his dissertation was on the dialectical method, he addressed himself as a philosopher of science. He gave lectures in symbolic logic and introductory philosophy of science at several universities. Students showed great interest in a completely new philosophy, but he could not draw many of them. He produced five MAs in philosophy of science through the 1960s.

Kim was a prolific writer. Among his many books four are worth mentioning. In the revised edition of his *Logic* (1954), he introduced symbolic logic for the first time. *Contemporary Philosophy* (1958) is a collection of papers in philosophy of science and *Introduction to Philosophy of Science* appeared in 1963. His later papers were assembled in *Studies in Philosophy and Logic* in 1975.

Some more MAs in philosophy of science came out of Korea University and Yonsei University. At Korea University there was Pak Hi-song who got a Ph.D with the thesis *Subjectivism and Intuition* at the University of Michigan in 1937. Chong Sok-hae at Yonsei was trained in Paris, Würzburg and Berlin. Both of them published very little and concentrated on teaching. *Skepticism and Truth*, the collection of Pak's papers was published posthumously in 1989. Chong translated Russell's *History of Western Philosophy* and Moore's *Principia Ethica*. Chong's intellectual biography by a former student was published in 1989 on his ninetieth birthday.

Philosophy of science, however, could not be prominent for more than twenty years in a climate where German idealism was dominant. Its level remained very low until the late 1970s. Many frustrated students went to the United States for proper training in philosophy of science. Some visible changes took place with their return.

In 1972 Kim Choon Sup, LeeHan-jo, Sonu Nam, Lee Cho-Sik and Song Sang-yong got together to have informal seminars which lasted for two years. With the increase of young philosophers, philosophy in Korea underwent a drastic turn in the 1970s. In 1976 ten subdivisions were organised within the Korean Philosophical Association. Division of Analytic Philosophy and Division of Logic were created by a handful of philosophers with Ph.Ds from the United States. The leading figures were Kim Yersu (Harvard), So Hung-Yul (Michigan), Chung Daihyun (Temple), Um Jung-Sik (Michigan State), and

Lee Myung-Hyun (Brown). They started with training of graduate students through seminars reading basic articles in analytic philosophy. Since then philosophy of science witnessed steady and remarkable progress. The provocative activities of the groups had a great impact on the whole philosophical community in Korea and quite a few philosophers in other fields became members. Thanks to the contributions of the groups, analytic philosophy articles came to outnumber those of German idealism in major philosophy journals after the 1980s.

Both divisions have more than one hundred members respectively and have had monthly meetings for twenty years. Wittgenstein was the centre of concern for some time. Out of five collections of papers published by the Society of Analytic Philosophy two are on Wittgenstein. Later it was expanded to include such philosophers as Quine, Kripke, Putnam, Davidson and J. Kim. The Society organises a special symposium or workshop twice a year. Scientific realism and philosophical naturalism were the topics of recent meetings.

Now I have to mention Jaegwon Kim, a foremost Korean philosopher of science living in the United States. He went to America forty years ago and graduated from Dartmouth College *summa cum laude.* Having got a Ph.D at Princeton under Carl Hempel, he had taught at Swarthmore, Cornell, Johns Hopkins, and Seoul National University. He was Roy Wood Sellars Professor at the University of Michigan and is now William Herbert Perry Faunce Professor at Brown University. His impressive achievements was a great stimulus to the Korean philosophy of science community. He trained many Korean philosophers of science both in Korea and in the United States. Last year Jaegwon Kim received *The Metaphysics of Supervenience*, a *Festschrift* on the occasion of his sixtieth birthday. The Korean version of his collected papers was also published. It shows that the philosophy of science in Korea has overcome the stage of enlightenment.

Korea has more than fifty Ph.Ds in philosophy from the United States, the majority of which are philosophers of science. Nearly one hundred Ph.Ds came from Europe. Some of them are also in philosophy of science. Among about three hundred Ph.Ds from Korean universities one fifth are in philosophy of science. According to a recent survey, out of six thousand published philosophical works including papers, dissertations, monographs and textbooks one fifth are in philosophy of science. There are now more than one hundred philosophers of science at the Ph.D level in Korea. Among them are small groups of great variety: German philosophy of nature, French epistemology, Marxism, Oriental philosophy, history and sociology of science, cognitive science, environmental philosophy, biomedical ethics, etc., etc. A new society for philosophy of science covering all of them is expected to emerge soon.

I would like to elaborate the activities of these groups a little further. Some philosophers trained in continental Europe view science with a different perspective from that of analytic tradition. Choi Jongduck represents German *Naturphilosophie* characterised by anti-positivism and anti-reductionism. The

French epistemology of Bachelard and Canguilhem was explored by Lee Jeong-woo. Foucault, Serres, Lacan, Lyotard and Thom draw special concern these days.

Until recently Marxism had been a taboo in South Korea. Young philosophers came to be much interested in Marxism during the democratisation movement under the repressive regimes. A Marxist-oriented social philosophy group evolved to the Korean Society for the Study of Philosophical Ideas in 1990. It is a national organisation and still active even after the fall of socialism. It has a study group in philosophy of nature which seeks for a philosophy of science from the viewpoint of dialectical and historical materialism. It has another group in *Chi* philosophy which concerns itself with traditional philosophy and medicine in the Orient.

Korea has a rapidly rising group of history of science. The Korean History of Science Society was established in 1960 and the Programme in History and Philosophy of Science at Seoul National University has been flourishing since 1984. Some historians of science who have lately come back from the United States and Europe are keenly interested in philosophy. Kim Mi Gyung, Im Gyeong Soon and Kim Kiyoon are some of them. They exert a considerable influence on the philosophy of science with the new achievements of their field such as social constructivism. It is expected that they will be in good collaboration with historically-minded philosophers of science like Jeong Byung-Hoon and Shin Joong Sop. There are also some sociologists of science. Kim Kyung-Man has an international reputation for criticising the strong programme.

The Korean Society for Cognitive Science was born in 1987 with three hundred members. About thirty philosophers including Kong Yong Hyun participated in it. It is a good example of an interdisciplinary study with psychology, artificial intelligence, neuroscience, linguistics, mathematics and social sciences. It has made impressive progress both in theoretical and practical approaches.

With the sudden industrialisation, environmental pollution became a serious problem in Korea. Nearly all ethicists and social philosophers are interested in environment and produce many papers. Biomedical ethics is becoming a very popular field too. It drew wide attention since Nicholas Fotion of Emory University came to Yonsei University as a visiting professor in medical ethics. Abortion has been a hot issue and brain death is a new problem at the moment.

In the 1980s Kim Yong-jun, a chemist, led two reading groups of scientists and philosophers which used Prigogine's *Order Out of Chaos* and Whitehead's *Process and Reality* as texts. These two merged into the Association of Science, Philosophy and Culture in 1988. It has held monthly meetings since then and publishes a journal *Science and Philosophy*. Zhang Hwe Ik, a physicist is noteworthy in this group. He wrote a highly original book *Science and Metascience* in 1990.

In conclusion, the philosophy of science in Korea showed an impressive development over a short span of time. In the first half, we achieved little, while in the latter half we were quite successful. In spite of this, I must confess that we have many weaknesses. For instance, we have no more than ten philosophers of science who do logic in its own right. There is no philosopher of biology. We have been very busy in absorbing the philosophy of science in the West. Ours is too thin to have a solid ground. But I hope that we will be able to produce original philosophy of science in the near future.

Hallym University
South Korea

CHENG-HUNG LIN, DAIWIE FU

PHILOSOPHY OF SCIENCE IN TAIWAN

Since modern science was introduced into China, many questions have been raised concerning the subject. Some scholars, e.g., Shih Hu and Wen-chiang Tin, considered science not only as systematic knowledge, but more importantly also as a way of thinking and knowing, or even as an attitude or way of life. Those who held this position usually rejected metaphysics, religion, traditional Chinese medicine and the traditional Chinese way of life as unscientific and hence harmful. They also claimed that the humanities and social sciences should also be studied with scientific method. On the other hand, some scholars insisted that the above scientific position is unacceptable. The controversy is still alive. Although most contemporary Chinese accept modern science without any hesitation, questions concerning the nature of science and the attitude towards science remain unsettled.

During the Japanese occupation (1895–1945), the scientific education and research in Taiwan were of a high quality. Some research on medical science and physical anthropology were truly excellent. However, it seems that there were very few, if any, studies in the philosophy or history of science. When the Chinese nationalist government moved to Taiwan in 1949, the abovementioned scientific viewpoint was also introduced into Taiwan. The first philosophy of science course in Taiwan was taught by Professor Hai-kuang Yin at National Taiwan University in around 1960. Professor Yin wrote several introductory articles on logical positivism. He also made use of the teachings of the logical positivists and the techniques of logical analysis in his severe criticisms of traditional Chinese philosophy, communism and the ideology of the Nationalist government. Thus he was disliked by both the Communist and the Nationalist Chinese governments. At one time, the Nationalist government in Taiwan considered logic, logical positivism and analytic philosophy as anti-government doctrine.

Since Professor Yin's untimely death in 1969, some of his former students have continued to work on logic, analytic philosophy and philosophy of science. Professor Shih-Chao Liu, the eldest of Professor Yin's students, had a paper 'On the Analytic and the Synthetic' published in *The Philosophical Review* (1956) and earned a Ph.D. degree in mathematical logic from the University of Wisconsin, Madison. To the best of my knowledge, Professor Liu was the first Taiwanese scholar to gain a Ph.D in logic and the abovementioned paper was the first article by a Taiwanese scholar to be published in a major philo-

M.L. Dalla Chiara et al. (eds.), Structures and Norms in Science, 487–489.

sophical journal. Later on, some other former students of Professor Yin and several younger scholars also obtained their doctorates in logic, philosophy of science or other related fields (e.g., cognitive science) from the United States or Canada. Most of them have returned to Taiwan and are teaching or conducting research in universities or academic institutes. Their research topics include those discussed by the logical positivists (e.g., Hempel), Popper and the so-called 'new philosophers of science' (e.g., Kuhn, Lakatos, Feyerabend, Laudan). As far as I know, no one has specialized in philosophy of mathematics, philosophy of physics or philosophy of biology.

The notion of a conceptual history of science is a difficult one for Chinese or Taiwanese historians of science. Considering the studies of history of science as a historical research activity which has close relationships with philosophy of science, with 'epistemology' in history of science of a French style, or at least with the 'internal logic' of the ancient science in question is a more or less foreign notion to Chinese or Taiwanese historians of science working before the 1960s. For them, to do history of science is mostly to select from Chinese ancient texts fragments of material that could be judged as 'scientific' from the viewpoint of contemporary Western sciences, but with very little concern with the internal coherence or even philosophies of nature within those texts. This old way of doing Chinese history of science (leaving aside the social history of science) was of course closely related to the social and intellectual situations of Chinese intellectuals at the beginnings of this century when the fate of China was at stake and the confidence of Chinese intellectuals towards China's cultural heritages was completely smashed. In order to build a new pride based on modern civilization, Westernized Chinese intellectuals elevated the contemporary Western sciences as the new standard by which ancient texts must be scanned in order to prove the existence and even occasional superiority over the West of the 'scientific mind' in Chinese culture.

A closer relationship between history of science and philosophy of science did not exist in Taiwan until, to my knowledge, the new studies led by Popper, Lakatos, Feyerabend, and of course those by Kuhn and Butterfield were introduced around the '70s. Before these introductions, the studies of philosophy of science in Taiwan were more concentrated on the logical structures of scientific statements or theories, studies that were closer to logic and American traditional analytic philosophy than to sciences and their histories. Perhaps T. Kuhn's *The Structure of Scientific Revolution*[2] was very important in challenging the assumptions of these two previously distinct disciplines of 'meta-'science, and thereby in bringing them closer and learning from each other. The interesting consequences of these impacts was that some historians of science in Taiwan gradually formed new conceptions of history of science and developed new lines of research in fields that are more or less foreign to Western history of science: history of Chinese science and Taiwan's conceptual history of science. Without falling into the trap of the history of 'ideas', the newly developed conceptual history of Chinese science gradually gained new grounds and motivations outside the old way of doing Chinese history of science. Similarly, Taiwan's conceptual history of science and knowledge, a field that obviously has no relation with China's cultural pride in its ancient 'science' and which interested no Chinese historians of science of the old school, broke new ground and developed only recently.

Department of Philosophy
National Taiwan University
Taipei
and
Institute of History
National Tsing-Hua University
Hsinchu

TABLE OF CONTENTS VOLUME I

140. C. Perelman, *The New Rhetoric and the Humanities.* Essays on Rhetoric and Its Applications. Translated from French and German. With an Introduction by H. Zyskind. 1979 ISBN 90-277-1018-X; Pb 90-277-1019-8
141. W. Rabinowicz, *Universalizability.* A Study in Morals and Metaphysics. 1979 ISBN 90-277-1020-2
142. C. Perelman, *Justice, Law and Argument.* Essays on Moral and Legal Reasoning. Translated from French and German. With an Introduction by H.J. Berman. 1980 ISBN 90-277-1089-9; Pb 90-277-1090-2
143. S. Kanger and S. Öhman (eds.), *Philosophy and Grammar.* Papers on the Occasion of the Quincentennial of Uppsala University. 1981 ISBN 90-277-1091-0
144. T. Pawlowski, *Concept Formation in the Humanities and the Social Sciences.* 1980 ISBN 90-277-1096-1
145. J. Hintikka, D. Gruender and E. Agazzi (eds.), *Theory Change, Ancient Axiomatics and Galileo's Methodology.* Proceedings of the 1978 Pisa Conference on the History and Philosophy of Science, Volume I. 1981 ISBN 90-277-1126-7
146. J. Hintikka, D. Gruender and E. Agazzi (eds.), *Probabilistic Thinking, Thermodynamics, and the Interaction of the History and Philosophy of Science.* Proceedings of the 1978 Pisa Conference on the History and Philosophy of Science, Volume II. 1981 ISBN 90-277-1127-5
147. U. Mönnich (ed.), *Aspects of Philosophical Logic.* Some Logical Forays into Central Notions of Linguistics and Philosophy. 1981 ISBN 90-277-1201-8
148. D. M. Gabbay, *Semantical Investigations in Heyting's Intuitionistic Logic.* 1981 ISBN 90-277-1202-6
149. E. Agazzi (ed.), *Modern Logic – A Survey.* Historical, Philosophical, and Mathematical Aspects of Modern Logic and Its Applications. 1981 ISBN 90-277-1137-2
150. A. F. Parker-Rhodes, *The Theory of Indistinguishables.* A Search for Explanatory Principles below the Level of Physics. 1981 ISBN 90-277-1214-X
151. J. C. Pitt, *Pictures, Images, and Conceptual Change.* An Analysis of Wilfrid Sellars' Philosophy of Science. 1981 ISBN 90-277-1276-X; Pb 90-277-1277-8
152. R. Hilpinen (ed.), *New Studies in Deontic Logic.* Norms, Actions, and the Foundations of Ethics. 1981 ISBN 90-277-1278-6; Pb 90-277-1346-4
153. C. Dilworth, *Scientific Progress.* A Study Concerning the Nature of the Relation between Successive Scientific Theories. 3rd rev. ed., 1994 ISBN 0-7923-2487-0; Pb 0-7923-2488-9
154. D. Woodruff Smith and R. McIntyre, *Husserl and Intentionality.* A Study of Mind, Meaning, and Language. 1982 ISBN 90-277-1392-8; Pb 90-277-1730-3
155. R. J. Nelson, *The Logic of Mind.* 2nd. ed., 1989 ISBN 90-277-2819-4; Pb 90-277-2822-4
156. J. F. A. K. van Benthem, *The Logic of Time.* A Model-Theoretic Investigation into the Varieties of Temporal Ontology, and Temporal Discourse. 1983; 2nd ed., 1991 ISBN 0-7923-1081-0
157. R. Swinburne (ed.), *Space, Time and Causality.* 1983 ISBN 90-277-1437-1
158. E. T. Jaynes, *Papers on Probability, Statistics and Statistical Physics.* Ed. by R. D. Rozenkrantz. 1983 ISBN 90-277-1448-7; Pb (1989) 0-7923-0213-3
159. T. Chapman, *Time: A Philosophical Analysis.* 1982 ISBN 90-277-1465-7
160. E. N. Zalta, *Abstract Objects.* An Introduction to Axiomatic Metaphysics. 1983 ISBN 90-277-1474-6
161. S. Harding and M. B. Hintikka (eds.), *Discovering Reality.* Feminist Perspectives on Epistemology, Metaphysics, Methodology, and Philosophy of Science. 1983 ISBN 90-277-1496-7; Pb 90-277-1538-6
162. M. A. Stewart (ed.), *Law, Morality and Rights.* 1983 ISBN 90-277-1519-X

163. D. Mayr and G. Süssmann (eds.), *Space, Time, and Mechanics.* Basic Structures of a Physical Theory. 1983 ISBN 90-277-1525-4
164. D. Gabbay and F. Guenthner (eds.), *Handbook of Philosophical Logic.* Vol. I: Elements of Classical Logic. 1983 ISBN 90-277-1542-4
165. D. Gabbay and F. Guenthner (eds.), *Handbook of Philosophical Logic.* Vol. II: Extensions of Classical Logic. 1984 ISBN 90-277-1604-8
166. D. Gabbay and F. Guenthner (eds.), *Handbook of Philosophical Logic.* Vol. III: Alternative to Classical Logic. 1986 ISBN 90-277-1605-6
167. D. Gabbay and F. Guenthner (eds.), *Handbook of Philosophical Logic.* Vol. IV: Topics in the Philosophy of Language. 1989 ISBN 90-277-1606-4
168. A. J. I. Jones, *Communication and Meaning.* An Essay in Applied Modal Logic. 1983 ISBN 90-277-1543-2
169. M. Fitting, *Proof Methods for Modal and Intuitionistic Logics.* 1983 ISBN 90-277-1573-4
170. J. Margolis, *Culture and Cultural Entities.* Toward a New Unity of Science. 1984 ISBN 90-277-1574-2
171. R. Tuomela, *A Theory of Social Action.* 1984 ISBN 90-277-1703-6
172. J. J. E. Gracia, E. Rabossi, E. Villanueva and M. Dascal (eds.), *Philosophical Analysis in Latin America.* 1984 ISBN 90-277-1749-4
173. P. Ziff, *Epistemic Analysis.* A Coherence Theory of Knowledge. 1984 ISBN 90-277-1751-7
174. P. Ziff, *Antiaesthetics.* An Appreciation of the Cow with the Subtile Nose. 1984 ISBN 90-277-1773-7
175. W. Balzer, D. A. Pearce, and H.-J. Schmidt (eds.), *Reduction in Science.* Structure, Examples, Philosophical Problems. 1984 ISBN 90-277-1811-3
176. A. Peczenik, L. Lindahl and B. van Roermund (eds.), *Theory of Legal Science.* Proceedings of the Conference on Legal Theory and Philosophy of Science (Lund, Sweden, December 1983). 1984 ISBN 90-277-1834-2
177. I. Niiniluoto, *Is Science Progressive?* 1984 ISBN 90-277-1835-0
178. B. K. Matilal and J. L. Shaw (eds.), *Analytical Philosophy in Comparative Perspective.* Exploratory Essays in Current Theories and Classical Indian Theories of Meaning and Reference. 1985 ISBN 90-277-1870-9
179. P. Kroes, *Time: Its Structure and Role in Physical Theories.* 1985 ISBN 90-277-1894-6
180. J. H. Fetzer, *Sociobiology and Epistemology.* 1985 ISBN 90-277-2005-3; Pb 90-277-2006-1
181. L. Haaparanta and J. Hintikka (eds.), *Frege Synthesized.* Essays on the Philosophical and Foundational Work of Gottlob Frege. 1986 ISBN 90-277-2126-2
182. M. Detlefsen, *Hilbert's Program.* An Essay on Mathematical Instrumentalism. 1986 ISBN 90-277-2151-3
183. J. L. Golden and J. J. Pilotta (eds.), *Practical Reasoning in Human Affairs.* Studies in Honor of Chaim Perelman. 1986 ISBN 90-277-2255-2
184. H. Zandvoort, *Models of Scientific Development and the Case of Nuclear Magnetic Resonance.* 1986 ISBN 90-277-2351-6
185. I. Niiniluoto, *Truthlikeness.* 1987 ISBN 90-277-2354-0
186. W. Balzer, C. U. Moulines and J. D. Sneed, *An Architectonic for Science.* The Structuralist Program. 1987 ISBN 90-277-2403-2
187. D. Pearce, *Roads to Commensurability.* 1987 ISBN 90-277-2414-8
188. L. M. Vaina (ed.), *Matters of Intelligence.* Conceptual Structures in Cognitive Neuroscience. 1987 ISBN 90-277-2460-1
189. H. Siegel, *Relativism Refuted.* A Critique of Contemporary Epistemological Relativism. 1987 ISBN 90-277-2469-5